T0212055

# Lecture Notes Editorial Policies

Lecture Notes in Statistics provides a format for the informal and quick publication of monographs, case studies, and workshops of theoretical or applied importance. Thus, in some instances, proofs may be merely outlined and results presented which will later be published in a different form.

Publication of the Lecture Notes is intended as a service to the international statistical community, in that a commercial publisher, Springer-Verlag, can provide efficient distribution of documents that would otherwise have a restricted readership. Once published and copyrighted, they can be documented and discussed in the scientific literature.

Lecture Notes are reprinted photographically from the copy delivered in camera-ready form by the author or editor. Springer-Verlag provides technical instructions for the preparation of manuscripts. Volumes should be no less than 100 pages and preferably no more than 400 pages. A subject index is expected for authored but not edited volumes. Proposals for volumes should be sent to one of the series editors or to Springer-Verlag in Santa Clara.

Authors of monographs receive 50 free copies of their book. Editors receive 50 free copies and are responsible for distributing them to contributors. Authors, editors, and contributors may purchase additional copies at the publisher's discount. No reprints of individual contributions will be supplied and no royalties are paid on Lecture Notes volumes. Springer-Verlag secures the copyright for each volume.

Series Editors:

Professor P. Bickel
Department of Statistics
University of California
Berkeley, California 94720
USA

Professor P. Diggle
Department of Mathematics
University of Lancaster
Lancaster LA1 4YL
England

Professor S. Fienberg
Department of Statistics
Carnegie Mellon University
Pittsburgh, Pennsylvania 15213
USA

Professor K. Krickeberg
3 Rue de L'Estrapade
75005 Paris
France

Professor I. Olkin
Department of Statistics
Stanford University
Stanford, California 94305
USA

Professor N. Wermuth
Department of Psychology
Johannes Gutenberg University
Postfach 3980
D-6500 Mainz
Germany

Professor S. Zeger
Department of Biostatistics
The Johns Hopkins University
615 N. Wolfe Street
Baltimore, Maryland 21205-2103
USA

# Lecture Notes in Statistics      **127**

Edited by P. Bickel, P. Diggle, S. Fienberg, K. Krickeberg,
I. Olkin, N. Wermuth, S. Zeger

# Springer

*New York*
*Berlin*
*Heidelberg*
*Barcelona*
*Budapest*
*Hong Kong*
*London*
*Milan*
*Paris*
*Santa Clara*
*Singapore*
*Tokyo*

Harald Niederreiter
Peter Hellekalek
Gerhard Larcher
Peter Zinterhof (Editors)

# Monte Carlo and
# Quasi-Monte Carlo Methods 1996

Proceedings of a conference at the
University of Salzburg, Austria,
July 9–12, 1996

 Springer

Harald Niederreiter
Institute of Information Processing
Austrian Academy of Sciences
A-1010 Vienna, Austria

Peter Hellekalek
Department of Mathematics
University of Salzburg
A-5020 Salzburg, Austria

Gerhard Larcher
Department of Mathematics
University of Salzburg
A-5020 Salzburg, Austria

Peter Zinterhof
Department of Mathematics
University of Salzburg
A-5020 Salzburg, Austria

Library of Congress Cataloging-in-Publication Data

Niederreiter, Harald, 1944-
    Monte Carlo and Quasi-Monte Carlo methods 1996 / Harald
Niederreiter, Peter Hellekalek, Gerhard Larcher.
        p.   cm. -- (Lecture notes in statistics ; 127)
    "This volume contains the refereed proceedings of the Second
International Conference on Monte Carlo and Quasi-Monte Carlo
Methods in Scientific Computing which was held at the University of
Salzburg (Austria) from July 9-12, 1996"--Preface.
    Includes bibliographical references.
    ISBN 0-387-98335-X (softcover : alk. paper)
    1. Science--Data processing--Congresses.  2. Monte Carlo method-
-Congresses.  I. Hellekalek, Peter.  II. Larcher, Gerhard, 1946-
. III. International Conference on Monte Carlo and Quasi-Monte
Carlo Methods in Scientific Computing (2nd : 1996 : Salzburg,
Austria) IV. Title.  V. Series: Lecture notes in statistics
(Springer-Verlag) ; v. 127.
Q183.9.N54  1997
519.2'82'028551--dc21                                    97-34133

Printed on acid-free paper.

Camera ready copy provided by the editors.
Printed and bound by Braun-Brumfield, Ann Arbor, MI.
Printed in the United States of America.

9 8 7 6 5 4 3 2 1

ISBN 0-387-98335-X Springer-Verlag New York Berlin Heidelberg  SPIN 10557821

# Preface

This volume contains the refereed proceedings of the Second International Conference on Monte Carlo and Quasi-Monte Carlo Methods in Scientific Computing which was held at the University of Salzburg (Austria) from July 9-12, 1996. The program of this conference was arranged by a committee consisting of Luc Devroye (McGill University), Henri Faure (University of Provence), Bennett L. Fox (University of Colorado at Denver), Peter Hellekalek (University of Salzburg), Gerhard Larcher (University of Salzburg), Pierre L'Ecuyer (University of Montreal), Gary L. Mullen (Pennsylvania State University), Harald Niederreiter (Austrian Academy of Sciences, chair), Jerome Spanier (Claremont Graduate School), and Peter Zinterhof (University of Salzburg). The local organization was in the hands of Peter Hellekalek, Gerhard Larcher, and Peter Zinterhof. The first conference of this type took place at the University of Nevada, Las Vegas, in June 1994 and the proceedings of that conference were published as Volume 106 of the *Lecture Notes in Statistics*.

Ever since their invention in the 1940s, Monte Carlo methods have been successfully employed to solve difficult problems in scientific computing that cannot be approached by other methods. The important steps in a Monte Carlo method are the setting up of an appropriate stochastic model and the design of a numerical approximation scheme based on random sampling. Monte Carlo methods had their origin in computational physics and were soon applied to computational problems in the other sciences, in engineering, and in econometrics. Today Monte Carlo methods form a basic staple of scientific computing and their applications become ever more diverse.

Quasi-Monte Carlo methods are deterministic versions of Monte Carlo methods, in the sense that random samples are replaced by suitably chosen deterministic point sets and sequences, and they were developed fairly soon after the introduction of Monte Carlo methods. In fact, the term "quasi-Monte Carlo method" already appeared in a Los Alamos technical report of R.D. Richtmyer from 1951. For a long time they were the province of specialists, but in the last ten years their appeal has broadened significantly since it was found that in certain types of computational problems they systematically outperform Monte Carlo methods. This development has gone hand in hand with accelerated progress in the theory of quasi-Monte Carlo methods. Spectacular success stories in exciting new applications such as to mathematical finance have brought quasi-Monte Carlo methods into the limelight in the last few years.

The conference was a showcase for recent developments in Monte Carlo and quasi-Monte Carlo methods and also provided the opportunity of discussing important applications of these methods. These proceedings contain all invited papers presented at the conference and a selection of the submitted contributed papers. The papers were filtered twice: first, the abstracts of the talks were screened by the Program Committee, and then the full papers were subjected to a strict peer-review process. The topics covered here range from theoretical issues in Monte Carlo and simulation methods, low-discrepancy point sets and sequences, lattice rules, and pseudorandom number generation to applications such as numerical integration, numerical linear algebra, integral equations, binary search, global optimization, computational physics, mathematical finance, and computer graphics.

We would like to thank IBM Austria, the Austrian Computer Society, and the Austrian Ministry for Science and Communications for financial contributions. Essential support for the conference in terms of logistics and manpower was provided by the Department of Mathematics of the University of Salzburg. For special thanks we want to single out Wolfgang Ch. Schmid and Karl Entacher who assisted us in many tasks with enormous energy and enthusiasm. We are greatly indebted to the referees of the papers who provided their time and expertise to help us ensure high quality standards.

We express our gratitude to Springer-Verlag for publishing this volume in their *Lecture Notes in Statistics*, and in particular to John Kimmel for the unfailing support he has given along the way. We are pleased to announce that the next conference in this series is scheduled to be held in June 1998 at the Claremont Graduate School (by then Claremont Graduate University) in California.

Harald Niederreiter
Peter Hellekalek
Gerhard Larcher
Peter Zinterhof

# Contents

viii

# Conference Participants

**Lothar Afflerbach**
Economics Department
FH Lausitz
D-01968 Senftenberg, Germany
afflerbach@fh-lausitz.de

**Kurt Binder**
Institut für Physik
Johannes Gutenberg Universität Mainz
D-55099 Mainz, Germany
binder@chaplin.physik.uni-mainz.de

**Patrick Cheridito**
Departement Mathematik
ETH Zürich, HG G 34
CH-8092 Zürich, Switzerland
dito@math.ethz.ch

**Ibrahim Coulibaly**
Laboratoire de Mathématiques
Université de Savoie
F-73376 Le Bourget du Lac Cedex, France
Ibrahim.Coulibaly@univ-savoie.fr

**Luc Devroye**
School of Computer Science
McGill University
P.Q. H3A 2A7 Montréal, Canada
luc@kriek.cs.mcgill.ca

**Ivan T. Dimov**
Center f. Inf. and Comp. Technology
Bulgarian Academy of Sciences
1113 Sofia, Bulgaria
ivdimov@amigo.acad.bg

**Karl Entacher**
Department of Mathematics
University of Salzburg
A-5020 Salzburg, Austria
charly@random.mat.sbg.ac.at

**Karin Frank**
Fachbereich Informatik
Universität Kaiserslautern
D-67653 Kaiserslautern, Germany
frank@informatik.uni-kl.de

**Seiji Fujino**
Faculty of Information Sciences
Hiroshima City University
731-31 Hiroshima, Japan
fujino@ce.hiroshima-cu.ac.jp

**Jürgen Bierbrauer**
Department of Mathematical Sciences
Michigan Technological University
MI 49931-1295 Houghton, U.S.A.
jbierbra@mtu.edu

**Stéphane Bulteau**
Irisa
Campus de Beaulieu
F-35042 Rennes Cédex, France
sbulteau@irisa.fr

**At Compagner**
Laboratory of Applied Physics
University of Technology
2600 GA Delft, The Netherlands
ac@dutncp8.tudelft.nl

**Gerhard Derflinger**
Institut für Statistik
Wirtschaftsuniversität Wien
A-1090 Wien, Austria
Gerhard.Derflinger@wu-wien.ac.at

**Ulrich Dieter**
Institut für Statistik
Technische Universität Graz
A-8010 Graz, Austria
dieter@stat.tu-graz.ac.at

**Jürgen Eichenauer-Herrmann**
Fachbereich Mathematik
Technische Hochschule Darmstadt
D-64289 Darmstadt, Germany

**Henri Faure**
Centre de Mathématiques et d'Informatique
Université de Provence
F-13453 Marseille Cedex 13, France
faure@gyptis.univ-mrs.fr

**Michel Fromm**
LMN - UFR des
Sciences et des Techniques
F-25030 Besancon cedex, France
mfromm@utinam.univ-fcomte.fr

**Jessica G. Gaines**
Department of Mathematics & Statistics
University of Edinburgh
EH9 3JZ Edinburgh, Scotland, UK
jessica@maths.ed.ac.uk

**Paul Glasserman**
Graduate School of Business
Columbia University
NY 10027 New York, U.S.A.
pglasser@research.gsb.columbia.edu

**Helge Hagenauer**
Department of Computer Science
University of Salzburg
A-5020 Salzburg, Austria
hagenau@cosy.sbg.ac.at

**Erika Hausenblas**
Department of Mathematics
University of Salzburg
A-5020 Salzburg, Austria
Erika.Hausenblas@sbg.ac.at

**Carole Hayakawa**
Mathematics Department
The Claremont Graduate School
CA 91711-3988 Claremont, U.S.A.

**Peter Hellekalek**
Department of Mathematics
University of Salzburg
A-5020 Salzburg, Austria
Peter.Hellekalek@sbg.ac.at

**Jiri K. Hoogland**
NIKHEF-H
P.O. Box 41882
NL-1012 RK Amsterdam, The Netherlands
t96@nikhef.nl

**Fred James**
CERN
CH-1211 Geneva 23, Switzerland
F.James@cern.ch

**Ronald Kleiss**
Science Department
University of Nijmegen
NL-6500 GL Nijmegen, The Netherlands
kleiss@sci.kun.nl

**Marcin Kotulski**
Hugo Steinhaus Center for Stochastic Methods
Technical University of Wroclaw
50-370 Wroclaw, Poland
kotulski@im.pwr.wroc.pl

**Raymond Lacey**
Derivative Investment Advisers
9A Madeira Road
SW16 2DB London, England
raymondo@connemara.win-uk.net

**Hermann Haaf**
Südwestdeutsche Landesbank
Am Hauptbahnhof 2
D-70173 Stuttgart, Germany

**David C. Handscomb**
Computing Laboratory
Oxford University
OX1 3QD Oxford, England
dch@comlab.ox.ac.uk

**Mike Hawrylycz**
Center for Adaptive Systems Applications
Los Alamos National Laboratory
NM 87544 Los Alamos, U.S.A.
mjh@lacasa.com

**Philip Heidelberger**
IBM T.J. Watson Research Center
PO Box 704
NY 10598 Yorktown Heights, U.S.A.
berger@watson.ibm.com

**Eva Herrmann**
Fachbereich Mathematik
Technische Hochschule Darmstadt
D-64289 Darmstadt, Germany
eherrmann@mathematik.th-darmstadt.de

**Wolfgang Hörmann**
Institut für Statistik
Wirtschaftsuniversität Wien
A-1090 Wien, Austria
whoer@statrix2.wu-wien.ac.at

**Alexander Keller**
Fachbereich Informatik
Universität Kaiserslautern
D-67653 Kaiserslautern, Germany
keller@informatik.uni-kl.de

**Bernhard Klinger**
Institut für Mathematik
Technische Universität Graz
A-8010 Graz, Austria
klinger@fstghp05.tu-graz.ac.at

**Pierre L'Ecuyer**
Département d'Informatique
Université de Montréal
H3C 3J7 Montréal, Canada
lecuyer@iro.umontreal.ca

**Tim N. Langtry**
School of Mathematical Sciences
University of Technology, Sydney
NSW 2007 Broadway, Australia
tim@maths.uts.edu.au

**Gerhard Larcher**
Department of Mathematics
University of Salzburg
A-5020 Salzburg, Austria
Gerhard.Larcher@sbg.ac.at

**Christian Lécot**
Laboratoire de Mathématiques
Université de Savoie
F-73376 Le Bourget du Lac Cedex, France
Christian.Lecot@univ-savoie.fr

**Josef Leydold**
Institut f. Statistik
Wirtschaftsuniversität Wien
A-1090 Wien, Austria
Josef.Leydold@wu-wien.ac.at

**Pascal Meyer**
LMN - UFR des
Sciences et des Techniques
F-25030 Besancon cedex, France
pmeyer@utinam.univ-fcomte.fr

**William Morokoff**
Department of Mathematics
University of California
CA 90095-1555 Los Angeles, U.S.A.
morokoff@math.ucla.edu

**Rein D. Nobel**
Department of Econometrics
Vrije Universiteit
1081 HV Amsterdam, The Netherlands
rnobel@econ.vu.nl

**Giray Ökten**
Mathematics Department
The Claremont Graduate School
CA 91711-3988 Claremont, U.S.A.
okteng@cgs.edu

**Helmut J. Pradlwarter**
Institute of Engineering Mechanics
University of Innsbruck
A-6020 Innsbruck, Austria
Helmut.Pradlwarter@uibk.ac.at

**Igor Radovic**
Zeillergasse 52
A-8020 Graz, Austria
radovic@fstghp05.tu-graz.ac.at

**Klaus Ritter**
Mathematisches Institut
Univ. Erlangen-Nürnberg
D-91054 Erlangen, Germany
ritter@mi.uni-erlangen.de

**Rainer Lassahn**
FB WOW
UniBW - Hamburg
D-22039 Hamburg, Germany
rainer.lassahn@unibw-hamburg.de

**Hannes Leeb**
Department of Mathematics
University of Salzburg
A-5020 Salzburg, Austria
leeb@random.mat.sbg.ac.at

**Peter Mathé**
Weierstrass Institute
for Applied Analysis and Stochastics
D-10117 Berlin, Germany
mathe@wias-berlin.de

**Jochen Michels**
Inst. f. Physikal. u. Theoret. Chemie
Universität Bonn
D-53115 Bonn, Germany
michels@sunc.thch.uni-bonn.de

**Harald Niederreiter**
Institute of Information Processing
Austrian Academy of Sciences
A-1010 Vienna, Austria
niederreiter@oeaw.ac.at

**Shigeyoshi Ogawa**
Department of Mathematics
Kyoto Institute of Technology
Kyoto 606, Japan
ogawa@ipc.kit.ac.jp

**Werner Pohlmann**
Department of Computer Science
University of Salzburg
A-5020 Salzburg, Austria
pohlmann@cosy.sbg.ac.at

**Dean Prichard**
Center for Adaptive Systems Applications
Los Alamos National Laboratory
NM 87544 Los Alamos, U.S.A.
dp@lacasa.com

**Herbert Rief**
Joint Research Centre
Ispra Establishment
I-21020 Ispra (Va.), Italy
herbert.rief@jrc.it

**Gerardo Rubino**
Irisa
Campus de Beaulieu
F-35042 Rennes Cédex, France
rubino@irisa.fr

**Alfred Scheerhorn**
Competence Center Informatik
Lohberg 10
D-49702 Meppen, Germany
scheerhorn@cci.de

**Walter Schwaiger**
Inst. für Finanzwirtschaft und Controlling
Universität Innsbruck
A-6020 Innsbruck, Austria
Walter.Schwaiger@uibk.ac.at

**Jerome Spanier**
Mathematics Department
The Claremont Graduate School
CA 91711-3988 Claremont, U.S.A.
spanierj@cgs.edu

**Shu Tezuka**
IBM Japan, Ltd.
1623-14, Shimotsuruma
Kanagawa-ken 242, Japan
tezuka@trlvm.vnet.ibm.com

**Bruno Tuffin**
Irisa
Campus de Beaulieu
F-35042 Rennes Cédex, France
tuffin@irisa.fr

**Stefan Wegenkittl**
Department of Mathematics
University of Salzburg
A-5020 Salzburg, Austria
ste@random.mat.sbg.ac.at

**Peter Winker**
SFB 178
Universität Konstanz
D-78434 Konstanz, Germany
Peter.Winker@uni-konstanz.de

**Reinhard Wolf**
Department of Mathematics
University of Salzburg
A-5020 Salzburg, Austria
Reinhard.Wolf@sbg.ac.at

**Anatoly A. Zhigljavsky**
Faculty of Mathematics and Mechanics
Sankt-Petersburg State University
198904 Sankt Petersburg, Russia
zh@stat.math.lgu.spb.su

**Wolfgang Ch. Schmid**
Department of Mathematics
University of Salzburg
A-5020 Salzburg, Austria
Wolfgang.Schmid@sbg.ac.at

**Thomas Siegl**
Institut für Mathematik
Technische Universität Graz
A-8010 Graz, Austria
siegl@opt.math.tu-graz.ac.at

**Sibylle Strandt**
Fachbereich Mathematik
Technische Hochschule Darmstadt
D-64289 Darmstadt, Germany
strandt@mathematik.th-darmstadt.de

**Alev Topuzoglu**
Department of Mathematics
Middle East Technical University
06531 Ankara, Turkey
alevt@rorqual.cc.metu.edu.tr

**Wolfgang Wagner**
Weierstrass Institute
for Applied Analysis and Stochastics
D-10117 Berlin, Germany
wagner@wias-berlin.de

**Kalman Wilner**
Optronics Department
TAAS - Israel Industries Ltd.
47100 Ramat Hasharon, Israel

**Reinhard Winkler**
Kommission für Mathematik
Österr. Akademie der Wissenschaften
A-1010 Wien, Austria
reinhard.winkler@oeaw.ac.at

**Yi-Jun Xiao**
CERMICS - ENPC
Cité Descartes - Champs-sur-Marne
F-77455 Marne-la-Vallée, France
xy@cermics.enpc.fr

**Peter Zinterhof**
Department of Mathematics
University of Salzburg
A-5020 Salzburg, Austria
zinterhof@edvz.sbg.ac.at

# A Comparison of Some Monte Carlo and Quasi Monte Carlo Techniques for Option Pricing

Peter Acworth, Mark Broadie, and Paul Glasserman

ABSTRACT This article compares the performance of ordinary Monte Carlo and quasi Monte Carlo methods in valuing moderate- and high-dimensional options. The dimensionality of the problems arises either from the number of time steps along a single path or from the number of underlying assets. We compare ordinary Monte Carlo with and without antithetic variates against Sobol', Faure, and Generalized Faure sequences and three constructions of a discretely sampled Brownian path. We test the standard random walk construction with all methods, a Brownian bridge construction proposed by Caflisch and Morokoff with Sobol' points and an alternative construction based on principal components analysis also with Sobol' points. We find that the quasi Monte Carlo methods outperform ordinary Monte Carlo; the Brownian bridge construction generally outperforms the standard construction; and the principal components construction generally outperforms the Brownian bridge construction and is more widely applicable. We interpret both the Brownian bridge and principal components constructions in terms of orthogonal expansions of Brownian motion and note an optimality property of the principal components construction.

## 1  Introduction

The pricing of financial instruments is one of the fastest growing areas of application of Monte Carlo methods. Financial firms are among the leading consumers of high-end computers, and the computational demands of valuing portfolios of options, mortgage-backed securities, and complex derivatives push the limits of even the fastest computers currently available. The need for efficient simulation has generated particular interest in the financial industry in the use of quasi Monte Carlo methods. The extent of this interest in evinced by articles in the popular press (*The Economist*, August 12, 1995; *The New York Times*, September 25, 1995), and numerous articles in the practitioner literature, especially *Risk* magazine. The problems arising in financial applications are often of very high dimension (e.g., from 50 to several hundred) and therefore pose new challenges for research in this area.

The purpose of this article is to compare the performance of Monte Carlo (MC) and some quasi Monte Carlo (QMC) techniques in option pricing over a range of dimensions and computational budgets. In addition to comparing the straightforward implementation of QMC sequences, we examine the effectiveness of their use with a construction described by William Morokoff (see Caflisch and Morokoff [6], Caflisch and Moskowitz [7]) at the Salzburg workshop from which this volume originates. Motivated by the impressive performance of this construction, we propose an alternative which has some theoretical advantages and make a preliminary assessment of its effectiveness. Finally, since this article is intended primarily for a Monte Carlo audience, we hope it will provide a useful introduction to applications in option pricing. A more extensive overview of Monte Carlo methods for option pricing can be found in Boyle, Broadie, and Glasserman [3].

Section 2 of this article discusses option pricing in general and the specific cases we study. Section 3 describes a numerical study and Section 4 proposes and evaluates an alternative construction.

## 2 Option Pricing

In this section, we briefly review some theoretical background leading to the representation of option prices as expectations which are in turn amenable to Monte Carlo. For a complete and far more general treatment of this material, see Duffie [8]. The theory sketched here has grown out of the early work of Harrison and Kreps [11] and Harrison and Pliska [12].

Uncertainty is represented by a probability space $(\Omega, \mathcal{F}, P)$ supporting a $k$-dimensional Brownian motion $W_t = (W_t^1, \ldots, W_t^k)$. Each $W_t^i$ has drift zero and variance parameter 1, and the $i$th and $j$th components have correlation $\rho_{ij}$; i.e., $E[W_t^i] = 0$, $\mathrm{Var}[W_t^i] = t$, and

$$\mathrm{Cov}[W_t^i, W_t^j] = \rho_{ij}t.$$

We consider an economy with $k$ risky assets with prices $S_t = (S_t^1, \ldots, S_t^k)$ at time $t$. The prices satisfy

$$dS_t^i = \mu_i S_t^i \, dt + \sigma_i S_t^i \, dW_t^i, \quad i = 1, \ldots, k \tag{1}$$

for some mean return parameters $\mu_1, \ldots, \mu_k$ and positive volatility parameters $\sigma_1, \ldots, \sigma_k$. Thus, each asset price is a geometric Brownian motion process and its instantaneous returns $dS_t^i/S_t^i$ follow ordinary Brownian motion. The parameters $\mu_i$ and $\sigma_i$ could, in general, be both time-varying and stochastic, but we restrict attention to the constant parameter case. We complete the description of the economy by assuming a constant continuously compounded risk-free interest rate $r > 0$. The present value, as of time $t$, of a payoff to be received at time $T > t$ is the product of the payoff and the discount factor $\exp(-r(T - t))$. We fix an interval $[0, T]$ and let $S_t^{k+1} = \exp(-r(T - t))$ denote the price at time $t$ of a bond with a certain payoff of 1 at time $T$.

A trading strategy is a process $\pi_t = (\pi_t^1, \ldots, \pi_t^{k+1})$ in which $\pi_t^i$ represents the number of shares held of the $i$th asset at time $t$. If we think of $\pi_t$ as a row vector and $S_t$ as a column vector, then the value at time $t$ of the portfolio described by $\pi_t$ is given by the scalar product $\pi_t S_t$. The trading gains under $\pi$ over an interval $[0, t]$ are given by the Itô integral

$$\int_0^t \pi_u \, dS_u.$$

The trading strategy $\pi$ is called *self-financing* if

$$\pi_t S_t = \pi_0 S_0 + \int_0^t \pi_u \, dS_u, \quad \text{for all } t \in [0, T].$$

This says that the value $\pi_t S_t$ at time $t$ equals the initial investment $\pi_0 S_0$ plus the gains from trading over $[0, t]$ — no additional infusion of wealth is required to sustain the strategy. An *arbitrage* is a self-financing trading strategy with $\pi_0 S_0 < 0$ and $\pi_T S_T \geq 0$, a.s.; in other words, this is a strategy that requires a negative initial investment yet guarantees a nonnegative terminal value.

An *equivalent martingale measure* is a measure $Q$, equivalent (in the sense of measures) to $P$, under which the discounted asset prices

$$\tilde{S}_t^i = e^{-rt} S_t^i, \quad i = 1, \ldots, k,$$

are martingales. The existence of an equivalent martingale measure precludes arbitrage. For if the $\tilde{S}_t^i$ are martingales under $Q$, then (under integrability conditions on $\pi$) $\int \pi_u \, d\tilde{S}_u$ is a mean-zero martingale, and hence for any self-financing trading strategy

$$\pi_0 \tilde{S}_0 = E_Q \left[ \pi_T \tilde{S}_T - \int_0^T \pi_t d\tilde{S}_t \, dt \right] = E[\pi_T \tilde{S}_T].$$

It is therefore impossible to have $\pi_T \tilde{S}_T \geq 0$ if $\pi_0 \tilde{S}_0 < 0$. It can be shown that the self-financing property is unaffected by the change of measure to complete the argument.

Consider, next, a process $X_t$ on $[0, T]$ with $X_T$ interpreted as the payoff at time $T$ of some additional security. The payoff $X_T$ is *attainable* in the original economy if there exists a self-financing trading strategy $\pi$ for which $\pi_T S_T = X_T$. In this case, to augment the economy with the new security without introducing an arbitrage opportunity we must have $X_t = \pi_t S_t$ for all $t \in [0, T]$. Under the measure $Q$, each discounted asset price is a martingale and therefore so too is $X_t$; in particular,

$$X_0 = E_Q[e^{-rT} X_T]. \tag{2}$$

This is the fundamental result of pricing by arbitrage: the value today of an attainable payoff to be received at a future date is the expected present value of the payoff, the expectation taken under the equivalent martingale measure $Q$.

It remains to specify which payoffs are attainable. Set

$$\Sigma_{ij} = \sigma_i \sigma_j \rho_{ij}, \quad i,j = 1, \ldots, k,$$

with the $\sigma_i$ and the $\rho_{ij}$ as defined above. We assume that $\Sigma$ has full rank, a condition known as market completeness. In the single-asset case, this reduces to $\sigma \equiv \sigma_1 > 0$. In a complete market, every square-integrable payoff depending only on the history of $(W_t^1, \ldots, W_t^k)$ for $t \in [0, T]$ is attainable.

In the simple setting we are considering of a complete market in which asset prices follow geometric Brownian motions, the martingale measure $Q$ is unique. Moreover, changing to the measure $Q$ is equivalent to replacing each $\mu_i$ with $r$ to get

$$dS_t^i = rS_t^i \, dt + \sigma_i S_t^i \, dW_t^i. \tag{3}$$

With this change in the drift parameters, each $e^{-rt} S_t^i$ becomes a martingale. If, now, $X_T = g(S_T^1, \ldots, S_T^k)$ is the payoff of security derived from the underlying assets, the pricing rule developed above yields

$$X_0 = e^{-rT} E[g(S_T^1, \ldots, S_T^k)].$$

(We have dropped the subscript $Q$ from the expectation with the understanding that (1) has been replaced by (3).) Thus, the problem of pricing the derivative security reduces to one of computing an expectation — a problem well-suited to Monte Carlo. In fact, we will also consider payoffs that depend on the paths of the underlying assets, and not just their values at the terminal time $T$. The same pricing mechanism applies to path-dependent options.

We conclude this section with a simple example. Suppose $k = 1$ and write $S_t$ for $S_t^1$. Let $g(s) = \max(0, s - K)$ with $K$ a constant. Then $g(S_T)$ is the payoff of a call option expiring at $T$ with a strike price of $K$; that is, an option that grants the holder the right to buy the underlying asset at time $T$ at a price of $K$. The price of the call at time 0 is $e^{-rT} E[\max(0, S_T - K)]$, with $S_T$ determined by (3). Because $S_T$ has a lognormal distribution, this expectation can be evaluated in closed form. The result is the famous Black-Scholes formula.

# 3 Comparisons

## 3.1 Option Examples and First Constructions

There is an extensive literature on specific examples of option payoffs that allow closed-form pricing and at least as large a literature on deterministic numerical methods for computing prices that cannot be evaluated analytically. (See, e.g., Hull [13] or Jarrow and Turnbull [15] for an introduction or Rogers and Talay [22] for a survey and recent developments.) But for high-dimensional problems, option prices can generally be evaluated only by some form of Monte Carlo. There are essentially two considerations that contribute to high dimensionality: the option payoff may depend on many points along the path of a set of underlying assets, or the number of underlying assets may be large. In this section, we compare the performance of some Monte Carlo and quasi Monte Carlo methods on both types of problems.

We consider the following general classes of options:

(a) *Discrete down-and-out call.* This is a *barrier* option on a single underlying asset with payoff

$$C_B = \max(S_T - K, 0)\mathbf{1}_{\{\min_{1 \le i \le m} S_{t_i} > H\}},$$

and price $e^{-rT} E[C_B]$. Here $K$ and $H$ are constants and $t_1, \ldots, t_m$ are points in $[0, T]$. This is a standard call option struck at $K$ except that if the underlying asset is below the barrier $H$ at any of the monitoring dates $t_1, \ldots, t_m$ the option is *knocked out* and the holder receives nothing. In our examples we take evenly spaced monitoring dates $t_i = i(T/m)$, $i = 0, 1, \ldots, m$. With continuous monitoring (i.e., with the min in the indicator taken over all $t \in [0, T]$) this option can be priced in closed form (see, e.g., p.462 of [13]). The discrete version can be priced quickly and accurately using a numerical method of Broadie, Glasserman, and Kou [5]. We use this method to compare results obtained by simulation.

(b) *Discrete average rate option.* Perhaps the simplest frequently encountered option for which Monte Carlo is a primary pricing method has as its payoff

$$C_A = \max\left(0, \frac{1}{m}\sum_{i=1}^{m} S_{t_i} - K\right).$$

This is an option on the average price of the underlying asset over a fixed set of monitoring dates. Its intractability arises from the fact that the sum of lognormal random variables does not, in general, admit a closed-form distribution. As a surrogate for this option we work with

$$C_G = \max\left(0, \left(\prod_{i=1}^{m} S_{t_i}\right)^{1/m} - K\right).$$

The geometric average of the $S_{t_i}$ appearing in this payoff is itself lognormally distributed, so $e^{-rT} E[C_G]$ can be evaluated exactly through a simple modification of the Black-Scholes formula; see p.466 of [13].

(c) *A Multiasset option.* The examples we consider have payoff

$$C_M = \max\left(0, \left(\prod_{i=1}^{m} S_T^i\right)^{1/m} - K\right),$$

where $S_T^1, \ldots, S_T^m$ are the terminal prices of $m$ correlated assets, as in Section 2. Again, the geometric average appearing in the payoff leads to a closed-form solution against which to compare simulation results. Minor variations in the payoff function, however, lead to pricing problems with no closed-form solution.

Expressed as an integral, the price of each of the options just described is $m$-dimensional, though in (a) and (b) the different dimensions correspond to different points along a one-dimensional path. We are therefore able to examine the effect of dimension by varying $m$. To price the options by simulation, we use the fact that the solution to (3) is given by

$$S_t^i = S_0^i \exp((r - \tfrac{1}{2}\sigma_i^2)t + \sigma_i W_t^i).$$

Thus, to simulate $S_{t_1}, \ldots, S_{t_m}$ for a single asset or $S_T^1, \ldots, S_T^m$ for multiple assets it suffices to generate discrete increments of Brownian motion.

For the one-dimensional case, the standard construction sets $W_0 = 0$ and

$$W_{t_{i+1}} = W_{t_i} + \sqrt{t_{i+1} - t_i}\, Z_{i+1}, \quad i = 0, \ldots, m-1, \tag{4}$$

where $Z_1, \ldots, Z_m$ are independent standard normal random variates. For the $m$-asset case, we may take

$$\begin{pmatrix} W_T^1 \\ \vdots \\ W_T^m \end{pmatrix} = \sqrt{T} A \begin{pmatrix} Z_1 \\ \vdots \\ Z_m \end{pmatrix}, \tag{5}$$

where $A$ is the Cholesky matrix obtained from $\Sigma$; i.e., a lower triangular matrix satisfying $AA' = \Sigma$. In fact, (4) amounts to multiplying the vector of $Z_i$ by a lower triangular matrix all of whose nonzero entries in column $i$ are $\sqrt{t_i - t_{i-1}}$. This is the Cholesky matrix for the covariance matrix with $ij$-entry equal to $\min(t_i, t_j)$, the covariance matrix of the Brownian path.

The constructions (4) and (5) yield the required points from a Brownian path, given $Z_i$ as inputs. To generate the $Z_i$ by Monte Carlo, we apply the inverse of the cumulative normal distribution to $U_i$ that are uniformly distributed on $[0, 1]$. We compute the inverse normal numerically using the routine of Moro [17]. To use quasi Monte Carlo with the standard construction, we generate points $x = (x_1, \ldots, x_m)$ in $[0, 1)^m$, apply the inverse normal to each $x_i$ to get a value $z_i$ and then substitute $z_1, \ldots, z_m$ for $Z_1, \ldots, Z_m$ in (4) and (5).

Caflisch and Morokoff [6] have proposed the use of quasi Monte Carlo with an alternative construction of a discretely sampled one-dimensional Brownian path. Their construction is based on the Brownian bridge property that, for $u < s < t$, the distribution of $W_s$ conditional on $W_u = a$ and $W_t = b$ is normal with mean

$$\left(\frac{t-s}{t-u}\right) a + \left(\frac{s-u}{t-u}\right) b \tag{6}$$

and variance

$$\frac{(s-u)(t-s)}{(t-u)}. \tag{7}$$

If, for example, $m = 2^n$ for some integer $n$, the discretely sampled Brownian path can be generated by consecutively determining its values at the times $T, T/2, T/4, 3T/4, \ldots, T/m, \ldots, (m-1)T/m$. Specifically, we may set

$$W_T = \sqrt{T} z_1$$
$$W_{T/2} = \frac{1}{2} W_T + \frac{\sqrt{T}}{2} z_2$$
$$\vdots$$
$$W_{(m-1)T/m} = \frac{1}{2} (W_{(m-2)T/m} + W_T) + \sqrt{\frac{T}{2m}} z_m.$$

This construction fills in the path at an increasingly fine grid of points. With ordinary Monte Carlo, it has the same effect as the standard construction. Indeed, the Brownian bridge construction simply replaces the Cholesky matrix implicit in (4) with another matrix $B$ for which $BB'$ is the covariance matrix of the discretely sampled path. (We return to this point in the next section.) Since the joint distribution of the $W_{t_i}$ is unchanged, the choice of construction has no effect on the variance of a Monte Carlo estimate. However, it is well known that the initial coordinates of a low discrepancy sequence typically have better equidistribution properties than the later coordinates. The standard construction (4) does not account for this. In contrast, the Brownian bridge construction uses the best coordinates from each point to determine most of the structure of a path and reserves the later coordinates to fill in fine detail. See Fox [10] for a related construction of Poisson processes.

## 3.2 Numerical Comparison

We now turn to the design of our numerical study.

*Options.* We use four classes of options: the discrete barrier and geometric average options with payoffs $C_B$ and $C_G$ defined in (a) and (b) above; and the multiasset option with payoff $C_M$ in which either all underlying assets are independent or all pairs of underlying assets have correlation 0.3.

*Methods.* We compare the performance of ordinary Monte Carlo, Monte Carlo with antithetic variates, Sobol' points, Faure points, Generalized Faure (GFaure) points, all based on the standard constructions (4) and (5). For the two single-asset options, we also tested the combination of Sobol' points with the Brownian bridge construction. The use of Sobol' points for this case was essentially arbitrary, though based in part on preliminary evidence suggesting that Sobol' points outperformed Faure points. (For a general treatment of low discrepancy sequences, see Niederreiter [18].) For the Brownian bridge construction we follow the steps in Section 2 for the largest power of two smaller than $m$ and then fill in the remaining points using (6) and (7).

*Algorithms.* For standard Monte Carlo we use the generator RAN1 from Press et al. [21]. The GFaure points are based on Section 6.1.2 of Tezuka [23]. Sobol' points can be generated using Bratley and Fox [4]. To generate both GFaure and the Sobol' points we use the FINDER software of Papageorgiou, Paskov, and Traub [19, 20]. This code is publicly available for noncommercial purposes through Columbia University. We generate Faure points using an implementation of Ken Seng Tan of the University of Waterloo.

A well-known property of standard low discrepancy sequences is that they tend to perform better if an initial portion of the sequence is deleted. For the Sobol' sequence, the number of points we skip in each case is the largest power of two smaller than the number of points to be used in the run. With the Faure sequence the number skipped is the fourth power of the prime base (following Fox [9]), and with the GFaure sequence it is always 200,000 (a recommendation from Anargyros Papageorgiou).

*Dimensions.* We test the methods on problems with $m = 10$, 50, and 100. Problems of each size are run for $n = 1.25$ thousand, 5 thousand, 20 thousand, and 80 thousand points.

*Figure of merit.* Because simulation by quasi Monte Carlo is purely deterministic, it offers no simple basis for comparison with ordinary Monte Carlo. To compare methods, we therefore compare their root mean square (RMS) relative errors over 250 randomly generated problem instances. This is a measure of the typical error using the various methods.

We generate the problem instances as follows. In all cases, the initial asset prices $S_0^i$ are set equal to 100, the expiration date $T$ is 1 year, annual volatilities ($\sigma_i$) are uniform between 10% and 60% (independent for different assets), and the interest rate is uniform between 0 and 10%. For the barrier option, the barrier $H$ is uniform between 70 and 95 and the strike $K$ is uniform between $1.1H$ and 130. For the average rate options, the strike is uniform between 80 and 120, and for the multiasset options

| | MC | MC_Anti | Sobol' | Faure | GFaure | BBridge |
|---|---|---|---|---|---|---|
| $m = 10$ | | | Barrier Options | | | |
| $n = 1{,}250$ | 5.09 | 3.63 | 1.32 | 1.93 | 1.32 | 0.78 |
| 5,000 | 2.28 | 1.96 | 0.75 | 0.77 | 0.62 | 0.41 |
| 20,000 | 1.29 | 1.13 | 0.48 | 0.35 | 0.54 | 0.53 |
| 80,000 | 0.73 | 0.70 | 0.47 | 0.44 | 0.51 | 0.47 |
| | | | Average Rate Options | | | |
| $n = 1{,}250$ | 5.29 | 4.09 | 2.14 | 0.73 | 1.47 | 0.71 |
| 5,000 | 2.79 | 2.46 | 0.18 | 0.41 | 0.50 | 0.24 |
| 20,000 | 1.39 | 1.15 | 0.08 | 0.38 | 0.18 | 0.08 |
| 80,000 | 0.60 | 0.58 | 0.03 | 0.14 | 0.08 | 0.03 |
| $m = 50$ | | | Barrier Options | | | |
| $n = 1{,}250$ | 4.73 | 4.09 | 7.10 | 4.59 | 2.72 | 1.14 |
| 5,000 | 2.59 | 1.93 | 1.10 | 2.15 | 0.98 | 0.87 |
| 20,000 | 1.19 | 1.00 | 0.30 | 0.35 | 0.40 | 0.25 |
| 80,000 | 0.61 | 0.50 | 0.22 | 0.11 | 0.17 | 0.12 |
| | | | Average Rate Options | | | |
| $n = 1{,}250$ | 5.20 | 5.04 | 4.24 | 2.85 | 2.92 | 0.53 |
| 5,000 | 2.60 | 2.08 | 0.61 | 3.02 | 1.07 | 0.16 |
| 20,000 | 1.38 | 1.18 | 0.24 | 0.82 | 0.44 | 0.05 |
| 80,000 | 0.61 | 0.59 | 0.06 | 0.03 | 0.18 | 0.03 |
| $m = 100$ | | | Barrier Options | | | |
| $n = 1{,}250$ | 4.61 | 4.07 | 9.83 | 5.98 | 3.25 | 1.32 |
| 5,000 | 2.57 | 2.17 | 1.70 | 2.21 | 1.74 | 0.91 |
| 20,000 | 1.32 | 0.93 | 0.62 | 0.99 | 0.65 | 0.23 |
| 80,000 | 0.65 | 0.51 | 0.19 | 0.15 | 0.41 | 0.09 |
| | | | Average Rate Options | | | |
| $n = 1{,}250$ | 5.04 | 4.51 | 10.12 | 11.12 | 3.42 | 0.63 |
| 5,000 | 2.80 | 2.07 | 1.27 | 2.97 | 1.56 | 0.18 |
| 20,000 | 1.15 | 1.21 | 0.24 | 1.28 | 0.73 | 0.04 |
| 80,000 | 0.62 | 0.58 | 0.05 | 0.37 | 0.30 | 0.03 |

TABLE 1. RMS Relative errors (in percent) for single-asset options. $m$ is the problem dimension (number of points on the path) and $n$ is the number of paths.

it is uniform between 70 and 110. These parameters give a broad range for comparison. As explained above, each of these options can be valued without Monte Carlo using either a formula or a numerical procedure; we use the true value to compute RMS errors. In all cases, if the true value is below 0.5 the instance generated is discarded and replaced with an independently generated one: very small option prices provide less interesting examples and lead to less reliable estimates of RMS relative error.

In practice the true option values will not be known. In this case, the standard errors available with Monte Carlo provide an advantage that is not captured in the RMS figure of merit.

Table 1 shows results for the single-asset instances. (Generating Tables 1 and 2 required more than 10 days on a dedicated 200 MHz Pentium PC.) The QMC methods outperform ordinary MC with and without antithetics except in a few cases when the dimension is high and the number of points is relatively small. (The Sobol' results would probably be better with $n = 1,024$ — a power of two — than with $n = 1,250$.) This is broadly consistent with results reported by others (e.g., Berman [2]) using different types of problems and a different basis for comparison. Among the QMC methods, none is uniformly superior, but the combination of the Brownian bridge construction with Sobol' points most often has the lowest RMS relative error and is never appreciably inferior to the others. Among the straightforward QMC results the comparison is more difficult, though Sobol' would appear to be the most effective overall and Faure the least effective.

The Brownian bridge construction is not directly applicable to options depending on the prices of multiple assets at a single date, so Table 2 compares only standard constructions. Again, with few exceptions, the QMC methods outperform the standard MC methods. The Sobol' errors are smallest in 16 out of the 24 cases and in all but two cases with $n \geq 5,000$. The GFaure errors are smaller than the Faure errors in all but two cases.

It is important to emphasize that the performance of ordinary MC can often be very substantially improved through the use of variance reduction techniques that exploit problem structure. In the case of barrier options, for example, Boyle et al. [3] show that importance sampling can produce dramatic variance reduction when the underlying asset is far from the barrier. Berman [2] finds that MC does as well as QMC in option pricing when used with a variance reduction technique. Thus, Tables 1 and 2 compare only black-box implementations of the various methods. This comparison is still relevant since in practice financial institutions may need to value a large number of very distinct options simultaneously, so developing special methods for each may not be feasible.

# 4  A Principal Components Construction

The effectiveness of the Brownian bridge construction appears to lie in the fact that it uses the best coordinates of each point to determine most of the structure of a path. With this intuition in mind, it is natural to ask whether this construction uses the best coordinates optimally, or whether even more of the structure of a path can be determined by the initial coordinates of a point. A related question is whether this type of construction can be extended to apply to simulating a single step of multiple assets. We now develop an alternative construction that addresses these questions.

| | MC | MC_Anti | Sobol' | Faure | GFaure |
|---|---|---|---|---|---|
| $m = 10$ | | No Correlation | | | |
| $n = 1,250$ | 4.24 | 2.94 | 1.20 | 0.73 | 1.21 |
| 5,000 | 2.23 | 1.69 | 0.37 | 0.75 | 0.56 |
| 20,000 | 1.00 | 0.99 | 0.19 | 0.59 | 0.15 |
| 80,000 | 0.48 | 0.40 | 0.06 | 0.39 | 0.08 |
| | | Correlation 0.3 | | | |
| $n = 1,250$ | 3.94 | 3.24 | 1.03 | 0.43 | 0.83 |
| 5,000 | 2.11 | 1.30 | 0.17 | 0.37 | 0.26 |
| 20,000 | 1.00 | 0.76 | 0.06 | 0.14 | 0.11 |
| 80,000 | 0.52 | 0.36 | 0.04 | 0.10 | 0.04 |
| $m = 50$ | | No Correlation | | | |
| $n = 1,250$ | 3.44 | 2.44 | 3.55 | 1.97 | 1.66 |
| 5,000 | 1.38 | 1.39 | 0.50 | 2.13 | 0.57 |
| 20,000 | 0.80 | 0.61 | 0.18 | 0.24 | 0.23 |
| 80,000 | 0.43 | 0.35 | 0.08 | 0.09 | 0.08 |
| | | Correlation 0.3 | | | |
| $n = 1,250$ | 3.37 | 2.23 | 1.58 | 4.69 | 1.26 |
| 5,000 | 1.73 | 1.21 | 0.21 | 2.87 | 0.37 |
| 20,000 | 0.87 | 0.57 | 0.05 | 0.29 | 0.13 |
| 80,000 | 0.43 | 0.28 | 0.03 | 0.11 | 0.05 |
| $m = 100$ | | No Correlation | | | |
| $n = 1,250$ | 1.79 | 1.63 | 3.18 | 2.66 | 1.07 |
| 5,000 | 1.11 | 0.76 | 0.53 | 0.92 | 0.47 |
| 20,000 | 0.55 | 0.44 | 0.13 | 0.25 | 0.17 |
| 80,000 | 0.24 | 0.15 | 0.07 | 0.10 | 0.10 |
| | | Correlation 0.3 | | | |
| $n = 1,250$ | 2.82 | 2.04 | 2.15 | 3.59 | 1.44 |
| 5,000 | 1.72 | 0.97 | 0.34 | 1.83 | 0.46 |
| 20,000 | 0.80 | 0.44 | 0.05 | 0.56 | 0.16 |
| 80,000 | 0.37 | 0.23 | 0.03 | 0.29 | 0.05 |

TABLE 2. RMS Relative errors (in percent) for multi-asset options. $m$ is the problem dimension (number of assets) and $n$ is the number of replications.

## 4.1 Discrete Paths

For any $0 = t_0 < t_1 < \cdots < t_m = T$ (not necessarily evenly spaced) let $V$ be the $m \times m$ matrix $V_{ij} = \min(t_i, t_j)$, $i, j = 1, \ldots, m$; this is the covariance matrix of $W_{t_1}, \ldots, W_{t_m}$. As noted in Section 3, the Brownian bridge and standard constructions correspond to multiplying a vector of $m$ independent normals by a matrix $M$ satisfying $MM' = V$; the standard construction uses the lower triangular matrix with this property. A standard notion in statistics of the variability *explained* by just the first $k$ standard normals in such a representation is the sum of the squared norms

| Construction | Cumulative Explained Variability (%) | | | | |
|---|---|---|---|---|---|
| Standard | 2.0 | 3.9 | 5.9 | 7.8 | 9.7 |
| Brownian bridge | 67.2 | 83.6 | 87.7 | 91.8 | 92.8 |
| Principal components | 81.6 | 90.1 | 93.3 | 95.0 | 96.0 |

TABLE 3. Cumulative explained variability from the first five dimensions using the standard (Cholesky) construction, the Brownian bridge construction, and the principal components construction, based on dimension $m = 64$.

of the first $k$ columns of the matrix. *Principal components analysis* (which is discussed in most textbooks on multivariate statistics) is based on the matrix $M$ maximizing the explained variability for each $k = 1, \ldots, m$ subject to the constraint that $MM'$ equal a specified covariance matrix. The solution is given by $M = QD^{1/2}$ where the columns of $Q$ are unit-length eigenvectors of the covariance matrix and $D$ is a diagonal matrix of eigenvalues. The eigenvalues appear in decreasing order of magnitude along the diagonal of $D$, and the $i$th column of $Q$ has as its eigenvalue the $i$th element along the diagonal of $D$.

Table 3 compares the cumulative variability explained by the first five columns of the Cholesky matrix, the Brownian bridge matrix, and the principal components matrix. These values were computed with $m = 64$, but they are rather insensitive to the choice of $m$. They suggest that the Brownian bridge construction is a substantial improvement over the standard construction in allocating importance to the initial dimensions, and also that the principal components matrix offers appreciable further improvement.

Using the principal components matrix $M$, we generate a path by multiplying $M$ by a vector $(Z_1, \ldots, Z_m)'$ of independent standard normals (using Monte Carlo) or a transformed low discrepancy point $(z_1, \ldots, z_m)'$ (using quasi Monte Carlo). With standard Monte Carlo, this offers no advantage. But because $M$ maximizes the variability explained by any initial set of $z_i$, it would seem to make optimal use of the initial coordinates of a QMC sequence. Moreover, this construction can be applied to any covariance matrix (with appropriate modifications in the singular case) and can therefore be used for both the single-asset path-dependent options and the multiple-asset path-independent options considered in the previous section. In the multi-asset case, the variability explained by the first dimension increases with the correlation across assets, so we expect the effectiveness of the method to increase as well. (In the uncorrelated case, the principal components construction merely reorders the assets, ranking them by their volatilities.)

Finding the eigenvalues and eigenvectors of an $m \times m$ covariance matrix to build the matrix $M$ is an $O(m^3)$ operation. In practice, this computation takes a small fraction of the total simulation time for even moderately large values of $m$ — certainly for $m$ up to 250, say. Moreover, this compu-

| | Single Asset | | Multiple Asset | |
|---|---|---|---|---|
| | BBridge | PrinComp | Sobol' | PrinComp |
| $m = 10$ | Barrier | | Corr. 0.0 | |
| $n = 1{,}250$ | 0.78 | 0.97 | 1.20 | 1.01 |
| 5,000 | 0.41 | 0.49 | 0.37 | 0.50 |
| 20,000 | 0.53 | 0.50 | 0.19 | 0.20 |
| 80,000 | 0.47 | 0.47 | 0.06 | 0.03 |
| | Average Rate | | Corr. 0.3 | |
| $n = 1{,}250$ | 0.71 | 0.32 | 1.03 | 0.23 |
| 5,000 | 0.24 | 0.11 | 0.17 | 0.06 |
| 20,000 | 0.08 | 0.02 | 0.06 | 0.02 |
| 80,000 | 0.03 | 0.01 | 0.04 | 0.01 |
| $m = 50$ | Barrier | | Corr. 0.0 | |
| $n = 1{,}250$ | 1.14 | 1.18 | 3.55 | 2.45 |
| 5,000 | 0.87 | 0.59 | 0.50 | 0.34 |
| 20,000 | 0.25 | 0.31 | 0.18 | 0.08 |
| 80,000 | 0.12 | 0.08 | 0.08 | 0.04 |
| | Average Rate | | Corr. 0.3 | |
| $n = 1{,}250$ | 0.53 | 0.33 | 1.58 | 0.16 |
| 5,000 | 0.16 | 0.11 | 0.21 | 0.05 |
| 20,000 | 0.05 | 0.02 | 0.05 | 0.02 |
| 80,000 | 0.03 | 0.01 | 0.04 | 0.01 |
| $m = 100$ | Barrier | | Corr. 0.0 | |
| $n = 1{,}250$ | 1.32 | 1.41 | 3.18 | 3.59 |
| 5,000 | 0.91 | 0.46 | 0.53 | 0.56 |
| 20,000 | 0.23 | 0.28 | 0.13 | 0.10 |
| 80,000 | 0.10 | 0.11 | 0.07 | 0.02 |
| | Average Rate | | Corr. 0.3 | |
| $n = 1{,}250$ | 0.63 | 0.33 | 2.15 | 0.16 |
| 5,000 | 0.18 | 0.11 | 0.34 | 0.04 |
| 20,000 | 0.04 | 0.02 | 0.06 | 0.02 |
| 80,000 | 0.03 | 0.01 | 0.03 | 0.00 |

TABLE 4. RMS relative errors (in percent) illustrating the performance of the principal components construction. $m$ is the problem dimension (number of points on the path) and $n$ is the number of quasi Monte Carlo points.

tation can be done off-line because it is independent of the problem data. Generating a path using the principal components matrix $M$ entails a full matrix-vector multiplication, so generating $n$ paths requires $O(nM^2)$ effort. In contrast, the Brownian bridge construction, if implemented recursively rather than as a matrix multiplication, is $O(nM)$. This certainly represents an advantage to the Brownian bridge method. We thank Russell Caflisch and William Morokoff for pointing this out.

Table 4 reports RMS relative errors illustrating the performance of the principal components construction combined with Sobol' points. For the single-asset options, we compare against the Brownian bridge construction (also combined with Sobol' points) on the same set of 250 randomly generated options as before. The Brownian bridge construction is not applicable to the multi-asset options so there we compare against a standard implementation of Sobol' points. (The Brownian bridge and Sobol' results are repeated from the previous tables to facilitate comparison.) The principal components construction outperforms the standard Sobol' in almost all cases. In the single-asset results, principal components outperforms Brownian bridge in 17 of the 24 cases; it never does appreciably worse than Brownian bridge and in several cases does much better.

## 4.2 Continuous Paths

The apparent effectiveness of both the Brownian bridge and principal components constructions motivates a deeper look at the relation between these constructions and Brownian motion. For simplicity, we set $T = 1$ and consider standard Brownian motion on $[0, 1]$.

The generalization of principal components analysis for general Gaussian processes is the Karhounen-Loève expansion, based on the eigenfunctions of the covariance kernel. In the case of Brownian motion, these are functions $\psi$ satisfying

$$\int_0^1 \min(s, t)\psi(s)\, ds = \lambda\psi(t)$$

for some $\lambda$. The solutions and the corresponding eigenvalues are known to be

$$\psi_n(t) = \sqrt{2}\sin\left(\frac{(2n+1)\pi s}{2}\right), \quad \lambda_n = \left(\frac{2}{(2n+1)\pi}\right)^2.$$

The corresponding Karhounen-Loève expansion of Brownian motion is

$$W_t = \sum_{n=0}^{\infty} \sqrt{\lambda_n}\psi_n(t)Z_n,$$

where $Z_0, Z_1, \ldots$ are independent standard normals; see, e.g., Adler [1]. The eigenvalues $\lambda_n$ decrease with $n$ and the eigenfunctions $\psi_n$ are of increasing frequency, so the $Z_n$ fill in increasingly fine detail in this representation of the Brownian path. For small values of $n$, the $\psi_n$ provide a good approximation to the eigenvectors of a discrete path of $m$ points, even for $m$ of moderate size — 30, say.

The counterpart of the discrete Brownian bridge construction for continuous paths is the Lévy-Ciesilski expansion (e.g., Janicki and Weron [14], Karlin and Taylor [16], pp.373-377) which we describe next. Define the Haar functions on $[0, 1]$ by setting $H_1 \equiv 1$; $H_2(t) = 1$ on $[0, 1/2)$ and $-1$

on $[1/2, 1]$;

$$H_{2^n+1}(t) = \begin{cases} 2^{n/2}, & 0 \le t < 2^{-(n+1)}, \\ -2^{n/2}, & 2^{-(n+1)} \le t \le 2^{-n}, \\ 0, & \text{otherwise}; \end{cases}$$

and

$$H_{2^n+j}(t) = H_{2^n+1}\left(t - \frac{j-1}{2^n}\right), \quad j = 1, \ldots, 2^n - 1.$$

The Haar functions form a complete orthonormal basis in $L^2[0,1]$ and, interestingly, play a role in quasi Monte Carlo through the analysis of digital nets. From the Haar functions define the Schauder functions

$$F_m(t) = \int_0^t H_m(u)\, du.$$

The Brownian bridge construction of an $m$-step path with $m = 2^n$ can be represented exactly as

$$W_t^{(m)} = \sum_{i=1}^m F_i(t) Z_i;$$

and the continuous Brownian path can be represented as

$$W_t = \sum_{i=1}^\infty F_i(t) Z_i.$$

Thus, viewed in continuous time, the Brownian bridge and principal components constructions correspond to expansions in terms of two different sets of functions. The Karhounen-Loève expansion and principal components construction are optimal among all such representations in that they allocate maximum variability to each initial portion of the driving sequence $Z_1, Z_2, \ldots$. Moreover, they immediately suggest extensions to other Gaussian processes. Finding Karhounen-Loève expansions explicitly is difficult, but a numerical solution suffices for the application proposed here.

# 5  Concluding Remarks

We have compared the use of ordinary Monte Carlo and quasi Monte Carlo in the pricing of moderate- and high-dimensional options. In addition to testing the straightforward implementation of quasi Monte Carlo we have examined the performance of a Brownian bridge construction proposed by Caflisch and Morokoff [6] and a related new construction based on principal components — i.e., based on the eigenvectors of a covariance matrix. Briefly, we summarize our numerical results as follows: all the QMC methods outperfom ordinary Monte Carlo; Sobol' points appear to give the best results among the QMC sequences tested; the Brownian bridge construction usually does better than a straightforward construction and sometimes much better; the principal components construction usually outperforms the Brownian bridge construction. Moreover, the principal components construction is applicable to essentially any covariance matrix whereas the Brownian bridge construction is specific to the covariance matrix of a discretely sampled Brownian path.

As always, we must be careful in extrapolating from a limited set of numerical results. Different models for the underlying assets and different payoffs could lead to different conclusions. It is also important to keep in mind that none of the methods examined here exploits knowledge of the integrand. The performance of ordinary Monte Carlo, in particular, could in many cases be made superior to that of quasi Monte Carlo when this type of knowledge can be used to develop an effective variance reduction technique.

*Acknowledgments:* We thank Anargyros Papageorgiou and Joseph Traub for providing and discussing the FINDER code.

## 6 References

[1] ADLER, R.J., *An Introduction to Continuity, Extrema, and Related Topics for General Gaussian Processes*, Institute of Mathematical Statistics, Hayward, California, 1990.

[2] BERMAN, L., Comparison of Path Generation Methods for Monte Carlo Valuation of Single Underlying Derivative Securities, Research Report RC20570, IBM Research Division, Yorktown Heights, New York.

[3] BOYLE, P., M. BROADIE, AND P. GLASSERMAN, Monte Carlo Methods for Security Pricing, *J. Economic Dynamics and Control*, to appear.

[4] BRATLEY, P., AND B.L. FOX, ALGORITHM 659: Implementing Sobol's Quasirandom Sequence Generator, *ACM Trans. Mathematical Software* **14**, 88-100, 1988.

[5] BROADIE, M., P. GLASSERMAN, AND S. KOU, Connecting Discrete and Continuous Path-Dependent Options, Working Paper, Columbia Business School, New York, 1996.

[6] CAFLISCH, R.E., AND W. MOROKOFF, Valuation of Mortgage Backed Securities Using the Quasi-Monte Carlo Method, Working Paper, Department of Mathematics, UCLA, 1996.

[7] CAFLISCH, R.E., AND B. MOSKOWITZ, Modified Monte Carlo Methods Using Quasi-Random Sequences. In *Monte Carlo and Quasi-Monte Carlo Methods in Scientific Computing*, (H. Niederreiter and P.J.-S. Shiue, eds.), Springer-Verlag, New York, 1995.

[8] DUFFIE, D., *Dynamic Asset Pricing Theory*, Second Edition, Princeton University Press, 1996.

[9] FOX, B.L., ALGORITHM 647: Implementation and Relative Efficiency of Quasi-Random Sequence Generators, *ACM Trans. Mathematical Software* **12**, 362–376, 1986.

[10] FOX, B.L., Generating Poisson Processes by Quasi-Monte Carlo. Working Paper, SIM-OPT Consulting, Boulder, Colorado, 1996.

[11] HARRISON, J.M., AND D. KREPS, Martingales and Arbitrage in Multiperiod Securities Markets, *J. Economic Theory* **20**, 381–408, 1979.

[12] HARRISON, J.M., AND S. PLISKA, Martingales and Stochastic Integrals in the Theory of Continuous Trading, *Stochastic Processes and their Applications* **11**, 215–260, 1981.

[13] HULL, J.C., *Options, Futures, and Other Derivatives*, Third Edition, Prentice-Hall, Englewood Cliffs, New Jersey, 1997.

[14] JANICKI, A., AND A. WERON, *Simulation and Chaotic Behavior of $\alpha$-Stable Stochastic Processes*, Marcel-Dekker, New York, 1994.

[15] JARROW, R., AND S. TURNBULL, *Derivative Securities*, South-Western College Publishing, Cincinnati, Ohio, 1996.

[16] KARLIN, S., AND H. TAYLOR, *A First Course in Stochastic Processes*, Academic Press, New York, 1975.

[17] MORO, B., The Full Monte, *RISK* **8**, 57–58, February 1995.

[18] NIEDERREITER, H., *Random Number Generation and Quasi-Monte Carlo Methods*, SIAM, Philadelphia, 1992.

[19] PAPAGEORGIOU, A., AND J. TRAUB, Beating Monte Carlo, *Risk* **9**, 63–65, June 1996.

[20] PASKOV, S., AND J. TRAUB, Faster Valuation of Financial Derivatives, *J. Portfolio Management*, 113–120, Fall 1995.

[21] PRESS, W.H., S.A. TEUKOLSKY, W.T. VETTERLING, AND B.P. FLANNERY, *Numerical Recipes in C: The Art of Scientific Computing*, Second Edition, Cambridge University Press, 1992.

[22] ROGERS, C., AND D. TALAY, EDS., *Numerical Methods in Financial Mathematics*, Cambridge University Press, 1997.

[23] TEZUKA, S., *Uniform Random Numbers: Theory and Practice*, Kluwer Academic Publishers, Norwell, Massachusetts, 1995.

Peter A. Acworth
Doctoral Office, 808 Uris Hall
Columbia Business School
New York, NY 10027
pa73@columbia.edu

Mark Broadie
415 Uris Hall
Columbia Business School
New York, NY 10027
mbroadie@research.gsb.columbia.edu

Paul Glasserman
403 Uris Hall
Columbia Business School
New York, NY 10027
pglasser@research.gsb.columbia.edu

# Monte Carlo Methods: a powerful tool of statistical physics

Kurt Binder
Institut für Physik, Johannes Gutenberg-Universität Mainz
D-55099 Mainz, Germany

email:binder@chaplin.physik.uni-mainz.de

## Abstract:

Statistical mechanics of condensed matter systems (solids, fluids) tries to express macroscopic equilibrium properties of matter as averages computed from a Hamiltonian that expresses interactions of an atomistic many body system. While analytic methods for most problems involve crude and uncontrolled approximations, the Monte Carlo computer simulation method allows a numerically exact treatment of this problem, apart from "statistical errors" which can be made as small as desired, and the systematic problem that a system of finite size is treated rather than the thermodynamic limit. However, the simulations of phase transitions then elucidate how a symmetry breaking arises via breaking of ergodicity, if the Monte Carlo sampling is interpreted as a time average along a stochastic trajectory in phase space, in the thermodynamic limit. These concepts are illustrated for the transition paramagnet-ferromagnet of the Ising model, and unmixing transitions in polymer mixtures. As an example of the application of Monte Carlo to clarify questions about dynamic processes, simulations of interdiffusion in lattice models of alloys are discussed.

# 1. Introduction: Purpose and scope of simulations in statistical thermodynamics

Monte Carlo simulation is a tool of science that complements both analytical theory and experiment [1, 2, 3, 4, 5]. Computer simulation can clarify research problems in physics, when the direct comparison between theory and experiment is inconclusive.

A good example illustrating this statement is the interdiffusion in solid mixture (Fig. 1). This is a thermally activated hopping process, and typically the jump-rates of the two species may be orders of magnitude different, $\Gamma_A \ll \Gamma_B$. Now the description of the processes in terms of analytical theory is rather controversial, since very different conclusions are reached, depending on the exact nature of the assumptions and approximations. The "slow mode theory" [6, 7] suggests that interdiffusion is controlled by the selfdiffusion constant of the slow species, i.e. the interdiffusion constant $D_{int} \propto D_A^{self} \propto \Gamma_A$, while the "fast mode theory" suggests the contrary [8], $D_{int} \propto D_B^{self} \propto \Gamma_B$. Surprisingly, experimental evidence has been claimed [9] for both theories, and since these theories contradict each other, the comparison between theory and experiment obviously is problematic! Now such a comparison can be inconclusive since inadequacies of the simple model of Fig. 1 (such as the neglect of A-B-interactions) obscure inadequacies of the various approximations made. In contrast, the simulation [10] can study precisely the same model on which the theory is based. All parameters of the theory (e.g. "Onsager coefficients" in our case [6, 7, 8, 9] can be independently estimated [10], see Sec. 4.

So simulations do not suffer from uncontrolled approximations, but there are other caveats, of course: one must be aware of statistical errors [1, 2, 3, 4, 5, 11]; in addition, there are "controllable" systematic errors. As their name suggests, Monte Carlo methods [1, 2, 3, 4] rely on the use of random numbers, but the concept of "randomness" often is an idealization [12]. In particular, the "pseudo-random numbers" supplied by built-in computer routines use deterministic formulas and hence are never truly random in the sense of perfectly uncorrelated [13, 14]. Examples where this lack of randomness hampers applications in statistical physics have again and again been documented in the literature (see e.g. [15]). However, this problem is outside of consideration here, and rather we emphasize two other problems that computer simulations have.

(i) Finite size effects: simulations deal mostly with cubic boxes of size $L \times L \times L$ with periodic boundary conditions, and have typically between $N = 10^2$ to $10^6$ degrees of freedom. This situation is not identical with the thermodynamic limit ($L \to \infty$ and $N \to \infty$), usually considered in statistical mechanics, and thus systematic deviations from this limit must be carefully considered [16, 17, 18, 19]. This problem is obvious for second-order phase transitions, where the order parameter correlation length $\xi$ diverges — the growth of $\xi$ is limited by the system size $L$. The "finite size scaling" theory [16, 17, 18, 19] developed for this problem has in fact

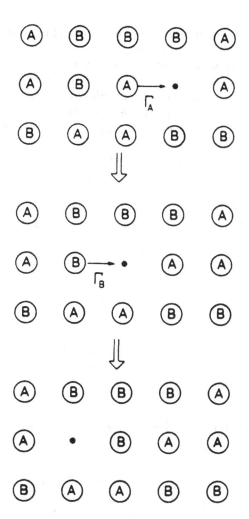

**Fig. 1:** Schematic description of interdiffusion in a randomly mixed crystal, where two different kinds of atoms (A,B) may occupy the sites of a regular lattice. Occupied sites are indicated by circles. Diffusion proceeds by the vacancy mechanism: A-atoms may jump to vacant sites with a jump rate $\Gamma_A$, B-atoms with a jump rate $\Gamma_B$. (For simplicity, it is assumed that corresponding pairwise interaction energies are zero and hence these jump rates do not depend on the occupation of neighboring lattice sites.)

become a powerful tool for the analysis of bulk critical phenomena, so this is no longer a serious limitation. But finite size effects are still a problem for weak first-order transitions — for problems such as the nearest neighbor Ising antiferromagnet in a magnetic field on a face-centered cubic lattice a volume of 1 million Ising spins is not enough [20]. There are many size effects unrelated to critical phenomena: e.g. a pathintegral Monte Carlo study of $Ar$ crystals at low temperatures did not yield the expected Debye law for the specific heat, $C \propto T^3$, but rather $C$ vanishes exponentially $\{C \propto \exp(-\Delta/T)\}$ [21]. This gap $\Delta$ arises because no acoustic phonons with wavelengths $\lambda$ larger than the box size $L$ are present.

(ii) The second problem is the finite "observation time" during which a simulated system evolves and the "dynamic correlations" that develop. We adopt here the dynamic interpretation of the Monte Carlo sampling [1, 2, 3, 22]: just as the Molecular Dynamics method [5] where by solving Newton's equation of motion for a many-body system one generates a deterministic trajectory through phase space, Monte Carlo methods generate a stochastic trajectory. This notion gets a precise meaning when the importance sampling Monte Carlo process [23] is considered as numerical realization of the associated Markovian master equation [22]. This yields a basis for the study of diffusion processes and relaxation phenomena by dynamic Monte Carlo methods [1, 2, 3, 4, 5], as well as for a discussion of the "statistical errors" [22] and systematic errors resulting from the finite length of the Monte Carlo trajectory [24].

## 2. Monte Carlo methods for equilibrium statistical mechanics

The basic task is to compute thermal averages $(\langle \ldots \rangle)$ of suitable observables from an atomistic many-body Hamiltonian. E.g., magnetic properties of anisotropic (uniaxial) solids may be described by the Ising model, with Hamiltonian

$$\mathcal{H}_{\text{Ising}} = -J \sum_{\langle i,j \rangle} S_i S_j - H \sum_i S_i, \qquad S_i = \pm 1. \qquad (1)$$

We assume a ferromagnetic interaction ($J > 0$) between nearest neighbors (n.n.) only ($\langle i, j \rangle$ means over n.n. is summed once), and $H$ is the magnetic field, $i$ labels the $N$ lattice sites. The phase space $\{\vec{x}\}$ is the set of all possible spin orientations, $\{S_1, S_2, \ldots, S_i, \ldots, S_N\}$. Another example is a (classical) fluid, $\{\vec{x}\}$ is the space of all coordinates $\{\vec{r}_1, \vec{r}_2, \ldots, \vec{r}_i, \ldots, \vec{r}_N\}$ of the atoms in a volume $V$, and the Hamiltonian is the sum of (pairwise) interactions that depend on the interatomic distance $r \equiv |\vec{r}_i - \vec{r}_j|$, e.g. the Lennard-Jones potential $\{v(r) = 4\epsilon[(\sigma/r)^{12} - (\sigma/r)^6]$, where $\epsilon$ sets the energy scale and $\sigma$ describes the potential range$\}$. In this canonical ensemble for the magnet temperature $T$, field $H$ and particle number $N$ are the independently given variables (for the fluid, one may have instead the NVT ensemble), and averages are then defined as follows

$$\langle A \rangle = \int d\vec{X}\, P_{eq}(\vec{X}) A(\vec{X}), \qquad P_{eq}(\vec{X}) = \exp[-\mathcal{H}(\vec{X})/k_B T]/Z, \quad (2)$$

$$Z = \int d\vec{x}\, \exp[-\mathcal{H}(\vec{X})/k_B T] = \text{partition function}, \qquad (3)$$

$k_B$ being Boltzmann's constant.

Now it is well known that problems as described by Eqs.(1)-(3) can be solved exactly in rare cases only, and simple approximations that neglect statistical fluctuations such as the mean field approximation (MFA) are inadequate. The basic idea of the Monte Carlo method hence is to sample these fluctuations by a numerical integration over phase space,

$$\langle A \rangle \approx \bar{A} = (1/M) \sum_\nu P_{eq}(\vec{X}_\nu) A(\vec{X}_\nu) \qquad (4)$$

The difficulty, of course, is that the integration space is very high-dimensional, and it is a problem to choose the sample of $M$ points $\vec{X}_\nu$ appropriately. It is easy to see that a regular grid of points $\vec{X}_\nu$ does not work, but a uniform random sampling of phase space ("simple sampling") does not work either in most cases. The reason is that for thermal averages (and $N$ large) it is only a very small region of phase space where the important contributions come from. This is best seen for specific examples, e.g. the Ising ferromagnet, Eq.(1): sampling each spin configuration completely at random, the probability to find a nonzero magnetization per spin $(m)$ is very small, namely

$$P_N^{ss}(m) \propto \exp(-m^2 N/2), \qquad m = (1/N) \sum_T S_i. \qquad (5)$$

(It is an elementary exercise of probability theory to show that a gaussian of width $1/\sqrt{N}$ centered at zero results.) On the other hand, the actual distribution is sharply peaked but in general at another value (e.g., for $T < T_c$ and $H = 0$ it is peaked at the "spontaneous magnetization $M_{\text{sp}}$, see Fig. 2 [17]). Actually, the distribution $P_L(m)$ is symmetric around $m = 0$ for $H = 0$, thus another peak occurs at $m = -M_{\text{sp}}$ that is not shown here.

It is clear that for large $N$ (and large $L$) almost no states would be generated from simple sampling in the region where $P_L(m)$ has its peak: most of the computational effort would be wasted for exploring a completely uninteresting part of the phase space. The same argument holds for other variables as well, e.g. the energy per spin $E = \langle \mathcal{H} \rangle /N$.

Thus a method is needed that leads us automatically in the important region of phase space, sampling points preferentially from the region which yields the peaks of distributions such as $P_L(m)$, $P_L(E)$, etc. Such a method actually exists, namely the importance sampling scheme of Metropolis et al. [23] chooses the states $\vec{X}_\nu$ with a probability proportional to the Boltzmann factor, $P_{eq}(\vec{X})$ {Eq.(2)}.

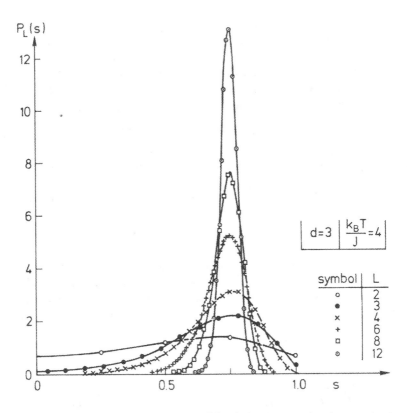

**Fig. 2:** Probability distribution $P_L(s)$ of the magnetization per spin of $L \times L \times L$ subsystems of a simple cubic Ising lattice with $N = 24^3$ spins and periodic boundary conditions for zero magnetic field and temperature $k_B T / J = 4.0$ {the critical temperature occurs at $k_B T_c / J \approx 4.51$, there $M_{\mathrm{sp}}$ vanishes as $M_{\mathrm{sp}} \propto (1 - T/T_c)^\beta$ with $\beta \approx 0.32$}. $P_L(s)$ is defined as $P_L(s) = \int d\vec{x}\, \sigma\, (s - L^{-d} \sum' S_i)\, P_{eq}(\vec{X})$, the $\sum'$ being extended over all possible subsystems of the $d$- dimensional volume $L^d$. From Binder [17].

In this scheme one generates a sequence of states $\vec{X}_\nu \to \vec{X}_{\nu+1} \to \vec{X}_{\nu+2} \to$ ... recursively one from the other, with a properly choosen transition probability $W(\vec{X} \to \vec{X}')$. The "move" $\vec{X} \to \vec{X}'$ may be chosen as is convenient for the considered model: for the Ising model, it may be a single spin flip, a spin exchange, or overturning of a large cluster of spins [25]. For a fluid, the simplest choice of a move is a random displacement of a randomly chosen particle, $\vec{r}_i' = \vec{r}_i + \delta\vec{r}_i$.

Now one can show that the states $\vec{X}_\nu$ are distributed according to the Boltzmann factor $P_{eq}(\vec{X})$, if the transition probability satisfies the condition of detailed balance,

$$P_{eq}(\vec{X})W(\vec{X} \to \vec{X}') = P_{eq}(\vec{X}')W(\vec{X}' \to \vec{X}) \qquad (6)$$

For example, one may choose (the constant $\tau_o$ is introduced to set a time scale)

$$W(\vec{X} \to \vec{X}') = \begin{cases} \tau_0^{-1}, & \text{if } \delta\mathcal{H} \equiv \mathcal{H}(\vec{X}') - \mathcal{H}(\vec{X}) < 0 \\ \tau_0^{-1} \exp(-\delta\mathcal{H}/k_BT) & \text{if } \delta\mathcal{H} > 0 \end{cases} \qquad (7)$$

This "time" unit $\tau_o$ is fixed such that 1 Monte Carlo step (MCS) per degree of freedom is the unit of this "Monte Carlo time". This notion makes sense since the Monte Carlo sampling can be interpreted [22] as "time averaging" along stochastic trajectories in phase space, described by a master equation for the probability $P(\vec{X}, t)$ that a state $\vec{X}$ occurs at time $t$,

$$\frac{dP(\vec{X},t)}{dt} = -\sum_{\vec{X}'} W(\vec{X} \to \vec{X}')P(\vec{X},t) + \sum_{\vec{X}'} W(\vec{X}' \to \vec{X})P(\vec{X}',t) \qquad (8)$$

This equation is a rate equation, describing the balance between the loss of probability by all processes $\vec{X} \to \vec{X}'$ leading away from the considered state and the gain of probability due to the inverse processes. Obviously, $dP(\vec{X},t)/dt = 0$ if $P(\vec{X},t) = P_{eq}(\vec{X}) =$ equilibrium is the stationary solution of the master equation.

Of course, the dynamical properties of a system described by such a stochastic trajectory (Monte Carlo is a realization of a Markov process) differs in general from dynamic properties derived from a deterministic trajectory (remember that the Molecular Dynamics [MD] method amounts to solve Newton's equation of motion, and this method is in fact the only reasonable way to proceed if one wants to know problems such as dynamic properties of Lennard-Jones fluids). But Monte Carlo (MC) is a reasonable method for describing dynamic properties of systems where the considered degrees of freedoms are a slow subset of all degrees of freedom, weakly coupled to the fast ones, that then act like a heat bath. This is true for Ising magnets, as well as for diffusion in solid alloys (Fig. 1), where the phonons of the crystal act like a heat bath [1, 2, 3]. There are also problems, like the slow Brownian motion of polymer chains in melts, where either MC or MD can be applied [4].

Now a few words are in order to describe how the algorithm defined by Eq.(7) is realized in practice: one chooses a random number $n$ that is uniformly distributed in the unit interval $[0,1]$, and compares $W(\vec{X} \to \vec{X}')$ with $n$: if $W \geq n$, the trial configuration $\vec{X}'$ is accepted as a new configuration and counted in the average. If $W < n$, one rejects $\vec{X}'$, counts the old configuration $\vec{X}$ once more; and this process is repeated again and again. It then follows that the "time average" $\bar{A}$ converges to the correct ensemble average $\langle A \rangle$ {Eq.(2)},

$$\bar{A} = \frac{1}{M - M_o} \sum_{\nu = M_o + 1}^{M} A(\vec{X}_\nu) = \frac{1}{t - t_o} \int_{t_o}^{t} dt' A(\vec{X}, t) \qquad (9)$$

The first $M_o$ configurations are omitted from the average in Eq.(8), to diminish the influence of the arbitrary starting configuration $\vec{X}_1$ which is not characteristic of equilibrium (e.g., for the Ising model one may start from a state with all spins up, or a random spin configuration). The "times" $t, t', t_o$ in Eq.(8) are defined as $t = M/N$, $t' = \nu/N$, $t_o = M_o/N$, invoking the time unit MCS mentioned above. One can show that for $M \to \infty$ the states $\{\vec{X}_\nu\}$ are indeed distributed proportional to $P_{eq}(\vec{X})$ irrespective of the starting condition (the state $\vec{X}_1$).

# 3. An application example: studying the phase transition of the Ising model with MC

For lattices of dimensionality $d \geq 2$, the model Eq.(1) has at zero field $H = 0$ a second-order phase transition at a temperature $T = T_c$ from a paramagnetic state at $T > T_c$ to a ferromagnetic state for $T < T_c$. This critical temperature implies critical singularities: the spontaneous magnetization vanishes according to a power law, $M \propto (1 - T/T_c)^\beta$ where $\beta$ is the critical exponent of this "order parameter" and the susceptibility diverges: $\chi \equiv (\partial M/\partial H)_T \propto |1 - T/T_c|^{-\gamma}$. For $T < T_c$, one observes a first-order transition at $H = 0$ when $H$ is varied from positive to negative fields = the magnetization {first derivative of the free energy $F$, $M = -(\partial F/\partial H)_T$} exhibits a jump, and $\chi$ has then a delta-function singularity at $H = 0$. However, all these phase transitions occur in the thermodynamic limit ($L \to \infty$) only, for finite lattice linear dimension $L$ the critical singularities are rounded and shifted [16, 17, 18, 19]. Qualitatively this finite size rounding can be understood considering the order parameter distribution $P_L(m)$ and its moments (Fig. 3). In the paramagnetic region, for sizes $L \gg \xi$, the correlation length of order parameter fluctuations ($\xi \propto |1 - T/T_c|^{-\nu}$), one has a gaussian of width $\sqrt{k_B T \chi/L^d}$. As $T$ approaches $T_c$, $\xi$ grows until $\xi$ becomes comparable to $L$: then the width is of order $(\xi^{\gamma/\nu}/L^d)^{1/2}$ $(L^{\gamma/\nu - d})^{1/2} = L^{-\beta/\nu}$, invoking the hyperscaling relation ($d\nu = \gamma + 2\beta$) between the critical exponents. In the critical region near $T_c$, $P_L(m)$ is strongly non-gaussian and actually develops gradually a double-peak structure. For $T < T_c$, the correlation length shrinks again,

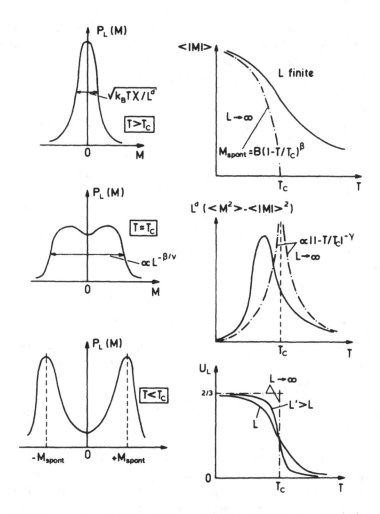

**Fig. 3:** Schematic variation of the probability distribution $P_L(m)$ to find a magnetization $m$ in a finite Ising system at $H = 0$, implying a cubic box of linear dimension $L$ with periodic boundary conditions (left part). The associated temperature variation of the average order parameter $\langle|m|\rangle$, the "susceptibility" $k_B T \chi' = L^d(\langle m^2 \rangle - \langle|m|\rangle^2)$ and reduced fourth-order cumulant $U_L = 1 - \langle m^4 \rangle/[3\langle m^2 \rangle^2]$ are also shown (right part).

and the two peaks get more and more separated, finally one gets two well-separated gaussians, one centered at $+M_{sp}$, the other centered at $-M_{sp}$. In this way one can understood how it happens that a true spontaneous magnetization results in the thermodynamic limit, while $\langle m \rangle \equiv 0$ for $H = 0$ for finite systems: for any finite system there is always a nonzero probability to move from the right peak to the left peak or vice versa.

The principle of finite size scaling theory [16, 17, 18, 19] thus is a comparison of characteristic lengths in the system: $L$ scales with $\xi$, the order parameter correlation length. Since this length diverges as $\xi \propto (1 - T/T_c)^{-\nu}$, the right scaling variable is $L/\xi$ or $L(1 - T/T_c)^{\nu} \equiv Lt^{\nu}$ or $tL^{1/\nu}$. A technique to extrapolate via finite size scaling to the thermodynamic limit is "data collapsing": e.g., one has $m = L^{-\beta/\nu}\tilde{m}(t\,L^{1/\nu})$, where $\tilde{m}$ is a "scaling function", and thus all data for $m = m(T, L)$ should collapse on a single curve when plotted as $mL^{\beta/\nu}$ vs. $t\,L^{1/\nu}$ (provided $T_c$, $\nu$ and $\beta$ are known). As an example, Fig. 4 shows such an analysis for the square lattice Ising antiferromagnet with nearest and next nearest neighbor exchange $(J_{nnn}/J_{nn} = 1)$ [26]. In this case one has an order-disorder transition from the paramagnetic to the "super-antiferromagnetic" structure (one-dimensional ferromagnetically aligned rows of up spins alternate with rows of down spins). Critical exponents for this problem are not those of the exactly solved [27] nearest neighbor problem $(\beta = 1/8, \nu = 1.0)$ but were unknown. Choosing the "wrong" exponents (upper part of Fig. 4) the points systematically deviate from a unique function, while the "true" exponents (lower part of Fig. 4) are found $(\beta \approx 0.10, \nu \approx 0.85)$ by the condition of optimal "collapse" on a unique function, which must be a straight line $\{\ln \tilde{m}(\zeta) = \beta \ln \zeta + \text{const}\}$ for large $\zeta = tL^{1/\nu}$.

We emphasize here that Fig. 4 is only one of the first examples where previously unknown critical properties were extracted from Monte Carlo simulations [26]. Meanwhile the techniques of analysis have been enormously refinded, and since much larger lattices can be simulated than used in Fig. 4 (and much better statistics is possible, due to much faster computers) widespread applications are possible. As an example, Fig. 5 shows the distribution of the order parameter $P(m)$ for a symmetric mixture of flexible polymers (chain lengths $N_A = N_B = N = 128$). Here the various pairs of monomers interact with energies $\epsilon_{AB} = -\epsilon_{AA} = -\epsilon_{BB} = \epsilon$ if their distance on the simple cubic lattice does not exceed $\sqrt{6}$ lattice spacings. $P(m)$ indeed changes smoothly with temperature, as anticipated in the schematic figure 3. Note that in this case the two peaks of $P(m)$ for $T < T_c$ correspond to phase coexistenc between a B-rich phase ($m$ negative) and an A-rich phase ($m$ positive) [28].

Returning to the Ising model again, we now consider the first-order transition for $T < T_c$ as function of the field, where in the thermodynamic limit a delta-function singularity is expected for the susceptibility $\chi(H)$ at $H = 0$. Recording $\chi(H)$ from magnetization fluctuations $\{\chi_L \equiv L^d(\langle m^2 \rangle - \langle m^2 \rangle^2)/k_B T\}$ we see that this peak is again rounded off for finite lattices (Fig.6 [29]). This behavior again can be understood in

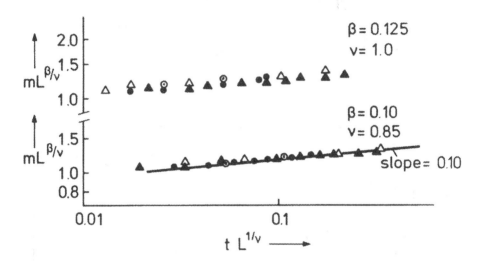

**Fig. 4:** Finite size scaling plot of the order parameter $m = \sqrt{\psi_1^2 + \psi_2^2}$ of the Ising antiferromagnet with nearest and next nearest neighbor exchange ($J_{nnn} = J_{nn}$) on the square lattice. Here $\psi_1$, $\psi_2$ are the two order parameter components $\{\psi_1 = 1, \psi_2 = 0$ if ferromagnetic rows alternate in sign ($\uparrow\downarrow$) along $x$ direction, $\psi_1 = 0$, $\psi_2 = 1$ if they alternate in $y$ direction on the lattice$\}$. Data are for $L = 10$ (full triangles), 20 (full circles), 40 (open triangles), and 60 (open circles). From Binder and Landau [26].

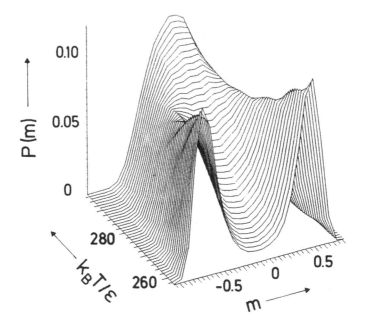

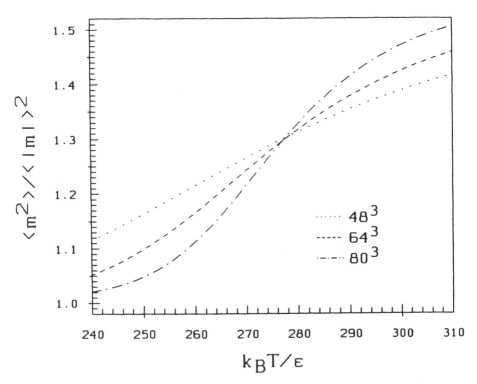

**Fig.5 :** Distribution of the order paramter $P(m)$ as function of reduced temperature for a polymer mixture (A,B). The system is a cubic box of linear dimension $L = 80$ containing $n = n_A + n_B = 250$ chains. The order

**Fig.5 ff :** parameter of the mixture is $m = (n_A - n_B)/(n_A + n_B)$. Recording ratios of moments (such as $\langle m^2 \rangle / \langle |m|^2 \rangle$ or the cumulant $U_L$ in Fig.3) vs. temperature for different sizes $L$ one finds $T_c$ from the intersection of these curves (lower part). From Deutsch and Binder [28].

terms of the double gaussian approximation for $P_L(m)$ at $T < T_c$ (Fig. 3), noting that for $H = 0$ the peaks have equal weight (and then the fluctuation of $m$ is maximal) while for $H \neq 0$ the weights of the peaks are given by factors proportional to $\exp\left(\pm \frac{mL^dH}{k_BT}\right)$, i.e. Boltzmann factors involving the Zeemann energy. Working this idea [29] out one finds again a scaling description for the finite size rounding at first-order transitions, namely [29] for $d = 2$

$$\frac{\chi_L}{L^2} = \frac{\chi_\infty}{L^2} + \frac{M_{sp}^2}{k_BT}/\cosh^2(HM_{sp}L^2/k_BT) \tag{10}$$

Using the exact result for $M_{sp}$ [27] one calculates the curve shown in Fig. 7 (neglecting the correction $\chi_\infty/L^2$) which indeed nicely describes the Monte Carlo results without any adjustable parameters. Thus the problem of size effects at phase transitions is rather well understood, and can be taken into account in the analysis of Monte Carlo simulations [1, 2, 3, 4, 5].

# 4. An application to problems in nonequilibrium statistical thermodynamics: interdiffusion in mixtures

We now return to the problem of Fig. 1, keeping in mind that Monte Carlo methods can simulate dynamic processes when the considered degrees of freedom are weakly coupled to much faster ones that form a heatbath. This is the case for a lattice model of binary alloys, the considered degrees of freedom being occupation variables $c_i^A$, $c_i^B$ of lattice sites $i$ which are unity if the site $i$ is taken by an A atom or B atom, respectively, and zero else (phonons of the crystal are then thought to induce random hops with jump rates $\Gamma_A$, $\Gamma_B$ to vacant lattice sites. Many such random hops lead to a description in terms of diffusion).

Now the framework of nonequilibrium statistical thermodynamics is not atomistic but rather phenomenological. One postulates "constitutive equations" for the current densities $\vec{J}_A$, $\vec{J}_B$ of A,B atoms, namely linear relations between them and the driving forces, the gradients of chemical potential differences between $A(B)$ atoms and vacancies $V$ $(\mu_A - \mu_V, \mu_B - \mu_V)$:

$$\vec{J}_A = -(\wedge_{AA}/k_BT)\,\nabla\,(\mu_A - \mu_V) - (\wedge_{AB}/k_BT)\,\nabla\,(\mu_B - \mu_V), \tag{11}$$

$$\vec{J}_B = -(\wedge_{BA}/k_BT)\,\nabla\,(\mu_A - \mu_V) - (\wedge_{BB}/k_BT)\,\nabla\,(\mu_B - \mu_V). \tag{12}$$

Here $\wedge_{AA}$, $\wedge_{AB} = \wedge_{BA}$, $\wedge_{BB}$ are the "Onsager coefficients". Eqs.(11,12) are at best approximately valid of course (nonlinearities and fluctuations are neglected).

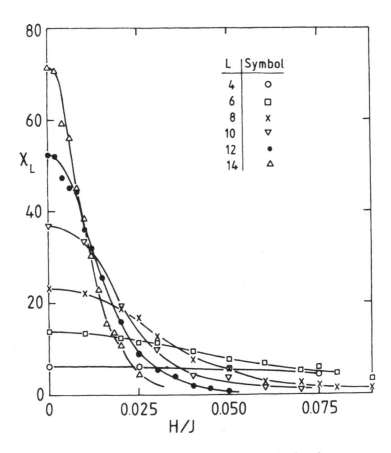

**Fig. 6:** Susceptibility $\chi_L$ vs. reduced magnetic field for the nearest neighbor Ising square lattice at a temperature $k_B T/J = 2.1$ (where the correlation length $\xi$ is about 3 lattice spacings; note $k_B T_c/J \equiv 2.269$ [27]). Since for this model $\chi_L(-H) = \chi_L(H)$, there is no shift of the peak with finite size, and only positive fields are shown. From Binder and Landau [29].

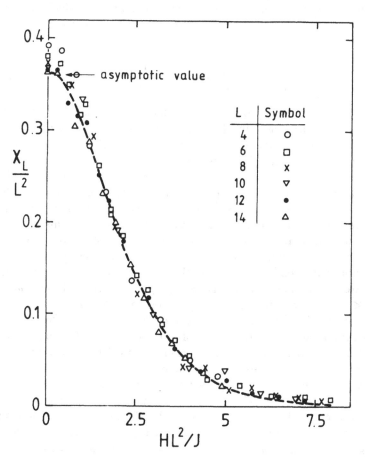

**Fig. 7:** Scaling plot of the susceptibility as function of the scaled field, using the data of fig.6, for the two-dimensional Ising nearest neighbor square lattice at $k_B T/J = 2.1$. Curve shows Eq.(10). From Binder and Landau [29].

Taken together with the continuity equations that express the conservation laws for the local concentrations $c_A(\vec{r}, t)$, $c_B(\vec{r}, t)$

$$\partial c_A(\vec{r},t)/\partial t + \nabla \cdot \vec{J}_A = 0, \qquad \partial c_B(\vec{r},t)/\partial t + \nabla \cdot \vec{J}_B = 0, \qquad (13)$$

it is a matter of simple algebra [10] to obtain a complete description of the interdiffusion process. E.g. the interdiffusion constant $D_i$ (which describes how a weak concentration difference between A and B spreads out) is given by [10]

$$D_i = \frac{\Lambda_{AA} \, \Lambda_{BB} - \Lambda_{AB}^2}{\Lambda_{AA} + 2\,\Lambda_{AB} + \Lambda_{BB}} \left( \frac{1}{c_A} + \frac{1}{c_B} \right), \qquad c_V \to 0. \qquad (14)$$

However, there are many problems with such a treatment: Where do the Onsager coefficients come from? How are they related to the atomistic rates $\Gamma_A$, $\Gamma_B$ (Fig.1)? Is the "mean field" character of Eqs.(11),(12) an accurate description? etc. In particular, it is common [6, 7, 8, 9] to neglect the off-diagonal coefficient $\Lambda_{AB}$ (since nothing is known about it!) in comparison with the diagonal ones, but it is questionable whether this is reasonable. All the questions can be answered by "taylored" computer experiments: by imposing chemical potential gradients either on the A-atoms or on the B-atoms one can create steady state currents in the system (particles leaving the box at one boundary reenter at the opposite one, because of the periodic boundary conditions!) Thus the Onsager coefficients can simply be measured from their definitions, Eqs.(11),(12). Fig. 8 shows that for $\Gamma_A/\Gamma_B \ll 1$ it is wrong to neglect $\Lambda_{AB}$ in comparison to $\Lambda_{AA}$. However, using the determined Onsager coefficients in the formula, Eq. (14) provides an accurate description of interdiffusion, as Fig. 9 shows. Fig. 9 also illustrates that another concept of statistical physics can be implemented directly in simulations, namely linear response: we apply a wavevector-dependent chemical potential difference $\Delta\mu(\vec{k})$ $\{k = 2\pi/\lambda\}$ to the system, to prepare an initial state of the model where a concentration wave with wavelength $\lambda$ is present. Typically the amplitude is chosen such that [10] $\delta c_A(t = 0) = \delta c_B(t = 0) = 0.02$. At time $t = 0$, this perturbation $\Delta\mu(\vec{k})$ is suddenly switched off, and then one simply watches the decay of the concentration wave with time. Different wavelengths (Fig.9) are used to check that one is actually in the long wavelength limit. And while the full mean field treatment {Eq.(14)} based on the actual Onsager coefficients works well, approximations [6, 7, 8, 9] where the Onsager coefficients are somehow related to selfdiffusion coefficients fail badly [10].

## 5. Conclusions

In this brief view, it was shown that in physics the technique of Monte Carlo simulation plays a crucial role, complementing both analytical theory and experiment. Comparison between analytical theory and experiment often is inconclusive: analytical theory needs both a model (e.g. a Hamiltonian)

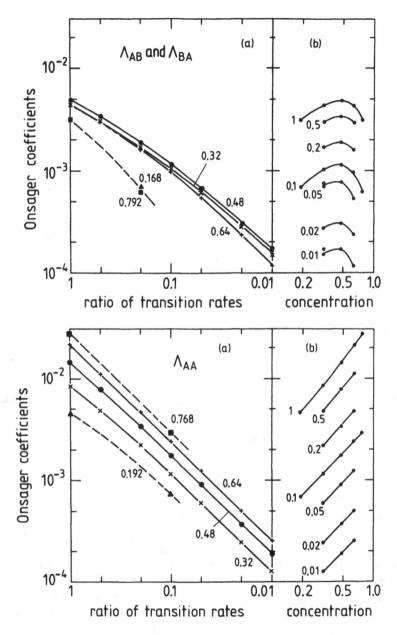

**Fig. 8:** Onsager coefficients $\wedge_{AB}$ (upper part) and $\wedge_{AA}$ (lower part of a two-dimensional noninteracting random alloy model (ABV mdoel) plotted vs. $\Gamma_A/\Gamma_B$ (left part) or concentration $c_A$ (right part), with $c_A$ as parameter (a) or $\Gamma_A/\Gamma_B$ as parameter (b). All data were obtained from $L \times L$ square lattices with $L = 80$ and $c_V = 0.004$. Curves are guides to the eye only. From Kehr et al. [10].

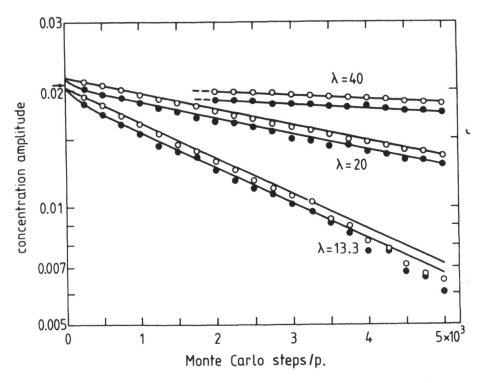

**Fig. 9:** Amplitudes of concentration waves with wavelengths $\lambda$ as a function of time $t$ (in units of Monte Carlo steps (MCS) per particle), after a chemical potential variation with that wavelength has been shut off at $t = 0$. Open circles represent A-atoms, full dots B-atoms, for a lattice of $L^3$ sites, $L = 80$, $c_A = c_B = 0.48$, $c_V = 0.04$, $\Gamma_A/\Gamma_B = 0.1$. Three different wavelengths $\lambda$ are shown (the arrow indicates the initial concentration amplitude for $\lambda = 40$). Note that one must choose $\lambda\nu = L$ with $\nu$ integer, to comply with the periodic boundary conditions. The curves represent theoretical predictions, based on the use of actual Onsager coefficients (cf. Fig.8) in a mean-field theory based on Eqs.(11)-(13) {for $c_V \to 0$ this theory predicts a single exponential decay proportional to $\exp[-D_i(2\pi/\lambda)^2 t]$ with $D_i$ given by Eq.(14)}. From Kehr et al. [10].

with undetermined parameters and mathematical approximations. The adjustable parameters obscure the comparison with experiment — often an acceptable fit of a wrong theory to experiment results; and also experiments have their problems — often the physical system is incompletely characterized, and the incomplete information may even be obscured by inessential details. Comparing computer simulation and experiment, on the other hand, one learns whether a model captures the essential physics. Comparing simulation and analytical theory one can perform stringent tests of approximations, since simulations use simple well-characterized models, there are no approximations, and full microscopic detail is available that allows to check intermediate steps of a theoretical derivation.

But one must be aware of technical difficulties of simulations (such as problems of "bad" random numbers, finite size effects, etc.) and thus the successful strategy of a simulation needs careful consideration. A few "rules of thumb" are: simplify the model as much as possible! Only very efficient computer programs yield conclusive answers, otherwise statistical errors may still be too large. And most important is: think about the problem first, and tell the computer what to measure! Finally it is useful to first test the model by simple limiting cases.

## Acknowledgements:

This overview is based on joint research with H.-P. Deutsch, K.W. Kehr, D.P. Landau and S. Reulein. Support is also acknowledged from the Deutsche Forschungs-Gemeinschaft (grants No. Bi314/3 and SFB 262/D1).

## References

[1] K. Binder (ed.) *Monte Carlo Methods in Statistical Physics, 2nd ed.* (Springer, Berlin 1986)

[2] K. Binder (ed.) *Applications of the Monte Carlo Method in Statistical Physics, 2nd ed.* (Springer, Berlin 1987)

[3] K. Binder (ed.) *The Monte Carlo Method in Condensed Matter Physics, 2nd ed.* (Springer, Berlin 1995

[4] K. Binder (ed.) *Monte Carlo and Molecular Dynamics Simulations in Polymer Science* (Oxford University Press, New York 1995)

[5] K. Binder and G. Ciccotti (ed.) *Monte Carlo and Molecular Dynamics of Condensed Matter Systems* (Societá Italiana di Fisica, Bologna 1996)

[6] F. Brochard, J. Jouffray, and P. Levinson, Macromolecules **16**, 2638 (1983)

38

[7] K. Binder, J. Chem. Phys. **79**, 6387 (1983); Colloid & Polymer Sci. **265**, 273 (1987)

[8] E. J. Kramer, P. Green, and C. J. Palmstrom, Polymer **25**, 437 (1984)

[9] For a review, see e.g. K. Binder and H. Sillescu, in *Encyclopedia of Polymer Science and Engineering, Suppl. Vol.* (H. Mark, ed.) p. 297 (Wiley, New York 1979)

[10] K. W. Kehr, K. Binder, and S. M. Reulein, Phys. Rev. **B 39**, 4891 (1989)

[11] M. P. Allen and D. J. Tildesley, *Computer Simulation of Liquids* (Clarendon, Oxford 1987)

[12] A. Compagner, Am. J. Phys. **59**, 700 (1991)

[13] D. Knuth, *The Art of Computer Programming, Vol. 2* (Addison-Wesley, Reading MA, 1969)

[14] F. James, Computer Phys. Commun. **60**, 329 (1990)

[15] A. M. Ferrenberg, D. P. Landau, and Y. J. Wong, Phys. Rev. Lett. **69**, 3382 (1992)

[16] M. N. Barber, in *Phase Transitions and Critical Phenomena, Vol. 8* (C. Domb and J. L. Lebowitz, eds.) Chap. 2 (Academic Press, New York 1983)

[17] K. Binder, Z. Physik **B 43**, 119 (1981)

[18] V. Privman (ed.) *Finite Size Scaling and Numerical Simulation of Statistical Systems* (World Scientific, Singapore 1990)

[19] K. Binder, in *Computational Methods in Field Theory* (H. Gausterer and C. B. Lang, eds.) p. 59 (Springer, Berlin 1992)

[20] S. Kämmerer, B. Dünweg, M. d'Onorio de Meo and K. Binder, Phys. Rev. **B 53**, 2345 (1996)

[21] M. H. Müser, P. Nielaba and K. Binder, Phys. Rev. **B 51**, 2723 (1995)

[22] H. Müller-Krumbhaar and K. Binder, J. Stat. Phys. **8**, 1 (1973)

[23] N. Metropolis, A. W. Rosenbluth, M. N. Rosenbluth, A. M. Teller and E. Teller, J. Chem. Phys. **21**, 1087 (1953)

[24] A. M. Ferrenberg, D. P. Landau and K. Binder, J. Stat. Phys. **63**, 867 (1991)

[25] R. H. Swendsen, J. S. Wang and A. M. Ferrenberg, in Ref.3, Chapter 4

[26] K. Binder and D. P. Landau, Phys. Rev. **B 21**, 1941 (1980)

[27] L. Onsager, Phys. Rev. **65**, 117 (1944); N. Yang, Phys. Rev. **85**, 808 (1952)

[28] H. P. Deutsch and K. Binder, Macromolecules **25**, 6214 (1992)

[29] K. Binder and D. P. Landau, Phys. Rev. **30**, 1477 (1984)

# Binary search trees based on Weyl and Lehmer sequences

Luc Devroye
School of Computer Science
McGill University
Montreal, Canada H3A 2K6
luc@cs.mcgill.ca

## 1 Introduction

This paper is based upon the presentation at the meeting in Salzburg. As a courtesy to those who attended the meeting, I will try to faithfully reproduce—with minor omissions and additions—what I said at that meeting. There are two basic background references for mathematical details, Devroye (1987) and Devroye and Goudjil (1996).

The purpose of the study is to amass even more evidence than already exists regarding the inherent dangers of using certain random number generators in simulations. For example, linear congruential generators (based upon the recurrences $x_n = (ax_n + b) \bmod m$ for $a, b, m, x_n$ all integer) derive their popularity from easily understood properties of Weyl and Lehmer sequences. As such generators produce integer sequences that are necessarily periodic, they can strictly speaking not be compared with aperiodic sequences such as $\{n\theta\}$ for $\theta$ irrational. However, they are fundamental akin to Weyl sequences, which may be obtained by the recurrence

$$x_{n+1} = \{x_n + \theta\} , \ x_0 = 0 ,$$

and Lehmer sequences, which may be obtained by the recurrence

$$x_{n+1} = \{ax_n\} , x_0 \in (0, 1) ,$$

We study binary search trees formed by consecutive (standard) insertions of numbers $\{\theta\}, \{2\theta\}, \{3\theta\}, \ldots$, where $\theta \in (0, 1)$ is an irrational number, and $\{.\}$ denotes "mod 1". The sequence in question is called the <u>Weyl sequence</u> for $\theta$, after Weyl (but see also Bohl, 1909), who showed that for all irrational $\theta$ the sequence is equi-distributed. We recall that a sequence $x_n$, $n \geq 1$, is _equi-distributed_ if for all $0 \leq a \leq b \leq 1$,

$$\lim_{n \to \infty} \frac{1}{n} \sum_{i=1}^{n} I_{x_i \in [a,b]} = b - a$$

(see Freiberger and Grenander (1971), Hlawka (1984) or Kuipers and Niederreiter (1974)).

The equi-distribution property makes Weyl sequences, or suitable generalizations of them, prime candidates for pseudo-random number generation. Of course, various regularities in the sequence make them rather unsuitable for most purposes (see, e.g., Knuth, 1981). Let $\mathcal{T}_n(\theta)$ be the binary search tree based upon the first $n$ numbers in the Weyl sequence for $\theta$. This tree, called the _Weyl tree_, captures a lot of refined information regarding the structure of the Weyl sequence, and is a fundamental tool for the analysis of algorithms involving Weyl sequences in the input stream. Computer scientists are mostly concerned with the following structural qualities: the average depth of a node (the depth is the path distance from a node to the root), the height (the maximal depth), and the number of leaves (the number of nodes with no children). The discussion in this paper focuses on these quantities. We relate various properties of the structure of these trees to the continued fraction expansion of $\theta$. If $H_n$ is the height of the tree with $n$ nodes when $\theta$ is chosen at random and uniformly on $[0, 1]$, then we show that in probability, $H_n \sim (12/\pi^2) \log n \, \log \log n$. We conclude with some properties of binary search trees for Lehmer sequences $\{x_0 a^n\}$ where $x_0$ is a real number and $a > 1$ is an integer.

## 2 Random binary search trees

Binary search trees are binary trees in which each datum is associated with a node, and each node has the search tree property, that is, all nodes in its left subtree have smaller values, and all nodes in its right subtree have larger values. Nodes have two possible children. There are actual children (which are nodes) and potential children (places for future placement of nodes). Potential children are called external nodes. A binary tree with $n$ nodes has $n + 1$ external nodes. Nodes without children are called leav s. Standard insertion of datum $x$ proceeds by finding the unique external node that could accept $x$, given the binary search tree property, and placing $x$ there. If a tree is constructed in this manner from an i.i.d. sequence $X_1, \ldots, X_n$ (drawn from a uniform distribution on $[0, 1]$), or from a random permutation of $\{1, \ldots, n\}$, it is called a random binary search tree, and will be denoted by $\mathcal{R}_n$. These trees may easily be drawn by placing $X_i$ at $(X_i, -i)$, as is done below. In this manner, one easily sees the binary search tree property in action.

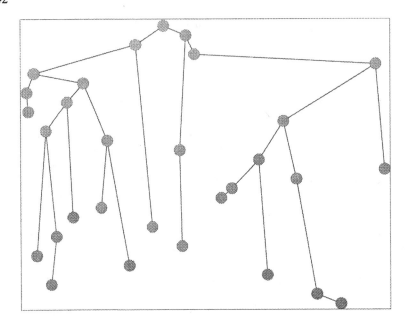

FIGURE 1. A random binary search tree in which $X_i$ is placed at $(X_i, -i)$ in the plane. As the projections of the points show the data in order, the binary search tree is a tool for sorting and for unraveling permutations.

Mostly everything is known about the behavior of $\mathcal{R}_n$ (see Mahmoud, 1992, for a survey). For example, the depth $D_n$ of $X_n$ (that is, the path distance to the root) satisfies

$$\frac{D_n}{2\log n} \to 1 \text{ in probability}$$

(Lynch, 1965, and Knuth, 1973). In fact, $(D_n - 2\log n)/\sqrt{2\log n} \overset{\mathcal{L}}{\to}$ normal $(0, 1)$, where $\overset{\mathcal{L}}{\to}$ denotes convergence in distribution (Devroye, 1988). Therefore, most nodes may be found at distance $2\log n \pm O(\sqrt{\log n})$ from the root. The height $H'_n$ of $\mathcal{R}_n$ satisfies

$$\frac{H'_n}{\log n} \to 4.31107\ldots \text{ almost surely}$$

(Robson, 1979, 1982; Devroye, 1986, 1987). All levels of the tree are full up to about level $0.3711\ldots\log n$ (Devroye, 1986). The constants given above are the two solutions of the equation $c\log(2e/c) = 1$. These properties may be used for the purpose of comparison, as they exemplify ideal behavior. Departures of randomness will very likely result in trees with radically different behavior. All of the above properties are summarized in the figure below.

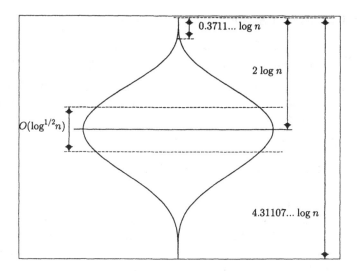

FIGURE 2. The properties of a random binary search tree are summarized. The width of the blob is proportional to the number of nodes at a given level.

## 3 The structure of Weyl trees

The following notation will be used. The height of $T_n(\theta)$ is $H_n(\theta)$. The set of leaves of $T_n(\theta)$ is $\mathcal{L}_n(\theta)$. The collection of $n+1$ possible positions for a new node to be added to $T_n(\theta)$ is called the set of external nodes, and is denoted by $\mathcal{E}_n(\theta)$. When $\theta$ is understood, the suffix $(\theta)$ will be dropped from the notation. The collection $\mathcal{E}_n$ may be split into $\mathcal{E}_n^R$ and $\mathcal{E}_n^L$, where $\mathcal{E}_n^R$ has those nodes that are right children, and $\mathcal{E}_n^L$ collects all left children in $\mathcal{E}_n$.

In this section, we look at the structure of a Weyl tree for fixed irrational $\theta$. Crucial connections are made with the continued fraction for $\theta$. Well-known properties of continued fractions then permit us to deduce results about $H_n$ and $|\mathcal{L}_n|$ without too much trouble.

Let $1 = T_1 < T_2 < \cdots$ be the <u>record times</u>, i.e., the times at which $x_n = \{n\theta\}$ is minimum or maximum among $x_1, \ldots, x_n$. The times of occurrence of a minimum or maximum are denoted by $L_n$ and $R_n$, and the indices of these sequences are synchronized with the $T_i$'s as follows:

$$(L_n, R_n) = \begin{cases} (L_{n-1}, T_n) & \text{if at } T_n \text{ there is a maximum;} \\ (T_n, R_{n-1}) & \text{if at } T_n \text{ there is a minimum.} \end{cases}$$

As it turns out, there is a lot of structure in these sequences. The fundamental property in this respect is the following.

**Lemma 1 (Ellis and Steele, 1981).** *We have*

$$(L_n, R_n) = \begin{cases} (L_{n-1}, L_{n-1} + R_{n-1}) & \text{if at } T_n \text{ there is a maximum;} \\ (L_{n-1} + R_{n-1}, R_{n-1}) & \text{if at } T_n \text{ there is a minimum.} \end{cases}$$

*Let $k$ be the smallest integer such that $n < L_k + R_k$. Then, if $x_{(1)} < \ldots < x_{(n)}$ denotes the ordered sequence for $x_1, \ldots, x_n$, then the indices $(1), \ldots, (n)$ coincide with*

$$\{L_k, 2L_k, 3L_k, \ldots, \} \pmod{(L_k + R_k)} \cap \{1, \ldots, n\} .$$

*Also, $(L_n, R_n)$ are relatively prime for all $n$.*

A quick verification: if $n = L_k + R_k - 1$, then the index of the maximum is $(L_k + R_k - 1)L_k \pmod{(L + k + R_k)} = -L_k \pmod{(L_k + R_k)} = R_k$, as was expected. This Lemma says that at $n = L_k + R_k - 1$, the shape of the binary search tree for $x_1, \ldots, x_n$ is entirely determined by the two numbers $L_k$ and $R_k$. In fact, then, there are only $O(n^2)$ possible Weyl search trees with $n$ elements, even though there are $\frac{1}{n+1}\binom{2n}{n} = \Theta(4^n/n^{3/2})$ possible binary search trees on $n$ nodes. As the simplest example, consider the five binary search trees on 3 nodes. Two of these are impossible to obtain as Weyl trees (the ones in which the root has one child and the child has one child but of different polarity). This fact was used by Ellis and Steele to derive a method that would sort any Weyl sequence using comparisons only (thus, without being capable of numerically inspecting entries) in $O(\log n)$ comparisons.

There is a natural way of looking at the growth of the Weyl search tree in layers. The $(i + 1) - st$ layer consists of all $x_j$ with $T_i \leq j \leq T_{i+1} - 1$. A special role is played also by the ancestor tree $T_{T_i - 1}$. A layer can be considered as a new coat of leaves painted on the ancestor tree. Each layer adds one and just one coat, as the next Lemma explains.

**Lemma 2 (Devroye and Goudjil, 1996).** *All nodes in the $(i + 1)$-st layer are leaves, and all leaves of $T_{T_{i+1}-1}$ are in the $(i + 1)$-st layer. All nodes in the $(i + 1) - st$ layer are either right children or left children, but not both. In fact,*

$$|\mathcal{E}^L_{T_{i+1}-1}| = R_i , |\mathcal{E}^R_{T_{i+1}-1}| = L_i ,$$

*and*

$$|T_{T_{i+1}-1}| = T_{i+1} - 1 = L_i + R_i - 1 .$$

**Lemma 3.** *We have $|\mathcal{L}_{T_{i+1}-1}| = \min(L_i, R_i)$, and $H_{T_{i+1}-1} = i$. Put differently, $k - 1 \leq H_n \leq k$ if $k$ is the unique integer with $T_k \leq n < T_{k+1}$.*

PROOF. The first statement is an immediate corollary of Lemma 2. Also, as each layer destroys all the leaves of the ancestor tree, it is clear by induction

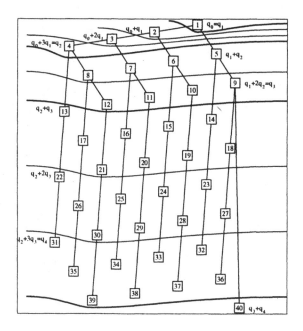

FIGURE 3. This figure shows the Weyl tree for $\theta = \sqrt{77} = [8; 1, 3, 2, 3, \ldots]$. The nodes are at positions $(x_i, -i)$ and $i$ is shown in the boxes that represent the nodes. Layers are separated by wiggly lines. Thicker lines separate layers of different polarity. Note that just before an extremum, all leaves may be found in the last layer. Using Lemma 1, can the user guess who the parent will be of $x_{41}$? And how many nodes are there in the layer started with $x_{40}$? The heads of the layers have indices that satisfy a certain recurrence; the integers in this recurrence are $q_0 = 1, q_1 = 1, q_2 = 4, q_3 = 9, q_4 = 31$.

that the height of the tree is exactly equal to the number of layers minus one. □

The study of the height and of the number of leaves reduces to the study of the sequence $(L_i, R_i)$. For the height, the growth of $T_k$ as a function of $k$ is important. This is closely related to the continued fraction expansion

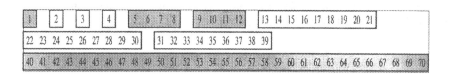

FIGURE 4. The figure shows the break-up of the Weyl sequence for $\theta = \sqrt{77}$ into layers. The shaded layers are all leaves in the tree of right polarity. Each layer covers the existing tree with a new layer of leaves.

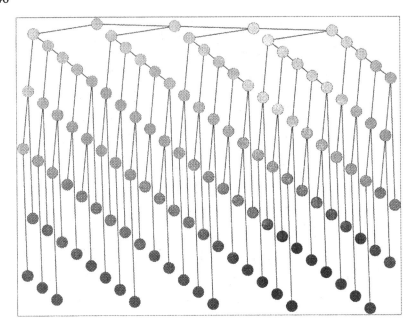

FIGURE 5. The Weyl tree $\mathcal{T}_n(\sqrt{1311})$ in which each $X_i$ is placed at $(X_i, -i)$ exhibits an obvious orchard pattern. This phenomenon is observed for all $\theta$. Note that $\sqrt{1311} = [36; (4, 1, 4, 2, 1, 2, 4, 1, 4, 72)^*]$.

of $\theta$. To understand the rest of the paper, we recall a few basic facts from the theory of continued fractions.

## 4 Continued fractions

Let $\theta$ be irrational, and define the Weyl sequence with $n$-th term $x_n = \{n\theta\}$, $n \geq 1$, where $\{.\}$ denotes the "modulo 1" operator: $\{u\} = u - \lfloor u \rfloor$. Denote the <u>continued fraction</u> expansion of $\theta$ by

$$\theta = [a_0; a_1, a_2, \ldots] \,,$$

where the $a_i$'s are the <u>partial quotients</u>, $a_i \geq 1$ for $i \geq 1$ (see Lang (1966) or LeVeque (1977)). Thus, we have

$$\theta = a_0 + \cfrac{1}{a_1 + \cfrac{1}{a_2 + \cfrac{1}{a_3 + \cdots}}} \,,$$

with $a_0 = \lfloor \theta \rfloor$. The <u>$i$-th convergent</u> of $\theta$ is

$$r_i = [a_0; a_1, \ldots, a_i] \,.$$

It can be computed recursively as $r_i = \frac{p_i}{q_i}$, where $\gcd(p_i, q_i) = 1$, and

$$p_{-2} = 0, p_{-1} = 1, p_i = a_i p_{i-1} + p_{i-2} , \ i \geq 0 ,$$

and

$$q_{-2} = 1, q_{-1} = 0, q_i = a_i q_{i-1} + q_{i-2} , \ i \geq 0 .$$

Note that $r_0 = a_0$ and $r_1 = a_0 + 1/a_1$. The $r_i$'s alternately underestimate and overestimate $\theta$. The denominators $q_i$ of the convergents play a special role as

$$1 = q_0 \leq q_1 \leq q_2 \leq \cdots$$

and

$$\left| \theta - \frac{p_i}{q_i} \right| \leq \frac{1}{q_i q_{i+1}} , \ i > 0 .$$

To study the number of records and the the evolution of the layers, the following result is essential. It extends a theorem of Lang (1966).

**Lemma 4 (Boyd and Steele, 1978).** *In a Weyl sequence for an irrational $\theta$ with partial quotients $a_n$, and convergents $p_n/q_n$, there are $a_1$ right extrema, followed by $a_2$ left extrema, $a_3$ right extrema and so forth. These extrema occur when $n$ is in the following list:*

$a_1$ right extremes $q_{-1} + q_0 , q_{-1} + 2q_0 , \ldots , q_{-1} + a_1 q_0 = q_1$

$a_2$ left extremes $q_0 + q_1 , q_0 + 2q_1 , \ldots , q_0 + a_2 q_1 = q_2$

$a_3$ right extremes $q_1 + q_2 , q_1 + 2q_2 , \ldots , q_1 + a_3 q_2 = q_3$

$a_4$ left extremes $q_2 + q_3 , q_2 + 2q_3 , \ldots , q_2 + a_4 q_3 = q_4$

$\cdots$

*The list above is the list of indices of the first elements in each layer.*

Lemma 4 shows that we start with $a_1$ right extremes, followed by $a_2$ left extremes, then $a_3$ right extremes, and so forth. This description, together with Lemma 1 and Lemma 2 should suffice to completely describe the tree. In fact, it suffices to show the right and left extrema in groups of sizes $a_1$, $a_2$, and so forth, and let the user paint in the layers of leaves one by one. In obvious notation, we may represent $\theta$ as $R^{a_1} L^{a_2} R^{a_3} L^{a_4} \ldots$, where $L$ and $R$ stand for left extremes and right extremes respectively. We note that this is precisely a representation of an infinite path in the Stern-Brocot tree ($L$ means that you should take a left turn and $R$ a right turn), which was found in the nineteenth century and is described in detail in Graham, Knuth and Patashnik (1989). Basically, the Stern-Brocot tree is an infinite binary tree which holds all possible convergents so that irrational numbers correspond to infinite paths.

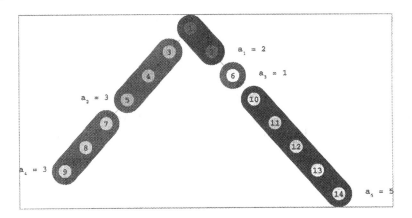

FIGURE 6. The groups of right and left extrema are shown. To construct the Weyl tree, first paint in $a_1$ layers of right leaves, then $a_2$ layers of left leaves, the $a_3$ layers of right leaves, and so forth. The "guts" of the tree are not shown in the figure.

## 5 Height of random Weyl trees

In this section, we look at random Weyl trees, that is, binary search trees for $\theta = U$, where $U$ is a uniform $[0, 1]$ random variable. This study allows us to make statements that are true for almost all $\theta$. The probabilistic setting comes in handy for the purpose of analysis. The main result of the next three sections, for example, shows that

$$\frac{H_n}{\log n \, \log \log n} \to \frac{12}{\pi^2} \text{ in probability,}$$

where $c > 0$ is a universal constant. This shows that $\mathcal{T}_n$ differs greatly from $\mathcal{R}_n$.

From Lemma 3 and Lemma 4, we easily determine the relationship between height and partial quotients.

**Proposition 1.** *Let $\theta$ be irrational. Let $k \geq 2$. If $n = q_k - 1$, then there are exactly $\sum_{i=1}^{k} a_i - 1$ full layers, and the Weyl tree $\mathcal{T}_n$ has height $H_n = \sum_{i=1}^{k} a_i - 2$ . In general, if $q_k \leq n < q_{k+1}$ , then*

$$\sum_{i=1}^{k} a_i \leq H_n + 2 \leq \sum_{i=1}^{k+1} a_i \ .$$

**Remark (Discrepancy).** There is another field in which the behavior of the partial sums $S_n = \sum_{i=1}^{n} a_i$ matters. In quasi-random number generation, the notion of discrepancy is important. In general, the discrepancy for a sequence $x_1, \ldots, x_n$ is

$$D_n = \sup_{A \in \mathcal{A}} \left| \frac{\sum_{i=1}^{n} I_{x_i \in A}}{n} - \lambda(A) \right| ,$$

where $\lambda(.)$ denotes Lebesgue measure, and $\mathcal{A}$ is a suitable subclass of the Borel sets. For example, if we take the intervals, then (Schmidt, 1972; Béjian, 1982)

$$D_n \geq \frac{0.12 \log n}{n}$$

infinitely often. From Niederreiter (1992, p. 24), we note that for a Weyl sequence for irrational $\theta$,

$$D_n \leq \frac{1}{n} \sum_{i=1}^{l(n)+1} a_i = \frac{S_{l(n)+1}}{n} ,$$

where $l(n)$ is the unique integer with the property that

$$q_{l(n)} \leq n < q_{l(n)+1} .$$

For example, Niederreiter's bound implies that if $\theta$ is such that $\sum_{i=1}^{m} a_i = O(m)$ (as when all $a_i$'s are bounded), then

$$D_n = O\left( \frac{\log n}{n} \right) .$$

Thus, Weyl sequences with small partial quotients behave well in this sense. We will see that the same is true for random search trees based on Weyl sequences.

## 6 Partial quotients of random irrationals

Replace $\theta$ by a uniform $[0, 1]$ random variable. Its continued fraction has been studied in depth by Khintchine (1963), Philipp (1970) and the references found there.

**Lemma 5 (Borel-Bernstein theorem; see Khintchine, 1924).** *For almost all $\theta$, $a_n \geq \varphi(n)$ infinitely often if and only if $\sum_n 1/\varphi(n) = \infty$. (Thus, if $\theta$ is uniform $[0, 1]$, then with probability one, $a_n \geq n \log n \log \log n$ infinitely often, for example.)*

This shows that the $a_n$'s necessarily have large oscillations. The result can also be used to show that certain subclasses of $\theta$'s have zero measure. Examples include:

a. The $\theta$'s with bounded partial quotients. The extreme example here is $\theta = \frac{1+\sqrt{5}}{2}$, which has $a_0 = a_1 = a_2 = \cdots = 1$.

b. The $\theta$'s that are quadratic irrationals (solutions of quadratic equations). It is known that the $a_i$'s are eventually periodic and thus bounded (in fact, the periodicity characterizes the quadratic irrationals, see Khintchine, 1963, p. 56).

**Lemma 6 (Kusmin, 1928; Lévy, 1929).** *Let $z_n$ denote the value of the continued fraction*

$$[0; a_{n+1}, a_{n+2}, \ldots] \ .$$

*(That is, $z_n = r_n - a_n = \{r_n\}$, where*

$$r_n = [a_n; a_{n+1}, a_{n+2}, \ldots] \ .)$$

*Then, if $\theta$ is uniform $[0, 1]$, then $z_n$ tends in distribution to the so-called Gauss-Kusmin distribution with distribution function*

$$F(x) = \log_2(1 + x) \ , \ 0 \le x \le 1 \ .$$

This limit theorem is easy to interpret if we consider convergents. Indeed, $r_0 = \theta$, and in general, $a_{n+1} = \lfloor 1/z_n \rfloor$. Thus, Lemma 6 also gives an accurate description of the limit law for $a_n$. In fact, as a corollary, one obtains another result of Lévy (1929), which states that the proportion of $a_i$'s taking value $k$ tends for almost all $\theta$ to a finite constant only depending upon $k$. If $\theta$ is uniform $[0, 1]$, then $a_1 = \lfloor 1/\theta \rfloor$ is a discretized version of a uniform $[0, 1]$ random variable. As $n$ grows, the distribution gradually shifts to a discretized version of one over a Gauss-Kusmin random variable. As the latter law has a density $f(x) = 1/((1 + x) \log 2)$ on $[0, 1]$ which varies monotonically from $1/\log 2$ to $1/\log 4$, for practical purposes, it is convenient to think of the $a_n$'s as having a law close to that of $1/U$. For example, the Borel-Bernstein law holds also for the sequence $1/U_n$ where $U_1, U_2, \ldots$ are i.i.d. uniform $[0, 1]$.

There is stability if we start the process with $\theta$ having the Gauss-Kusmin law, just if we were firing up a Markov chain by starting with the stationary distribution: if $\theta$ has the Gauss-Kusmin law, then all $z_n$'s have the Gauss-Kusmin law, and all the $a_n$'s have the same distribution (however, they are not independent.)

We must master the dependence between the $a_n$'s. This was done by Philipp in 1970:

**Lemma 7 (Philipp, 1970).** *Let $\theta$ have the Gauss-Kusmin distribution. Let $\mathcal{M}_{u,v}$ be the smallest $\sigma$-algebra with respect to which the coefficients $a_u, \ldots, a_v$ are measurable. Then for any sets $A \in \mathcal{M}_{1,t}$ and $B \in \mathcal{M}_{t+n,\infty}$,*

$$|\mathbf{P}\{AB\} - \mathbf{P}\{A\}\mathbf{P}\{B\}| < c\rho^n \mathbf{P}\{A\}\mathbf{P}\{B\} \ ,$$

*where $\rho \in (0, 1)$ and $c$ is a constant.*

This result states that in effect the $a_n$'s are almost independent, with the dependence decreasing in an exponential fashion. Next we consider the behavior of partial sums of the partial quotients of a random Weyl sequence, and obtain a limit law. More precisely, we study the behavior of

$$S_n = \sum_{i=1}^{n} a_i$$

when $\theta$ is replaced by $U$, a uniform $[0, 1]$ random variable. Proposition 2 below was obtained without the help of modern probability theoretical tools such as Lemma 7 by Khintchine (1935). A short proof of it and of Theorem 1 below is given in Devroye and Goudjil (1996).

**Proposition 2.** *If $\theta$ is Gauss-Kusmin distributed, then*

$$\frac{\sum_{i=1}^{n} a_i}{n \log_2 n} \to 1$$

*in probability.*

Proposition 2 may be rephrased as follows, if $A_n$ denotes the collection of all $\theta$'s on $[0, 1]$ with $|\sum_{i=1}^{n} a_i/(n \log_2 n) - 1| > \epsilon$:

$$\lim_{n \to \infty} \mathbf{P}\{\theta \in A_n\} = 0 .$$

**Theorem 1.** *If $\theta$ has a distribution with a density on $[0, 1]$, then*

$$\frac{\sum_{i=1}^{n} a_i}{n \log_2 n} \to 1$$

*in probability.*


# 7 The behavior of the denominator of the convergents

**Lemma 8 (Khintchine, 1935, and Lévy, 1937; see Khintchine, 1963, p. 75).** *There exists a universal constant $\gamma = \pi^2/12 \ln 2 \approx 1.186569111$ such that for almost all $\theta$,*

$$q_n = e^{(\gamma+o(1))n} .$$

Lemma 8 is related to the property (Khintchine, 1963, p. 101) that

$$\left( \prod_{i=1}^{n} a_i \right)^{1/n} \to c \overset{\text{def}}{=} \prod_{j=1}^{\infty} \left( 1 + \frac{1}{j(j+2)} \right)^{\frac{\ln j}{\ln 2}}$$

for almost all $\theta$. Indeed, to get this intuition, recall from the recurrences for the $q_n$'s that

$$q_{n+1} = a_{n+1}q_n + q_{n-1} \leq (a_{n+1} + 1)q_n \ ,$$

so that

$$q_n \leq \prod_{j=1}^{n}(1 + a_j) \leq 2^n \prod_{j=1}^{n} a_j \ . \ \Box$$

We also note that $q_n$ must grow faster than a Fibonacci sequence, as $q_{n+1} \geq q_n + q_{n-1}$. This implies that $q_n \geq \rho^{n-1}$ for all $n$, where $\rho = (1+\sqrt{5})/2$ is the golden ratio. Another simple lower bound is $q_n \geq 2^{(n-1)/2}$ (Khintchine, 1963, p. 18).

**Theorem 2 (Devroye and Goudjil, 1996).** *If $\theta$ has any density on $[0, 1]$, then*

$$\frac{H_n}{\frac{1}{\gamma} \log n \log_2 \log n} = \frac{H_n}{\frac{12}{\pi^2} \log n \log \log n} \to 1$$

*in probability. Note that $12/\pi^2 \approx 1.215854203$.*

PROOF. By Theorem 1, as $k \to \infty$,

$$\sum_{i=1}^{k} a_i \sim k \log_2 k$$

in probability. Next, $\log q_k \sim \gamma k$ in probability. The latter fact implies that in probability, $k \sim (1/\gamma) \log n$ if $k$ is the unique integer such that $q_k \leq n < q_{k+1}$. But Theorem 1 and Proposition 1 then imply that

$$\frac{H_n}{k \log_2 k} \sim \frac{H_n}{(1/\gamma) \log n \log_2 \log n} \to 1$$

in probability. $\qquad\Box$

This theorem does not describe the behavior as $n \to \infty$ for a single $\theta$ (the "strong" behavior). Rather, it refers to a metric property and takes for each $n$ a cross-section of $\theta$'s that give a height in the desired range, and confirms that the measure (probability) of these $\theta$'s tends to one. For oscillations and strong behavior, a bit more is required. By the Borel-Bernstein theorem, with probability one,

$$a_n \geq n \log n \log \log n$$

infinitely often (the Borel-Bernstein theorem yields a better lower bound, but the one given here suffices to make our point). Since with probability one, $q_k^{1/k} \to e^{\gamma}$ as $k \to \infty$, we see from Proposition 1 that with probability one,

$$H_n \geq (1/\gamma) \log n \log \log n \log \log \log n$$

infinitely often. Thus, Theorem 2 cannot be strengthened to almost sure convergence, as the oscillations are too wide.

It is of interest to bound the oscillations in the strong behavior as well. Also, again by the Borel-Bernstein theorem, with probability one, for all but finitely many $n$, $a_n \le n \log n \log^{1+\epsilon} \log n$ for $\epsilon > 0$. This implies that with probability one, for all but finitely many $n$, $\sum_{j=1}^n a_j \le n^2 \log n \log^{1+\epsilon} \log n$. But then, by Proposition 1 and Lemma 8, with probability one, for all but finitely many $n$,

$$H_n \le \frac{2}{\gamma^2} (\log n)^2 (\log \log n)(\log \log n \log n)^{1+\epsilon} .$$

## 7.1 Very good trees

From the inequality of Proposition 1, we recall that $H_n = O(\log n)$ if $\sum_{i=1}^n a_i = O(n)$. Such irrationals have zero probability. As the most prominent member with the smallest partial sums of partial quotients, we have the golden ratio ($a_n \equiv 1$ for $n \ge 0$). Indeed, as for these sequences, $q_n \le \prod_{i=1}^n (1 + a_i) \le \exp(\sum_{i=1}^n a_i) = \exp(O(n))$, we have the claimed result on $H_n$ without further ado. In fact, for the golden ratio, we have $q_n \sim c\rho^n$, where $\rho = (1 + \sqrt{5})/2$ and $c > 0$ is a constant. As $\sum_{i=1}^n a_i \equiv n$, we see that $H_n \sim \frac{\log n}{\log \rho}$. The Weyl tree is simply not high enough compared to typical random Weyl trees or truly random binary search trees.

If $a_n \equiv a$ for all $n$, then $q_n = aq_{n-1} + q_{n-2}$ for all $n$. This simple linear recurrence has a solution $q_n \sim c \left( \frac{a+\sqrt{a^2+4}}{2} \right)^n$ for some constant $c$. As $\sum_{i=1}^n a_i = an$, we see that

$$H_n \sim \frac{a}{\log \left( \frac{a+\sqrt{a^2+4}}{2} \right)} \log n .$$

Note that the coefficient can be made as large as desired by picking $a$ large enough.

Quadratic algebraic numbers $\theta$ are solutions of quadratic equations with integer-valued coefficients. Such numbers have periodic partial quotients, and thus satisfy the boundedness property discussed here. If the period length is $p$, and the sum of the partial quotients over a period is $s$, then $\sum_{i=1}^k a_i \sim sk/p$ as $k \to \infty$. Furthermore, $q_k = (\rho+o(1))^k$ where $\rho$ depends upon the periodic elements only. This implies that

$$H_n \sim \frac{s \log n}{p \log \rho} .$$

## 7.2 Very bad trees

We first show that Weyl trees can be almost of arbitrary height.

**Theorem 3.** *Let $h_n$ be a monotone sequence of numbers decreasing from 1 to 0 at any slow rate. Then there exists an irrational $\theta$ such that for the Weyl tree, $H_n \geq n h_n$ infinitely often.*

PROOF. We exhibit a monotonically increasing sequence $a_n$ of partial quotients to describe $\theta$. The inequality will be satisfied at instants when the tree size $n = q_k$ for some $k$. Thus, we will have for all $k$ large enough,

$$H_{q_k} \geq q_k h_{q_k} .$$

Now, $H_{q_k} \geq \sum_{i=1}^{k} a_i \geq a_k$, and

$$a_k \leq q_k \leq 2^k \prod_{i=1}^{k} a_i \leq 2^k a_k (a_{k-1})^{k-1} .$$

Thus,

$$\frac{H_{q_k}}{q_k} \geq \frac{1}{2^k (a_{k-1})^{k-1}} \geq h_{a_k} \geq h_{q_k}$$

by choosing $a_k$ large enough (note that $k$ and $a_{k-1}$ are fixed). $\qquad\square$

EXAMPLE 1. Take $a_k = 2^k$. Then

$$2^{k(k+1)/2} \leq q_k \leq 2^{k+k(k+1)/2} ,$$

so that $k = \sqrt{2 \log_2 n} - K - o(1)$, where $K \in [1/2, 3/2]$ and possibly depends upon $n$. As $\sum_{i=1}^{k} a_i = 2^{k+1} - 1$, we have at those times when $n = q_k$ for some $k$,

$$H_n = 2^{k+1} - 1 = \Theta\left(2^{\sqrt{2 \log_2 n}}\right) .$$

This grows much faster than any power of the logarithm.

EXAMPLE 2. If we set $a_k = 2^{2^k}$, then $q_k \leq 2^k \prod_{i=1}^{k} a_i \leq 2^{k+2^{k+1}-1} \leq \log_2(a_k) a_k^2 / 2$. Combine this with $H_{q_k} \geq a_k$, and note that when $n = q_k$ for some $k$,

$$H_n \geq \sqrt{\frac{2n}{\log_2 H_n}} ,$$

and therefore,

$$H_n \geq (1 + o(1)) \sqrt{\frac{4n}{\log_2 n}} .$$

EXAMPLE 3. By considering $a_k = b^{b^k}$ for integer $b$, the height increases at least as $(n / \log_2 n)^{1-1/b}$.

## 7.3 Trees for a few selected transcendental numbers

The partial quotients are known for just a few transcendental numbers. For example

$$\tan(1/2) = [0; 1, 1, 4, 1, 8, 1, 12, 1, 16, \ldots] .$$

Thus, $a_{2k} = 1$, $a_{2k+1} = 4k$, $k \geq 1$. From $q_{2k+1} = 4kq_{2k} + q_{2k-1}$ and $q_{2k} = q_{2k-1} + q_{2k-2}$, one can show (see Boyd and Steele, 1978, p. 57) that

$$4^k k! < q_{2k+1} < 8^k (k+1)!$$

and

$$q_{2k+1} \approx q_{2k+2} \approx (ck)^k$$

for some constant $c$. In fact, then, we see that the $k$ for Theorem 2 satisfies

$$k \sim \frac{\log n}{\log \log n} .$$

But then

$$H_n \sim \sum_{j=1}^{k/2} (4j) \sim \frac{k^2}{2} \sim \frac{\log^2 n}{2 \log^2 \log n} .$$

The Weyl tree is much higher than that of a typical random Weyl tree.

In a second example, consider

$$e = [2; 1, 2, 1, 1, 4, 1, 1, 6, 1, 1, 8, \ldots]$$

so that $a_0 = 2$, $a_{3m} = a_{3m-2} = 1$ and $a_{3m-1} = 2m$ for $m \geq 1$. Then (Lang, 1966, p. 74) there exist constants $C_1$ and $C_2$ such that

$$C_1 4^n \Gamma(n + 3/2) \leq q_{3n+1} \leq C_2 4^n \Gamma(n + 3/2) .$$

This shows that $k \sim \log n / \log \log n$. Thus,

$$H_n \sim \frac{\log^2 n}{9 \log^2 \log n} .$$

Again, the Weyl tree has an excessive height.

# 8 Sorting Weyl sequences

Ellis and Steele (1981) have shown that the first $n$ elements of any Weyl sequence can be sorted with the aid of $O(\log(n))$ comparisons only, even though these sequences too are equi-distributed for any irrational $\theta$. This shows that such sequences possess a lot of structure. Of course, the fact that discrete random Weyl sequences and random Lehmer sequences are imperfect is because they can be "described" very simply by a small number of bits. The randomness of a sequence has been related by several authors to the length of the descriptors (see e.g. Martin-Löf (1966), Knuth (1973), Bennett (1979)). For surveys and discussions on the topic of uniform random variate generation, one could consult Niederreiter (1991) or L'Ecuyer (1989).

It is well-known that the number of comparisons needed in quicksort is equal to the sum of the depths of all the nodes in the binary search tree constructed from the data by ordinary insertion. As this sum is bounded from below by $H_n(H_n + 1)/2$ (just by summing over the path leading to the furthest node), we see that the number of comparisons in quicksort is infinitely often at least equal to $nh_n(nh_n + 1)/2$ for any sequence $h_n$ decreasing to zero, and some irrational $\theta$. Yet, for i.i.d. data drawn from the same nonatomic distribution, the expected number of comparisons is asymptotic to $2n \log n$ (Sedgewick, 1977). Therefore, Weyl sequences are not appropriate for generating test data for sorting algorithms. With a uniform $[0, 1]$ $\theta$, the expected number of comparisons grows as $n \log n \, \log \log n$. In fact, we have the following.

**Proposition 3 (Devroye and Goudjil, 1996).** *Let $\theta$ be uniform $[0, 1]$. For any constant $C$, with probability one, the number of comparisons for quicksort-ing the first $n$ numbers of a random Weyl sequence is infinitely often larger than*

$$Cn \log n \log \log n \log \log \log n \ .$$

# 9 Lehmer sequences

When $a$ is integer, the sequence

$$x_n = \{a^n x_0\}$$

will be called a Lehmer sequence. Lehmer sequences are equi-distributed for all integer $a > 1$ and almost all $x_0 \in (0, 1)$ (see Freiberger and Grenander (1971), Hlawka (1984) or Kuipers and Niederreiter (1974)). The behavior of these sequences models the behavior of multiplicative sequences on infinite-wordsize machines. In particular, if Lehmer sequences don't have a certain desirable property, then it is just about impossible that multiplicative sequences possess that property. In addition, the multiplier $a$ in the Lehmer sequence is inherited from the multiplicative generator, because it determines the same geometric increase in the sequence of values (before truncation due to the $\{.\}$ operation). In fact, the "complexity" of the generator can be measured by $a$. This continuous model is at the same time the genesis and an abstraction of the discrete finite wordsize sequences. It was employed by many to explain certain properties, gain insight, and advance the understanding of certain generators.

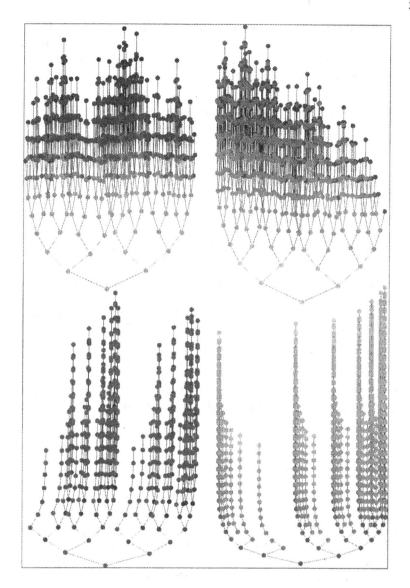

FIGURE 7. Four Weyl trees with $\theta$, from top left to bottom right ranging from $\sqrt{2}$, $\sqrt{3}$, $\tan(1/2)$ and $\pi$. Colors (gray-levels) represent the various layers of paint. The points are slightly perturbed so that one gets a better impression of the number of nodes at each level.

For $a$ not integer, the sequence obtained by $x_{n+1} = \{ax_n\}$ does not coincide with $x_n = \{x_0 a^n\}$. In fact, while the latter sequence is equi-distributed for almost all $x_0$, it is easy to show that the former sequence is not equi-distributed for any value $x_0$.

EXAMPLE 7 ($x_0$ IS RATIONAL). In computers, we can only store rational numbers. If we begin with a rational $x_0 = k/m$ for integers $k$ and $m$ that are relatively prime, then $x_0$ has a periodic decomposition base $a$:

$$\{x_0\} = \sum_{i=1}^{\infty} \frac{x_{0i}}{a^i} ,$$

where $x_{0i} \in \{0, 1, \ldots, a-1\}$, and the sequence $\{x_{0i}\}$ is periodic for all $i \geq N$ and some $N$. Therefore, the sequence $x_N, x_{N+1}, \ldots$ is periodic, so that $\{x_n, n \geq 1\}$ is not equi-distributed.

EXAMPLE 8 (POOR CHOICES FOR $x_0$). Decompose $x_0$ base $a$ as in the previous example. Let $N_{0,n}, \ldots, N_{a-1,n}$ denote the number of values in $\{x_{01}, \ldots, x_{0n}\}$ equal to $0, 1, \ldots, a-1$ respectively. Equi-distribution implies

$$\lim_{n \to \infty} \frac{N_{i,n}}{n} = \frac{1}{a} , \quad 0 \leq i < a .$$

All integers in the decomposition must occur equally frequently. In fact, equi-distribution occurs if and only if for all integer $k \geq 1$, and all $b = (b_1, \ldots, b_k) \subseteq \{0, \ldots, a-1\}^k$,

$$\lim_{n \to \infty} \frac{1}{n} \sum_{i=1}^{n} I_{(x_{0,i+1}, \ldots, x_{0,i+k})=b} = \frac{1}{a^k} .$$

Thus, all pairs, triples, and so forth must occur equally often. Thus, $x_0$ itself is very restricted. If the bits in a binary expansion of $x_0$ are independent Bernoulli ($p$), then the only acceptable $p$ is $1/2$. It is also the only value for which $x_0$, the random variable with this bit expansion, has a density (the uniform density). For all other values, $x_0$ has a singular distribution. Considered in this manner, we see that the vast majority of $x_0$'s will in fact be poor. It is easy to construct an irrational $x_0$ for which there is no equi-distribution.

A random Lehmer sequence has an integer multiplier $a$ and a uniform $[0, 1]$ random seed $X_0$. How such a seed is obtained is of course problematic. In fact, having $X_0$ is like having an infinite i.i.d. sequence of uniform $[0, 1]$ random variables $X_1, X_2, \ldots$. To see this, consider the bit representation of $X_0$:

$$X_0 = 0.X_{01} X_{02} X_{03} \cdots ,$$

and distribute the bits in a triangular fashion among the $X_i$'s: $X_{01}$ goes to $X_1$, $X_{02}$ and $X_{03}$ go to $X_1$ and $X_2$ respectively, $X_{04}$ through $X_{06}$ go to $X_1$ through $X_3$ and so forth. This establishes a one-to-one mapping between the bits of $X_0$ and all the bits of all $X_i$'s. But since an i.i.d. sequence of bits (with 1's and 0's occurring with equal probability), when considered as a fraction of a real number, defines a uniform $[0, 1]$ random variable, we see that $X_1, X_2, \ldots$ are indeed i.i.d. uniform $[0, 1]$ as claimed. This perfect generator is, however, not of the form $X_{n+1} = f(X_n)$, so that we need not celebrate prematurely.

A random Lehmer sequence has obvious imperfections. For example, the pairs $(X_n, X_{n+1})$ fall on one of many parallel lines in the plane because $X_{n+1} - aX_n$ is integer-valued. What we would like to discuss here is how the sequence behaves globally. Knuth (1973, p. 88) spends some time on the present model. He shows for example that while $\mathbf{P}(X_{n+1} < X_n) = 1/2$,

$$\mathbf{P}(X_{n+2} < X_{n+1} < X_n) = \frac{1}{6} + \frac{1}{6a} + O(a^{-2}).$$

This value should be $1/6$ and is thus wrong. Starting with $X_0$, a monotone run of length $k$ occurs if $X_0 \geq X_1 \geq \cdots \geq X_{k-1} < X_k$. The expected length of such a run is shown by Knuth to be

$$\left(\frac{a}{a-1}\right)^a - \frac{a}{a-1} = e - 1 + \frac{e/2 - 1}{a} + O(a^{-2}),$$

which is too big—for an i.i.d. sequence the answer is $e - 1$.

## 10 Binary search trees for Lehmer sequences

A Lehmer tree is a binary search tree for a Lehmer sequence, and thus has two parameters, an integer multiplier $a$ and a real-valued seed $x_0$. A random Lehmer tree is a Lehmer tree in which $x_0$ is uniformly chosen on $[0, 1]$ (and thus has one parameter, $a$). We will point out a few flaws of particular Lehmer trees. For example, in view of the fact that the height of $\mathcal{R}_n$ has expected value $\Theta(\log n)$, it is puzzling that even with good multipliers, binary search trees based upon Lehmer sequences have heights that can take any value $o(n)$ for all $n$ large enough.

**Theorem 4.** *Let $a > 1$ be an arbitrary integer. Let $g_n$ be a monotonically increasing sequence of positive integers with $g_n = o(n)$. Then there exist infinitely many seeds $x_0$ for which the Lehmer sequence $\{x_0 a^n\}$ is equidistributed, yet $H_n \geq g_n$ for all $n$ large enough, where $H_n$ is the height of the binary search tree with $n$ nodes obtained by inserting the first $n$ numbers of the Lehmer sequence.*

PROOF. We proceed by showing that for a specially constructed random seed $x_0$, the Theorem is true with probability one, and thus, that there must

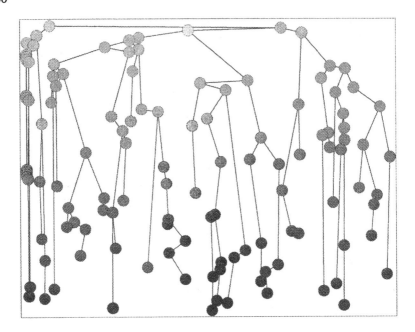

FIGURE 8. A random Lehmer tree with multiplier $a$, in which each $X_i$ is placed at $(X_i, -i)$. The orchard pattern from Weyl trees has disappeared.

exist infinitely many deterministic seeds $x_0$ for which the Theorem is true. Let $x_0$ be a uniformly distributed random variable on $[0, 1]$. It is well-known that its fraction has independent digits that are uniformly distributed on $\{0, \dots, a-1\}$. With probability one, all $k$-tuples occur equally often in the limit (see above), and the Lehmer sequence is thus equi-distributed. Define $n_i = 2^i$. Define an integer sequence $h_{n_i}$ recursively as follows: $h_{n_0} = 0$, and for $i \geq 0$,

$$h_{n_{i+1}} = h_{n_i} + 2i + g_{n_{i+2}} .$$

Observe that $h_{n_i}$ is increasing, and that $h_{n_i} = o(n_i)$. That follows from telescoping the recurrence relation and noting that

$$h_{n_i} = g_{n_{i+1}} + g_{n_i} + \dots + g_{n_2} + i(i-1) .$$

Replace $h_{n_{i+1}}$ digits starting at digit $n_i + 1$ by 0, for $i = 1, 2, \dots$. Among the first $n_i$ digits, at most $h_{n_2} + \dots + h_{n_i}$ digits have been thus modified. Clearly, this is $o(n_i)$ as $i \to \infty$. Therefore, all $k$-tuples still occur equally often in the limit, and the Lehmer sequence for the new $x_0$ is equi-distributed.

In a sequence of i.i.d. random digits $X_1, \dots, X_n$, with $\mathbf{P}\{X_i = i\} = 1/a$, $0 \leq i < a$, let $L_n$ be the length of the longest run of zeros. Clearly,

$$\mathbf{P}\{L_n \geq k\} \leq \frac{n}{a^k} .$$

Let $L_n$ now denote the same thing for the first $n$ digits of $x_0$. We have added a few sequences of consecutive zeros, but obvious calculations show that

$$\mathbf{P}\{L_{n_i} \geq k + h_{n_i}\} \leq \frac{n_i}{a^k} .$$

The infinite string starting at position $n_i + 1$ starts with $h_{n_{i+1}}$ zeros. It and the next $h_{n_{i+1}} - h_{n_i} - k$ infinite strings form a father/right child chain in the binary search tree if $L_{n_i} < k + h_{n_i}$. Therefore,

$$H_{n_i + h_{n_{i+1}}} \geq h_{n_{i+1}} - h_{n_i} - k$$

with probability at least $1 - n_i/a^k$. We assume without loss of generality that $n_i + h_{n_{i+1}} < n_{i+1}$. Thus,

$$H_{n_{i+1}} \geq h_{n_{i+1}} - h_{n_i} - 2i$$

with probability at least $1 - n_i/a^{2i}$. Let $A$ be the event that all these inequalities hold except finitely often. Then, by the Borel-Cantelli lemma, $\mathbf{P}\{A\} = 1$ if

$$\sum_{i=1}^{\infty} \frac{n_i}{a^{2i}} < \infty .$$

As $n_i = 2^i$, this is indeed the case. Therefore, with probability one, for all $i$ large enough,

$$H_{n_{i+1}} \geq h_{n_{i+1}} - h_{n_i} - 2i .$$

Assume $i$ is such a large integer. Take $n \in [n_{i+1}, n_{i+2})$. Then

$$H_n \geq H_{n_{i+1}}$$
$$\geq h_{n_{i+1}} - h_{n_i} - 2i$$
$$= g_{n_{i+2}} \quad \text{(by definition of } h\text{)}$$
$$\geq g_n \quad \text{(by monotonicity of } g\text{)}$$

so that $H_n \geq g_n$ for all $n$ large enough. This concludes the proof. □

## 11 Virtual trees

The random trees introduced in this note may also be used as crude compact representations of real-life trees for possible use in graphics. One could take each tree and associate a branch with each node, and add nodes at the end of the leaf branches. If the branches are randomly rotated with respect to their parent branches, and if the branches have a direction that is somehow forced to grow towards the sun (north, up), and if we attach

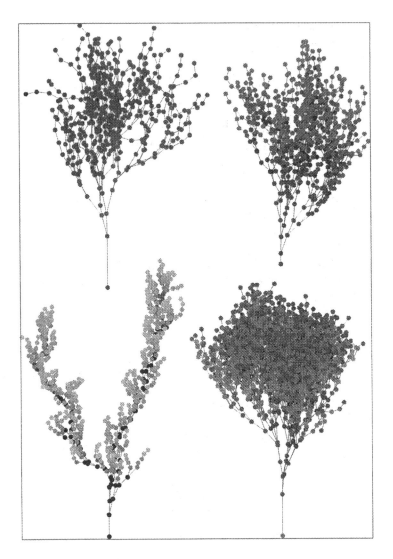

FIGURE 9. Four Weyl trees with $\theta$, from top left to bottom right ranging from $e$, $\tan(1/2)$, $e^{2\pi}$ and $\sqrt{7}$. Colors (gray-levels) represent the various layers of paint.

random branch lengths, very simple but sometimes interesting tree shapes occur, as is apparent from Figure 9.

## *References*

Aho, A.V., Hopcroft, J.E., and Ullman, J.D., 1983. *Data Structures and Algorithms*. Reading, Mass.: Addison-Wesley.

Beck, J., 1991. Randomness of $n\sqrt{2}$ (mod 1) and a Ramsey property of the hyperbola. *Colloquia Mathematica Societatis János Bolyai* 60. Budapest.

Béjian, R., 1982. Minoration de la discrépance d'une suite quelconque sur $T$. *Acta Arithmetica* 41. 185–202.

Bennett, C.H., 1979. *On random and hard-to-describe numbers*. IBM Watson Research Center Report RC 7483 (no 32272). Yorktown Heights, N.Y.

Bohl, P., 1909. Über ein in der Theorie der säkularen Störungen vorkommendes Problem. *Jl. Reine und Angewandte Mathematik* 135. 189–283.

Boyd, D.W. and Steele, J.M., 1978. Monotone subsequences in the sequence of fractional parts of multiples of an irrational. *Jl. Reine und Angewandte Mathematik* 204. 49–59.

Chatterji, S.D., 1966. Masse, die von regelmässigen Kettenbrüchen induziert sind. *Mathematische Annalen* 164. 113–117.

Chow, Y.S., and Teicher, H., 1978. *Probability Theory*. New York, N.Y.: Springer-Verlag.

del Junco, A. and Steele, J.M., 1979. Hammersley's law for the van der Corput sequence: an instance of probability theory for pseudorandom numbers. *Annals of Probability* 7. 267–275.

Devroye, L., 1986. A note on the height of binary search trees. *Journal of the ACM* 33. 489–498.

Devroye, L., 1987. Branching processes in the analysis of the heights of trees *Acta Informatica* 24. 277–298.

Devroye, L., 1988. Applications of the theory of records in the study of random trees. *Acta Informatica* 26. 123–130.

Devroye, L., and Goudjil, A., 1996. *A study of random Weyl trees*. Technical Report, School of Computer Science, McGill University, Montreal.

Diaconis, P., 1977. The distribution of leading digits and uniform distribution mod 1. *Annals of Probability* 5. 72–81.

Ellis, M.H., and Steele, J.M., 1981. Fast sorting of Weyl sequences using comparisons. *SIAM Journal on Computing* 10. 88–95.

Franklin, J.N., 1963. Deterministic simulation of random sequences. *Mathematics of Computation* 17. 28–59.

Freiberger, W., and Grenander, U., 1971. *A Short Course in Computational Probability and Statistics*. New York: Springer-Verlag.

Fushimi, M., and Tezuka, S., 1983. The $k$-distribution of generalized feedback shift register pseudorandom numbers. *Communications of the ACM* 26. 516–523.

Fushimi, M., 1988. Designing a uniform random number generator whose subsequences are $k$-distributed. *SIAM Journal on Computing* 17. 89–99.

Galambos, J., 1972. The distribution of the largest coefficient in continued fraction expansions. *Quarterly Journal of Mathematics Oxford Series* 23. 147–151.

Graham, R.L., Knuth, D.E., and Patashnik, O., 1989. *Concrete Mathematics—A Foundation for Computer Science*. Reading, MA: Addison-Wesley.

Hlawka, E., 1984. *The Theory of Uniform Distribution*. Berkhamsted, U.K.: A B Academic.

Kesten, H., 1960. Uniform distribution mod 1. *Annals of Mathematics* 71. 445–471.

Khintchine, A., 1924. Einige Sätze über Kettenbrüche, mit Anwendungen auf die Theorie der Diophantischen Approximationen, *Mathematische Annalen* 92. 115–125.

Khintchine, A., 1935. Metrische Kettenbruchprobleme. *Compositio Mathematica* 1. 361–382.

Khintchine, A., 1936. Metrische Kettenbruchprobleme. *Compositio Mathematica* 2. 276–285.

Khintchine, A., 1963. *Continued Fractions*. Groningen: Noordhoff.

Knuth, D.E., 1973. *The Art of Computer Programming, Vol. 3 : Sorting and Searching*. Reading, MA: Addison-Wesley.

Knuth, D.E., 1981. *The Art of Computer Programming, Vol. 2, 2nd Ed.*. Reading, Mass.: Addison-Wesley.

Kuipers, L., and Niederreiter, H., 1974. *Uniform Distribution of Sequences*. New York: John Wiley.

Kusmin, R.O., 1928. On a problem of Gauss. *Reports of the Academy of Sciences* A. 375–380.

Lang, S., 1966. *Introduction to Diophantine Approximations*. Reading, MA: Addison-Wesley.

L'Ecuyer, P. and Blouin, F., 1988. Linear congruential generators of order $k > 1$. *Proceedings of the 1988 Winter Simulation Conference*, ed. M.Abrams, P.Haigh and J.Comfort. 432–439. ACM.

L'Ecuyer, P., 1989. A tutorial on uniform variate generation. *Proceedings of the 1989 Winter Simulation Conference*, ed. E.A. MacNair, K.J. Musselman and P. Heidelberger. 40–49. ACM.

L'Ecuyer, P., 1990. Random numbers for simulation. *Communications of the ACM* 33. 85–97.

LeVeque, W.J., 1977. *Fundamentals of Number Theory*. Reading, MA: Addison-Wesley.

Lévy, P., 1929. Sur les lois de probabilité dont dépendent les quotients complets et incomplets d'une fraction continue. *Bulletin de la Société Mathématique de France* 57. 178–193.

Lévy, P., 1937. *Théorie de l'addition des variables aléatoires*. Paris.

Lynch, W.C., 1965. More combinatorial problems on certain trees. *Computer Journal* 7. 299–302.

Mahmoud, H.M., 1992. *Evolution of Random Search Trees*. New York: John Wiley.

Martin-Löf, P., 1966. The definition of random sequences. *Information and Control* 9. 602–619.

Niederreiter, H., 1977. Pseudo-random numbers and optimal coefficients. *Advances in Mathematics* 26. 99–181.

Niederreiter, H., 1978. Quasi-Monte Carlo methods and pseudo-random numbers. *Bulletin of the American Mathematical Society* 84. 957–1042.

Niederreiter, H., 1991. Recent trends in random number and random vector generation. *Annals of Operations Research* 31. 323–346.

Niederreiter, H., 1992. *Random Number Generation and Quasi-Monte Carlo Methods* 63. SIAM CBMS-NSF Regional Conference Series in Applied Mathematics. Philadelphia: SIAM.

Peskun, P.H., 1980. Theoretical tests for choosing the parameters of the general mixed linear congruential pseudorandom number generator. *Journal of Statistical Computation and Simulation* 11. 281–305.

Philipp, W., 1969. The central limit problem for mixing sequences of random variables. *Zeitschrift für Wahrscheinlichkeitstheorie und verwandte Gebiete* 12. 155–171.

Philipp, W., 1970. Some metrical theorems in number theory II. *Duke Mathematical Journal* 38. 477–485.

Robson, J.M., 1979. The height of binary search trees. *The Australian Computer Journal* 11. 151–153.

Robson, J.M., 1982. The asymptotic behaviour of the height of binary search trees. *Australian Computer Science Communications* 88–88.

Schmidt, W.M., 1972. Irregularities of distribution. *Acta Arithmetica* 21. 45–50.

Sedgewick, R., 1977. The analysis of quicksort programs. *Acta Informatica* 4. 327–355.

# A survey of quadratic and inversive congruential pseudorandom numbers

JÜRGEN EICHENAUER–HERRMANN, EVA HERRMANN,
AND STEFAN WEGENKITTL

**Abstract.** This review paper deals with nonlinear methods for the generation of uniform pseudorandom numbers in the unit interval. The emphasis is on results of the theoretical analysis of quadratic congruential and (recursive) inversive congruential generators, which are scattered over a fairly large number of articles. Additionally, empirical results of some sample generators in a two–level overlapping serial test are given.

## 1. Introduction and basic concepts

Uniform pseudorandom numbers in the interval $[0, 1)$ are basic ingredients of any stochastic simulation. Their quality is of fundamental importance for the success of the simulation, since the outcome of a typical stochastic simulation strongly depends on the structural and statistical properties of the underlying pseudorandom number generator. Therefore, the selection of a suitable one is a crucial task in any simulation project, which should not be left to chance, and one would be well advised to follow Donald E. Knuth's proposal 'to run each Monte Carlo program at least twice using quite different sources of random numbers, before taking the answers of the program seriously' [73, p. 173]. This state of affairs provided the motivation for studying several *nonlinear* methods during the last ten years. These approaches were not introduced in order to replace the classical methods, such as linear congruential and shift–register generators, but they were designed as possible alternatives in the sense of Knuth's advice. The theoretical and empirical analysis of nonlinear methods is the currently most active research area in uniform pseudorandom number generation. There is already a wide variety of expository literature on different nonlinear methods. The reader is referred to the excellent monograph of Harald Niederreiter [86] and to the survey articles [18, 22, 33, 40, 66, 75, 82 – 85, 87 – 89, 91, 92]. Results of the extensive empirical testing of uniform pseudorandom numbers generated by different nonlinear methods in direct comparison with linear methods can be found in [76] and the references given there.

**Uniform pseudorandom numbers.** A sequence of *uniform pseudorandom numbers* for stochastic simulations is generated by a deterministic algorithm and should simulate a sequence of random variables which are independent and uniformly distributed in the unit interval $[0, 1)$, i.e., it should have the same 'relevant statistical properties' as a sequence of realizations of such random variables. The standard algorithms for generating sequences of uniform pseudorandom numbers are based on recursive procedures and yield sequences that are (purely) periodic. The following basic requirements on uniform pseudorandom numbers are usually put forward. The generated sequences of pseudorandom numbers should have a sufficiently large *period length*, the pseudorandom numbers within a period should be *equidistributed* in the interval $[0, 1)$, successive terms in the generated sequences of pseudorandom numbers should have reasonable *statistical independence* properties, and the generated points of $s$ successive pseudorandom numbers should not show any (strong) *regular structure* in their distribution in $[0, 1)^s$. Additionally, there should be a reasonably fast (and easy) computer implementation of the generation algorithm. This computational requirement of efficiency was considered very important in the early days of computing, but with the increasing power of modern computers it becomes more and more insignificant. Moreover, pseudorandom number generation is not the bottleneck in a typical stochastic simulation, since the computer time taken to generate the pseudorandom numbers is usually only a small part of the effort in the simulation project.

**Discrepancy.** Equidistribution and statistical independence properties of uniform pseudorandom numbers in the interval $[0, 1)$ can be analysed based on the distribution behaviour of $s$–tuples of successive terms in the generated sequences. The notion of discrepancy provides a reliable measure for distribution properties of the corresponding point sets in $[0, 1)^s$. For $N$ arbitrary points $t_0, t_1, \ldots, t_{N-1} \in [0, 1)^s$, the *(extreme) discrepancy* is defined by

$$D_N = D_N(t_0, t_1, \ldots, t_{N-1}) = \sup_J |F_N(J) - V(J)|,$$

where the supremum is extended over all subintervals $J$ of $[0, 1)^s$, $F_N(J)$ is $N^{-1}$ times the number of points among $t_0, t_1, \ldots, t_{N-1}$ falling into $J$, and $V(J)$ denotes the $s$–dimensional volume of $J$. Besides the extreme discrepancy, the *star discrepancy* $D_N^* = D_N^*(t_0, t_1, \ldots, t_{N-1})$ is also of interest, which is defined in the same way except that the supremum is extended only over all subintervals $J$ of $[0, 1)^s$ with one vertex at the origin. The star discrepancy $D_N^*$ is related to the extreme discrepancy $D_N$ by the inequalities $D_N^* \leq D_N \leq 2^s D_N^*$. Note that $D_N^*$ is just the statistical test quantity for the $s$–dimensional two–sided *Kolmogorov test* in nonparametric statistics. In the context of pseudorandom numbers, this theoretical test for equidistribution $(s = 1)$ and statistical independence $(s \geq 2)$ is also called *uniformity test* and $s$–dimensional *serial test*, respectively. For a sequence $t_0, t_1, \ldots$ of independent random variables uniformly distributed in $[0, 1)^s$, the star discrepancy $D_N^* = D_N^*(t_0, t_1, \ldots, t_{N-1})$ follows a *law of the iterated logarithm*, namely, $\limsup_{N \to \infty} N^{1/2}(\log \log N)^{-1/2} D_N^* = 1/\sqrt{2}$ with probability 1 with respect to the $s$–dimensional Lebesgue measure. This result was shown by Kai-Lai Chung [6] for $s = 1$ and by Jack Kiefer [72] for $s \geq 2$. Therefore, the order of magnitude $N^{-1/2}(\log \log N)^{1/2}$ will serve as a benchmark in comparison with the asymptotic behaviour of the discrepancy of $N$ pseudorandom points from $[0, 1)^s$.

**Notation.** The following abbreviations will be used throughout the paper. For any positive integer $n$, let $\mathbb{Z}_n = \{0, 1, \ldots, n-1\}$ and write $\mathbb{Z}_n^*$ for the subset of integers of $\mathbb{Z}_n$ which are relatively prime to $n$. Note that $\mathbb{Z}_n^*$ consists exactly of the odd elements of $\mathbb{Z}_n$, whenever $n$ is a power of two. For a prime $p$, the sets $\mathbb{Z}_p$ and $\mathbb{Z}_p^* = \mathbb{Z}_p \setminus \{0\}$ can be identified with the finite field of order $p$ and its multiplicative group, respectively.

**Empirical analysis.** Empirical testing is an important complement to the theoretical investigation of pseudorandom numbers. In contrast to the theoretical analysis, empirical tests will treat the pseudorandom number generator as a 'black box' and do not directly analyse the underlying algorithm. This makes it possible to study the performance for arbitrary parts of the period with varying length. Furthermore, empirical tests can be regarded as prototypes of stochastic simulations. In order to be able to compare the test results for the nonlinear congruential generators of this paper with empirical results for different linear and nonlinear methods in [76], the so–called 'Load Test' design has been chosen.

For a sequence $(x_n)_{n \geq 0}$ of uniform pseudorandom numbers in the interval $[0, 1)$, the *'Load Test'* uses overlapping $s$–tuples of integers defined by the four leading bits of every pseudorandom number, i.e., these $s$–tuples are given by $z_n = (\lfloor 16x_n \rfloor, \lfloor 16x_{n \oplus 1} \rfloor, \ldots, \lfloor 16x_{n \oplus (s-1)} \rfloor) \in \mathbb{Z}_{16}^s$ for $0 \leq n < N$, where $N$ is a fixed *sample size*, $\lfloor x \rfloor$ denotes the greatest integer less than or equal to $x$, and $\oplus$ means addition modulo $N$ in the set $\mathbb{Z}_N$. For each $z \in \mathbb{Z}_{16}^s$, let $c_N(z)$ be the number of occurences of the $s$–tuple $z$ in the sample $z_0, z_1, \ldots, z_{N-1}$. In the case $s \geq 2$, let $\tilde{c}_N$ denote the corresponding counter for $(s-1)$–dimensional tuples, i.e., $\tilde{c}_N(\tilde{z})$ can be obtained as the sum of $c_N(\tilde{z}, z_s)$ over all $z_s \in \mathbb{Z}_{16}$ for any $\tilde{z} \in \mathbb{Z}_{16}^{s-1}$. The statistical test quantity $T_N = T_N(x_0, x_1, \ldots, x_{N-1})$ for the $s$–dimensional *overlapping serial* (or *frequency*) *test* of these counters is defined by

$$T_N = \sum_{z \in \mathbb{Z}_{16}^s} \frac{(c_N(z) - N/16^s)^2}{N/16^s} - \sum_{\tilde{z} \in \mathbb{Z}_{16}^{s-1}} \frac{(\tilde{c}_N(\tilde{z}) - N/16^{s-1})^2}{N/16^{s-1}},$$

where the second sum is left out for $s = 1$. For a sequence $t_0, t_1, \ldots$ of independent random variables uniformly distributed in $[0, 1)$, the test quantity $T_N = T_N(t_0, t_1, \ldots, t_{N-1})$ is asymptotically $\chi^2$–distributed with $16^s - 16^{s-1}$ degrees of freedom. Note that in this case $N/16^s$ and $N/16^{s-1}$ are just the expectations of each counter $c_N(z)$ and $\tilde{c}_N(\tilde{z})$, respectively. A detailed proof of the above result and further information about the overlapping serial test can be found in [94, Ch. 5] and the references given there. The 'Load Test' combines the overlapping serial test with a two–sided *Kolmogorov test*. To this end, a sample $T_N^{(1)}, \ldots, T_N^{(32)}$ of 32 repetitions of the test quantity $T_N$ in the overlapping serial test is calculated, where $T_N^{(k)}$ depends on the pseudorandom numbers $x_{(k-1)N}, x_{(k-1)N+1}, \ldots, x_{kN-1}$ for $1 \le k \le 32$. Let $K_{32} = K_{32}(T_N^{(1)}, \ldots, T_N^{(32)})$ be $\sqrt{32}$ times the corresponding test quantity of the two–sided Kolmogorov test with respect to the asymptotic distribution of $T_N$, namely the $\chi^2$–distribution with $16^s - 16^{s-1}$ degrees of freedom. The rejection area of this test at a level of significance of 0.01 is approximated by $[1.59, \infty)$.

In this paper, the 'Load Test' will be applied to three nonlinear congruential generators, each time for varying dimension $s \in \{1, 2, \ldots, 5\}$ and sample size $N \in \{2^{18}, 2^{19}, \ldots, 2^{26}\}$. Note that the generator is restarted at the same initial value for each of these 45 combinations and that at most $2^{31}$ elements of the generated sequence are used. The computations have been carried out using the PLAB–package which is available at http://random.mat.sbg.ac.at.

## 2. Quadratic congruential method

The earliest nonlinear congruential method for generating uniform pseudorandom numbers is Donald E. Knuth's *quadratic congruential method* which is already mentioned in the first edition of his famous book [73] from 1969. The theoretical analysis of this rather simple approach is surprisingly involved. Consequently, one is still far away from a complete understanding of the quadratic congruential method, but the current knowledge indicates that the generated pseudorandom numbers have several attractive properties, at least, if the parameters in the underlying quadratic recursion are selected carefully.

Let $m \ge 2$ be a (large) integer, called the *modulus*, and let $a, b, c \in \mathbb{Z}_m$ be three *parameters*. For an *initial value* $y_0 \in \mathbb{Z}_m$, a *quadratic congruential sequence* $(y_n)_{n \ge 0}$ of elements of $\mathbb{Z}_m$ is defined recursively by

$$y_{n+1} \equiv a y_n^2 + b y_n + c \ (\mathrm{mod}\, m), \quad n \ge 0,$$

and the corresponding sequence $(x_n)_{n \ge 0}$ of *quadratic congruential pseudorandom numbers* in the interval $[0, 1)$ is obtained by $x_n = y_n/m$ for $n \ge 0$. These sequences are purely periodic if and only if $a \equiv 0 \ (\mathrm{mod}\, p)$ and $b \not\equiv 0 \ (\mathrm{mod}\, p)$ for any prime $p$ dividing the modulus $m$ (except that the weaker condition $a + b \equiv 1 \ (\mathrm{mod}\, 2)$ is necessary whenever $p = 2$ but not 4 divides $m$). This implies that in the quadratic congruential method a prime modulus (or more generally, a squarefree modulus) is uninteresting.

The following presentation concentrates on the case of a power of two modulus $m = 2^\omega$ for some integer $\omega \ge 5$. This choice of the modulus is the most convenient for implementation on a digital computer. The case of an odd prime power modulus is briefly mentioned at the end of this section. The analysis of the quadratic congruential method with an arbitrary composite modulus is still at an early stage. These generators can be identified with compound quadratic congruential generators which show some computational advantages. The reader is referred to the very interesting contribution of Sibylle Strandt in this volume which presents the current knowledge of the quadratic congruential method with composite modulus.

It follows from [73, p. 34] that quadratic congruential sequences in the case of a power of two modulus are purely periodic with the maximum possible period length $m$ if and only if the conditions $a \equiv 0 \pmod{2}$, $b \equiv a + 1 \pmod{4}$, and $c \equiv 1 \pmod{2}$ are satisfied. Then the generated pseudorandom numbers $x_0, x_1, \ldots, x_{m-1}$ within a period run through all rationals in $[0, 1)$ with denominator $m$. Therefore, the full period shows a perfect equidistribution in the interval $[0, 1)$.

**Distribution of pairs.** The distribution of tuples of successive pseudorandom numbers strongly depends on the maximal power of two that divides the parameter $a$ in the quadratic congruential method. The conditions for the maximum possible period length imply that $\gcd(a, m) = 2^{\alpha}$ for some integer $\alpha \in \{1, \ldots, \omega\}$. In particular, the choice $a \equiv 0 \pmod{m/2}$, i.e., $\alpha \geq \omega - 1$, reduces the underlying quadratic recursion to a linear one, whereas $a \equiv 2 \pmod{4}$, i.e., $\alpha = 1$, yields the most 'quadratic' behaviour. From now on, only the latter case is discussed, i.e., it is always assumed that the parameters in the quadratic congruential method satisfy the conditions $a \equiv 2 \pmod{4}$, $b \equiv 3 \pmod{4}$, and $c \equiv 1 \pmod{2}$.

Subsequently, *overlapping pairs* $\mathbf{x}_n = (x_n, x_{n+1}) \in [0, 1)^2$ of quadratic congruential pseudorandom numbers are considered, and the abbreviation $D_m^{(2)} = D_m(\mathbf{x}_0, \mathbf{x}_1, \ldots, \mathbf{x}_{m-1})$ is used for their discrepancy over the full period. The upper bound for $D_m^{(2)}$ in Theorem 2.1 improves earlier results in [49, Theorem 1; 34, Theorem 6] and follows at once from the upper bound in Theorem 2.2 and an application of [74, Ch. 2, Theorem 2.6; 39, Lemma 3]. The lower bound for $D_m^{(2)}$ is cited from [49, p. 249; 34, Theorem 7].

THEOREM 2.1. *The discrepancy $D_m^{(2)}$ of all overlapping pairs satisfies*

$$D_m^{(2)} < \frac{2\sqrt{2} + 8}{7\pi^2} m^{-1/2} (\log m)^2 - (0.0791) m^{-1/2} \log m + (0.3173) m^{-1/2} + 4 m^{-1}$$

*and*

$$D_m^{(2)} \geq \frac{1}{3(\pi + 2)} m^{-1/2}$$

*for any parameters $a \equiv 2 \pmod{4}$, $b \equiv 3 \pmod{4}$, and $c \equiv 1 \pmod{2}$.*

The distribution of overlapping pairs $(x_n, x_{n+1})$ is illustrated in Figure 2.1 by a plot of all generated points in the small subregion $[0, 0.002)^2$ of the unit square. The underlying quadratic congruential generator has modulus (and period length) $m = 2^{31} = 2\,147\,483\,648$ and parameters $a = 67\,483\,458$, $b = 212\,709\,107$, and $c = 557\,316\,979$. These specific values of the parameters have been chosen arbitrarily among all those satisfying the conditions in Theorem 2.1. It turned out that the number of points within a period falling into the subregion $[0, 0.002)^2$ is $8\,493$; the expected number of points under the uniform distribution is about $8\,590$. Figure 2.1 shows that there is no obvious structure within the generated points. Similar plots are ob-

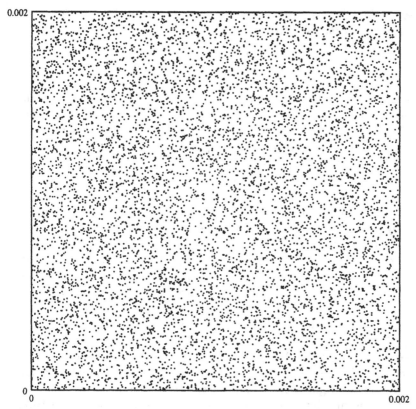

FIGURE 2.1. Distribution of overlapping pairs for $m = 2^{31}$ and parameters $a = 67\,483\,458$, $b = 212\,709\,107$, and $c = 557\,316\,979$.

tained in many other subregions of the unit square.

In the present case of a power of two modulus, it is also of interest to study the distribution of the two disjoint subsets of *nonoverlapping pairs* $\tilde{\mathbf{x}}_n = (x_{2n}, x_{2n+1}) \in [0,1)^2$. Let $\tilde{D}_{m/2}^{(2)} = D_{m/2}(\tilde{\mathbf{x}}_0, \tilde{\mathbf{x}}_1, \ldots, \tilde{\mathbf{x}}_{(m/2)-1})$ denote their discrepancy over the full period. The upper bound for $\tilde{D}_{m/2}^{(2)}$ in Theorem 2.2 refines an earlier result in [31, Theorem 1]; its proof is omitted, since the desired upper bound follows from [39, Lemmas 1 and 2(b)] by straightforward calculations and similar arguments as in [31, Proof of Theorem 1]. The lower bound for $\tilde{D}_{m/2}^{(2)}$ is taken from [31, Theorem 2].

THEOREM 2.2. *The discrepancy* $\tilde{D}_{m/2}^{(2)}$ *of all nonoverlapping pairs satisfies*

$$\tilde{D}_{m/2}^{(2)} < \frac{2\sqrt{2}+8}{7\pi^2}m^{-1/2}(\log m)^2 - (0.0791)m^{-1/2}\log m + (0.3173)m^{-1/2} + 4m^{-1}$$

*and*

$$\tilde{D}_{m/2}^{(2)} \geq \frac{2}{B(\pi+2)}m^{-1/2}$$

72

for any parameters $a \equiv 2 \pmod 4$, $b \equiv 3 \pmod 4$, and $c \equiv 1 \pmod 2$, where the constant $B$ in the lower bound is given by

$$B = \begin{cases} 1 & \text{for } y_0 \equiv (b+1)/4 \pmod 2), \\ 3 & \text{for } y_0 \equiv (b+5)/4 \pmod 2). \end{cases}$$

Theorems 2.1 and 2.2 show that the discrepancy of both overlapping and nonoverlapping pairs is always of an order of magnitude between $m^{-1/2}$ and $m^{-1/2}(\log m)^2$, provided the conditions $a \equiv 2 \pmod 4$, $b \equiv 3 \pmod 4$, and $c \equiv 1 \pmod 2$ on the parameters in the quadratic congruential method are satisfied. Note that it is in this range of magnitudes where one also finds the discrepancy of $m$ true random points from $[0, 1)^2$ according to Jack Kiefer's probabilistic law of the iterated logarithm.

The distribution of nonoverlapping pairs $(x_{2n}, x_{2n+1})$ is illustrated in Figure 2.2 by plots of the two subsets of all generated points in the small subregion $[0, 0.001)^2$ of the unit square. The underlying quadratic congruential generator is the same as in Figure 2.1, namely the one with modulus (and period length) $m = 2^{31} = 2\,147\,483\,648$ and parameters $a = 67\,483\,458$, $b = 212\,709\,107$, and $c = 557\,316\,979$. Hence, Figure 2.2 corresponds to the left lower subsquare of the plot of all overlapping pairs in Figure 2.1. The two disjoint subsets of nonoverlapping pairs belong to an even initial value (left plot) and an odd initial value (right plot), respectively. It turned out that the subregion $[0, 0.001)^2$ in Figure 2.2 contains $2\,186$ overlapping pairs in all, which split up into $1\,101$ nonoverlapping pairs on the left and $1\,085$ ones on the right; both times the expected number of points under the uniform distribution is about $1\,074$. Figure 2.2 shows that the distribution of the two subsets of nonoverlapping pairs looks truly random. Similar distributions appear in many other subregions of the unit square.

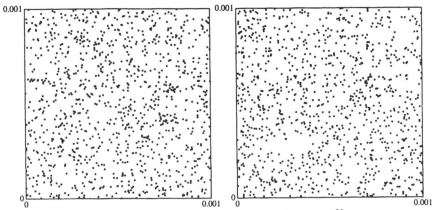

FIGURE 2.2. Distribution of nonoverlapping pairs for $m = 2^{31}$ and parameters $a = 67\,483\,458$, $b = 212\,709\,107$, and $c = 557\,316\,979$.

**Distribution of lagged pairs.** Besides the distribution of pairs of successive pseudorandom numbers, one is also interested in the distribution of more general types of pairs, since their distribution properties provide information on the autocorrelation structure of the generated sequence. Recently, *lagged pairs* $\mathbf{x}_n = (x_n, x_{n+2}) \in [0,1)^2$ of quadratic congruential pseudorandom numbers have been studied. Their discrepancy over the full period is denoted by $D_{m;a,b,c}^{[1,3]} = D_m(\mathbf{x}_0, \mathbf{x}_1, \ldots, \mathbf{x}_{m-1})$ in order to emphasize the dependence on the parameters $a, b, c$ in the underlying quadratic recursion. The discrepancy analysis for lagged pairs is much more involved than in the case of (ordinary) pairs discussed above. The upper bound for the average value of the discrepancy $D_{m;a,b,c}^{[1,3]}$ over the parameter $c$ in Theorem 2.3 slightly improves [37, Theorem 6]; its proof is left out, since the desired upper bound can be established by similar arguments as in [37, Proof of Theorem 6] and by an application of [39, Lemma 2(a)]. The lower bound for $D_{m;a,b,c}^{[1,3]}$ in Theorem 2.4 is cited from [37, Theorem 8].

**THEOREM 2.3.** *The average value of the discrepancy $D_{m;a,b,c}^{[1,3]}$ of all lagged pairs over the parameter $c \in \mathbb{Z}_m^*$ satisfies*

$$\frac{2}{m} \sum_{c \in \mathbb{Z}_m^*} D_{m;a,b,c}^{[1,3]} < \frac{4\sqrt{2}+2}{7\pi^2} m^{-1/2} (\log m)^2 + (0.0977)m^{-1/2} \log m$$
$$- (0.1753)m^{-1/2} + 2m^{-1}$$

*for any parameters $a \equiv 2 \pmod 4$ and $b \equiv 3 \pmod 4$.*

**THEOREM 2.4.** *Let $\nu \geq 1$ and $\mu \in \{0,1,2\}$ be integers with $\omega = 3\nu + \mu + 2$. Let $a \equiv 2 \pmod 4$, $b \equiv 3 \pmod 4$, and $c \equiv 1 \pmod 2$ be parameters with*

$$4ac \equiv (b-1)^2 - 28 + 2^{2\nu+4} \pmod{2^{\omega-\nu+1}}.$$

*Then the discrepancy $D_{m;a,b,c}^{[1,3]}$ of all lagged pairs satisfies*

$$D_{m;a,b,c}^{[1,3]} \geq \frac{2^{(\mu-1)/3}}{27(\pi+2)} m^{-1/3}.$$

The distribution of lagged pairs $(x_n, x_{n+2})$ is illustrated in Figures 2.3 and 2.4 by plots of all generated points in the small subregion $[0, 0.002)^2$ of the unit square. It turned out that the number of points within a period falling into this subregion is 8 551 and 8 750, respectively; both times the expected number of points under the uniform distribution is about 8 590. The underlying quadratic congruential generator in Figure 2.3 is the same as in Figures 2.1 and 2.2, namely the one with modulus (and period length) $m = 2^{31} = 2\,147\,483\,648$ and parameters $a = 67\,483\,458$, $b = 212\,709\,107$, and $c = 557\,316\,979$. Figure 2.3 shows a rather typical distribution of lagged pairs without particular structure. Similar plots can be obtained in other subregions of the unit square and for other arbitrary choices of the parameters in the quadratic congruential method. This behaviour of lagged pairs corresponds to the upper bound for their average discrepancy in Theorem 2.3, which is of the reasonable order of magnitude $m^{-1/2}(\log m)^2$. For $m = 2^{31}$, the crucial condition on the parameters $a, b, c$ in Theorem 2.4 takes the form $4ac \equiv (b-1)^2 + 4\,194\,276 \pmod{2^{23}}$, which is violated by the

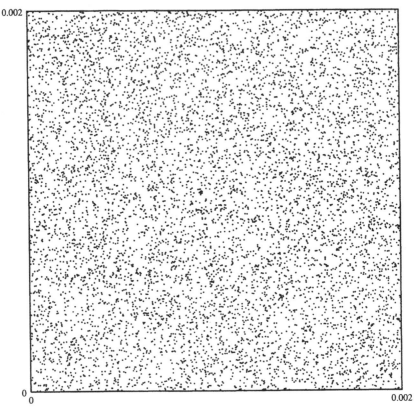

FIGURE 2.3. Distribution of lagged pairs for $m = 2^{31}$ and parameters $a = 67\,483\,458$, $b = 212\,709\,107$, and $c = 557\,316\,979$.

quadratic congruential generator of Figure 2.3 (and by most others).

The quadratic congruential generator of Figure 2.4 has modulus (and period length) $m = 2^{31} = 2\,147\,483\,648$ and parameters $a = 67\,483\,458$, $b = 212\,709\,107$, and $c = 557\,316\,981$, i.e., compared with the generator of Figure 2.3, only the parameter $c$ has been increased by 2. Hence, the distribution of (ordinary) pairs in the unit square remains almost the same; all generated points are merely shifted upwards by $2/m \approx 9.3 \cdot 10^{-10}$. Now, however, the crucial condition on the parameters $a, b, c$ in Theorem 2.4 is satisfied, and Figure 2.4 actually shows that a strongly pronounced line structure appears in the distribution of lagged pairs. It consists exactly of all $2^{22} = 4\,194\,304$ lagged pairs $(x_n, x_{n+2}) = (y_n, y_{n+2})/m$ with $y_n \equiv 452 \pmod{512}$. These lagged pairs satisfy the linear equation $27y_n + y_{n+2} \equiv 81\,212\,342 \pmod{2^{31}}$, i.e., they are lying on only 28 equidistant parallel lines in the unit square. The subregion $[0, 0.002)^2$ in Figure 2.4 contains a small part of one such line with 310 points. The proof of Theorem 2.4 in [37] reveals that the (large) order of magnitude $m^{-1/3}$ of the lower bound for the discrepancy $D_{m;a,b,c}^{[1,3]}$ of lagged pairs always stems from

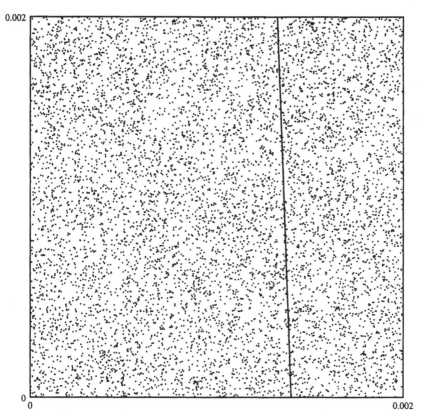

FIGURE 2.4. Distribution of lagged pairs for $m = 2^{31}$ and parameters
$a = 67\,483\,458$, $b = 212\,709\,107$, and $c = 557\,316\,981$.

the concentration of about $m^{2/3}$ points on 28 parallel lines in the unit square.

**Distribution of triples.** In the following section, *(overlapping) triples*
$\mathbf{x}_n = (x_n, x_{n+1}, x_{n+2}) \in [0, 1)^3$ of quadratic congruential pseudorandom
numbers are considered, and $D^{(3)}_{m;a,b,c} = D_m(\mathbf{x}_0, \mathbf{x}_1, \ldots, \mathbf{x}_{m-1})$ is written
for their discrepancy over the full period in order to stress again that it de-
pends on the parameters $a, b, c$ in the underlying quadratic recursion. The
discrepancy analysis for triples shows several similarities to the situation
of lagged pairs discussed above, and up to now, almost the same kind of
results could be established. The upper bound for the average value of the
discrepancy $D^{(3)}_{m;a,b,c}$ over the parameter $c$ in Theorem 2.5 can be deduced
from [34, Theorem 8 and its proof]. The (generally valid) lower bound for
$D^{(3)}_{m;a,b,c}$ in Theorem 2.6 follows at once from the corresponding result for
the discrepancy $D^{(2)}_m$ of all overlapping pairs in Theorem 2.1. The (particularly large) lower bound in Theorem 2.7(a) is taken from [29, Theorem;
34, Theorem 10], and the one in Theorem 2.7(b) is an immediate conse-

quence of the corresponding lower bound for the discrepancy $D_{m;a,b,c}^{[1,3]}$ of all lagged pairs in Theorem 2.4.

THEOREM 2.5. *The average value of the discrepancy* $D_{m;a,b,c}^{(3)}$ *of all triples over the parameter* $c \in \mathbb{Z}_m^*$ *satisfies*

$$\frac{2}{m} \sum_{c \in \mathbb{Z}_m^*} D_{m;a,b,c}^{(3)} < \frac{24\sqrt{2} + 68}{31\pi^3} m^{-1/2} (\log m)^3 + (0.8427) m^{-1/2} (\log m)^2$$
$$+ (2.0927) m^{-1/2} \log m + (1.6495) m^{-1/2} + 3m^{-1}$$

*for any parameters* $a \equiv 2 \pmod 4$ *and* $b \equiv 3 \pmod 4$.

THEOREM 2.6. *The discrepancy* $D_{m;a,b,c}^{(3)}$ *of all triples satisfies*

$$D_{m;a,b,c}^{(3)} \geq \frac{1}{3(\pi + 2)} m^{-1/2}$$

*for any parameters* $a \equiv 2 \pmod 4$, $b \equiv 3 \pmod 4$, *and* $c \equiv 1 \pmod 2$.

THEOREM 2.7. *Let* $\nu \geq 1$ *and* $\mu \in \{0, 1, 2\}$ *be integers with* $\omega = 3\nu + \mu + 2$.
(a) *Let* $a \equiv 2 \pmod 4$, $b \equiv 3 \pmod 4$, *and* $c \equiv 1 \pmod 2$ *be parameters with*

$$4ac \equiv (b-1)^2 - 4 \pm 8 + 2^{2\nu+4} \pmod{2^{\omega - \nu + 1}}.$$

*Then the discrepancy* $D_{m;a,b,c}^{(3)}$ *of all triples satisfies*

$$D_{m;a,b,c}^{(3)} \geq \frac{2^{(\mu-1)/3}}{4(\pi^2 + 3\pi + 3)} m^{-1/3}.$$

(b) *Let* $a \equiv 2 \pmod 4$, $b \equiv 3 \pmod 4$, *and* $c \equiv 1 \pmod 2$ *be parameters with*

$$4ac \equiv (b-1)^2 - 28 + 2^{2\nu+4} \pmod{2^{\omega - \nu + 1}}.$$

*Then the discrepancy* $D_{m;a,b,c}^{(3)}$ *of all triples satisfies*

$$D_{m;a,b,c}^{(3)} \geq \frac{2^{(\mu-1)/3}}{27(\pi + 2)} m^{-1/3}.$$

Theorems 2.5 and 2.6 show that the average discrepancy of triples (over the parameter $c$) is always of an order of magnitude between $m^{-1/2}$ and $m^{-1/2}(\log m)^3$, provided the conditions $a \equiv 2 \pmod 4$ and $b \equiv 3 \pmod 4$ on the parameters in the quadratic congruential method are satisfied. On the other hand, Theorem 2.7 implies that there exist corresponding parameters $a, b, c$ such that the discrepancy $D_{m;a,b,c}^{(3)}$ of triples is of an order of magnitude at least $m^{-1/3}$.

The proof of Theorem 2.7(a) in [29] shows that the (large) order of magnitude $m^{-1/3}$ of the lower bound for the discrepancy $D_{m;a,b,c}^{(3)}$ of triples stems from a concentration of about $m^{2/3}$ points on only 6 equidistant parallel planes in the unit cube.

**Empirical results.** The 'Load Test' as described in the previous section has been applied to the quadratic congruential generator with modulus (and period length) $m = 2^{31} = 2\,147\,483\,648$ and parameters $a = 67\,483\,458$, $b = 212\,709\,107$, and $c = 557\,316\,979$, which is the same generator as in Figures 2.1–2.3. The empirical results of the 'Load Test' for the initial value $y_0 = 0$ are summarized in Figure 2.5. The three–dimensional bar plot on the left reports the values of $K_{32} = K_{32}(T_N^{(1)}, \ldots, T_N^{(32)})$ (censored at 2) in the two–sided Kolmogorov test for the 45 analysed combinations of dimension $s$ and sample size $N$. Black bars indicate values in the rejection area $[1.59, \infty)$. The corresponding grey–scale plot on the right contains for each of these 45 bars an illustration of the underlying sample $T_N^{(1)}, \ldots, T_N^{(32)}$ of the test quantity in the overlapping serial test, which provides additional information on the behaviour of the generator and reveals possible reasons for a critical value under the 'Load Test'. A small value of $T_N^{(k)}$ in a quantile scale is represented by a white rectangle, whereas a black rectangle belongs to a large value. The resulting 32 rectangles are arranged in four rows of eight ones each, where the numbering is from left to right and from bottom to top. The quadratic congruential generator of Figure 2.5 performs quite well under the 'Load Test'. Only for the largest sample size $N = 2^{26}$, a somewhat suspicious behaviour appears, which can be attributed to the fact that in this case the period length $m = 2^{31}$ is just exhausted by the 32 repetitions of the overlapping serial test. The empirical results in [76] show that this generator behaves more robust under the 'Load Test' than linear congruential generators of comparable period length.

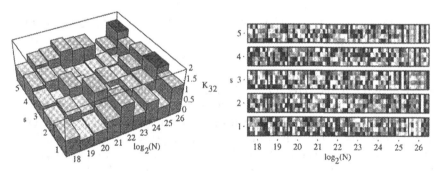

FIGURE 2.5. Empirical results of the 'Load Test' with initial value $y_0 = 0$ for $m = 2^{31}$ and parameters $a = 67\,483\,458$, $b = 212\,709\,107$, and $c = 557\,316\,979$.

Notes. Equidistribution properties of parts of the period in the quadratic congruential method with arbitrary composite modulus are analysed in an interesting paper of Frank Emmerich [63]. It follows from [63, Theorem 5] that, for relatively large parts of the period, the distribution of the pseudorandom numbers tends to be more regular than the one of true random numbers, which reflects the perfect equidistribution of the full period.

The lattice structure of quadratic congruential pseudorandom numbers with arbitrary composite modulus is studied in [11, 25]. It is shown that, in any dimension $s \geq 2$, the set of all $s$–tuples $(x_n, x_{n+1}, \ldots, x_{n+s-1}) \in [0, 1)^s$

of successive pseudorandom numbers can be described by a superposition of shifted lattices. The corresponding bases and shift vectors of these lattices are explicitly given. In the case $m = 2^\omega$ and $a \equiv 2 \pmod 4$ discussed above, there exist $2^{-\delta/2}m^{1/2}$ shifted lattices, each consisting of $2^{\delta/2}m^{1/2}$ generated points in $[0, 1)^s$, where $\delta \in \{0, 1\}$ with $\delta \equiv \omega \pmod 2$.

Statistical independence properties of quadratic congruential pseudorandom numbers have also been analysed in the case of an odd prime power modulus. The discrepancy of pairs is studied in [49, 55], and results on the distribution of triples can be found in [16, 38].

# 3. Inversive congruential method with $2^\omega$–modulus

The *inversive congruential method with power of two modulus* for generating uniform pseudorandom numbers was introduced in [12] based on a corresponding suggestion of Donald E. Knuth given in a personal communication. The theoretical analysis of this method is still in progress, but the current knowledge already indicates that the generated pseudorandom numbers have several attractive properties, at least, if the parameters in the underlying inversive recursion are selected carefully.

Let $m = 2^\omega$ for some integer $\omega \geq 6$ be the *modulus*, and let $a, b \in \mathbb{Z}_m$ be two *parameters* with $a + b \equiv 1 \pmod 2$. For an *initial value* $y_0 \in \mathbb{Z}_m^*$, an *inversive congruential sequence* $(y_n)_{n \geq 0}$ of elements of $\mathbb{Z}_m^*$ is defined recursively by

$$y_{n+1} \equiv ay_n^{-1} + b \pmod m, \quad n \geq 0,$$

where $z^{-1}$ denotes the multiplicative inverse of $z \in \mathbb{Z}_m^*$. An algorithm for its efficient calculation is described in the next section. The corresponding sequence $(x_n)_{n \geq 0}$ of *inversive congruential pseudorandom numbers* in the interval $[0, 1)$ is obtained by $x_n = y_n/m$ for $n \geq 0$. These sequences are purely periodic if and only if $a \equiv 1 \pmod 2$ and $b \equiv 0 \pmod 2$.

It follows from [12, Theorem; 86, Theorem 8.9] that inversive congruential sequences are purely periodic with the maximum possible period length $m/2$ if and only if the conditions $a \equiv 1 \pmod 4$ and $b \equiv 2 \pmod 4$ are satisfied. Then the generated pseudorandom numbers $x_0, x_1, \ldots, x_{(m/2)-1}$ within a period run through all rationals in $[0, 1)$ with odd numerator and denominator $m$. Hence, the full period shows a perfect equidistribution in the interval $[0, 1)$.

**Distribution of pairs.** Subsequently, the distribution of *overlapping pairs* $\mathbf{x}_n = (x_n, x_{n+1}) \in [0, 1)^2$ of inversive congruential pseudorandom numbers is studied. Let $D_{m/2}^{(2)} = D_{m/2}(\mathbf{x}_0, \mathbf{x}_1, \ldots, \mathbf{x}_{(m/2)-1})$ denote their discrepancy over the full period. The upper bound for $D_{m/2}^{(2)}$ in Theorem 3.1 is cited from [39, Theorem 1] and improves an earlier result in [80, Theorem 2]. The lower bound for $D_{m/2}^{(2)}$ in Theorem 3.2 is taken from [50, Theorems 1 and 2].

THEOREM 3.1. *The discrepancy $D_{m/2}^{(2)}$ of all overlapping pairs satisfies*

$$D_{m/2}^{(2)} < \frac{8\sqrt{2}+4}{7\pi^2}m^{-1/2}(\log m)^2 - (0.4191)m^{-1/2}\log m + (0.6328)m^{-1/2} + 8m^{-1}$$

*for any parameters $a \equiv 1 \pmod 4$ and $b \equiv 2 \pmod 4$.*

THEOREM 3.2. *Let $c \in \{1, 5\}$ and $A(t) = (4-t^2)/(8-t^2)$ for some $t \in (0, 2]$. Then there exist more than $A(t)m/8$ parameters $a \in \mathbb{Z}_m$ with $a \equiv c \pmod 8$ such that the discrepancy $D_{m/2}^{(2)}$ of all overlapping pairs satisfies*

$$D_{m/2}^{(2)} \geq \frac{t}{B(\pi+2)}m^{-1/2}$$

*for any parameter $b \equiv 2 \pmod 4$, where the constant $B$ in the lower bound is given by $B = 1$ for $c = 1$ and $B = 3$ for $c = 5$.*

The distribution of overlapping pairs $(x_n, x_{n+1})$ is illustrated in Figure 3.1 by a plot of all generated points in the small subregion $[0, 0.003)^2$ of the unit square. The underlying inversive congruential generator has modu-

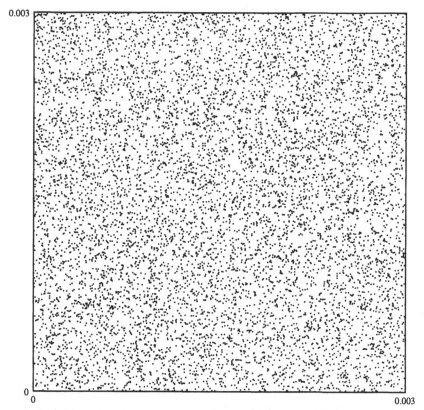

FIGURE 3.1. Distribution of overlapping pairs for $m = 2^{31}$ and parameters $a = 67\,483\,461$ and $b = 212\,709\,106$.

lus (and double period length) $m = 2^{31} = 2\,147\,483\,648$ and parameters $a = 67\,483\,461$ and $b = 212\,709\,106$. These values of the parameters are almost the same as in the examples of quadratic congruential generators in the previous section. It turned out that the number of points within a period falling into the subregion $[0, 0.003)^2$ is $9\,541$; the expected number of points under the uniform distribution is about $9\,664$. Figure 3.1 shows that there is no obvious structure within the generated points. Similar plots are obtained in many other subregions of the unit square.

In the present case of a power of two modulus, it is again of additional interest to study the distribution of the two disjoint subsets of *nonoverlapping pairs* $\tilde{\mathbf{x}}_n = (x_{2n}, x_{2n+1}) \in [0, 1)^2$. Let $\tilde{D}^{(2)}_{m/4} = D_{m/4}(\tilde{\mathbf{x}}_0, \tilde{\mathbf{x}}_1, \ldots, \tilde{\mathbf{x}}_{(m/4)-1})$ denote their discrepancy over the full period. The upper bound for $\tilde{D}^{(2)}_{m/4}$ in Theorem 3.3 is taken from [39, Theorem 2] and improves an earlier result in [51, Theorem 1]; the lower bound for $\tilde{D}^{(2)}_{m/4}$ is cited from [51, Theorem 2].

THEOREM 3.3. *The discrepancy $\tilde{D}^{(2)}_{m/4}$ of all nonoverlapping pairs satisfies*

$$\tilde{D}^{(2)}_{m/4} < \frac{8\sqrt{2}+4}{7\pi^2} m^{-1/2}(\log m)^2 - (0.4191)m^{-1/2}\log m + (0.6328)m^{-1/2} + 8m^{-1}$$

*and*

$$\tilde{D}^{(2)}_{m/4} \geq \frac{2\sqrt{2}}{B(\pi+2)}m^{-1/2}$$

*for any parameters $a \equiv 1 \pmod 4$ and $b \equiv 2 \pmod 4$, where the constant $B$ is given by $B = 1$ for $a \equiv 1 \pmod 8$ and $B = 3$ for $a \equiv 5 \pmod 8$.*

Apart from the weaker result on the lower bound for the discrepancy of overlapping pairs in Theorem 3.2, it follows from Theorems 3.1–3.3 that the discrepancy of both overlapping and nonoverlapping pairs is of

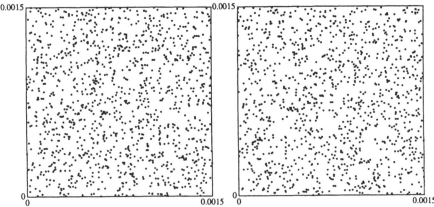

FIGURE 3.2. Distribution of nonoverlapping pairs for $m = 2^{31}$ and parameters $a = 67\,483\,461$ and $b = 212\,709\,106$.

an order of magnitude between $m^{-1/2}$ and $m^{-1/2}(\log m)^2$ for all parameters $a \equiv 1 \pmod 4$ and $b \equiv 2 \pmod 4$ in the inversive congruential method. Note that it is in this range of magnitudes where one also finds the discrepancy of $m/2$ true random points from $[0,1)^2$ according to the probabilistic law of the iterated logarithm.

The distribution of nonoverlapping pairs $(x_{2n}, x_{2n+1})$ is illustrated in Figure 3.2 by plots of the two subsets of all generated points in the small subregion $[0, 0.0015)^2$ of the unit square. The underlying inversive congruential generator is the same as in Figure 3.1, namely the one with modulus (and double period length) $m = 2^{31} = 2\,147\,483\,648$ and parameters $a = 67\,483\,461$ and $b = 212\,709\,106$. Hence, Figure 3.2 corresponds to the left lower subsquare of the plot of all overlapping pairs in Figure 3.1. The two disjoint subsets of nonoverlapping pairs belong to an initial value $y_0 \equiv 1 \pmod 4$ (left plot) and an initial value $y_0 \equiv 3 \pmod 4$ (right plot), respectively. It turned out that the subregion $[0, 0.0015)^2$ in Figure 3.2 contains $2\,473$ overlapping pairs in all, which split up into $1\,266$ nonoverlapping pairs on the left and $1\,207$ ones on the right; both times the expected

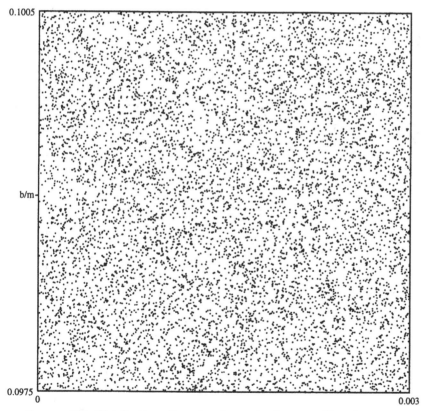

FIGURE 3.3. Distribution of overlapping pairs for $m = 2^{31}$ and parameters $a = 67\,483\,461$ and $b = 212\,709\,106$.

82

number of points under the uniform distribution is about 1 208. Figure 3.2 shows that the distribution of the two subsets of nonoverlapping pairs looks truly random. Similar distributions appear in many other subregions of the unit square.

**Hyperbolic structures.** The current knowledge indicates that there exist two subregions of the unit square which are particularly critical with respect to hyperbolic structures in the distribution of pairs $(x_n, x_{n+1})$ of inversive congruential pseudorandom numbers, namely the regions around the points $(0, b/m)$ and $(1, b/m)$, respectively. The situation around the point $(0, b/m)$ is illustrated in Figures 3.3 and 3.4 by plots of all generated points in the small subregion $[0, 0.003) \times [0.0975, 0.1005)$ of the unit square. The two underlying inversive congruential generators have modulus (and double period length) $m = 2^{31} = 2\,147\,483\,648$ and parameter $b = 212\,709\,106$, but different values of the parameter $a$. This implies that the ratio $b/m \approx 0.09905$ is the same. It turned out that the number of points within a period falling into the subregion $[0, 0.003) \times [0.0975, 0.1005)$

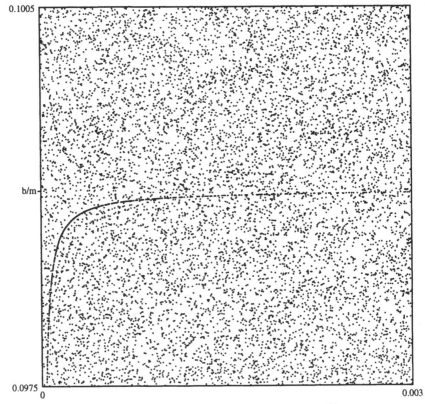

FIGURE 3.4. Distribution of overlapping pairs for $m = 2^{31}$ and parameters $a = 69\,282\,925$ and $b = 212\,709\,106$.

is 9 419 and 10 306, respectively; both times the expected number of points under the uniform distribution is about 9 664. The inversive congruential generator of Figure 3.3 has parameter $a = 67\,483\,461$, i.e., this generator is the same as in Figures 3.1 and 3.2. Figure 3.3 shows a rather typical distribution of pairs around the point $(0, b/m)$ without remarkable hyperbolic structures. Similar plots can be obtained for many other arbitrary choices of the parameters in the inversive congruential method.

The underlying inversive congruential generator in Figure 3.4 has parameter $a = 69\,282\,925$. Now, a strongly pronounced hyperbolic structure appears in the distribution of pairs. It consists exactly of all 1 154 pairs $(x_n, x_{n+1}) = (y_n, y_{n+1})/m$ with the property that $y_n$ is an (odd) divisor of $125m - a = 268\,366\,173\,075 = 3^2 \cdot 5^2 \cdot 7^2 \cdot 11 \cdot 13 \cdot 17^2 \cdot 19 \cdot 31$ which is greater than $(125m - a)/b \approx 1\,261.658$ and less than $m$. These pairs satisfy the hyperbolic equation $y_{n+1} = -(125m - a)/y_n + b$, i.e., they are lying on one hyperbola in the unit square. The subregion $[0, 0.003) \times [0.0975, 0.1005)$ in Figure 3.4 contains a small but significant part of this curve with 487 points.

**Distribution of triples.** In the following section, *(overlapping) triples* $\mathbf{x}_n = (x_n, x_{n+1}, x_{n+2}) \in [0, 1)^3$ of inversive congruential pseudorandom numbers are considered, and $D^{(3)}_{m/2;a,b} = D_{m/2}(\mathbf{x}_0, \mathbf{x}_1, \ldots, \mathbf{x}_{(m/2)-1})$ is written for their discrepancy over the full period in order to stress that it depends on the parameters $a, b$ in the underlying inversive recursion. Again, the discrepancy analysis for triples is much more involved than the one for pairs discussed above. Only very recently, the following results have been established. The upper bound for the average value of the discrepancy $D^{(3)}_{m/2;a,b}$ over the parameter $a$ (with parameter $b \equiv ac \pmod m$ for some fixed integer $c \equiv 2 \pmod 4$) in Theorem 3.4 can be deduced from [56, Theorem 1 and its proof]. The (generally valid) lower bound for $D^{(3)}_{m/2;a,b}$ in Theorem 3.5 is cited from [56, Theorem 3]. The (particularly large) lower bound in Theorem 3.6 follows at once from [57, Corollary].

THEOREM 3.4. *The average value of the discrepancy $D^{(3)}_{m/2;a,b}$ of all triples over the parameter $a \in \mathbb{Z}_m$ with $a \equiv 1 \pmod 4$ and $b \equiv ac \pmod m$ satisfies*

$$\frac{4}{m} \sum_{\substack{a \in \mathbb{Z}_m \\ a \equiv 1 \,(\mathrm{mod}\,4)}} D^{(3)}_{m/2;a,ac} < \frac{68\sqrt{2} + 48}{31\pi^3} m^{-1/2}(\log m)^3 + (0.579)m^{-1/2}(\log m)^2$$
$$+ (0.3106)m^{-1/2}\log m + (0.0333)m^{-1/2} + 6m^{-1}$$

*for any integer $c \equiv 2 \pmod 4$.*

THEOREM 3.5. *The discrepancy $D^{(3)}_{m/2;a,b}$ of all triples satisfies*

$$D^{(3)}_{m/2;a,b} \geq \frac{2}{\pi + 2} m^{-1/2}$$

*for any parameters $a \equiv 1 \pmod 4$ and $b \equiv 2 \pmod 4$.*

THEOREM 3.6. *Let $\nu \geq 2$ and $\mu \in \{0, 1, 2\}$ be integers with $\omega = 3\nu + \mu$. Let $a \equiv 1 \pmod{4}$ and $b \equiv 2 \pmod{4}$ be parameters with*

$$4a \equiv b^2 + 2^{2\nu+2} \pmod{2^{\omega-\nu+1}}.$$

*Then the discrepancy $D^{(3)}_{m/2;a,b}$ of all triples satisfies*

$$D^{(3)}_{m/2;a,b} \geq \frac{2^{\mu/3}}{2(\pi^2 + 3\pi + 3)} m^{-1/3}.$$

Theorems 3.4 and 3.5 show that the average discrepancy of triples is of an order of magnitude between $m^{-1/2}$ and $m^{-1/2}(\log m)^3$. On the other hand, Theorem 3.6 implies that there exist parameters $a, b$ such that the discrepancy $D^{(3)}_{m/2;a,b}$ of triples is of an order of magnitude at least $m^{-1/3}$.

The proof of Theorem 3.6 in [57] shows that the (large) order of magnitude $m^{-1/3}$ of the lower bound for the discrepancy $D^{(3)}_{m/2;a,b}$ of triples stems from a concentration of about $m^{2/3}$ points on only 4 equidistant parallel planes in the unit cube. By the way, the crucial condition on the parameters $a, b$ in Theorem 3.6 is violated by the inversive congruential generators of Figures 3.1–3.4.

**Empirical results.** The 'Load Test' as described above has been applied to the inversive congruential generator with modulus (and double period length) $m = 2^{31} = 2\,147\,483\,648$ and parameters $a = 67\,483\,461$ and $b = 212\,709\,106$, which is the same generator as in Figures 3.1–3.3. The empirical results of the 'Load Test' for the initial value $y_0 = 1$ are summarized in Figure 3.5. These two plots are organized as in Figure 2.5 of the previous section. This inversive congruential generator shows some deficiencies under the 'Load Test'. For a fair comparison with the generators of Figures 2.5 and 4.4, it must be taken into account that this inversive congruential generator has only period length $m/2 = 2^{30}$, which explains the poor behaviour and the stripy patterns for the two largest sample sizes $N = 2^{25}$ and $N = 2^{26}$. It should be observed that there are several suspicious values for smaller sample sizes, too. However, the empirical results

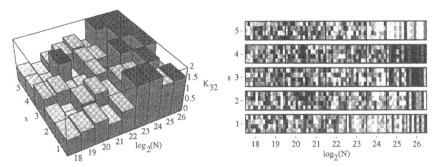

FIGURE 3.5. Empirical results of the 'Load Test' with initial value $y_0 = 1$ for $m = 2^{31}$ and parameters $a = 67\,483\,461$ and $b = 212\,709\,106$.

in [76] show that this generator behaves still more robust under the 'Load Test' than linear congruential generators of comparable period length.

**Notes.** Equidistribution properties of parts of the period in the inversive congruential method are analysed in [35]. It follows from [35, Theorem 1] that, for relatively large parts of the period, the distribution of the pseudorandom numbers tends to be more regular than the one of true random numbers from $[0, 1)$, which reflects the perfect equidistribution of the full period.

The lattice structure of inversive congruential pseudorandom numbers is studied in [25, 44]. It is shown that, in any dimension $s \geq 2$, the set of all $s$-tuples $(x_n, x_{n+1}, \ldots, x_{n+s-1}) \in [0, 1)^s$ of successive pseudorandom numbers can be described by a superposition of shifted lattices. The corresponding bases and shift vectors of these lattices are explicitly given. There exist $2^{-\delta/2}(m/2)^{1/2}$ shifted lattices, each consisting of $2^{\delta/2}(m/2)^{1/2}$ generated points in $[0, 1)^s$, where $\delta \in \{0, 1\}$ with $\delta \equiv \omega + 1 \pmod 2$.

The autocorrelation structure of inversive congruential pseudorandom numbers is studied in [19]. It follows from [19, Theorem 3] that, for any odd integer $\ell \geq 3$, the discrepancy of all lagged pairs $(x_n, x_{n+\ell}) \in [0, 1)^2$ over the full period is less than $(0.2217)m^{-1/2}((\log m)^2 + (5.052)\log m + 6.09) + 4m^{-1}$ for any parameters $a \equiv 1 \pmod 4$ and $b \equiv 2 \pmod 4$.

Recently, the following generalized inversive congruential method with power of two modulus was introduced by Takashi Kato, Li-Ming Wu, and Niro Yanagihara in [69]. Let $a, b, c \in \mathbb{Z}_m$ be three parameters with $a + b + c \equiv 1 \pmod 2$. For an initial value $y_0 \in \mathbb{Z}_m^*$, a generalized inversive congruential sequence $(y_n)_{n \geq 0}$ of elements of $\mathbb{Z}_m^*$ is defined recursively by $y_{n+1} \equiv a y_n^{-1} + b + c y_n \pmod m$ for $n \geq 0$. The corresponding sequence $(x_n)_{n \geq 0}$ of pseudorandom numbers in the interval $[0, 1)$ is obtained again by $x_n = y_n/m$ for $n \geq 0$. These sequences are purely periodic if and only if $a + c \equiv 1 \pmod 2$ and $b \equiv 0 \pmod 2$. It follows from [69, Theorem] that they have the maximum possible period length $m/2$ if and only if $a + c \equiv 1 \pmod 4$ and $b \equiv 2 \pmod 4$. Note that the choice $c \equiv 0 \pmod{m/2}$ yields the ordinary inversive congruential method. The distribution of pairs of successive pseudorandom numbers strongly depends on the maximal power of two that divides the parameter $a$. In particular, the choice $a \equiv 0 \pmod{m/8}$ reduces the underlying recursion to a linear one, whereas an odd value of the parameter $a$ yields the most 'inversive' behaviour. Only the latter case is briefly discussed below. It follows from [70, Theorem 1(I)] that the discrepancy $D_{m/2}^{(2)}$ of all overlapping pairs $(x_n, x_{n+1}) \in [0, 1)^2$ over the full period is less than $(0.2217)m^{-1/2}((\log m)^2 + (5.052)\log m + 6.09) + 4m^{-1}$ for any parameters $a \equiv 1 \pmod 2$, $b \equiv 2 \pmod 4$, and $c \equiv a - 1 \pmod 4$. On the other hand, it can be deduced from [70, Theorem 2 and its proof] that, for any $t \in (0, 2]$, there exist more than $((4 - t^2)/(8 - t^2))m/8$ integers $d \in \mathbb{Z}_m$ with $d \equiv 1 \pmod 8$ such that $D_{m/2}^{(2)} \geq t(\pi + 2)^{-1}m^{-1/2}$ for any parameters $a \equiv 1 \pmod{2}$, $b \equiv 2 \pmod{4}$, and $c \equiv a - 1 \pmod{8}$ with $a(c + 1) \equiv d \pmod{m}$. The lattice structure of generalized inversive congruential pseudorandom numbers is studied in [71], and similar results as in the (ordinary) inversive congruential method are established.

In the following, a modified inversive congruential method with power of two modulus is briefly described, which was introduced in [43]. Let $(\cdot)^- : \mathbb{Z}_m \to \mathbb{Z}_m$ denote a generalized multiplicative inverse modulo $m$ defined by $(0)^- = 0$ and $(2^\alpha z)^- \equiv 2^\alpha z^{-1} \pmod{m}$ for integers $z \in \mathbb{Z}_{2^{\omega - \alpha}}^*$ and $\alpha \in \{0, 1, \ldots, \omega - 1\}$. For two parameters $a, b \in \mathbb{Z}_m$ and an initial value $y_0 \in \mathbb{Z}_m$, a modified inversive congruential sequence $(y_n)_{n \geq 0}$ of elements of $\mathbb{Z}_m$ is defined recursively by $y_{n+1} \equiv a(y_n)^- + b \pmod{m}$ for $n \geq 0$. The corresponding sequence $(x_n)_{n \geq 0}$ of pseudorandom numbers in the interval $[0, 1)$ is obtained again by $x_n = y_n/m$ for $n \geq 0$. These sequences are purely periodic if and only if the parameter $a$ is odd. It is shown in [43, Theorem 1] that they have the maximum possible period length $m$ if and only if $a \equiv 1 \pmod{4}$ and $b \equiv 1 \pmod{2}$. Hence, the obvious advantage of this modified inversive congruential method is that the period length can attain the larger value $m$, as opposed to the maximum possible period length $m/2$ in the original method. It follows from [43, Theorem 2] that the discrepancy $D_m^{(2)}$ of all overlapping pairs $(x_n, x_{n+1}) \in [0, 1)^2$ over the full period is less than $(1.5136)m^{-1/2}((\log m)^2 + (3.77) \log m + 3.553) + 2m^{-1}$ for any parameters $a \equiv 1 \pmod{4}$ and $b \equiv 1 \pmod{2}$. On the other hand, it can be deduced from [43, Theorem 3 and the remark after its proof] that, for any $t \in (0, 1]$ and $c \in \{1, 5\}$, there exist more than $((1 - t^2)/(8\sqrt{2} + 12 - t^2))m/8$ parameters $a \in \mathbb{Z}_m$ with $a \equiv c \pmod{8}$ such that $D_m^{(2)} > (t/B)(\pi + 2)^{-1}m^{-1/2}$ for any parameter $b \equiv 1 \pmod{2}$, where the constant $B$ is given by $B = 1$ for $c = 1$ and $B = 3$ for $c = 5$.

The inversive congruential method has also been studied with a prime power modulus $m = p^\omega$ for some odd prime $p \geq 3$ and an integer $\omega \geq 2$. This approach was introduced in [59]. The parameters $a, b \in \mathbb{Z}_m$ in the underlying inversive recursion $y_{n+1} \equiv ay_n^{-1} + b \pmod{m}$ and the initial value $y_0 \in \mathbb{Z}_m^*$ have to be chosen in such a way that each generated element $y_{n+1}$ belongs to $\mathbb{Z}_m^*$. The generated sequences are purely periodic if $a \not\equiv 0 \pmod{p}$, and the period length is at most $\varphi(m) = (p - 1)p^{\omega - 1}$. Periodicity properties are studied in [3, 17, 59, 68], and it follows from these results that the maximum possible period length is $((p + 1)/2)p^{\omega - 1}$ (which is less than $\varphi(m)$ for all primes $p \geq 5$). Statistical independence properties, based on the discrepancy of pairs and triples of successive pseudorandom numbers, are analysed in [15, 23], and the lattice structure of these pseudorandom numbers is studied in [25]. The reader is referred to [18, Sec. 3.2] for a brief survey of the inversive congruential method with odd prime power modulus.

There exist a few explicit variants of the inversive congruential method with prime power modulus, the earliest one was introduced and studied in [46]. Some results on the properties of these explicit inversive congruential pseudorandom numbers can be found in [32, 36, 46].

# 4. Inversive congruential method with prime modulus

The earliest *inversive congruential method* for generating uniform pseudo-random numbers is the one with *prime modulus* which was introduced in [10]. It belongs to the currently most attractive nonlinear congruential methods, although a complete understanding of this approach has of course not been reached (and probably never will be). Its theoretical analysis is rather demanding, and significant applications of number theory play an important role in this theory.

Let $p \geq 5$ be a (large) prime, called the *modulus*, and let $a, c \in \mathbb{Z}_p^*$ be two *parameters*. For an *initial value* $y_0 \in \mathbb{Z}_p$ (with $y_0 \equiv cz_0 \pmod{p}$ for some fixed $z_0 \in \mathbb{Z}_p$), an *inversive congruential sequence* $(y_n)_{n \geq 0}$ of elements of $\mathbb{Z}_p$ is defined recursively by

$$y_{n+1} \equiv ac^2 y_n^{-1} + c \pmod{p}, \quad n \geq 0,$$

where $z^{-1}$ denotes the multiplicative inverse of $z \in \mathbb{Z}_p^*$ and $0^{-1} = 0$. Note that $y_n \equiv cz_n \pmod{p}$ for $n \geq 0$, where the sequence $(z_n)_{n \geq 0}$ of elements of $\mathbb{Z}_p$ satisfies the inversive recursion with $c = 1$, i.e., $z_{n+1} \equiv az_n^{-1} + 1 \pmod{p}$ for $n \geq 0$. The efficient calculation of the multiplicative inverse of $z \in \mathbb{Z}_p^*$ can be based on the Euclidean algorithm with the integers $z$ and $p$. To this end, put $u_0 = p$, $u_1 = z$, $v_0 = 0$, $v_1 = 1$, and $i = 1$. Now, while $u_i > 1$ put $u_{i+1} \equiv u_{i-1} \pmod{u_i}$ with $u_{i+1} \in \mathbb{Z}_{u_i}$, $v_{i+1} = v_{i-1} - v_i(u_{i-1} - u_{i+1})/u_i$, and increase $i$ by 1. Then the algorithm terminates after $O(\log p)$ steps with $z^{-1} \equiv v_i \pmod{p}$. The corresponding sequence $(x_n)_{n \geq 0}$ of *inversive congruential pseudorandom numbers* in the interval $[0, 1)$ is obtained by $x_n = y_n/p$ for $n \geq 0$. These sequences are always purely periodic.

**Period length.** The period length of inversive congruential sequences has been studied in [2, 4, 10, 65]. Obviously, the property that an inversive congruential sequence has the maximum possible period length $p$ only depends on the parameter $a \in \mathbb{Z}_p^*$, but not on the specific value of the parameter $c \in \mathbb{Z}_p^*$. In this case, the generated pseudorandom numbers $x_0, x_1, \ldots, x_{p-1}$ within a period run through all rationals in $[0, 1)$ with denominator $p$ and show a perfect equidistribution in the interval $[0, 1)$. Let $\mathbb{M}_p$ be the corresponding set of all $a \in \mathbb{Z}_p^*$ which belong to sequences with period length $p$. It can be deduced from [2, Theorem 3] that the set $\mathbb{M}_p$ contains exactly $\varphi(p+1)/2$ elements, where $\varphi$ denotes Euler's totient function. These elements are characterized in [65, Theorem 1] by an algebraic property of the polynomial $z^2 - z - a \in \mathbb{Z}_p[z]$, namely that the ratio of its two roots (in the finite field with $p^2$ elements) has multiplicative order $p + 1$. It is a handy sufficient condition for $a \in \mathbb{M}_p$ that $z^2 - z - a$ is a primitive polynomial over the finite field $\mathbb{Z}_p$, which is already mentioned in [10, p. 321; 86, Theorem 8.4]. The following elementary algorithmic characterization of the elements of $\mathbb{M}_p$ can be deduced from [10, Algorithm]. For $a \in \mathbb{Z}_p^*$, let $A = \begin{pmatrix} 1 & a \\ 1 & 0 \end{pmatrix} \in \mathbb{Z}_p^{2 \times 2}$ be a corresponding matrix. Let $(B)_{11}$ denote the entry $b_{11}$ of a matrix $B = (b_{ij})_{1 \leq i,j \leq 2}$. Then $a \in \mathbb{M}_p$ if and only if $(A^p)_{11} \equiv 0 \pmod{p}$ and $(A^{((p+1)/d)-1})_{11} \not\equiv 0 \pmod{p}$ for any prime divisor $d$ of $p+1$. Note that the powers of the matrix $A$ can be calculated efficiently by using the standard square-and-multiply technique.

The situation is illustrated for the prime $p = 2^{31} - 1 = 2\,147\,483\,647$. In this case, the set $\mathbb{M}_p$ contains $\varphi(p+1)/2 = 2^{29}$ elements and the conditions for $a \in \mathbb{M}_p$ reduce to $(A^p)_{11} \equiv 0 \pmod{p}$ and $(A^{(p-1)/2})_{11} \not\equiv 0 \pmod{p}$. Some selected parameters $a \in \mathbb{M}_p$ are given in Table 4.1, namely the five smallest ones and twenty arbitrary values, which have been chosen at random among all elements of $\mathbb{M}_p$.

| 1  | 46 648 502  | 701 979 605   | 1 174 954 052 | 1 552 129 765 |
|----|-------------|---------------|---------------|---------------|
| 13 | 175 937 607 | 779 922 292   | 1 320 564 073 | 1 626 175 446 |
| 16 | 240 135 714 | 957 368 305   | 1 425 460 855 | 1 701 029 437 |
| 17 | 410 752 179 | 1 077 355 727 | 1 462 827 631 | 1 893 375 808 |
| 18 | 521 230 544 | 1 116 658 433 | 1 488 863 516 | 2 053 672 816 |

TABLE 4.1. Elements of $\mathbb{M}_p$ for $p = 2^{31} - 1$.

**Distribution of tuples over the full period.** Subsequently, the distribution of *(overlapping) s–tuples* $\mathbf{x}_n = (x_n, x_{n+1}, \ldots, x_{n+s-1}) \in [0,1)^s$ of inversive congruential pseudorandom numbers is studied for any dimension $s \geq 2$. Let $D_p^{(s)} = D_p(\mathbf{x}_0, \mathbf{x}_1, \ldots, \mathbf{x}_{p-1})$ denote their discrepancy over the full period. The upper bound for $D_p^{(s)}$ in Theorem 4.1 can be deduced from [86, Theorem 8.7 and its proof]. Note that the slight improvement stems from a smaller upper bound for a prime denominator $p \geq 5$ in [7, Theorem 1], which can be applied in the proof of [86, Corollary 3.11]. Theorem 4.1 improves an earlier upper bound in [80, Theorem 1]. The lower bound for $D_p^{(s)}$ in Theorem 4.2 is cited from [27, Result 2] and improves an earlier result in [81, Theorem 2].

THEOREM 4.1. *The discrepancy $D_p^{(s)}$ of all s–tuples satisfies*

$$D_p^{(s)} < (2 + p^{-1/2})(s-1)p^{-1/2}\left(\frac{4}{\pi^2}\log p + 1.51\right)^s + sp^{-1}$$

*for all dimensions $s \geq 2$ and any parameters $a \in \mathbb{M}_p$ and $c \in \mathbb{Z}_p^*$.*

THEOREM 4.2. *Let $A(t) = (((p-3)/(p-1)) - t^2)/((2 + p^{-1/2})^2 - t^2)$ for some $t \in (0, \sqrt{(p-3)/(p-1)}]$. Let the parameter $a \in \mathbb{M}_p$ be fixed. Then there exist more than $A(t)(p-1)$ parameters $c \in \mathbb{Z}_p^*$ such that the discrepancy $D_p^{(s)}$ of all s–tuples satisfies*

$$D_p^{(s)} \geq \frac{t}{2(\pi+2)}p^{-1/2}$$

*for all dimensions $s \geq 2$.*

Theorem 4.1 shows that the discrepancy of $s$–tuples is of an order of magnitude at most $p^{-1/2}(\log p)^s$ for any dimension $s \geq 2$ and all parameters $a \in \mathbb{M}_p$ and $c \in \mathbb{Z}_p^*$ in the inversive congruential method. On the other

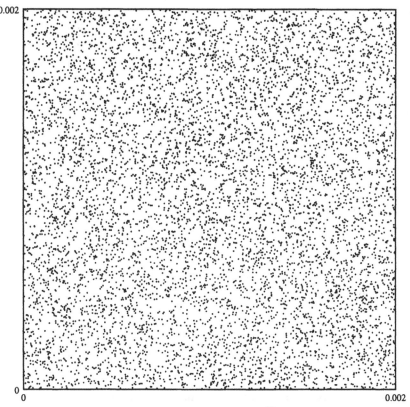

FIGURE 4.1. Distribution of pairs for $p = 2^{31} - 1$ and parameters
$a = 957\,368\,305$ and $c = 212\,709\,106$, i.e., $ac^2 \equiv 67\,483\,458 \pmod{p}$.

hand, it follows from Theorem 4.2 that, for a positive fraction of these parameters, the discrepancy of $s$-tuples is of an order of magnitude at least
$p^{-1/2}$ for all dimensions $s \geq 2$. It is in this range of magnitudes where one
also finds the discrepancy of $p$ true random points from $[0, 1)^s$ according to
the probabilistic law of the iterated logarithm.

The distribution of pairs $(x_n, x_{n+1})$ is illustrated in Figure 4.1 by a plot
of all generated points in the small subregion $[0, 0.002)^2$ of the unit square.
The underlying inversive congruential generator has modulus (and period
length) $p = 2^{31} - 1 = 2\,147\,483\,647$ and parameters $a = 957\,368\,305$ and
$c = 212\,709\,106$. Since $ac^2 \equiv 67\,483\,458 \pmod{p}$, the values of multiplier
and increment in the inversive recursion are almost the same as in the first
example of the previous section. It turned out that the number of points
within a period falling into the subregion $[0, 0.002)^2$ is $8\,492$; the expected
number of points under the uniform distribution is about $8\,590$. Figure 4.1
shows that there is no obvious structure within the generated points. Similar plots are obtained in many other subregions of the unit square.

**Distribution of tuples over parts of the period.** In the following section, for an arbitrary dimension $s \geq 2$, *nonoverlapping s–tuples* $\mathbf{x}_n = (x_{sn}, x_{sn+1}, \ldots, x_{sn+s-1}) \in [0,1)^s$ of inversive congruential pseudorandom numbers are considered. The discrepancy of the first $N$ generated $s$-tuples is abbreviated by $D_{N;a,c}^{(s)} = D_N(\mathbf{x}_0, \mathbf{x}_1, \ldots, \mathbf{x}_{N-1})$ in order to emphasize the dependence on the parameters $a, c$ in the underlying inversive recursion. Over the full (prime) period, the set of all nonoverlapping $s$-tuples (with $s < p$) coincides with the set of all overlapping $s$-tuples discussed above, so that the discrepancy bounds of Theorems 4.1 and 4.2 remain valid for the discrepancy $D_{p;a,c}^{(s)}$. The discrepancy analysis over parts of the period is much more complicated. Recently, an upper bound for the average value of the discrepancy $D_{N;a,c}^{(s)}$ over the parameter $c$ could be established. Theorem 4.3 is a slightly improved special version of the corresponding more general result in [41, Theorem 1]; its proof is omitted, since it follows at once from the results in [41].

THEOREM 4.3. *The average value of the discrepancy $D_{N;a,c}^{(s)}$ of nonoverlapping s–tuples over the parameter $c \in \mathbb{Z}_p^*$ satisfies*

$$\frac{1}{p-1} \sum_{c \in \mathbb{Z}_p^*} D_{N;a,c}^{(s)} < \sqrt{2s-1}\, N^{-1/2} \left( \frac{2}{\pi} \log p + \frac{7}{5} \right)^s$$

*for $1 \leq N < p$, all dimensions $s \geq 2$, and any parameter $a \in \mathbb{M}_p$.*

Theorem 4.3 shows that the average value of the discrepancy of nonoverlapping $s$–tuples over parts of the period is of an order of magnitude at most $N^{-1/2}(\log p)^s$ for any dimension $s \geq 2$, which is basically in accordance with the probabilistic law of the iterated logarithm for the discrepancy of $N$ true random points from $[0,1)^s$.

**Hyperbolic structures.** The analysis of hyperbolic structures in the distribution of pairs $(x_n, x_{n+1})$ of inversive congruential pseudorandom numbers is a subject of current research by the third author (marginally assisted by the first one). The results obtained so far indicate that there exist two particularly critical subregions of the unit square, namely the regions around the points $(0, c/p)$ and $(1, c/p)$, respectively. The situation around the point $(0, c/p)$ is illustrated in Figures 4.2 and 4.3 by plots of all generated points in the small subregion $[0, 0.002) \times [0.098, 0.1)$ of the unit square. The two underlying inversive congruential generators have modulus (and period length) $p = 2^{31} - 1 = 2\,147\,483\,647$ and second parameter $c = 212\,709\,106$ and $c = 212\,709\,105$, respectively, which implies that the ratio $c/p \approx 0.09905$ is (nearly) the same. It turned out that the number of points within a period falling into the subregion $[0, 0.002) \times [0.098, 0.1)$ is $8\,364$ and $8\,598$, respectively; both times the expected number of points under the uniform distribution is about $8\,590$. The inversive congruential generator of Figure 4.2 has first parameter $a = 957\,368\,305$, i.e., $ac^2 \equiv 67\,483\,458 \pmod{p}$.

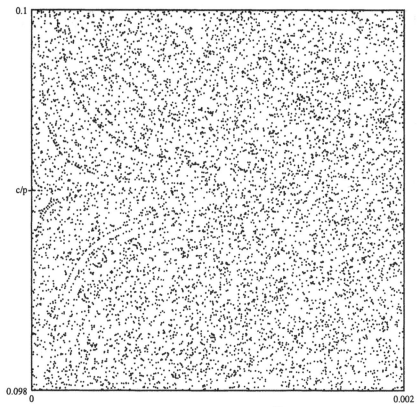

FIGURE 4.2. Distribution of pairs for $p = 2^{31} - 1$ and parameters $a = 957\,368\,305$ and $c = 212\,709\,106$, i.e., $ac^2 \equiv 67\,483\,458 \pmod{p}$.

Hence, this generator is the same as in Figure 4.1. Figure 4.2 shows a rather typical distribution of pairs around the point $(0, c/p)$ with several clearly visible but not very pronounced hyperbolic structures. Similar plots can be obtained for many other arbitrary choices of the parameters in the inversive congruential method.

The underlying inversive congruential generator in Figure 4.3 has first parameter $a = 1\,893\,375\,808$, i.e., $ac^2 \equiv 66\,784\,045 \pmod{p}$. Now, a strongly pronounced hyperbolic structure appears. It consists exactly of all 2\,746 pairs $(x_n, x_{n+1}) = (y_n, y_{n+1})/p$ with the property that $y_n$ is a divisor of $365p + 66\,784\,045 = 783\,898\,315\,200 = 2^6 \cdot 3^2 \cdot 5^2 \cdot 7^2 \cdot 11 \cdot 13 \cdot 17 \cdot 457$ which is greater than $(365p + 66\,784\,045)/(p - c) \approx 405.163$ and less than $p$. These pairs satisfy the hyperbolic equation $y_{n+1} = (365p + 66\,784\,045)/y_n + c$, i.e., they are lying on one hyperbola in the unit square. The subregion $[0, 0.002) \times [0.098, 0.1)$ in Figure 4.3 contains a small but significant part of this curve with 569 points.

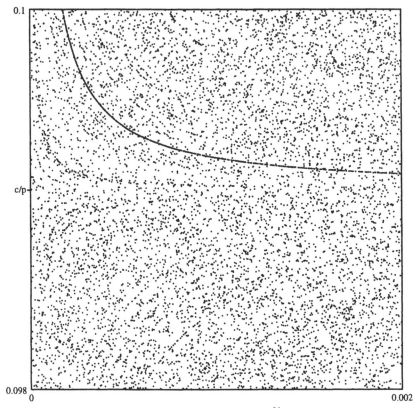

FIGURE 4.3. Distribution of pairs for $p = 2^{31} - 1$ and parameters
$a = 1\,893\,375\,808$ and $c = 212\,709\,105$, i.e., $ac^2 \equiv 66\,784\,045 \pmod{p}$.

**Empirical results.** The 'Load Test' as described above has been applied to the inversive congruential generator with modulus (and period length) $p = 2^{31} - 1 = 2\,147\,483\,647$ and parameters $a = 957\,368\,305$ and

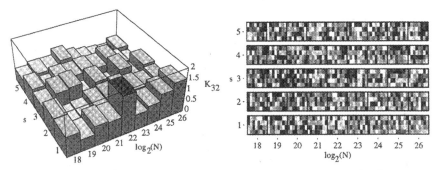

FIGURE 4.4. Empirical results of the 'Load Test' with initial value
$y_0 = 0$ for $p = 2^{31} - 1$ and parameters $a = 957\,368\,305$ and
$c = 212\,709\,106$, i.e., $ac^2 \equiv 67\,483\,458 \pmod{p}$.

$c = 212\,709\,106$, i.e., $ac^2 \equiv 67\,483\,458 \pmod{p}$. Hence, this generator is the same as in Figures 4.1 and 4.2. The empirical results of the 'Load Test' for the initial value $y_0 = 0$ are summarized in Figure 4.4. Again, these two plots are organized as in Figures 2.5 and 3.5 of the previous sections. This inversive congruential generator shows an excellent behaviour under the 'Load Test', even for the largest sample size $N = 2^{26}$, which is quite typical for inversive congruential generators with prime modulus [76].

**Notes.** Equidistribution properties of parts of the period in nonlinear congruential methods are analysed in [48]. It follows from [48, Theorem 1] that, for relatively large parts of the period, the distribution of inversive congruential pseudorandom numbers tends to be more regular than the one of true random numbers from $[0, 1)$, which reflects the perfect equidistribution of the full period.

The following remarkable nonlinearity property of the inversive congruential method with prime modulus was proved in [14, Theorem]: Any hyperplane in the $s$–dimensional affine space $\mathbb{Z}_p^s$ with $s \geq 2$ contains at most $s$ of the $s$–tuples $(y_n, y_{n+1}, \ldots, y_{n+s-1})$ generated within a period which additionally satisfy $y_n \cdots y_{n+s-2} \neq 0$. Note that the last condition eliminates only $s - 1$ of the generated $s$–tuples. A simplified proof of this result can be found in [86, Proof of Theorem 8.6].

The (very favourable) performance of inversive congruential pseudorandom numbers under Marsaglia's lattice test is studied in [5, 8, 65, 78]. However, an example in [13] illustrates the weakness of this criterion.

A generalization of the inversive congruential method based on higher-order recursions was introduced in [9], but apart from some periodicity properties no further results could be established up to now.

A compound version of the inversive congruential method was introduced in [28]; it corresponds to the special case of a squarefree modulus in the generalized inversive congruential method with arbitrary modulus of Klaus Huber [68]. In the compound approach, distinct primes $p_1, \ldots, p_r$ and corresponding sequences $(x_n^{(i)})_{n \geq 0}$ of inversive congruential pseudorandom numbers with prime modulus $p_i$ for $1 \leq i \leq r$ are considered. A sequence $(x_n)_{n \geq 0}$ of compound inversive congruential pseudorandom numbers in the interval $[0, 1)$ is defined by $x_n \equiv x_n^{(1)} + \cdots + x_n^{(r)} \pmod{1}$ for $n \geq 0$. The compound approach allows to reach a large period length $p_1 \cdots p_r$ and an excellent discretization of the interval $[0, 1)$, although exact integer calculations have to be performed only with respect to relatively small primes $p_1, \ldots, p_r$. Additionally, this technique has the obvious advantage that it is particularly suitable for parallelization. Statistical independence properties of compound inversive congruential pseudorandom numbers are analysed in [28, 41]. The reader is referred to [40, Sec. 4] for a review of these very promising results.

Recently, a digital inversive method was introduced and studied in [54], which is based on inversive recursions over finite fields with $p^k$ elements. The binary case ($p = 2$) of the digital inversive method is particularly in-

94

teresting, since it allows a fast computer implementation. A brief survey of this approach can be found in [91, Sec. 3.4].

Inversive recursions over finite fields of order $p^k$ can also be used to generate uniform pseudorandom vectors in $[0, 1)^k$. This stimulating idea is due to Harald Niederreiter, who introduced and studied the inversive method for pseudorandom vector generation in [82, 85, 90]. Related results can be found in [2, 4, 61, 62, 64]. The reader is also referred to the nice survey article of Frank Emmerich [60].

There also exists an explicit variant of the inversive congruential method with prime modulus, which was introduced in [24]. An explicit inversive congruential sequence in $\mathbb{Z}_p$ is explicitly defined by $y_n \equiv (an + b)^{-1} \pmod{p}$ for $n \geq 0$ with two parameters $a \in \mathbb{Z}_p^*$ and $b \in \mathbb{Z}_p$. Pseudorandom numbers are obtained in the usual way by $x_n = y_n/p$ for $n \geq 0$. It is an obvious advantage of this approach that any choice of the parameters leads to a purely periodic sequence with the maximum possible period length $p$. By the way, an explicit inversive congruential sequence can also be described recursively, namely by $y_0 \equiv b^{-1} \pmod{p}$ and $y_{n+1} \equiv (y_n^{-1} + a)^{-1} \pmod{p}$ for $n \geq 0$, which shows that the parameter $b$ plays the role of an initial value. There is even another close connection to the recursive inversive congruential method. Let $(z_n)_{n \geq 0}$ be an inversive congruential sequence in $\mathbb{Z}_p$ with $z_0 \neq 1$ and $z_{n+1} \equiv -z_n^{-1} + 2 \pmod{p}$ for $n \geq 0$. Then it follows from [10, Theorem (ii)] that $(z_n)_{n \geq 0}$ is purely periodic with period length $p - 1$, and hence $\{z_0, z_1, \ldots, z_{p-2}\} = \mathbb{Z}_p \setminus \{1\}$. Taking into account that the successor of 0 is 2, a purely periodic sequence $(\tilde{z}_n)_{n \geq 0}$ in $\mathbb{Z}_p$ with period length $p$ can be defined by inserting the missing element 1 every time between 0 and 2 in the sequence $(z_n)_{n \geq 0}$. Then a short calculation shows that the choice $z_0 \equiv ab^{-1} + 1 \pmod{p}$ yields $(an + b)^{-1} \equiv a^{-1}(\tilde{z}_n - 1) \pmod{p}$ for $n \geq 0$, i.e., any explicit inversive congruential sequence $(y_n)_{n \geq 0}$ (with $b \neq 0$) can be identified with a suitable linear transformation of $(\tilde{z}_n)_{n \geq 0}$. (The case $b = 0$ can be treated analogously.) A lot of results on the explicit inversive congruential method are already available [21, 24, 30, 42, 52, 84, 89], and they show that these pseudorandom numbers have very attractive properties. The reader is referred to recent surveys in [33, Sec. 3; 40, Sec. 3; 91, Sec. 3.3] for more details on the properties of explicit inversive congruential generators.

Other nonlinear congruential methods are studied in the articles [8, 20, 26, 30, 45, 47, 48, 53, 58, 78, 79], and recent reviews are given in [33, Sec. 2; 40, Sec. 2; 91, Sec. 3.1].

## Acknowledgements

The authors would like to thank Frank Emmerich, Peter Hellekalek, Gerhard Larcher, Harald Niederreiter, Sibylle Strandt, and the referee for their comments and criticism.

# References

[1] C. Alexopoulos, K. Kang, W.R. Lilegdon, and D. Goldsman, Eds., *Proceedings of the 1995 Winter Simulation Conference*, IEEE Press, Piscataway, NJ, 1995.

[2] W.-S. Chou, On inversive maximal period polynomials over finite fields, *Appl. Algebra Engrg. Comm. Comput.* **6** (1995) 245–250.

[3] ——, The period lengths of inversive congruential recursions, *Acta Arith.* **73** (1995) 325–341.

[4] ——, The period lengths of inversive pseudorandom vector generations, *Finite Fields Appl.* **1** (1995) 126–132.

[5] W.-S. Chou and H. Niederreiter, On the lattice test for inversive congruential pseudorandom numbers, in: [93], 186–197.

[6] K.-L. Chung, An estimate concerning the Kolmogoroff limit distribution, *Trans. Amer. Math. Soc.* **67** (1949) 36–50.

[7] T. Cochrane, On a trigonometric inequality of Vinogradov, *J. Number Theory* **27** (1987) 9–16.

[8] J. Eichenauer, H. Grothe, and J. Lehn, Marsaglia's lattice test and non–linear congruential pseudo random number generators, *Metrika* **35** (1988) 241–250.

[9] J. Eichenauer, H. Grothe, J. Lehn, and A. Topuzoğlu, A multiple recursive non–linear congruential pseudo random number generator, *Manuscripta Math.* **59** (1987) 331–346.

[10] J. Eichenauer and J. Lehn, A non–linear congruential pseudo random number generator, *Statist. Papers* **27** (1986) 315–326.

[11] ——, On the structure of quadratic congruential sequences, *Manuscripta Math.* **58** (1987) 129–140.

[12] J. Eichenauer, J. Lehn, and A. Topuzoğlu, A nonlinear congruential pseudorandom number generator with power of two modulus, *Math. Comp.* **51** (1988) 757–759.

[13] J. Eichenauer and H. Niederreiter, On Marsaglia's lattice test for pseudorandom numbers, *Manuscripta Math.* **62** (1988) 245–248.

[14] J. Eichenauer–Herrmann, Inversive congruential pseudorandom numbers avoid the planes, *Math. Comp.* **56** (1991) 297–301.

[15] ——, On the discrepancy of inversive congruential pseudorandom numbers with prime power modulus, *Manuscripta Math.* **71** (1991) 153–161.

[16] ——, A remark on the discrepancy of quadratic congruential pseudorandom numbers, *J. Comput. Appl. Math.* **43** (1992) 383–387.

[17] ——, Construction of inversive congruential pseudorandom number generators with maximal period length, *J. Comput. Appl. Math.* **40** (1992) 345–349.

[18] ——, Inversive congruential pseudorandom numbers: a tutorial, *Internat. Statist. Rev.* **60** (1992) 167–176.

[19] ——, On the autocorrelation structure of inversive congruential pseudorandom number sequences, *Statist. Papers* **33** (1992) 261–268.

[20] ——, Equidistribution properties of nonlinear congruential pseudorandom numbers, *Metrika* **40** (1993) 333–338.

[21] ——, Explicit inversive congruential pseudorandom numbers: the compound approach, *Computing* **51** (1993) 175–182.

[22] ——, Inversive congruential pseudorandom numbers, *Z. Angew. Math. Mech.* **73** (1993) T644–T647.

[23] ——, On the discrepancy of inversive congruential pseudorandom numbers with prime power modulus, II, *Manuscripta Math.* **79** (1993) 239–246.

[24] ——, Statistical independence of a new class of inversive congruential pseudorandom numbers, *Math. Comp.* **60** (1993) 375–384.

[25] ——, The lattice structure of nonlinear congruential pseudorandom numbers, *Metrika* **40** (1993) 115–120.

[26] ——, Compound nonlinear congruential pseudorandom numbers, *Monatsh. Math.* **117** (1994) 213–222.

[27] ——, Improved lower bounds for the discrepancy of inversive congruential pseudorandom numbers, *Math. Comp.* **62** (1994) 783–786.

[28] ——, On generalized inversive congruential pseudorandom numbers, *Math. Comp.* **63** (1994) 293–299.

[29] ——, On the discrepancy of quadratic congruential pseudorandom numbers with power of two modulus, *J. Comput. Appl. Math.* **53** (1994) 371–376.

[30] ——, A unified approach to the analysis of compound pseudorandom numbers, *Finite Fields Appl.* **1** (1995) 102–114.

[31] ——, Discrepancy bounds for nonoverlapping pairs of quadratic congruential pseudorandom numbers, *Arch. Math.* **65** (1995) 362–368.

[32] ——, Nonoverlapping pairs of explicit inversive congruential pseudorandom numbers, *Monatsh. Math.* **119** (1995) 49–61.

[33] ——, Pseudorandom number generation by nonlinear methods, *Internat. Statist. Rev.* **63** (1995) 247–255.

[34] ——, Quadratic congruential pseudorandom numbers: distribution of triples, *J. Comput. Appl. Math.* **62** (1995) 239–253.

96

[35] ——, Equidistribution properties of inversive congruential pseudorandom numbers with power of two modulus, *Metrika* **44** (1996) 199–205.

[36] ——, Modified explicit inversive congruential pseudorandom numbers with power of two modulus, *Statistics and Computing* **6** (1996) 31–36.

[37] ——, Quadratic congruential pseudorandom numbers: distribution of lagged pairs, *J. Comput. Appl. Math.* **79** (1997) 75–85.

[38] ——, Quadratic congruential pseudorandom numbers: distribution of triples, II, *J. Comput. Appl. Math.* (to appear).

[39] ——, Improved upper bounds for the discrepancy of pairs of inversive congruential pseudorandom numbers with power of two modulus, Preprint.

[40] J. Eichenauer–Herrmann and F. Emmerich, A review of compound methods for pseudorandom number generation, in: [67], 5–14.

[41] ——, Compound inversive congruential pseudorandom numbers: an average–case analysis, *Math. Comp.* **65** (1996) 215–225.

[42] J. Eichenauer–Herrmann, F. Emmerich, and G. Larcher, Average discrepancy, hyperplanes, and compound pseudorandom numbers, *Finite Fields Appl.* (to appear).

[43] J. Eichenauer–Herrmann and H. Grothe, A new inversive congruential pseudorandom number generator with power of two modulus, *ACM Trans. Modeling and Computer Simulation* **2** (1992) 1–11.

[44] J. Eichenauer–Herrmann, H. Grothe, H. Niederreiter, and A. Topuzoğlu, On the lattice structure of a nonlinear generator with modulus $2^{\alpha}$, *J. Comput. Appl. Math.* **31** (1990) 81–85.

[45] J. Eichenauer–Herrmann and E. Herrmann, Compound cubic congruential pseudorandom numbers, *Computing* (to appear).

[46] J. Eichenauer–Herrmann and K. Ickstadt, Explicit inversive congruential pseudorandom numbers with power of two modulus, *Math. Comp.* **62** (1994) 787–797.

[47] J. Eichenauer–Herrmann and G. Larcher, Average behaviour of compound nonlinear congruential pseudorandom numbers, *Finite Fields Appl.* **2** (1996) 111–123.

[48] ——, Average equidistribution properties of compound nonlinear congruential pseudorandom numbers, *Math. Comp.* **66** (1997) 363–372.

[49] J. Eichenauer–Herrmann and H. Niederreiter, On the discrepancy of quadratic congruential pseudorandom numbers, *J. Comput. Appl. Math.* **34** (1991) 243–249.

[50] ——, Lower bounds for the discrepancy of inversive congruential pseudorandom numbers with power of two modulus, *Math. Comp.* **58** (1992) 775–779.

[51] ——, Kloosterman–type sums and the discrepancy of nonoverlapping pairs of inversive congruential pseudorandom numbers, *Acta Arith.* **65** (1993) 185–194.

[52] ——, Bounds for exponential sums and their applications to pseudorandom numbers, *Acta Arith.* **67** (1994) 269–281.

[53] ——, On the statistical independence of nonlinear congruential pseudorandom numbers, *ACM Trans. Modeling and Computer Simulation* **4** (1994) 89–95.

[54] ——, Digital inversive pseudorandom numbers, *ACM Trans. Modeling and Computer Simulation* **4** (1994) 339–349.

[55] ——, An improved upper bound for the discrepancy of quadratic congruential pseudorandom numbers, *Acta Arith.* **69** (1995) 193–198.

[56] ——, Inversive congruential pseudorandom numbers: distribution of triples, *Math. Comp.* (to appear).

[57] ——, Lower bounds for the discrepancy of triples of inversive congruential pseudorandom numbers with power of two modulus, *Monatsh. Math.* (to appear).

[58] ——, Parallel streams of nonlinear congruential pseudorandom numbers, *Finite Fields Appl.* (to appear).

[59] J. Eichenauer–Herrmann and A. Topuzoğlu, On the period length of congruential pseudorandom number sequences generated by inversions, *J. Comput. Appl. Math.* **31** (1990) 87–96.

[60] F. Emmerich, New methods in pseudorandom vector generation, in: [67], 15–23.

[61] ——, *Pseudorandom Number and Vector Generation by Compound Inversive Methods*, Dissertation, Darmstadt, 1996.

[62] ——, Pseudorandom vector generation by the compound inversive method, *Math. Comp.* **65** (1996) 749–760.

[63] ——, Equidistribution properties of quadratic congruential pseudorandom numbers, *J. Comput. Appl. Math.* **79** (1997) 207–214.

[64] F. Emmerich and T. Müller–Gronbach, A law of the iterated logarithm for discrete discrepancies and its applications to pseudorandom vector sequences, Preprint.

[65] M. Flahive and H. Niederreiter, On inversive congruential generators for pseudorandom numbers, in: [77], 75–80.

[66] P. Hellekalek, Inversive pseudorandom number generators: concepts, results, and links, in: [1], 255–262.

[67] P. Hellekalek, G. Larcher, and P. Zinterhof, Eds., *Proceedings of the 1st Salzburg Minisymposium on Pseudorandom Number Generation and Quasi–Monte Carlo Methods*, Austrian Center for Parallel Computation, Vienna, 1995.

[68] K. Huber, On the period length of generalized inversive pseudorandom number generators, *Appl. Algebra Engrg. Comm. Comput.* **5** (1994) 255–260.

[69] T. Kato, L.-M. Wu, and N. Yanagihara, On a nonlinear congruential pseudorandom number generator, *Math. Comp.* **65** (1996) 227–233.

[70] ——, The serial test for a nonlinear pseudorandom number generator, *Math. Comp.* **65** (1996) 761–769.

[71] ——, On the lattice structure of the modified inversive congruential generator with modulus $2^\alpha$, Preprint.

[72] J. Kiefer, On large deviations of the empiric d.f. of vector chance variables and a law of the iterated logarithm, *Pacific J. Math.* **11** (1961) 649–660.

[73] D.E. Knuth, *The Art of Computer Programming*, Vol. 2: *Seminumerical Algorithms*, Addison–Wesley, Reading, MA, 2nd ed., 1981.

[74] L. Kuipers and H. Niederreiter, *Uniform Distribution of Sequences*, Wiley, New York, 1974.

[75] P. L'Ecuyer, Uniform random number generation, *Ann. Oper. Res.* **53** (1994) 77–120.

[76] H. Leeb and S. Wegenkittl, Inversive and linear congruential pseudorandom number generators in empirical tests, *ACM Trans. Modeling and Computer Simulation* (to appear).

[77] G.L. Mullen and P.J.-S. Shiue, Eds., *Finite Fields, Coding Theory, and Advances in Communications and Computing*, Dekker, New York, 1993.

[78] H. Niederreiter, Remarks on nonlinear congruential pseudorandom numbers, *Metrika* **35** (1988) 321–328.

[79] ——, Statistical independence of nonlinear congruential pseudorandom numbers, *Monatsh. Math.* **106** (1988) 149–159.

[80] ——, The serial test for congruential pseudorandom numbers generated by inversions, *Math. Comp.* **52** (1989) 135–144.

[81] ——, Lower bounds for the discrepancy of inversive congruential pseudorandom numbers, *Math. Comp.* **55** (1990) 277–287.

[82] ——, Finite fields and their applications, in: D. Dorninger, G. Eigenthaler, H.K. Kaiser, and W.B. Müller, Eds., *Contributions to General Algebra 7*, Teubner, Stuttgart, 1991, 251–264.

[83] ——, Recent trends in random number and random vector generation, *Ann. Oper. Res.* **31** (1991) 323–345.

[84] ——, New methods for pseudorandom number and pseudorandom vector generation, in: J.J. Swain, D. Goldsman, R.C. Crain, and J.R. Wilson, Eds., *Proceedings of the 1992 Winter Simulation Conference*, IEEE Press, Piscataway, NJ, 1992, 264–269.

[85] ——, Nonlinear methods for pseudorandom number and vector generation, in: G. Pflug and U. Dieter, Eds., *Simulation and Optimization*, Lecture Notes in Economics and Math. Systems 374, Springer, Berlin, 1992, 145–153.

[86] ——, *Random Number Generation and Quasi-Monte Carlo Methods*, SIAM, Philadelphia, PA, 1992.

[87] ——, Finite fields, pseudorandom numbers, and quasirandom points, in: [77], 375–394.

[88] ——, Pseudorandom numbers and quasirandom points, *Z. Angew. Math. Mech.* **73** (1993) T648–T652.

[89] ——, On a new class of pseudorandom numbers for simulation methods, *J. Comput. Appl. Math.* **56** (1994) 159–167.

[90] ——, Pseudorandom vector generation by the inversive method, *ACM Trans. Modeling and Computer Simulation* **4** (1994) 191–212.

[91] ——, New developments in uniform pseudorandom number and vector generation, in: [93], 87–120.

[92] ——, Some linear and nonlinear methods for pseudorandom number generation, in: [1], 250–254.

[93] H. Niederreiter and P.J.-S. Shiue, Eds., *Monte Carlo and Quasi-Monte Carlo Methods in Scientific Computing*, Lecture Notes in Statistics 106, Springer, New York, 1995.

[94] S. Wegenkittl, *Empirical Testing of Pseudorandom Number Generators*, Diplomarbeit, Salzburg, 1995 (Available at http://random.mat.sbg.ac.at/team/).

Jürgen Eichenauer-Herrmann, Eva Herrmann
Fachbereich Mathematik
Technische Hochschule Darmstadt
Schloßgartenstraße 7
D-64289 Darmstadt
eherrmann@mathematik.th-darmstadt.de

Stefan Wegenkittl
Institut für Mathematik
Universität Salzburg
Hellbrunner Straße 34
A-5020 Salzburg
ste@random.mat.sbg.ac.at

# A Look At Multilevel Splitting

Paul Glasserman
Philip Heidelberger
Perwez Shahabuddin
Tim Zajic

ABSTRACT
This paper gives a non-technical overview of our work on the efficiency of a simple multilevel splitting technique for estimating rare event probabilities. For a particular class of stochastic models, this method is asymptotically optimal provided the splitting factor is properly chosen. However, for more general classes of processes, the method fails to be asymptotically optimal unless the splitting is done in a manner consistent with the dominant way in which the rare event happens (the large deviations behavior). In the absence of such consistency, the method also exhibits a kind of bias in which the estimate is too small with high probability. We briefly contrast our approach with one used in nuclear physics.

## 1  Introduction

Requirements for low cell loss rates in telecommunications networks and for low system failure rates in fault-tolerant computing systems have emphasized the importance of analyzing rare events in stochastic models. For example, in ATM (Asynchronous Transfer Mode) communications networks, quality of service requirements may be specified in terms of a cell loss rate in the range of $10^{-9}$. Exact analysis is limited to models having special structure while numerical analysis is limited to models whose underlying state space is not too large. Asymptotic analysis, in the form of large deviations analysis, has become popular [Buc90]. Large deviations deals with the asymptotic probability of a rare event and the way in which the rare event occurs. However large deviations typically only gives the rate at which a rare event probability approaches zero. Furthermore, the class of problems amenable to large deviations analysis is also limited. For example, in the realm of queueing networks, large deviations results typically only exist for single node networks (e.g., [Cha94, GW94, Whi93] and the references therein to name only a few), and for multiple node networks having special structure (e.g., [Cha94]).

Simulation thus appears as an attractive alternative. However, efficient simulation techniques for estimating a rare event probability are essential since the run length requirement using standard simulation typically increases as the inverse of the probability being estimated. While many variance reduction techniques have been proposed, two seem particularly applicable to the rare event problem: importance sampling (e.g., [HH65, GI89]) and splitting (e.g., [KH51]).

A survey on the application of importance sampling to problems arising in queueing and reliability models may be found in [Hei95]. There is a relationship between large deviations and effective importance sampling. The zero variance importance sampling change of measure (page 58 of [HH65]) requires sampling from the conditional distribution of the rare event. Since the theory of large deviations describes the asymptotic distribution of the rare event, the change of measure suggested by large deviations appears to be a good candidate for use in importance sampling. Indeed this is the case in some simple cases [CHJS94, CFM83, Sad91, SB90] such as a multiple server queue. However, in more complicated models heuristics suggested by large deviations may not only fail to be (asymptotically) efficient, but may in fact lead to variance inflation and sometimes infinite variance [GK95, GW96]. Thus importance sampling needs to be applied with great caution in complex models, especially in complex queueing networks.

Various forms of splitting have existed in the literature since the early 1950's. It appears to be of widespread use in problems arising in nuclear physics, e.g., [Dub85, DGB86, Bur90]. (We will contrast the use and analysis of these physics problems with our approach in Section 5.) However, it has only recently come into use for queueing and reliability problems [VMMF94, VV91, VV94], under the name of RESTART. Some approximate analysis of RESTART's efficiency has been performed, but that analysis does not relate to the large deviations behavior of the process being simulated. In [EM95, Mel93, Mel94], particular forms of splitting and Russian Roulette are analyzed for a class of Markovian models. Near-optimal splitting factors are derived (they are given in terms of quantities related to the steady state distribution), but their focus is on neither rare events nor asymptotic analysis.

On the surface, splitting is an easy method to implement: one needs only to decide when to split and what the splitting factor should be. This seems easier to do than selecting an effective change-of-measure as required with importance sampling. This suggests that splitting should be a more broadly applicable procedure than importance sampling.

In this paper, we summarize our work on splitting: detailed expositions can be found in [GHSZ97, GHSZ96a, GHSZ96b]. To make this paper accessible to a larger audience, the description here will contain a minimum of notation, will leave a number of messy technical assumptions unstated, and will contain no proofs. Our particular form of splitting is described in Section 2. We then focus on when splitting is "asymptotically optimal." When a method is asymptotically optimal, its computing requirements grow slowly as the event becomes rarer. Indeed, as shown in Section 3, for a particular class of stochastic models, our form of splitting is asymptotically optimal provided the right splitting factor is selected. This optimal factor is identified for this class of problems. In Section 4, we relate the large deviations behavior of the process to the efficiency of splitting. We identify necessary conditions on the splitting factor and on the large deviations rate functions that must be satisfied in order for splitting to be asymptotically optimal. These conditions imply that the large deviations behavior of the process, i.e., the dominant way in which the rare event occurs, must be understood. Failure to be consistent with the process's large deviations behavior leads not only to a sub-optimal variance, but also to a form of bias in which the estimator may be orders of magnitude too low with high probability. (This type of bias may also occur with importance sampling if the wrong change-of-measure is selected.) Furthermore, even if the levels are consistent with the large deviations behavior, conditions on the rates of less likely paths to failure must also be satisfied to maintain asymptotic optimality. Such conditions, which have been shown to be violated for one of the simplest queueing network models, would be quite difficult to check for more complex models. Thus, as with importance sampling, the use of simple splitting techniques must be applied with caution.

## 2   Multilevel Splitting

We are interested in estimating $\gamma_k$ which is defined to be the probability that a stochastic process, starting in some set $A_0$, reaches a "rare" set $A_k$ before it returns to $A_0$. For example, in a queue, $A_k$ may be the set where the total queue length exceeds $k$ and efficient estimation of $\gamma_k$ is key when estimating the steady-state buffer overflow rate, e.g., [Hei95]. (Or more generally, $A_k$ may be the set where the total queue length exceeds some constant $q_k$.) As $k$ increases, $\gamma_k$ decreases and special simulation techniques become necessary. We assume throughout that the process being simulated is a countable state space Markov chain in either discrete or continuous time. We assume that there are intermediate sets (levels) $A_1$, ... $A_{k-1}$, constructed in such a way that the process must pass through $A_{j-1}$ before reaching $A_j$. In this way, we can write $\gamma_k = p_1 \cdots p_k$ where $p_j$ is the probability of reaching level $j$ before hitting $A_0$ given that level $j-1$ was reached. We assume that $\gamma_k$ approaches zero at rate $\rho^{k+o(k)}$ for some constant $0 < \rho < 1$, i.e., we assume

$$\lim_{k \to \infty} \log(\gamma_k)/k = \log(\rho). \tag{1.1}$$

Such decay behavior is common in queueing models.

To estimate $\gamma_k$, we use a simple multilevel splitting procedure as depicted in Figure 1. When the process first enters $A_1$, it splits into $R_2$ subpaths each of which is started from the entrance state of $A_1$. This procedure is repeated recursively whenever a subpath first enters the next highest level, i.e., upon first entrance to $A_j$, $R_{j+1}$ new subpaths are started. ($R_1$ is the number of trials from the starting level $A_0$.) An unbiased estimate of $\gamma_k$ is then

$$\hat{\gamma}_k = \frac{Z_k}{R_1 \cdots R_k} \tag{1.2}$$

where $Z_k$ is the total number of subpaths that reach level $k$. Note that $\{Z_j, j \geq 0\}$ can be viewed as a branching process. We are interested in

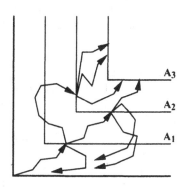

FIGURE 1. An example of multilevel splitting with two intermediate levels and a splitting factor of three. Two of the three subpaths from the first split reach the next level; four of their subpaths reach the final level.

the asymptotic efficiency, as $k \to \infty$, of the method. Two factors affect the efficiency: $w_k$, the expected work per sample, and $\sigma_k^2$, the variance per sample. (For simplicity, we assume that the computational cost is one unit per subpath.) The appropriate measure of efficiency combines these into the work-variance product $w_k \sigma_k^2$ [GW92, HH65]. Since, for any splitting factors $R_j$ and intermediate levels $A_j$ the mean is unchanged (i.e., $E[\hat{\gamma}_k] = \gamma_k$), we concentrate on $\eta_k = w_k \times E[\hat{\gamma}_k^2]$. With too little splitting, entrance to $A_k$ is still a rare event, while with too much splitting, the work per sample grows but the second moment cannot be decreased below its minimum possible value $\gamma_k^2$. In fact, since $E[\hat{\gamma}_k^2] \geq \gamma_k^2$ and $w_k \geq 1$, there is a best possible rate at which $\eta_k$ can approach zero and we will call the procedure asymptotically optimal if this best possible rate is achieved, i.e., if

$$\lim_{k \to \infty} \log(\eta_k)/k = 2 \log(\rho). \tag{1.3}$$

# 3   The Good News

We now describe a setting in which splitting is asymptotically optimal under simple conditions [GHSZ96a]. We assume that the process being simulated is a two dimensional Markov chain with a state space denoted by $(m, i)$. The first dimension denotes the process's level while the second dimension contains enough auxiliary information so as to render the process Markovian. For example, in an ATM queue fed by multiple Markov-modulated sources, $m$ could represent the total queue length and $i$ could represent an encoding that identifies the state of each source. We assume that there are only finitely many auxiliary states $i$. $A_k$ then represents all states with first component $\geq k$.

We assume that there is a transition matrix $\mathbf{P}_m$ where $P_m(i, j)$ denotes the probability that the process first enters $A_m$ in state $j$ (before entering $A_0$) given that it first entered level $m - 1$ in state $i$. Since the dynamics of the process may be very complicated, we do not assume that $\mathbf{P}_m$ is known in closed form, rather we only assume that samples can be simulated having this distribution. Because the probability of ever reaching the next highest level (before $A_0$) may be less than one, $\mathbf{P}_m$ may be a sub-stochastic matrix. We assume that a limiting transition matrix $\mathbf{P}$ exists, i.e., $\lim_{m \to \infty} P_m(i, j) = P(i, j)$, as is often the case in queueing models if $m$ denotes the queue length. We will refer to this class of models as "finite Markovian." Under appropriate irreducibility assumptions, (1.1) is satisfied with $\rho$ being the spectral radius of $\mathbf{P}$. Furthermore, if a constant splitting factor is used at each level, i.e., if $R_j = R$ for all $j$, then the method is asymptotically optimal if and only if

$$R = 1/\rho. \tag{1.4}$$

With this choice of $R$, $E[Z_k]$ remains roughly constant. Since $1/\rho$ is typically not an integer, the constant number of splits from level $j$ can be replaced by a random number of splits $R_j$, chosen independently from everything else, with mean $E[R_j]$. The estimator $\hat{\gamma}_k = Z_k / (E[R_1] \cdots E[R_k])$ is an unbiased estimate of $\gamma_k$ and is asymptotically optimal if $E[R_j] = 1/\rho$ (and $R_j$ is bounded).

Numerical results in [GHSZ96a] indicate that the method works extremely well when $R$ is properly chosen. For example, in an ATM model with on-off sources, $\gamma_{40} = 2.33 \times 10^{-20}$ can be estimated to within $\pm 13\%$ error in only 500 seconds of CPU time on an IBM RS/6000 workstation. (Here, the relative error is the half-width of a 99% confidence interval divided by $\gamma_k$.) The performance of the method degrades as $R$ deviates from $1/\rho$, although some deviation from optimality can be tolerated. In the above ATM example, the relative error increases to about $\pm 35\%$ when $R$ is 20% greater than optimal. However, the method becomes more sensitive to deviations from $1/\rho$ as $k$ increases and $\gamma_k$ decreases. These results suggest that the simple splitting scheme described in Section 2, augmented by pilot studies to estimate a good splitting factor, should be highly effective for this broad class of problems.

# 4 The Bad News

Not all processes fit within the finite Markovian framework of the previous section (infinite in one dimension, finite in the other). For example, in a queueing network, many queues may be simultaneously large. In the finite Markovian case, $R = 1/\rho$ is a necessary and sufficient condition for asymptotic optimality. In this more general setting, assuming (1.1) holds, the condition $R = 1/\rho$, is a necessary, but not sufficient, condition for splitting to be asymptotically optimal [GHSZ97].

Unlike the finite Markovian case, in which there is a natural set of intermediate levels and essentially only one way to reach $A_k$, there may now be many ways to reach $A_k$. Thus in designing the splitting procedure, not only must the splitting factors $R_j$ be chosen, the intermediate levels $A_j$ must also be chosen. As we will see, if these intermediate levels are not chosen properly, the method loses efficiency.

Consider the situation of Figure 2 in which some intermediate level $B$ and the final level $A$ are depicted. We assume that rare events in the process

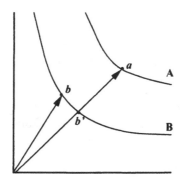

FIGURE 2. The most likely path to the intermediate set $B$ hits $B$ at $b$, but the most likely path to the final set $A$ hits $B$ at $b'$.

occur according to a large deviations principle. Informally, given that the process enters $B$, it does so by following the path that ends at the point $b$ with high probability. (By "following the path" we generally mean that an appropriately scaled version of the process stays close to such a deterministic path with probability approaching one.) Similarly, given that the process enters $A$, it does so by following the path that enters $B$ at $b'$ and ends at the point $a \in A$ with high probability.

Now consider the effect that such behavior has on the efficiency of splitting. Assume that the splitting factor is $R = 1/\rho$, in accordance with the necessary condition. By (1.1), (1.2) and the fact that $\hat{\gamma}_k$ is unbiased, $\log(E[Z_k])/k \to 0$. Almost all of the effort in simulating subpaths from level $B$ will be spent on subpaths that originate at (or near) $b$ since $b$ is the most likely hitting point of $B$. However, since the most likely path to

$A$ does not pass through $b$, much of that effort is, in essence, wasted since few of those subpaths will reach the final level $A$. Thus $Z_k = 0$ with high probability. Only rarely, will a subpath enter $B$ at the "correct" point $b'$. However, when such a subpath does enter $b'$ it will eventually generate a large expected number of subpaths that enter $A$. This phenomenon precludes asymptotic optimality. Thus a necessary condition for asymptotic optimality is that $b = b'$, i.e., the large deviations path to any intermediate level must coincide with the large deviations path to the final level.

Furthermore, if the situation shown in Figure 2 exists a kind of bias occurs. More formally, there is a growth rate for the sample size $R_1 = R_1(k)$ of replications from level 0 and a set $F_k$ such that $R_1(k) \to \infty$, $P\{F_k\} \to 1$ and

$$\lim_{k \to \infty} E[\hat{\gamma}_k 1_{F_k}]/\gamma_k = 0 \qquad (1.5)$$

where $1_{F_k}$ is the indicator of the set $F_k$. The interpretation of (1.5) is as follows: for a large sample size $R_1(k)$, $\hat{\gamma}_k$ appears to underestimate $\gamma_k$ by a large amount with high probability. Similar effects occur when the number of levels is fixed and the probabilities of moving to the next level approach zero.

However, even if the intermediate levels are consistent with large deviations as described above (i.e., $b = b'$), the method is not always guaranteed to be asymptotically optimal. Suppose the intermediate level $B$ is a fraction $\alpha$ of the way to $A$, i.e., $\log(P\{B\})/k \to \alpha \log(\rho)$. Let $x$ be any entrance point to $B$. Suppose the rate of hitting $A$ by passing through $x$ can be written as $\alpha \log(\delta_1(x)) + (1 - \alpha) \log(\delta_2(x))$, i.e., the probability of hitting $B$ near $x$ is approximately $\delta_1(x)^{\alpha k}$ and the probability of going from $x$ to $A$ is approximately $\delta_2(x)^{(1-\alpha)k}$. For any such $x$,

$$\alpha \log(\delta_1(x)) + (1 - \alpha) \log(\delta_2(x)) \leq \log(\rho) \qquad (1.6)$$

since the path through $x$ is not the only way to reach $A$. In order for splitting to be asymptotically optimal, an additional condition must be satisfied for every $x$ and every intermediate level:

$$\alpha \log(\delta_1(x)) + 2(1 - \alpha) \log(\delta_2(x)) \leq (2 - \alpha) \log(\rho). \qquad (1.7)$$

No matter how unlikely it is to enter $B$ at $x$, (1.7) states that it cannot be too "easy" to reach $A$ from $x$ since (1.7) may be violated if $\delta_2(x) \gg \rho$. If (1.7) is violated, only infrequently will a subpath enter $B$ at $x$, but when it does it will spawn an expected avalanche of subpaths that reach $A$.

The inequality (1.7) is a necessary, but not sufficient, condition for asymptotic optimality. Verification of (1.7) appears difficult in practice. However, it has been shown *not* to hold in a simple queueing network, even when the intermediate levels are consistent with large deviations. This network is a two queue tandem Jackson network (Poisson arrivals and exponential service time distributions) and the rare event of interest is when both queues are simultaneously large. Experimental results confirm that much slower convergence is obtained in this example as compared to finite Markovian examples. Also, in this example, if the intermediate levels are chosen in a way that is not consistent with large deviations, then the type of biasing predicted by (1.5) is observed experimentally: in one case $\hat{\gamma}_k$ is five orders of magnitude too low, even after hours of CPU time.

# 5 Contrast With Splitting in Physics

We now briefly contrast our approach to splitting with that described in [Dub85, DGB86, Bur90] for studying problems in nuclear physics. There, the methods and analysis seem highly tailored to the particular problem setting, specifically estimating the probability that a particle passing through a material reaches a detector. The material is divided into cells and when a particle enters a new cell it may split (or be randomly killed). When a particle is split, its weight is adjusted: the weight contains information about both the splitting procedure and the physics of the problem. The splitting factor may be cell-dependent. In some splitting procedures, certain transitions may be forced to occur in which case the weight is adjusted by the probability of the transition. An approximation for the variance, $\sigma^2$, is derived. This expression is approximate for a number of reasons including the fact that an infinite series is truncated and an approximation is used that considers only the average behavior of particles passing through a cell (the "surface-point approximation"). An approximate expression for the work per sample, $w$, is also derived. The expressions for $\sigma^2$ and $w$ are quite complicated and involve some unknown quantities that must be estimated via pilot studies. Such estimation is then embedded within a numerical optimization routine to minimize the estimated work-variance product $w\sigma^2$.

In our splitting approach, all subpaths reaching the final level carry the same weight, $1/R_1 \cdots R_k$, whereas in physics particles may carry different weights. This simplification makes asymptotic analysis possible. However, even with this simplification, in the finite Markovian setting our approach is as effective as *any* simulation algorithm can possibly be (in the sense of asymptotically minimizing the work times second moment). Furthermore, in this setting, the asymptotically optimal splitting factor can be simply identified using the mathematical structure of the problem, whereas in physics a numerical optimization procedure is applied. However, in more general settings the constant weight per subpath property of this simple approach may be a liability, since the method can easily fail to be asymptotically optimal. Whether or not large deviations analysis can be used to effectively assign state-dependent splitting factors (and thus weights) remains to be seen.

# 6 Acknowledgments

This work is supported by NSF grants DMI-94-57189, DMS-9508709 and DMI-96-25291.

# 7 References

[Buc90] J.A. Bucklew. *Large Deviation Techniques in Decision, Simulation and Estimation*. Wiley, New York, 1990.

[Bur90] K. Burn. Optimizing cell importances using an extension of the DSA – theory, implementation, preliminary results. *Progress in Nuclear Energy*, 24:39–54, 1990.

[CFM83] M. Cottrell, J.-C. Fort, and G. Malgouyres. Large deviations and rare events in the study of stochastic algorithms. *IEEE Trans. Automatic Control*, 28:907–920, 1983.

[Cha94] C.S. Chang. Stability, queue length and delay of deterministic and stochastic queueing networks. *IEEE Trans. Automatic Control*, 39:913–931, 1994.

[CHJS94] C.S. Chang, P. Heidelberger, S. Juneja, and P. Shahabuddin. Effective bandwidth and fast simulation of ATM intree networks. *Performance Evaluation*, 20:45–65, 1994.

[DGB86] A. Dubi, A. Goldfeld, and K. Burn. Application of the direct statistical approach on a multisurface splitting problem in Monte Carlo calculations. *Nuclear Science and Engineering*, 93:204–213, 1986.

[Dub85] A. Dubi. General statistical model for geometrical splitting in Monte Carlo – I. *Transport Theory and Statistical Physics*, 14:167–193, 1985.

[EM95] S.M. Ermakov and V.B. Melas. *Design and Analysis of Simulation Experiments*. Kluwer Academic Publishers, Dordrecht, 1995.

[GHSZ96a] P. Glasserman, P. Heidelberger, P. Shahabuddin, and T. Zajic. Multilevel splitting for estimating rare event probabilities. Technical Report RC 20478, IBM T.J. Watson Research Center, Yorktown Heights, NY, 1996.

[GHSZ96b] P. Glasserman, P. Heidelberger, P. Shahabuddin, and T. Zajic. Splitting for rare event simulation: Analysis of simple cases. In *1996 Winter Simulation Conference Proceedings*, pages 302 – 308, San Diego, CA, 1996. IEEE Computer Society Press.

[GHSZ97] P. Glasserman, P. Heidelberger, P. Shahabuddin, and T. Zajic. A large deviations perspective on the efficiency of multilevel splitting. Technical Report 20692, IBM T.J. Watson Research Center, Yorktown Heights, NY, 1997.

[GI89] P.W. Glynn and D.L. Iglehart. Importance sampling for stochastic simulations. *Management Science*, 35:1367–1392, 1989.

[GK95] P. Glasserman and S.G. Kou. Analysis of an importance sampling estimator for tandem queues. *ACM Trans. Modeling and Computer Simulation*, 5:22–42, 1995.

[GW92] P.W. Glynn and W. Whitt. The asymptotic efficiency of simulation estimators. *Operations Research*, 40:505–520, 1992.

[GW94] P.W. Glynn and W. Whitt. Logarithmic asymptotics for steady-state tail probabilities in a single server queue. *Journal of Applied Probability*, 31A:131–156, 1994.

[GW96] P. Glasserman and Y. Wang. Counterexamples in importance sampling for large deviations probabilities. Technical report, Columbia Business School, New York, NY, 1996. To appear in *Annals of Applied Probability*.

[Hei95] P. Heidelberger. Fast simulation of rare events in queueing and reliability models. *ACM Trans. Modeling and Computer Simulation*, 5:43 – 85, 1995.

[HH65] J.M. Hammersley and D.C. Handscomb. *Monte Carlo Methods*. Methuen and Co. Ltd., London, 1965.

[KH51] H. Kahn and T.E. Harris. Estimation of particle transmission by random sampling. *National Bureau of Standards Applied Mathematics Series*, 12:27–30, 1951.

[Mel93] V.B. Melas. Optimal simulation design by branching technique. In H.P. Wynn W.G. Muller and A.A. Zhigljavsky, editors, *Model Oriented Data Analysis*, pages 113–128. Physica-Verlag, Heidelberg, 1993.

[Mel94] V.B. Melas. Branching technique for Markov chain simulation (finite state case). *Statistics*, 25:159–171, 1994.

[Sad91] J.S. Sadowsky. Large deviations and efficient simulation of excessive backlogs in a GI/G/m queue. *IEEE Trans. Automatic Control*, 36:1383–1394, 1991.

[SB90] J.S. Sadowsky and J.A. Bucklew. On large deviations theory and asymptotically efficient Monte Carlo estimation. *IEEE Trans. Information Theory*, 36:579–588, 1990.

[VMMF94] M. Villén-Altamirano, A. Martínez-Marrón, J. Gamo, and F. Fernández-Cuesta. Enhancements of the accelerated simulation method RESTART by considering multiple thresholds. In J. Labetoulle and J.W. Roberts, editors, *International Telecommunications Conference 14*, pages 797–810, Amsterdam, 1994. Elsevier Science Publishers.

[VV91] M. Villén-Altamirano and J. Villén-Altamirano. RESTART: A method for accelerating rare event simulations. In J.W. Cohen and C.D. Pack, editors, *Queueing, Performance and Control in ATM*, pages 71–76. Elsevier Science Publishers, Amsterdam, 1991.

[VV94] M. Villén-Altamirano and J. Villén-Altamirano. RESTART: A straightforward method for fast simulation of rare events. In *1994 Winter Simulation Conference Proceedings*, pages 282–289, Orlando, FL, 1994.

[Whi93] W. Whitt. Tail probability with statistical multiplexing and effective bandwidths in multi-class queues. *Telecommunication Systems*, 2:71–107, 1993.

## Author's Addresses

Paul Glasserman
Columbia Business School
New York, New York 10027

Philip Heidelberger
IBM Research Division
T.J. Watson Research Center
Yorktown Heights, New York 10598

Perwez Shahabuddin, Tim Zajic
IEOR Department
Columbia University
New York, New York 10027

# On the Distribution of Digital Sequences

## Gerhard Larcher, Salzburg

*Acknowledgments:* Research supported by the Austrian Research Foundation (FWF), Project P 11009 MAT

ABSTRACT Low-discrepancy point sets are the basic tool for various quasi-Monte Carlo methods. The most important concept for the construction of low-discrepancy point sets in an $s$-dimensional unit cube is the concept of digital point sets, introduced by Niederreiter in 1987. We show that "almost all" digital sequences in a prime base are, in a certain sense, almost best possible distributed in the unit cube. Thereby we improve a result given in [LN95].

## 1 Introduction

Various quasi-Monte Carlo methods are based on the use of low-discrepancy point sets in an $s$-dimensional unit cube. So in quasi-Monte Carlo integration the use of low-discrepancy point sets for example is motivated by the - already "classical" - Koksma-Hlawka inequality which may be stated in the following form:

if $f : [0, 1)^s \longrightarrow R$ is a function of bounded variation $V(f)$ in the sense of Hardy and Krause, and if $\mathbf{x}_1, \ldots, \mathbf{x}_N$ is a point set in $[0, 1)^s$ with star discrepancy $D_N^*$ then

$$\left| \int_{[0,1)^s} f(\mathbf{x}) d\mathbf{x} - \frac{1}{N} \sum_{k=1}^{N} f(\mathbf{x}_k) \right| \leq V(f) \cdot D_N^*.$$

For a proof of the inequality and for details on the variation of a function see [KN74].

The star discrepancy $D_N^*$ of a point set $\mathbf{x}_1, \ldots, \mathbf{x}_N$ in $[0, 1)^s$ is defined by

$$D_N^* = \sup_{B \subseteq [0,1)^s} \left| \frac{| \{1 \leq n \leq N \mid \mathbf{x}_n \in B\} |}{N} - \lambda(B) \right|,$$

where the supremum is taken over all subintervals $B = \prod_{i=1}^{s} [0, b_i)$ with $0 < b_i \leq 1$ for $i = 1, \ldots, s$ of $[0, 1)^s$ and where $\lambda$ denotes the $s$-dimensional Lebesgue measure.

So the essential question is: how small can $D_N^*$ be, and how can we find $\mathbf{x}_1, \ldots, \mathbf{x}_N$ in $[0, 1)^s$ with small $D_N^*$ ?

Pure Monte Carlo integration essentially is based on Koksma-Hlawka type estimates and on the fact that on the average, for a random point set $\mathbf{x}_1, \ldots, \mathbf{x}_N$ in $[0, 1)^s$, a star discrepancy of about $N^{-\frac{1}{2}}$ can be expected. (For details see [Kie61].)

General lower bounds for the star discrepancy of point sets and of point sequences were given by K.F.Roth [Rot54]:

– for every dimension $s$ there is a constant $c_s > 0$ such that for every point set $\mathbf{x}_1, \ldots, \mathbf{x}_N$ in $[0, 1)^s$ we have

$$D_N^* \geq c_s \cdot \frac{(\log N)^{\frac{s-1}{2}}}{N},$$

and

– for every dimension $s$ there is a constant $c_s > 0$ such that for every sequence $\mathbf{x}_1, \mathbf{x}_2, \ldots$ in $[0, 1)^s$ we have for the star discrepancy $D_N^*$ of the first $N$ elements of the sequence

$$D_N^* \geq c_s \cdot \frac{(\log N)^{\frac{s}{2}}}{N}$$

for infinitely many $N$.

It is a famous conjecture in the theory of uniform distribution that the above estimates also hold true if the exponent $\frac{s-1}{2}$ is replaced by $s - 1$ and if $\frac{s}{2}$ is replaced by $s$ respectively. Until now it is only known that this conjecture is true for $s = 1$ and $s = 2$ in the first case, and if $s = 1$ in the second case.

Quasi-Monte Carlo methods now are based on the fact that there are point sets and point sequences in any $s$-dimensional unit cube which are best possible distributed with respect to the above conjecture. That means:

– for every dimension $s$ there is a constant $c_s'$ such that for every $N \geq 2$ there are $\mathbf{x}_1, \ldots, \mathbf{x}_N$ in $[0, 1)^s$ with star discrepancy

$$D_N^* \leq c_s' \cdot \frac{(\log N)^{s-1}}{N},$$

and

– for every dimension $s$ there is a constant $c_s'$ such that there are infinite sequences $\mathbf{x}_1, \mathbf{x}_2, \ldots$ in $[0, 1)^s$ such that for the star discrepancy $D_N^*$ of the first $N$ elements of the sequence we have

$$D_N^* \leq c_s' \cdot \frac{(\log N)^s}{N}$$

for all $N \geq 2$.

Classical examples are the Hammersley point sets and the Halton sequences. (For details see [Nie92b].) These point sets and sequences have the best possible order of discrepancy in $N$. However in all currently known discrepancy estimates the occurring constant $c_s'$ is of order about $s^s$, which is very large for applications in higher dimensions.

In 1967 Sobol [Sob67] and in 1982 Faure [Fau82] gave examples for point sets and sequences in $[0,1)^s$ with best possible order of discrepancy in $N$ and with an essentially smaller occurring constant $c_s'$.

Since 1987 in a series of papers Niederreiter introduced and studied in detail the concept of $(t,m,s)$-nets and $(t,s)$-sequences. (See for example [Nie87],[Nie88] and [Nie92b].) He introduced the general concept of digital $(t,m,s)$-nets and of digital $(t,s)$- and $(T,s)$-sequences (see [Nie92a] and [Nie92b]) and based on these general concepts he provided concrete point sets and sequences (e.g. Niederreiter sequences and Niederreiter-Xing sequences, see [NX96a] and [NX96b] and the references given there) with best possible order of discrepancy in $N$ and where - at least for $N$ large enough - the constants $c_s'$ in the discrepancy estimates are of order of about $\frac{1}{s^s}$. This of course is a drastical improvement on the previously known discrepancy estimates.

So the large classes of digital point sets and of digital sequences contain extremely well distributed examples of concrete point sets. So already from the point of view of the Koksma-Hlawka inequality, digital point sets are utmost fruitful for high-dimensional numerical integration. But moreover in recent years a new and very efficient integration method based on digital point sets was developed in analogy to the classical good lattice point method. (See for example [LT94], [LLNS96] or [LNS96].) It was shown that especially functions which are "smooth in some digital sense" (i.e. functions which are represented by rapidly converging Walsh series) can be excellently integrated by digital point sets. A survey on these "digital lattice rules" is given in [LS95].

So the search for digital sequences of high quality is highly motivated by quasi-Monte Carlo applications. We will show in this paper that, indeed, almost all digital sequences (over certain rings) are almost best possible distributed with respect to star discrepancy. We will make this statement more precise and we will prove it in Section 3.

First, in Section 2, we will recall the fundamental notions concerning digital point sets and digital sequences.

# 2  Digital point sets

To avoid too many technical notions, here we restrict to digital point sets and sequences defined over a residue class field $Z_q$ with $q$ prime. For a more general definition (over arbitrary finite commutative rings) see for example

[Nie92b] or [LNS96].

**Definition 1** *Let a dimension $s \geq 1$, a prime base $q \geq 2$, and an integer $m \geq 1$ be given. Let $C_1, \ldots, C_s$ be $m \times m$-matrices over $Z_q$. For $0 \leq n < q^m$ let $n = \sum_{k=0}^{m-1} a_k q^k$ be the q-adic representation of $n$ in base $q$. Consider the digits $a_0, \ldots, a_{m-1}$ as elements of $Z_q$. Let*

$$C_i \cdot (a_0, \ldots, a_{m-1})^T =: (y_1^{(i)}(n), \ldots, y_m^{(i)}(n))^T$$

*for $i = 1, \ldots, s$ and*

$$\mathbf{x}_n := (x_n^{(1)}, \ldots, x_n^{(s)}) \in [0, 1)^s$$

*with*

$$x_n^{(i)} := \sum_{k=1}^m \frac{y_k^{(i)}(n)}{q^k}.$$

*Then $\mathbf{x}_0, \ldots, \mathbf{x}_{q^m-1} \in [0, 1)^s$ is called a digital net over $Z_q$.*

**Definition 2** *Let $\mathbf{x}_0, \ldots, \mathbf{x}_{q^m-1} \in [0, 1)^s$ be a digital point set over $Z_q$ generated by the $m \times m$-matrices $C_1, \ldots, C_s$. By $c_i^{(j)}$ we denote the j-th row of $C_i$. Let an integer $t$, $0 \leq t \leq m$ be such that for all integers $0 \leq d_1, \ldots, d_s \leq m$, $d_1 + \cdots + d_s = m - t$ the set of vectors $\{c_i^{(j)} \mid j = 1, \ldots, d_i; i = 1, \ldots, s\}$ is linearly independent over $Z_q$. Then $\mathbf{x}_0, \ldots, \mathbf{x}_{q^m-1}$ is called a digital $(t, m, s)$-net over $Z_q$.*

The quantity $t$ is a quality parameter for the distribution of the point set $\mathbf{x}_0, \ldots, \mathbf{x}_{q^m-1}$ and it should be small. So for example the combination of Theorem 4.10 and of Theorem 4.28 in [Nie92b] gives, that for the star-discrepancy $D_N^*$ of every $(t, m, s)$-net $\mathbf{x}_0, \ldots, \mathbf{x}_{N-1}$ over $Z_q$ we have

$$D_N^* \leq \frac{B(s, q) \cdot q^t (\log N)^{s-1}}{N}$$

with a constant $B(s, q)$ depending only on $s$ and $q$.

**Definition 3** *Let a dimension $s \geq 1$ and a prime base $q \geq 2$ be given. Let $C_1, \ldots, C_s$ be $\infty \times \infty$-matrices over $Z_q$. For $n \geq 0$ let $n = \sum_{k=0}^{r-1} a_k q^k$ be the representation of $n$ in base $q$. Consider the digits $a_0, \ldots, a_{r-1}$ as elements of $Z_q$. Let*

$$C_i \cdot (a_0, \ldots, a_{r-1}, 0, 0, \ldots)^T =: (y_1^{(i)}(n), y_2^{(i)}(n), \ldots)^T$$

*for $i = 1, \ldots, s$ and*

$$\mathbf{x}_n := (x_n^{(1)}, \ldots, x_n^{(s)}) \in [0, 1)^s$$

*with*

$$x_n^{(i)} := \sum_{k=1}^{\infty} \frac{y_k^{(i)}(n)}{q^k}.$$

Then $\mathbf{x}_0, \mathbf{x}_1, \ldots$ *is called a digital sequence over* $Z_q$. *(Here the following condition is assumed to hold: for each* $n \geq 0$ *and* $1 \leq i \leq s$ *we have* $y_k^{(i)}(n) < b - 1$ *for infinitely many* $k$.)

**Definition 4** *Let* $\mathbf{x}_0, \mathbf{x}_1, \ldots$ *be a digital sequence over* $Z_q$ *generated by the matrices* $C_1, \ldots, C_s$. *For* $m \geq 1$ *let* $C_i^{(m)}$ *denote the left upper* $m \times m$-*submatrix of* $C_i$. *Let a function* $T : N \longrightarrow N$ *be such that* $T(m) \leq m$ *for all* $m$ *and such that* $C_1^{(m)}, \ldots, C_s^{(m)}$ *generate a digital* $(T(m), m, s)$-*net over* $Z_q$. *Then* $\mathbf{x}_0, \mathbf{x}_1, \ldots$ *is called a digital* $(T,s)$-*sequence over* $Z_q$.

A "small" quality function $T$ guarantees a good distribution of the corresponding digital sequence. For example a combination of Theorem 1 and of Lemma 3 in [LN95] gives the following result:
– let $\mathbf{x}_0, \mathbf{x}_1, \ldots$ be a digital $(T, s)$-sequence over $Z_q$, then for the star-discrepancy $D_N^*$ of the first $N$ elements of the sequence we have

$$D_N^* \leq B(q,s) \cdot \frac{1}{N} \sum_{m=1}^{k} q^{T(m)} m^{s-1}$$

for all $k \geq 1$ and all $q^k \leq N < q^{k+1}$. Especially if the sequence

$$(\frac{1}{k} \sum_{m=1}^{k} q^{T(m)})_{k=1,2,\ldots}$$

is bounded, then $D_N^* \leq B'(q,s) \cdot \frac{(\log N)^s}{N}$ for all $N \geq 2$.
($B(q,s)$ and $B'(q,s)$ are constants depending only on $q$ and on $s$.)

Note that matrices $C_1, \ldots, C_s$ generating a digital $(T, s)$-sequence over $Z_q$ with "small" $T$ provide for all $m \geq 1$ matrices $C_1^{(m)}, \ldots, C_s^{(m)}$ generating digital $(t(m), m, s)$-nets over $Z_q$ with "small" $t(m)$ and by the principle given in Lemma 4.22 in [Nie92b], even digital $(t(m), m, s+1)$-nets over $Z_q$. So in some sense it is sufficient to search for optimal digital sequences.

## 3 A metric result on the distribution of digital sequences

In the following we will show that digital sequences constructed over $Z_q$ ($q$ prime) show, on the average, an almost optimal distribution behavior. For this purpose we use a natural measure on the class of all $s$-dimensional digital sequences constructed over $Z_q$. The distribution properties of such a sequence depend only on the matrices $C_1, \ldots, C_s$. Therefore we identify the class of all $s$-dimensional digital sequences constructed over $Z_q$ with the set $M_s$ of all $s$-tuples $(C_1, \ldots, C_s)$ of infinite matrices over $Z_q$. We define the probability measure $\mu_s$ on $M_s$ as the product measure induced by a certain probability measure $\mu$ on the set $M$ of all infinite matrices over $Z_q$. We can view $M$ as the product of denumerably many copies of the sequence space $Z_q^\infty$ over $Z_q$, and so we define $\mu$ as the product measure induced by a certain probability measure $\mu^*$ on $Z_q^\infty$. We identify each $c = (c_1, c_2, \ldots) \in Z_q^\infty$ with the real number $\sum_{k=1}^\infty \frac{c_k}{q^k} \in [0, 1]$. This identification is up to a subset of $[0,1]$ of Lebesgue measure zero bijective. So we may define $\mu^*$ as the measure on $Z_q^\infty$ which with respect to the above identification coincides with the Lebesgue measure on $[0,1]$.

In [LN95] it was shown that for $\mu_s$-almost all $s$-dimensional sequences $S$ constructed over $Z_q$ the star-discrepancy $D_N^*(S)$ of the first $N$ elements of $S$ satisfies

$$D_N^*(S) = O(\frac{(\log N)^{2s+\epsilon}}{N})$$

for every $\epsilon > 0$.

We will improve this in the following form:

**Theorem 1** *Let $G : N \longrightarrow R^+$ be monotonically increasing and such that*

$$\sum_{m=1}^\infty \frac{m^{s-1} \log m}{G(m)} < \infty$$

*then for $\mu_s$ -almost all s-dimensional sequences $S$ constructed over $Z_q$ the star-discrepancy $D_N^*(S)$ of the first $N$ elements of $S$ satisfies*

$$D_N^*(S) = O(\frac{G(\log N)}{N}) .$$

*For example for $\mu_s$-almost all $S$ we have*

$$D_N^* = O(\frac{(\log N)^s \cdot (\log \log N)^{2+\epsilon}}{N})$$

*for all $\epsilon > 0$.*

The result in [LN95] was shown via two steps. First the discrepancy bound

$$D_N^*(S) \le B(q, s) \cdot \frac{1}{N} \sum_{m=1}^k q^{T(m)} m^{s-1}$$

for all $k \ge 1$ and all $q^k \le N < q^{k+1}$ for $(T, s)$-sequences $S$, and then the following result were used:
 – let $D : N \longrightarrow [0, \infty)$ be such that

$$\sum_{m=1}^\infty \frac{m^{s-1}}{q^{D(m)}} < \infty.$$

Then $\mu_s$-almost all $s$-dimensional digital sequences constructed over the finite field $Z_q$ ($q$ is prime !) are $(T, s)$-sequences with

$$T(m) \leq D(m) + O(1)$$

for all $m \geq 1$. (The implied constant may depend on the sequence.)

We will show in a forthcoming paper that this second result cannot be improved. Namely we have:

– if $D$ is such that

$$\sum_{m=1}^{\infty} \frac{m^{s-1}}{q^{D(m)}} = \infty,$$

then for all $T$ with $T(m) \leq D(m)$ for all $m \geq 1$, we have:

$\mu_s$-almost no $s$-dimensional digital sequence constructed over $Z_q$ is a $(T, s)$-sequence.

Consequently for a proof of Theorem 1 we have to refine the estimate for the discrepancy of digital $(t, m, s)$-nets and consequently for digital $(t, s)$-sequences. This essentially will be done in Lemma 1 and in Lemma 2.

We will need the following notations:

let a prime $q$, a dimension $s$ and a positive integer $m$ be given. Let the $m \times m$-matrices $C_1, \ldots, C_s$ over $Z_q$ be given and denote by $c_j^{(i)} \in Z_q^m$ with $j = 1, \ldots, m$ the rows of $C_i$ for $i = 1, \ldots, s$.

For $0 \leq w \leq s$ a $w$-tuple $(p_1, \ldots, p_w)$ of non-negative integers is called admissible with respect to $C_1, \ldots, C_s$, if the system $\{c_j^{(i)} : j = 1, \ldots, p_i; i = 1, \ldots, w\}$ is linearly independent over $Z_q$. For $w = 0$ we call the "zero-tuple" () admissible.

For $w \leq s - 1$ and $(p_1, \ldots, p_w)$ admissible we define

$$h(p_1, \ldots, p_w) := \max\{h \geq 0 \mid (p_1, \ldots, p_w, h) \text{ is admissible}\}.$$

For the proof we also will need the concept of a "digital translation net".

**Definition 5** *Let a dimension $s \geq 1$, a prime base $q \geq 2$, and an integer $m \geq 1$ be given. Let $C_1, \ldots, C_s$ be $m \times m$-matrices over $Z_q$ and let $W_1, \ldots, W_s$ be vectors from $Z_q^m$. For $0 \leq n < q^m$ let $n = \sum_{k=0}^{m-1} a_k q^k$ be the $q$-adic representation of $n$ in base $q$. Consider the digits $a_0, \ldots, a_{m-1}$ as elements of $Z_q$. Let*

$$C_i \cdot (a_0, \ldots, a_{m-1})^T + W_i =: (y_1^{(i)}(n), \ldots, y_m^{(i)}(n))^T$$

*for $i = 1, \ldots, s$ and*

$$\mathbf{x}_n := (x_n^{(1)}, \ldots, x_n^{(s)}) \in [0, 1)^s$$

*with*

$$x_n^{(i)} := \sum_{k=1}^{m} \frac{y_k^{(i)}(n)}{q^k}.$$

*Then* $\mathbf{x}_0, \dots, \mathbf{x}_{q^m-1} \in [0,1)^s$ *is called a digital translation net over* $Z_q$ *generated by* $C_1, \dots, C_s$ *and by the translation vectors* $W_1, \dots, W_s$.

Then we have

**Lemma 1** *Let* $D^*$ *denote the star discrepancy of the digital translation net* $\mathbf{x}_0, \dots, \mathbf{x}_{q^m-1}$ *over* $Z_q$ *generated by* $C_1, \dots, C_s$ *and the translation vectors* $W_1, \dots, W_s$. *Then*

$$D^* \le \sum_{w=0}^{s-1} (q-1)^w \sum_{\substack{p_1, \dots, p_w > 0 \\ (p_1, \dots, p_w) \, admissible}} q^{-(p_1 + \cdots + p_w + h(p_1, \dots, p_w))}.$$

**Proof.** We do not work out all details of the proof. For more details see the proof of Lemma 1 in [Lar96]. By the definition of the digital translation net, of admissible tuples, and by Theorem 4.26 in [Nie92b] (see also Lemma 1 in [SW96]) we have the following: let $(p_1, \dots, p_w)$ be admissible and let $B \subseteq [0,1]^s$ with

$$B = \prod_{i=1}^{w} [\frac{a_i}{q^{p_i}}, \frac{b_i}{q^{p_i}}) \times [0,1)^{s-w}$$

with integers $0 \le a_i < b_i \le q^{p_i}$. (We call an interval of such a form an admissible interval.) Then $B$ contains exactly

$$q^{m-(p_1 + \cdots + p_w)} \cdot \prod_{i=1}^{w} (b_i - a_i)$$

of the net points.

Let $E = \prod_{i=1}^{s} [0, \alpha_i) \subseteq [0,1)^s$ with $\alpha_i := \sum_{j=1}^{\infty} \frac{\alpha_j^{(i)}}{q^j}$ for $i = 1, \dots, s$ be taken arbitrarily. (If the representation of $\alpha_i$ in base $q$ is not unique then we take its infinite representation.) Then on the one hand we have:

$$\tilde{E} := \bigcup_{\substack{(p_1, \dots, p_s) \\ admissible \\ p_1, \dots, p_w > 0}} \prod_{i=1}^{s} [\sum_{j=1}^{p_i-1} \frac{\alpha_j^{(i)}}{q^j}, \sum_{j=1}^{p_i} \frac{\alpha_j^{(i)}}{q^j}) \subseteq E.$$

The intervals in the above union are pairwise disjoint and admissible. On the other hand it is easy to deduce (for example by induction on $s$, see [Lar96]) that:

$$E \subseteq \tilde{E} \cup \bigcup_{w=0}^{s-1} \bigcup_{\substack{(p_1, \dots, p_w) \\ admissible \\ p_1, \dots, p_w > 0}} \left( \prod_{i=1}^{w} [\sum_{j=1}^{p_i-1} \frac{\alpha_j^{(i)}}{q^j}, \sum_{j=1}^{p_i} \frac{\alpha_j^{(i)}}{q^j}) \right)$$

$$\times \ [ \ \sum_{j=1}^{h(p_1,\ldots,p_w)} \frac{\alpha_j^{(w+1)}}{q^j}, \ \sum_{j=1}^{h(p_1,\ldots,p_w)} \frac{\alpha_j^{(w+1)}}{q^j} + \frac{1}{q^{h(p_1,\ldots,p_w)}})$$

$$\times \ [0,1)^{s-w-1}).$$

Again all intervals in the second union above are admissible. Therefore

$$\left| \frac{| \{ 0 \le n \le N-1 \mid \mathbf{x}_n \in E \} |}{N} - \lambda(E) \right| \le$$

$$\sum_{w=0}^{s-1} (q-1)^w \cdot \sum_{\substack{(p_1,\ldots,p_w) \\ admissible \\ p_1,\ldots,p_w > 0}} q^{-(p_1+\cdots+p_w+h(p_1,\ldots,p_w))}$$

and the result follows. ◇

For the proof of the Theorem we further need the following result:

**Lemma 2** *Let $D_N^*(S)$ denote the star discrepancy of the first $N$ elements $\mathbf{x}_0,\ldots,\mathbf{x}_{N-1}$ of the digital sequence $S$ generated by the infinite matrices $C_1,\ldots,C_s$ over $Z_q$ ($q$ prime). Let $N = \sum_{k=0}^{t-1} a_k q^k$ with $0 \le a_k < q$, $a_{t-1} \ne 0$ and let $C_i(m)$ denote the left upper $m \times m$-submatrix of $C_i$. Then*

$$N D_N^*(S) \le ts(q-1) + \sum_{m=1}^{t-1} \sum_{w=0}^{s-1} (q-1)^{w+1} \sum_{\substack{p_1,\ldots,p_w > 0 \\ \mathbf{p}:=(p_1,\ldots,p_w) \\ admissible \ to \ m}} q^{m-(p_1+\cdots+p_w+h(\mathbf{p}))}.$$

*(Here "admissible to $m$" means "admissible to $C_1(m),\ldots,C_s(m)$" and also $h(p_1,\ldots,p_w)$ is determined with respect to these matrices.)*

**Proof.** We divide the sequence $\mathbf{x}_0,\ldots,\mathbf{x}_{N-1}$ into the subsequences $w_{m,b}$; $b = 0,\ldots,b_m - 1$; $m = 0,\ldots,t-1$ where $w_{m,b}$ is the subsequence

$$\mathbf{x}_n; \quad \sum_{k=m+1}^{t-1} b_k q^k + b q^m \le n < \sum_{k=m+1}^{t-1} b_k q^k + (b+1) q^m.$$

Let now $E_i(m)$ denote the left $\infty \times m$-submatrix of $C_i$, let $F_i(m)$ denote $C_i$ without $E_i(m)$ and let

$$U_i(m,b,N) := F_i(m) \cdot (b, b_{m+1},\ldots,b_{t-1},0,0,\ldots)^T \in Z_q^\infty.$$

If $n = \sum_{k=m+1}^{t-1} b_k q^k + b q^m + \sum_{k=0}^{m-1} a_k q^k$ and

$$\vec{n} := (a_0,\ldots,a_{m-1}, b, b_{m+1},\ldots,b_{t-1},0,0,\ldots)^T$$

then $C_i \cdot \vec{n} = E_i(m) \cdot (a_0 \ldots a_{m-1})^T + U_i(m,b,N)$. Let $V_i(m,b,N)$ denote the vector consisting of the first $m$ coordinates of $U_i(m,b,N)$. Let $\tilde{w}_{m,b}$

be the digital translation net generated by $C_1(m), \ldots, C_s(m)$ and by the translation vectors $V_1 := V_1(m, b, N), \ldots, V_s := V_s(m, b, N)$.

Denote the sequence $w_{m,b}$ by $\mathbf{y}_0, \ldots, \mathbf{y}_{q^m-1}$ and $\tilde{w}_{m,b}$ by $\mathbf{z}_0, \ldots, \mathbf{z}_{q^m-1}$ then $\|\mathbf{z}_j - \mathbf{y}_j\|_\infty \le \frac{1}{q^m}$ for $j = 0, \ldots, q^m - 1$. (Note that $C_i(m)$ is the $m \times m$-upper part of $E_i(m)$ !) Therefore for the star discrepancies $D_{m,b}^*(1)$ of $w_{m,b}$ and $D_{m,b}^*(2)$ of $\tilde{w}_{m,b}$ we have $\mid q^m D_{m,b}^*(1) - q^m D_{m,b}^*(2) \mid \le s$. This last inequality immediately follows from the fact that if we enlargen the sides of an $s$-dimensional interval $\prod_{i=1}^s [0, \alpha_i)$ on the "right hand side" by $\frac{1}{q^m}$ (i.e. if we consider $\prod_{i=1}^s [0, \alpha_i + \frac{1}{q^m})$), then the volume of the part of the new interval which is in the unit cube is at most by $\frac{s}{q^m}$ larger than the volume of the initial interval. Therefore by [KN74], Theorem 2.6, page 115 (the assertion of this theorem at once can be checked to be true in arbitrary dimension) we have

$$ND_N^* \le \sum_{m=0}^{t-1} \sum_{b=0}^{b_m-1} q^m D_{m,b}^*(1) \le q-1+(t-1)(q-1)s+\sum_{m=1}^{t-1}(q-1)q^m D_{m,b}^*(2)$$

and by Lemma 1 the result follows. $\diamond$

In Lemma 2 the value $t$ is at most $\frac{\log N}{\log q} + 1$. So for the proof of the Theorem it suffices to show that for every $w$ with $0 \le w \le s - 1$, for $\mu_s$-almost all $C_1, \ldots, C_{w+1}$ we have

$$\sum_{m=1}^{t-1} \sum_{\substack{p_1, \ldots, p_w > 0 \\ (p_1, \ldots, p_w) \, admissible \, to \, m}} q^{m-(p_1+\cdots+p_w+h(p_1,\ldots,p_w))} = O(G(t)).$$

Let now $w$, $0 \le w \le s - 1$ be fixed. For an arbitrary $w$-tuple of positive integers $\mathbf{p} := (p_1, \ldots, p_w)$ let $\bar{p} := p_1 + \cdots + p_w$.

Let $g = g(m)$ be a non-negative integral-valued function in the variable $m$. Let

$$M(g) := \{(C_1, \ldots, C_{w+1}) \quad | \quad \textit{There exist } m \ge 1 \textit{ and}$$
$$\mathbf{p} \textit{ admissible with respect to } m$$
$$\textit{such that } h(\mathbf{p}) < m - \bar{p} - g(m)\}.$$

(Here $C_1, \ldots, C_{w+1}$ are $\infty \times \infty$-matrices over $Z_q$ and "admissible with respect to m" means "admissible with respect to the matrices $C_1(m), \ldots, C_{w+1}(m)$".)

We show

**Lemma 3**

$$\mu_{w+1}(M(g)) \le \sum_{m=1}^{\infty} \frac{m^w}{q^{g(m)}}.$$

**Proof.** We have

$$\mu_{w+1}(M(g)) \leq \sum_{m=1}^{\infty} \sum_{\substack{\mathbf{p}=(p_1,\ldots,p_w) \\ \bar{p}<m-g(m)}} \sum_{\substack{\lambda:=(\lambda_1,\ldots,\lambda_{m-g(m)})\in Z_q^{m-g(m)} \\ \lambda_{\bar{p}+1},\ldots,\lambda_{m-g(m)} \text{ not all } 0}} \mu_{w+1}(M(\lambda,\mathbf{p},m)),$$

where

$$
\begin{aligned}
M(\lambda,\mathbf{p},m) \quad := \quad & \{C_1,\ldots,C_{w+1} \mid \mathbf{p} \text{ admissible with respect to} \\
& C_1(m),\ldots,C_{w+1}(m) \\
& \text{and} \\
& \lambda_1 c_1^{(1)}(m) + \cdots + \lambda_{p_1} c_1^{(p_1)}(m) + \\
& \lambda_{p_1+1} c_2^{(1)}(m) + \cdots + \lambda_{p_1+p_2} c_2^{(p_2)}(m) + \\
& \cdots + \\
& \lambda_{\bar{p}+1} c_{w+1}^{(1)}(m) + \cdots + \lambda_{m-g(m)} c_{w+1}^{(m-g(m)-\bar{p})}(m) \\
& = 0 \}
\end{aligned}
$$

(here $c_i^{(j)}(m)$ denotes the $j$-th row of $C_i(m)$).

Since one of $\lambda_{\bar{p}+1},\ldots,\lambda_{m-g(m)}$ - say $\lambda_j$ - is different from zero, the row $c_{w+1}^{j-\bar{p}}(m)$ is determined by the other rows.

So $\mu_{w+1}(M(\lambda,\mathbf{p},m)) = \frac{1}{q^m}$ and

$$\mu_{w+1}(M(g)) \leq \sum_{m=1}^{\infty} \sum_{\substack{p_1,\ldots,p_w \\ p_1+\cdots+p_w<m}} \sum_{\lambda_1,\ldots,\lambda_{m-g(m)}\in Z_q} \frac{1}{q^m} \leq \sum_{m=1}^{\infty} \frac{m^w}{q^{g(m)}}.$$

$\diamond$

## Proof of the Theorem.

The proof of the Theorem now runs quite similar to the proof of the Theorem in [Lar95].

For $0 \leq w \leq s-1$ fixed, let $M_{w+1}$ again denote the set of all $w+1$-tuples $C_1,\ldots,C_{w+1}$ of $\infty \times \infty$-matrices over $Z_q$, and for $g$ as in Lemma 3 fixed, let $\overline{M} := M_{w+1} \backslash M(g)$.

For a positive function $G(m)$ we then consider the integral

$$
\int := \int_{(C_1,\ldots,C_{w+1})\in\overline{M}} \sum_{m=1}^{t-1} \frac{1}{G(m)} \times \\
\sum_{\substack{\mathbf{P}=(p_1,\ldots,p_w) \text{ admissible to } m, \\ p_i>0}} q^{m-\bar{p}-h(\mathbf{P})} d\mu_{w+1}(C_1,\ldots,C_{w+1}).
$$

We have

$$
\int \leq \int_{\overline{M}} \sum_{m=1}^{t-1} \frac{1}{G(m)} \sum_{\substack{\mathbf{P} \text{ admissible to } m, \\ h(\mathbf{P})<m-\overline{\mathbf{P}}}} q^{m-\bar{p}-h(\mathbf{P})} d\mu_{w+1}
$$

$$+ \int_{\overline{M}} \sum_{m=1}^{t-1} \frac{1}{G(m)} \sum_{\substack{\mathbf{p} \text{ admissible to } m, \\ \bar{p} \leq m}} 1 \, d\mu_{w+1}$$

$$=: \int_1 + \int_2$$

We have

$$\int_2 \leq \sum_{m=1}^{t-1} \frac{m^w}{G(m)}.$$

Further

$$\int_1 \leq \int_{\overline{M}} \sum_{m=1}^{t-1} \frac{1}{G(m)} \cdot$$

$$\sum_{\mathbf{p} \text{ admissible to } m} \sum_{i=m-g(m)-\bar{p}+1}^{m-\bar{p}} \sum_{\lambda} q^{m-\bar{p}-i+1} d\mu_{w+1}$$

where summation in the last sum is over all

$$\lambda = (\lambda_1, \ldots, \lambda_{i+\bar{p}}) \in Z_q^{i+\bar{p}} \backslash \{0\}$$

for which

$$\lambda_1 c_1^{(1)}(m) + \cdots + \lambda_{p_1} c_1^{(p_1)}(m) +$$
$$+ \cdots +$$
$$\lambda_{\bar{p}+1} c_{w+1}^{(1)}(m) + \cdots + \lambda_{\bar{p}+i} c_{w+1}^{(i)}(m) = 0.$$

Therefore

$$\int_1 \leq \sum_{m=1}^{t-1} \frac{1}{G(m)} \sum_{\substack{\mathbf{p} \text{ admissible to } m \\ \bar{p} \leq m}} \sum_{i=m-g(m)-\bar{p}+1}^{m-\bar{p}}$$

$$\sum_{\lambda \in Z_q^{i+\bar{p}} \backslash \{0\}} q^{m-\bar{p}-i+1} \int d\mu_{w+1},$$

where integration in the last integral is over all $(C_1, \ldots, C_{w+1}) \in M_{w+1}$ for which
$$\lambda_1 c_1^{(1)}(m) + \cdots + \lambda_{\bar{p}+i} c_{w+1}^{(i)}(m) = 0.$$

This integral (by the same argument as in the proof of Lemma 3) is $\frac{1}{q^m}$.
Consequently

$$\int_1 \leq q \cdot \sum_{m=1}^{t-1} \frac{1}{G(m)} m^w g(m).$$

Let now for $L > 1$

$$g(m) := \lceil \frac{\log L}{\log q} + \frac{(w+2)\log m}{\log q} \rceil \geq 1,$$

then

$$\int \leq (q+1) \sum_{m=1}^{t-1} \frac{g(m)m^w}{G(m)}$$

and by Lemma 3

$$\mu_{w+1}(M(g)) \leq \frac{1}{L} \cdot \frac{\pi^2}{6}.$$

Let $G(m)$ be such that

$$\sum_{m=1}^{\infty} \frac{m^w \log m}{G(m)} < \infty,$$

then

$$(q+1) \sum_{m=1}^{\infty} \frac{g(m)m^w}{G(m)} < \infty.$$

Since the function

$$H_t(C_1, \ldots, C_{w+1}) := \sum_{m=1}^{t-1} \frac{1}{G(m)} \sum_{\substack{\mathbf{p}=(p_1,\ldots,p_w) \, admissible \, to \, m, \\ p_i > 0}} q^{m-\bar{p}-h(\mathbf{p})}$$

is monotonically increasing in t, and since G is monotonically increasing we get that for all L for almost all $C_1, \ldots, C_{w+1} \in \overline{M}$ we have:

$$\sum_{m=1}^{t-1} \sum_{\mathbf{p}=(p_1,\ldots,p_w) \, admissible \, to \, m} q^{m-\bar{p}-h(\mathbf{p})} = O(G(t)).$$

Therefore (and by Lemma 3) we get that for almost all $(C_1, \ldots, C_s) \in M_s$ we have

$$\sum_{m=1}^{t-1} \sum_{w=0}^{s-1} \sum_{\mathbf{p} \, admissible \, to \, m} q^{m-(\bar{p}+h(\mathbf{p}))} = O(G(t)),$$

and by Lemma 2 the result follows.

◇

# 4 References

[Fau82]   H. Faure.   Discrépance de suites associées à un système de numération (en dimension s). *Acta Arith.*, **41**:337–351, 1982.

122

[Kie61]    J. Kiefer. On large deviations of the empiric d.f. of vector chance variables and a law of the iterated logarithm. *Pacific J. Math.*, 11:649–660, 1961.

[KN74]    L. Kuipers and H. Niederreiter. *Uniform Distribution of Sequences.* John Wiley, New York, 1974.

[Lar95]    G. Larcher. On the distribution of an analog to classical Kronecker-sequences. *J. Number Theory*, 52:198–215, 1995.

[Lar96]    G. Larcher. A bound for the discrepancy of digital nets and its application to the analysis of certain pseudo-random number generators. Preprint. University Salzburg, 1996.

[LLNS96] G. Larcher, A. Lauß, H. Niederreiter, and W. Ch. Schmid. Optimal polynomials for $(t, m, s)$-nets and numerical integration of multivariate Walsh series. *SIAM J. Numer. Analysis*, 33-6, 1996.

[LN95]    G. Larcher and H. Niederreiter. Generalized $(t, s)$-sequences, Kronecker-type sequences, and diophantine approximations of formal Laurent series. *Trans. Amer. Math. Soc.*, 347:2051–2073, 1995.

[LNS96]    G. Larcher, H. Niederreiter, and W. Ch. Schmid. Digital nets and sequences constructed over finite rings and their application to quasi-Monte Carlo integration. *Monatsh. Math.*, 121:231–253, 1996.

[LS95]    G. Larcher and W. Ch. Schmid. Multivariate Walsh series, digital nets and quasi-Monte Carlo integration. In H. Niederreiter and P. J.-S. Shiue, editors, *Monte Carlo and Quasi-Monte Carlo Methods in Scientific Computing (Las Vegas, 1994)*, volume 106 of *Lecture Notes in Statistics*, pages 252–262. Springer, New York, 1995.

[LT94]    G. Larcher and C. Traunfellner. On the numerical integration of Walsh series by number-theoretic methods. *Math. Comp.*, 63:277–291, 1994.

[Nie87]    H. Niederreiter. Point sets and sequences with small discrepancy. *Monatsh. Math.*, 104:273–337, 1987.

[Nie88]    H. Niederreiter. Low-discrepancy and low-dispersion sequences. *J. Number Theory*, 30:51–70, 1988.

[Nie92a]    H. Niederreiter. Low-discrepancy point sets obtained by digital constructions over finite fields. *Czechoslovak Math. J.*, 42:143–166, 1992.

[Nie92b]   H. Niederreiter. *Random Number Generation and Quasi-Monte Carlo Methods*. Number **63** in CBMS–NSF Series in Applied Mathematics. SIAM, Philadelphia, 1992.

[NX96a]   H. Niederreiter and C. P. Xing. Low-discrepancy sequences and global function fields with many rational places. *Finite Fields Appl.*, **2**:241–273, 1996.

[NX96b]   H. Niederreiter and C. P. Xing. Quasirandom points and global function fields. In S. Cohen and H. Niederreiter, editors, *Finite Fields and Applications (Glasgow, 1995)*, volume 233 of *Lect. Note Series of the London Math. Soc.*, pages 269–296. Camb. Univ. Press, Cambridge, 1996.

[Rot54]   K. F. Roth. On irregularities of distribution. *Mathematika*, 1:73–79, 1954.

[Sob67]   I. M. Sobol'. The distribution of points in a cube and the approximate evaluation of integrals. *Ž. Vyčisl. Mat. i Mat Fiz.*, 7:784–802, 1967. (Russian).

[SW96]   W. Ch. Schmid and R. Wolf. Bounds for digital nets and sequences. *Acta Arith.*, 1996. To appear.

Institut für Mathematik
Universität Salzburg
Hellbrunnerstraße 34
A–5020 Salzburg
Austria
e-mail address: GERHARD.LARCHER@SBG.AC.AT

# Random Number Generators and Empirical Tests

## Pierre L'Ecuyer

ABSTRACT  We recall some requirements for "good" random number generators and argue that while the construction of generators and the choice of their parameters must be based on theory, a posteriori empirical testing is also important. We then give examples of tests failed by some popular generators and examples of generators passing these tests.

## 1  Introduction

Roughly, a random number generator is a simple program producing a periodic sequence of numbers on a computer. This sequence should behave, for practical purposes, as the realization of a sequence of independent and identically distributed (i.i.d.) random variables, say, uniformly over the interval $(0, 1)$, or $U(0, 1)$ for short. In the next section, we provide a more formal definition and describe basic desired properties for a generator. With these properties in view, the construction of a generator must be based on a good theoretical analysis of its structure. In this paper, we do not study the tools for such theoretical analysis (this is done elsewhere), but we focus rather on empirical tests that can be applied a posteriori. Section 3 discusses the idea of empirical statistical testing and what it can do for us. The theoretical distribution functions used in statistical tests are often known only asymptotically or approximately, and one must assess and control the approximation error for finite sample size. Section 3 also discusses how to do this. Sections 4 and 5 describe two specific classes of tests, based on close points in space and on discrete entropy, respectively. They summarize the content of the two papers [LCS96, LCC96]. Section 6 gives the results of a set of tests applied to a set of previously proposed generators. This is mainly to illustrate what the tests can do, and not to recommend any specific generator. Section 7 is a brief conclusion.

## 2  Random number generators

We define a *random number generator* as a structure comprised of a finite set of *states* $S$, a probability distribution $\mu$ over $S$, a *transition function* $T : S \to S$, a finite set of *output* symbols $U$, and an *output function* $G : S \to U$. The *initial state* (or *seed*) $s_0 \in S$ is generated (externally) from the probability distribution $\mu$, and then the state evolves according to $s_n = T(s_{n-1})$, for $n \geq 1$. At step $n$, the generator outputs $u_n = G(s_n) \in U$. These $u_n$ are called the *random numbers*. Typically, $U$ is the interval $[0, 1)$. Often, the seed $s_0$ is chosen deterministically, so $\mu$ is degenerate. Obviously, the sequence of states $s_n$ is periodic with period length $\rho$ not exceeding the cardinality of the set $S$. If $b$ bits of memory are used to represent the state, then $\rho \leq 2^b$, and we generally want $\rho$ to be near $2^b$.

The unreachable dream is that no statistical test (or computer program) could distinguish between the output sequence of the generator and the realization of an i.i.d. uniform sequence over $U$; e.g., cannot tell which is which better than by flipping a fair coin. This is obviously impossible, since one can always distinguish them if the sequence is observed long enough, because $\rho < \infty$.

A (perhaps) more reachable dream can be defined via the point of view of computational complexity. Consider a *family* of generators $\{\mathcal{G}_b = (S_b, \mu_b, T_b, U_b, G_b), b \geq 1\}$, where the parameter $b$ indicates the size of the memory used to represent the state. So, the state space $S_b$ of generator $\mathcal{G}_b$ has cardinality $2^b$. We assume that the functions $T_b$ and $G_b$ can be computed in polynomial time (in $b$) and restrict our attention to the statistical tests whose running time is also polynomial in $b$. Assuming that the period length increases as $2^b$, this excludes any statistical test that exhausts the period. The family $\{\mathcal{G}_b\}$ is called *polynomial-time perfect* if there is a constant $\epsilon > 0$ such that, for any polynomial-time statistical test trying to distinguish the output sequence $\{u_n\}$ from the realization of a sequence of i.i.d. $U(0, 1)$ random variables, the probability of the test guessing right is less than $1/2 + e^{-b\epsilon}$. This is equivalent to the requirement that no polynomial-time algorithm, after observing $u_0, \ldots, u_{n-1}$, could correctly predict any given bit of $u_n$ with probability exceeding $1/2 + e^{-b\epsilon}$. This computational complexity setup is well-known in the area of cryptology, where unpredictability is a key issue. For more details, see, e.g., [BBS86, Lag93, LP89]. The idea is that statistical tests not running in polynomial time "quickly" become too long to apply, as $b$ gets large enough, even with the fastest computers. So, polynomial-time perfect generators *provably* pass (essentially) all statistical tests when their size is large enough.

That sounds nice but how large is large enough? These unpredictability properties are only asymptotic and we do not know how safe we are for any given $b$. Moreover, there are generator families *conjectured* to be polynomial-time perfect, but none has been *proved* to be so. In fact, nobody knows for sure whether or not such a family *exists*. A nice instance is the BBS generator, introduced by Blum, Blum, and Shub [BBS86, LP89], which is proved polynomial-time perfect under the (unproven) assumption that factoring the product of two (unknown) $b$-bit random primes cannot be done in polynomial time in $b$.

The generators used in simulation are *not* polynomial-time perfect, but are generally much faster than those conjectured to be so, and some of them seem to have statistical behavior good enough for most simulation applications.

The sequence of states of a generator with period length $\rho$ can be put on a circle, and the generator goes around just like a runner circling a track with $\rho$ steps per lap. In general, the generator may have many disjoint subcycles, and will cycle around one of many "tracks", depending on its initial state $s_0$. Here, to simplify, we suppose that there is a single cycle. If the seed $s_0$ is fixed to a default value, as is sometimes the case, then the sequence is always exactly the same, so it is very easy to construct a statistical test that this generator will fail.

Suppose now that the starting point $s_0$ is picked randomly, uniformly over the $\rho$ values on the circle. For any given $n$ and fixed $t$, $1 < t \ll \rho$, consider the vector of $t$ successive output values $\boldsymbol{U}_n = (U_n, \ldots, U_{n+t-1})$. This vector $\boldsymbol{U}_n$ is now truly random, uniformly distributed (at best) over a set $\Psi$ of cardinality $\rho$. Ideally, we would like $\boldsymbol{U}_n$ to be uniformly distributed over the $t$-dimensional hypercube $U^t$. Strictly speaking, this is possible only if $|U|$, the cardinality of the set $U$, is finite and if $\rho$ is a multiple of $|U|^t$. This can occur only for small $t$. Otherwise, for $t$ such that $|U|^t > \rho$, one would (heuristically) like the set $\Psi = \{\boldsymbol{U}_n, 0 \leq n < \rho\}$ to be uniformly spread over $U^t$. Observe that we do not ask here for $\Psi$ to look like a random set of points, but rather to have a *super-uniform* (or very even) distribution over $U^t$. Then, a few points picked randomly from $\Psi$ should behave pretty much as i.i.d. uniform points over $U^t$, assuming that $\rho/t$ is several orders of magnitudes larger than the number of those points. This justifies huge period lengths, much larger than what is actually used in a simulation (e.g., $\rho > 2^{100}$ or more).

As an analogy, suppose one is asked to place $10^{10}$ balls of 10 different colors in a (huge) box. Every evening, the balls are well-mixed and 16 of them are drawn randomly, without replacement. The 16 successive colors $C_1, \ldots, C_{16}$ are noted (e.g., to determine the winning numbers of a lottery), then the 16 balls are replaced in the box for the next day. The question is: How many balls of each color to place in the box if we want the vector $(C_1, \ldots, C_{16})$ to be approximately uniformly distributed over the set of $10^{16}$ possible vectors? Obviously, the answer is to place $10^9$ balls of each color. Then, if we pick a large fraction of the balls from the box, we get superuniformity, but if we pick only a few (e.g., 16), we get a close approximation to uniformity. Our reasoning is similar in the case of random number generators. Another possibility would be that the colors of the $10^{10}$ balls placed in the box be independent uniform random variables, so the numbers of balls of each color in the box, say $(X_1, \ldots, X_{10})$, would be multinomial with parameters $n = 10^{10}$ and $p_1 = \cdots = p_{10} = 1/10$. In this case, the *a priori* distribution of $(C_1, \ldots, C_{16})$ is uniform, so this is a perfect strategy if these balls are used only once. However, the distribution of $(C_1, \ldots, C_{16})$ *conditional* on $(X_1, \ldots, X_{10})$ is generally nonuniform and is the same from day to day, so this scheme is not so good if the balls are to be used for several days. According to this argument, if a random number generator is to be constructed once and used for several applications, then, over the entire period of the generator, a superuniform point set might be preferred over a point set whose distribution resembles a typical realization of i.i.d. uniform random variables.

The real interval $[0, 1)$ can be approximated by the finite set $U = \{0, 1/m, 2/m, \ldots, 1 - 1/m\}$, where $m$ is a large integer. Suppose that $\rho = |\Psi| \approx m^k$. If $t \leq k$, then $U_n$ can be approximately uniformly distributed over $U^t$ and this is what we want. If $t > k$, then $|U^t|/|\Psi| \approx m^{t-k}$, so $\Psi$ can cover only a small fraction of $U^t$. In that case, following the above argument, we would like the points of $\Psi$ to be evenly distributed over the unit hypercube $[0, 1)^t$. If this holds for all $t$ up to some reasonably large value $t_0$ and if $\rho$ is large enough, then the vectors $U_n$ should behave pretty much as independent random points in the unit hypercube. This is similar to the "scanning ensemble" framework described by Compagner [Com91, Com95].

The above super-uniformity requirements provide precise theoretical guidelines for constructing specific generators. Their application, in turn, implies a thorough theoretical understanding of the equidistribution properties of the generators. Theoretical analysis is crucial for the design of good generators and has been the subject of several books and papers; see, e.g., [Knu81, L'E94, L'E97, Nie92] and other references given there. For generators based on linear recurrences, for example, this is often achieved by analyzing the underlying lattice structure (sometimes called the "spectral test"). Generators with well-understood structural properties may appear less random, but the less understood ones could hide worse defects.

After a specific generator has been constructed and implemented, it is a good idea to test it empirically. This is what we discuss in the next section. Good theoretical and statistical behavior are not the only desired properties. Speed, low memory usage, ease of implementation, and fast jump-ahead capabilities for splitting, are among the other major requirements; see [L'E94, L'E97] for further details.

## 3 Empirical statistical tests

Empirical tests view a generator as a black box and compute certain statistics from part of its output sequence in order to catch up significant defects.

From now on, we assume that $U = [0, 1)$ and we view the generator's output (before it is observed) as a sequence of random variables $U_0, U_1, U_2, \ldots$. We want to test the null hypothesis $H_0$: "The output sequence $\{U_0, U_1, U_2, \ldots\}$ is a sequence of i.i.d. $U(0, 1)$ random variables". We know in advance that $H_0$ is false, but the question is "how easily can we detect it?". A *statistical test* in our context is defined by any real-valued function $T$ of any finite number of output values, say $\{U_0, \ldots, U_{n-1}\}$. Since

$n$ has no upper bound, there is an infinite number of such tests. The random variable $T$ is called a *statistic* and its distribution under the null hypothesis $H_0$ (or a good approximation of this distribution) must be known to apply the test. In its simplest version, the test computes the value of $T$ from the output sequence of the generator and rejects $H_0$ if that value differs significantly from the "typical" values of $T$ that are to be expected under $H_0$.

Since there is an unlimited number of tests, which tests are the good ones? There is no general answer to this question. Different tests may detect different types of defects. As a general rule, we want to detect the defects that may have a bad impact on our simulation results and do not really care about the other types of defects. So, what we need is a statistic $T$ whose distribution under $H_0$ is known (or well-approximated) and whose behavior resembles very well that of the random variable $Y$ (say) of interest in our simulation. This is usually too hard to achieve in practice for two reasons:

(a) Typically, if we need simulation to estimate the properties of $Y$ (such as, e.g., its expectation), it is precisely because they are too complicated to be studied directly. In this case, it is unlikely that we can find a $T$ as just described.

(b) The random number generators provided in general purpose softwares cannot be tested this way, because the random variable $Y$ of interest to the user is unknown in advance and varies from user to user.

In some cases, in may be possible to find a $T$ whose behavior is close to that of $Y$. Then, reason (a) no longer applies. Few users, however, are ready to spend time testing the random number generators. It is therefore important that the generators proposed for general use be submitted to a wide variety of tests that can detect different types of correlations or other harmful structures. New classes of applications bring the need for new families of tests, based on, e.g., spatial statistics, random walks, and so on.

Random number generators are often tested via a *two-level* procedure as follows. Compute $N$ "independent" copies of $T$, denoted $T_1, \ldots, T_N$, from disjoint segments of the output sequence, and compute their empirical distribution $\hat{F}_N$. Then, compare this empirical distribution to the theoretical distribution of $T$ under $H_0$, say $F$. This comparison can be done by a standard goodness-of-fit test, such as those of Kolmogorov-Smirnov (KS), Anderson-Darling, etc. [Dur73, Knu81, Ste86a].

The KS test computes the KS statistics

$$D_N^+ = \sup_{0 \le u \le 1} (\hat{F}(u) - F(u)),$$

$$D_N^- = \sup_{0 \le u \le 1} (F(u) - \hat{F}(u)).$$

TABLE 1.1. Significance levels for $D_N^+$

| $N$ | 10 | $10^2$ | $10^3$ | $10^4$ | $10^5$ |
|---|---|---|---|---|---|
| $\alpha = 0.95$ | .410 | .134 | .043 | .013 | .004 |
| $\alpha = 0.99$ | .495 | .161 | .051 | .016 | .005 |

These two random variables take values, say, $d_N^+$ and $d_N^-$, respectively, with significance levels

$$\delta^+ = P[D_N^+ > d_N^+] \quad \text{and} \quad \delta^- = P[D_N^- > d_N^-],$$

where the latter probabilities are under $H_0$. The distribution functions of $D_N^+$ and $D_N^-$ under $H_0$ can be computed as explained in [Dur73], assuming that $F$ is continuous. Under $H_0$, $\delta^+$ and $\delta^-$ are (correlated) $U(0, 1)$ random variables. One rejects $H_0$ if $\delta^+$ or $\delta^-$ is far too close to 0 or 1. In case of doubt, one can repeat the test with other segments of the sequence and reject $H_0$ if $\delta^+$ or $\delta^-$ is consistently close to 0 or 1. When the confidence levels are "reasonable", this is somewhat reassuring, but does not prove that the generator will behave correctly for any arbitrary application. No statistical test can prove this. Nevertheless, generally speaking, one would better trust a generator which has passed an extensive and varied set of tests. For better confidence, it is also a good idea to re-run important simulations with totally different random number generators.

For many tests, including those described in the following sections, the exact theoretical distribution of the statistic $T$ under $H_0$, say $\tilde{F}$, is unknown and only an approximation $F$ (usually an asymptotic), is available. For large $N$, the test will eventually detect the lack of fit of the approximation and even the best generators will be rejected. We must therefore control the approximation error. Since we have a KS test in mind, the following error measures are convenient:

$$\Delta^+ \overset{\text{def}}{=} \sup_{0 \le u \le 1} (\tilde{F}(u) - F(u)),$$

$$\Delta^- \overset{\text{def}}{=} \sup_{0 \le u \le 1} (F(u) - \tilde{F}(u)).$$

To determine what error size is acceptable, the following table is useful. It gives the values of $d_N^+$ yielding significance levels $\alpha$ of 0.95 and 0.99, for different values of $N$. For $N = 100$, for example, rejection at the 99% level occurs only for $d_N^+ > 0.16$, so an error $\Delta^+$ of, say, less than 0.01 on the distribution has little effect on the test. The same holds for $\Delta^-$.

Generally, $\Delta^+$ and $\Delta^-$ cannot be computed exactly, but they can at least be estimated by simulation, using simply $\hat{\Delta}^+ = D_N^+$ and $\hat{\Delta}^- = D_N^-$ as estimators, with very large $N$. This also provides confidence intervals (using the KS distribution). This is what we did here, using different types of "robust" generators, such as G14 in Section 6, and the results agreed well across the different generators.

Knuth [Knu81] describes a "classical" set of tests. Marsaglia [Mar85, Mar96] proposes a more recent battery, called DIEHARD. Some of these tests are more "stringent" than the classical ones, in the sense that more popular generators tend to fail them. References to other statistical tests can be found, for instance, in [L'E92, L'E94, LCS96, LCC96, L'E96b, LW97, Ste86b]. Another extensive testing package called TestU01 [L'E96b], still under development, implements most of the tests proposed so far, as well as several classes of generators implemented in generic form. The next two sections describe two classes of statistical tests implemented in [L'E96b], but not belonging to earlier packages. Further details on these tests can be found in [LCS96, LCC96].

## 4   Tests based on close points

Suppose we generate $n$ points $U_i = (U_i, \ldots, U_{i+k-1})$, $i = k, 2k, 3k, \ldots$, in the $k$-dimensional hypercube $[0, 1)^k$. Under $H_0$, since these vectors are non-overlapping, the points are i.i.d. uniformly distributed in the hypercube. For each pair $i \neq j$, let

$$D_{n,i,j} = \|U_j - U_i\|_p$$

be the distance between $U_i$ and $U_j$ induced by the $L_p$-norm, defined by

$$\|(x_1, \ldots, x_k)\|_p = \begin{cases} (|x_1|^p + \cdots + |x_k|^p)^{1/p} & \text{for } 1 \leq p < \infty; \\ \max(|x_1|, \ldots, |x_k|) & \text{for } p = \infty. \end{cases}$$

Let $V_k$ be the volume of the unit ball $\{x \in \mathbf{R}^k \mid \|x\|_p \leq 1\}$, $\mu = V_k/(n(n-1)/2)$ and, for all $t \geq 0$, $Y_n(t)$ be the number of pairs $(i, j)$, with $i < j$, such that $D_{n,i,j}^k \leq \mu t$. The next proposition characterizes the stochastic process $Y_n$. For the proof and more details, see [LCS96, RS78, SB78, Rip87].

**Proposition 1** *Under $H_0$, for each $t_1 > 0$, as $n \to \infty$, the process $\{Y_n(t), 0 \leq t \leq t_1\}$ converges weakly to a Poisson process with unit rate.*

Let

$$\begin{aligned} T_{n,0} &= 0, \\ T_{n,i} &= \inf\{t \geq 0 \mid Y_n(t) \geq i\}, \quad i \geq 1, \\ \Delta_{n,i} &= T_{n,i} - T_{n,i-1}. \end{aligned}$$

The $\Delta_{n,i}$ are the times between the successive jumps of the process $Y_n$. Therefore, under $H_0$, they are approximately i.i.d. exponential with unit mean, and the transformed random variables

$$W_{n,i}^* = 1 - \exp[-\Delta_{n,i}]$$

are approximately i.i.d. $U(0,1)$. In particular, the random variable

$$W_n^* = W_{n,1}^* = 1 - \exp[-(D_n^*)^k/\mu],$$

where $D_n^* = \min_{i \neq j} D_{n,i,j}$ is the smallest interpoint distance, is approximately $U(0,1)$.

Based on this, we can define a statistical test using $T = W_n^*$ as a statistic. As usual, $N$ independent copies of $T$ are computed and their empirical distribution is compared with the uniform via a KS test. We call this the *nearest-pair test*.

We can also define a test based on $W_{n,1}^*, \ldots, W_{n,m}^*$ for some fixed $m > 1$. These $m$ values are supposed to be i.i.d. $U(0,1)$, so one possibility is to compare their empirical distribution to the uniform distribution with a KS test, which yields significance levels $\delta^+$ and $\delta^-$. This is replicated $N$ times, and gives rise to two (correlated) tests, the first one using this $\delta^+$ as its statistic $T$, the second one using $\delta^-$. At the outer level, the empirical distribution of the $N$ values of $\delta^+$ are compared with the uniform distribution, and similarly for $\delta^-$. We call this the *m-nearest-pairs test*.

These two tests are based on asymptotic results in $n$, which means that the errors $\Delta^+$ and $\Delta^-$ should decrease with $n$, so the question arises of how large $n$ should be for the error to be negligible. For the nearest-pair test, our experiments showed that $\Delta^+$ is negligible (probably 0), but $\Delta^-$ is quite significant. For $p = 2$ and $n = 1000$, we obtained estimates $\hat{\Delta}^-$ equal to .02, .06, and .18 in dimensions 4, 6, and 9, respectively. In dimension 9, we still have $\Delta^- \approx .12$ for $n = 10000$. Similar results were obtained for $p = 1$ and $p = \infty$. This can be explained by the boundary effects: To prove Proposition 1, one assumes that for each $i$ and "small" $r$, the number of points $U_j, j \neq i$, in the ball of radius $r$ centered at $U_i$ is roughly Poisson with mean $(n-1)$ times the volume of that ball. But for points at distance less than $r$ from the boundary, this is not true. As a result, $W_n^*$ is stochastically larger than a $U(0,1)$. The fraction of the hypercube volume at distance more than $r$ from the boundary is $(1 - 2r)^k$, which decreases to zero exponentially with $k$.

To eliminate the boundary effects, one can transform the hypercube $[0,1)^k$ into a torus, by identifying the opposite faces. In the torus, points facing each other on opposite sides or in opposite corners become close to each other. We found that in the torus, for $n = 1000$, the approximation error becomes practically negligible (less than 0.01) at least up to dimension 12. We obtained similar results for the $m$-nearest-pairs test.

# 5   Tests based on entropy

Let us write the binary expansion of each $U_i$ as

$$U_i = .b_{i1} b_{i2} b_{i3}....$$

Extract the following bit sequence:

$$b_{1,r+1}, \ldots, b_{1,r+s}, b_{2,r+1}, \ldots, b_{2,r+s}, b_{3,r+1}, \ldots, b_{3,r+s}, \ldots, \qquad (1.1)$$

where $r \geq 0$ and $s > 0$ are fixed integers. Extract from this sequence $n$ adjacent and disjoint *blocks* of $L$ bits each, thus constructing $n$ integers $X_1, \ldots, X_n$. Under $H_0$, the $X_i$ are i.i.d. uniform over $\{0, 1, \ldots, C-1\}$, where $C = 2^L$. For each $x$, let $N_x$ be the number of $X_i$ that are equal to $x$, and compute the *discrete empirical entropy*

$$\hat{H}_d(C, n) = - \sum_{x=0}^{C-1} (N_x/n) \log_2(N_x/n).$$

This is our statistic $T$. Repeat this $N$ times, let $T_1, \ldots, T_N$ be the $N$ values of $\hat{H}_d(C, n)$, and define the normalized values

$$S_i = \frac{T_i - E[T_i]}{\sqrt{\mathrm{Var}[T_i]}}.$$

Under $H_0$, we can obtain expressions for the exact mean and variance of $\hat{H}_d(C, n)$, and we also know that $S_i$ converges to a $N(0, 1)$ as $n \to \infty$ and $C \to \infty$ simultaneously, while $n/C$ converges to some positive constant (see [LCC96]). This justifies the *discrete entropy distribution test*, which compares the empirical distribution of the $S_i$ to the $N(0, 1)$. Of course, the quality of the normal approximation is an issue. As for the nearest-pair test, we have estimated the approximation errors $\Delta^+$ and $\Delta^-$, and obtained (for example) values smaller than 0.01 for $n = 2^L \geq 256$.

One can also test for the independence of the entropy across successive blocks. The *discrete entropy correlation test* computes the sample correlation $\hat{\rho}_N$ between the pairs $(S_i, S_{i+1})$ and tests for significance. Here, we would rather take $n$ small and $N$ very large, in which case $\sqrt{N}\hat{\rho}_N$ is approximately $N(0, 1)$ under $H_0$.

To squeeze out more information from the sequence, one may also construct the blocks of bits with overlap. Relabel the first $n$ bits of the sequence (1.1) as $b_1, b_2, \ldots, b_n$, and put them on a circle, thus defining $b_0 = b_n$ and $b_j = b_{j-n}$ for $j > n$. Let $X_i$ be the integer whose binary representation is $b_i \cdots b_{i+L-1}$, and define $C$, $N_x$, $\hat{H}_d(C, n)$, $T_1, \ldots, T_N$, and $S_1, \ldots, S_N$ as before. The exact mean and variance of the $T_i$ can be computed by brute force in time $O(2^n)$. The *overlapping average entropy test* computes the statistic

$$\frac{1}{\sqrt{N}} \sum_{i=1}^{N} S_i,$$

which is approximately $N(0, 1)$ for large $N$, under $H_0$. The *overlapping entropy correlation test* computes the sample correlation between the pairs $(S_i, S_{i+1})$ and tests for significance.

TABLE 1.2. List of selected generators.

| | |
|---|---|
| G1. | LCG with $m = 2^{31} - 1$ and $a = 742938285$ (see [FM86]). |
| G2. | LCG with $m = 2^{31} - 1$ and $a = 630360016$ (see [LK91]). |
| G3. | LCG with $m = 2^{31} - 1$ and $a = 16807$ (see [BFS87, LK91]). |
| G4. | LCG with $m = 2^{32}$, $a = 69069$, and $c = 1$ (see [LK91, Rip90]). |
| G5. | LCG with $m = 2^{31}$ and $a = 65539$ (RANDU, see [LK91]). |
| G6. | LCG with $m = 2^{31}$ and $a = 452807053$ (see [DvdMST95]). |
| G7. | LCG with $m = 2^{31}$, $a = 1103515245$, $c = 12345$ (see [Pla92]). |
| G8. | Implicit inv. with $m = 2^{31} - 1$, $a_1 = a_2 = 1$ (see [Eic92]). |
| G9. | Explicit inv. with $m = 2^{31} - 1$, $a = b = 1$ (see [Eic92, Hel95]). |
| G10. | Implicit inv. with $m = 2^{32}$, $a = b = 1$, $z_0 = 5$ (see [EG92]). |
| G11. | Explicit inv. of [EI94] with $m = 2^{32}$, $a = 6$, $b = 1$. |
| G12. | Modified explicit inv. of [Eic96] with $m = 2^{32}$, $a = 6$, $b = 1$. |
| G13. | Combined LCG in Fig. 3 of [L'E88]. |
| G14. | Combined MRG in Fig. 1 of [L'E96a]. |

# 6  Some test results

We selected an arbitrary set of generators previously proposed or widely used, and applied to them some instances of the tests described in the previous sections. The aim here is not to test all generators nor recommend specific ones, but just to illustrate the fact that many fail these tests. The generators are listed in Table 1.2.

G1 to G7 are linear congruential generators (LCG), based on the recurrence $x_i = (ax_{i-1} + c) \bmod m$, with output $u_i = x_i/m$, and well-known parameters. The next five are inversive generators modulo $m$, with output $u_i = z_i/m$, and where $z_i$ obeys a nonlinear recurrence. G8 is an *implicit* inversive generator, with recurrence $z_i = (a_1 + a_2 z_{i-1}^{-1}) \bmod m$. G9 is an *explicit* inversive generator of the form $x_i = (ai + b) \bmod m$, $z_i = x_i^{-1} \bmod m = x_i^{m-2} \bmod m$. G10 is an implicit inversive generator based on the recurrence $z_i = T(z_{i-1})$ where $T(2^\ell z) = (a_1 + 2^\ell a_2 z^{-1}) \bmod 2^{32}$ for odd $z$. G11 and G12 are explicit inversive generators, the latter with the recurrence $z_i = i(ai + c)^{-1} \bmod 2^{32}$. G13 and G14 are two combined linear generators, with period lengths of approximately $2^{61}$ and $2^{185}$, respectively.

We applied to these generators the nearest-pair test with the following parameters: $p = 2$ (Euclidean norm), $N = 20$ for $n = 10000, 100000$ and $k = 2, 4, 6, 9, 12$; then $N = 10$ with $n = 500000$ and $k = 2, 4, 6, 9$. We also applied the $m$-nearest-pairs test with the same parameters and $m = 10$. We just summarize the results here. For more details and more refined versions of these tests, see [LCS96].

Because of their well-known regular lattice structure [Knu81, L'E94, Rip87], we expected the LCGs G1 to G7 to fail the nearest-pair tests and they all did, spectacularly. Most of them passed the tests with $n = 10000$,

TABLE 1.3. Parameters for entropy distribution tests

| $n$ | $L$ | $r$ | $s$ |
|-----|-----|-----|-----|
| $2^{12}$ | 12 | 0 | 12 |
| $2^{12}$ | 12 | 0 | 4 |
| $2^{12}$ | 12 | 20 | 4 |
| $2^{16}$ | 8 | 0 | 8 |
| $2^{16}$ | 8 | 0 | 4 |
| $2^{16}$ | 8 | 20 | 4 |
| $2^{16}$ | 16 | 0 | 16 |
| $2^{16}$ | 16 | 0 | 4 |
| $2^{16}$ | 16 | 20 | 4 |

TABLE 1.4. Parameters for the entropy tests with overlapping

| $n$ | $r$ | $s$ |
|-----|-----|-----|
| $10^4$ | 0 | 30 |
| $10^5$ | 0 | 30 |
| $10^6$ | 0 | 30 |
| $10^7$ | 0 | 30 |
| $10^4$ | 20 | 3 |
| $10^5$ | 20 | 3 |
| $10^6$ | 20 | 3 |
| $10^7$ | 20 | 3 |

but for larger $n$, they all failed, in most cases with significance levels less than $10^{-10}$. When the points are very evenly distributed over the entire period of a generator, then (as a general rule) the period should be several orders of magnitude larger than the number of values used. This is not the case with these LCGs: Their period length is too small. The inversive generators, on the other hand, do not have their points so regularly spaced; They look more randomly scattered even over the entire period length. They all did well in the tests (all passed). The two combined generators G13 and G14 also passed.

For the entropy tests, we selected the parameters of Table 1.3 with $N = 1000$ for the non-overlapping discrete entropy distribution test, while for the overlapping average and correlation tests, we used the parameters in Table 1.4 with $n = 30$ and $L = 5$.

The results can be summarized as follows. All the generators with power-of-two modulus, namely G4, G5, G6, G7, G10, G11, G12, failed the entropy distribution tests with $r = 20$, and also failed most of the overlapping average entropy with $r = 20$ and many entropy correlation tests, at significance level usually less than $10^{-10}$. Those tests with $r = 20$ look at less significant bits of the generators, and it is thus no surprise that those with power-of-two modulus (linear or inversive) fail, because their least significant bits have short period length. G5 and G6 also fail most of the tests based on the most significant bits (with $r = 0$). Other suspicious values with $r = 0$ were observed for the entropy distribution test for G1 and G10 (significance levels near 0.001), and for the overlapping average entropy tests for G4 and G9. For the overlapping entropy correlation tests with the first four parameter sets ($r = 0$), all generators with power-of-two modulus failed some of those tests. G9 also failed the first two overlapping entropy correlation tests with significance levels less than $10^{-10}$. The explanation is that the first $n$ values produced by G9 are the inverses (modulo $2^{31} - 1$) of the first $n$ positive integers, and with this modulus, the inverses of small integers tend to have low entropy in their high-order bits. In fact, we tried changing the parameter $b$ of G9 to some arbitrary values larger than $10^5$, and it passed the tests in all cases. Further discussion and results of entropy tests can be found in [LCC96].

## 7 Conclusion

Random number generators should be designed based on sound theoretical analysis, and then be tested empirically. The author believes that generators which fail simple tests such as the ones presented in this paper should be discarded. Only generators G8, G13, and G14 passed all the tests reported in this paper. Of course, many other generators not considered here would also pass.

*Acknowledgments:* This work has been supported by NSERC-Canada grants # OGP0110050 and # SMF0169893, and FCAR-Québec grant # 93-ER-1654. I wish to thank Peter Hellekalek and Harald Niederreiter for the organization of this wonderful MC& QMC'96 conference in Salzburg. I also thank At Compagner, Jean-François Cordeau, and Richard Simard, who contributed to the ideas and implementations of the tests described in this paper.

## 8 References

[BBS86]    L. Blum, M. Blum, and M. Schub. A simple unpredictable pseudo-random number generator. *SIAM Journal on Computing*, 15(2):364–383, 1986.

[BFS87]    P. Bratley, B. L. Fox, and L. E. Schrage. *A Guide to Simulation*. Springer-Verlag, New York, second edition, 1987.

136

[Com91]    A. Compagner. The hierarchy of correlations in random binary sequences. *Journal of Statistical Physics*, 63:883–896, 1991.

[Com95]    A. Compagner. Operational conditions for random number generation. *Physical Review E*, 52(5-B):5634–5645, 1995.

[Dur73]    J. Durbin. *Distribution Theory for Tests Based on the Sample Distribution Function*, volume 9 of *SIAM CBMS-NSF Regional Conference Series in Applied Mathematics*. SIAM, Philadelphia, 1973.

[DvdMST95] E. J. Dudewicz, E. C. van der Meulen, M. G. SriRam, and N. K. W. Teoh. Entropy-based random number evaluation. *American Journal of Mathematical and Management Sciences*, 15:115–153, 1995.

[EG92]    J. Eichenauer-Herrmann and H. Grothe. A new inversive congruential pseudorandom number generator with power of two modulus. *ACM Transactions on Modeling and Computer Simulation*, 2(1):1–11, 1992.

[EI94]    J. Eichenauer-Herrmann and K. Ickstadt. Explicit inversive congruential pseudorandom numbers with power of two modulus. *Mathematics of Computation*, 62(206):787–797, 1994.

[Eic92]    J. Eichenauer-Herrmann. Inversive congruential pseudorandom numbers: A tutorial. *International Statistical Reviews*, 60:167–176, 1992.

[Eic96]    J. Eichenauer-Herrmann. Modified explicit inversive congruential pseudorandom numbers with power-of-two modulus. *Statistics and Computing*, 6:31–36, 1996.

[FM86]    G. S. Fishman and L. S. Moore III. An exhaustive analysis of multiplicative congruential random number generators with modulus $2^{31} - 1$. *SIAM Journal on Scientific and Statistical Computing*, 7(1):24–45, 1986.

[Hel95]    P. Hellekalek. Inversive pseudorandom number generators: Concepts, results, and links. In C. Alexopoulos, K. Kang, W. R. Lilegdon, and D. Goldsman, editors, *Proceedings of the 1995 Winter Simulation Conference*, pages 255–262. IEEE Press, 1995.

[Knu81]    D. E. Knuth. *The Art of Computer Programming, Volume 2: Seminumerical Algorithms*. Addison-Wesley, Reading, Mass., second edition, 1981.

[Lag93]    J. C. Lagarias. Pseudorandom numbers. *Statistical Science*, 8(1):31–39, 1993.

[LCC96]    P. L'Ecuyer, A. Compagner, and J.-F. Cordeau. Entropy-based tests for random number generators. Submitted. Also GERAD technical report number G-96-41, 1996.

[LCS96]    P. L'Ecuyer, J.-F. Cordeau, and R. Simard. Close-neighbor tests for random number generators. In preparation, 1996.

[L'E88]    P. L'Ecuyer. Efficient and portable combined random number generators. *Communications of the ACM*, 31(6):742–749 and 774, 1988. See also the correspondence in the same journal, 32, 8 (1989) 1019–1024.

[L'E92]    P. L'Ecuyer. Testing random number generators. In *Proceedings of the 1992 Winter Simulation Conference*, pages 305–313. IEEE Press, Dec 1992.

[L'E94]    P. L'Ecuyer. Uniform random number generation. *Annals of Operations Research*, 53:77–120, 1994.

[L'E96a]    P. L'Ecuyer. Combined multiple recursive generators. *Operations Research*, 44(5):816–822, 1996.

[L'E96b]    P. L'Ecuyer. TestU01: Un logiciel pour appliquer des tests statistiques à des générateurs de valeurs aléatoires. In preparation, 1996.

[L'E97]    P. L'Ecuyer. Random number generation. In Jerry Banks, editor, *Handbook on Simulation*. Wiley, 1997. To appear. Also GERAD technical report number G-96-38.

[LK91]    A. M. Law and W. D. Kelton. *Simulation Modeling and Analysis*. McGraw-Hill, New York, second edition, 1991.

[LP89]    P. L'Ecuyer and R. Proulx. About polynomial-time "unpredictable" generators. In *Proceedings of the 1989 Winter Simulation Conference*, pages 467–476. IEEE Press, Dec 1989.

[LW97]    H. Leeb and S. Wegenkittl. Inversive and linear congruential pseudorandom number generators in selected empirical tests. *ACM Transactions on Modeling and Computer Simulation*, 1997. To appear.

[Mar85]    G. Marsaglia. A current view of random number generators. In *Computer Science and Statistics, Sixteenth Symposium on the Interface*, pages 3–10, North-Holland, Amsterdam, 1985. Elsevier Science Publishers.

138

[Mar96]    G. Marsaglia. Diehard: A battery of tests of randomness.
           http://stat.fsu.edu/~geo/diehard.html, 1996.

[Nie92]    H. Niederreiter. *Random Number Generation and Quasi-
           Monte Carlo Methods*, volume 63 of *SIAM CBMS-NSF Re-
           gional Conference Series in Applied Mathematics*. SIAM,
           Philadelphia, 1992.

[Pla92]    P. J. Plauger. *The Standard C Library*. Prentice Hall, En-
           glewood Cliffs, New Jersey, 1992.

[Rip87]    B. D. Ripley. *Stochastic Simulation*. Wiley, New York, 1987.

[Rip90]    B. D. Ripley. Thoughts on pseudorandom number genera-
           tors. *Journal of Computational and Applied Mathematics*,
           31:153–163, 1990.

[RS78]     B. D. Ripley and B. W. Silverman. Quick tests for spatial
           interaction. *Biometrika*, 65(3):641–642, 1978.

[SB78]     B. Silverman and T. Brown. Short distances, flat triangles
           and Poisson limits. *Journal of Applied Probability*, 15:815–
           825, 1978.

[Ste86a]   M. S. Stephens. Tests based on EDF statistics. In R. B.
           D'Agostino and M. S. Stephens, editors, *Goodness-of-Fit
           Techniques*. Marcel Dekker, New York and Basel, 1986.

[Ste86b]   M. S. Stephens. Tests for the uniform distribution. In
           R. B. D'Agostino and M. S. Stephens, editors, *Goodness-
           of-Fit Techniques*, pages 331–366. Marcel Dekker, New York
           and Basel, 1986.

Pierre L'Ecuyer
Département d'informatique et de recherche opérationnelle,
Université de Montréal,
C.P. 6128, Succ. Centre-Ville, Montréal, H3C 3J7, Canada
http://www.iro.umontreal.ca/~lecuyer

# The Algebraic-Geometry Approach to Low-Discrepancy Sequences

Harald Niederreiter and Chaoping Xing

*In Saloniki kenn ich einen, der mich liest,*
*und auch in Bad Nauheim – das sind schon zwei.*
Günter Eich, *Zuversicht*

ABSTRACT We give a survey of recent work of the authors in which low-discrepancy sequences are constructed by new methods based on algebraic geometry. The most powerful of these methods employ algebraic curves over finite fields with many rational points. These methods yield significant improvements on all earlier constructions. In fact, we obtain $(t, s)$-sequences in an arbitrary base $b$ where for fixed $b$ the quality parameter $t$ has the least possible order of magnitude when considered as a function of the dimension $s$, namely $t$ grows linearly in $s$.

## 1. Introduction

Low-discrepancy sequences are the basic tools of quasi-Monte Carlo methods. The aim in the construction of low-discrepancy sequences is to obtain sequences of points in the $s$-dimensional unit cube $I^s = [0, 1]^s$ with as small a star discrepancy as possible. This is motivated, for instance, by the fact that the standard error bound in quasi-Monte Carlo integration, the Koksma-Hlawka bound, is proportional to the star discrepancy of the points used in the integration rule. Thus, points with smaller star discrepancy guarantee smaller error bounds. Very demanding applications of quasi-Monte Carlo methods, such as the recent ones in mathematical finance (see e.g. Ninomiya and Tezuka [19], Paskov [20], and Paskov and Traub [21]), lead to the challenge of finding better low-discrepancy sequences in high-dimensional unit cubes than off-the-shelf sequences. The gist of the work to be presented here is that new methods using algebraic geometry (more precisely, algebraic curves over finite fields) yield by far the best low-discrepancy sequences that are currently available.

The most powerful known methods for the construction of low-discrepancy sequences are based on the theory of $(t, s)$-sequences, which are sequences satisfying strong uniformity properties with respect to their distribution in $I^s$. The quality parameter $t$ measures these uniformity properties and should be as small as possible. The methods using algebraic geometry yield digital $(t, s)$-sequences in a prime-power base $q$, but together with standard devices (see [15, Section 5]) one can then derive $(t, s)$-sequences in an arbitrary base $b \geq 2$. One of the main results is that for $(t, s)$-sequences in a fixed base $b$, the values of $t$ (which depend on the dimension $s$) attain the least possible order of magnitude (namely linear growth in $s$) when we employ certain constructions using algebraic geometry.

Before we turn to a more concrete discussion of the topics above, we recall some basic notations and concepts. For a subinterval $J$ of $I^s$ and for a point set $P$ consisting of $N$ points $\mathbf{x}_0, \mathbf{x}_1, \ldots, \mathbf{x}_{N-1} \in I^s$ we write $A(J; P)$ for the number of integers $n$ with $0 \leq n \leq N - 1$ for which $\mathbf{x}_n \in J$, and we put

$$R(J; P) = \frac{A(J; P)}{N} - \mathrm{Vol}(J).$$

Then the *star discrepancy* $D_N^*(P)$ of $P$ is defined by

$$D_N^*(P) = \sup_J |R(J; P)|,$$

where the supremum is extended over all subintervals $J$ of $I^s$ with one vertex at the origin. For a sequence $S$ of points in $I^s$, the star discrepancy $D_N^*(S)$ is meant to be the star discrepancy of the first $N$ terms of $S$. We call $S$ a *low-discrepancy sequence* if

$$D_N^*(S) = O(N^{-1}(\log N)^s) \quad \text{for all } N \geq 2.$$

The first construction of low-discrepancy sequences for all dimensions $s$ was given by Halton [4], and for the $s$-dimensional Halton sequence $S$ we have

$$D_N^*(S) \leq A_s N^{-1}(\log N)^s + O(N^{-1}(\log N)^{s-1}) \quad \text{for all } N \geq 2,$$

with a coefficient $A_s$ tending to $\infty$ at a superexponential rate as $s \to \infty$ (see [10, Theorem 3.6]).

With regard to the theory of $(t, s)$-sequences we follow [10, Chapter 4], with the slight extensions described in [15, Section 2] and [16, Section 2].

**Definition 1.** For integers $b \geq 2$ and $0 \leq t \leq m$, a $(t, m, s)$-*net in base $b$* is a point set $P$ consisting of $b^m$ points in $I^s$ such that $R(J; P) = 0$ for every subinterval $J$ of $I^s$ of the form

$$J = \prod_{i=1}^{s} [a_i b^{-d_i}, (a_i + 1)b^{-d_i})$$

with integers $d_i \geq 0$ and $0 \leq a_i < b^{d_i}$ for $1 \leq i \leq s$ and of volume $\mathrm{Vol}(J) = b^{t-m}$.

For a base $b \geq 2$ we write $Z_b = \{0, 1, \ldots, b-1\}$ for the set of digits in base $b$. Given a real number $x \in [0, 1]$, let

$$x = \sum_{j=1}^{\infty} y_j b^{-j} \qquad \text{with all } y_j \in Z_b \tag{1}$$

be a $b$-adic expansion of $x$, where the case $y_j = b - 1$ for all but finitely many $j$ is allowed. For an integer $m \geq 1$ we define the truncation

$$[x]_{b,m} = \sum_{j=1}^{m} y_j b^{-j}.$$

It should be emphasized that this truncation operates on the *expansion* of $x$ and not on $x$ itself, since it may yield different results depending on which $b$-adic expansion of $x$ is used. If $\mathbf{x} = \left(x^{(1)}, \ldots, x^{(s)}\right) \in I^s$ and the $x^{(i)}, 1 \leq i \leq s$, are given by prescribed $b$-adic expansions, then we define

$$[\mathbf{x}]_{b,m} = \left(\left[x^{(1)}\right]_{b,m}, \ldots, \left[x^{(s)}\right]_{b,m}\right).$$

**Definition 2.** Let $b \geq 2$ and $t \geq 0$ be integers. A sequence $\mathbf{x}_0, \mathbf{x}_1, \ldots$ of points in $I^s$ is a $(t, s)$-*sequence in base* $b$ if for all integers $k \geq 0$ and $m > t$ the points $[\mathbf{x}_n]_{b,m}$ with $kb^m \leq n < (k+1)b^m$ form a $(t, m, s)$-net in base $b$. Here the coordinates of all points of the sequence are given by prescribed $b$-adic expansions of the form (1).

Detailed information on bounds for the star discrepancy of $(t, s)$-sequences is provided in [10, Chapter 4]. We mention here only the following simplified bound: for any $(t, s)$-sequence $S$ in base $b$ we have

$$D_N^*(S) \leq C_b(s, t)N^{-1}(\log N)^s + O\left(N^{-1}(\log N)^{s-1}\right) \qquad \text{for all } N \geq 2, \tag{2}$$

where the implied constant in the Landau symbol depends only on $b, s$, and $t$. Here

$$C_b(s, t) = \frac{b^t}{s}\left(\frac{b-1}{2\log b}\right)^s .$$

if either $s = 2$ or $b = 2, s = 3, 4$; otherwise

$$C_b(s, t) = \frac{b^t}{s!} \cdot \frac{b-1}{2\lfloor b/2 \rfloor}\left(\frac{\lfloor b/2 \rfloor}{\log b}\right)^s . \tag{3}$$

It is clear from the discrepancy bound (2) that small values of $t$ are preferable if one wants to obtain good low-discrepancy sequences. Note also that the value of $t$ has a concrete significance for the distribution of the points of the sequence in view of Definitions 1 and 2, with smaller values of $t$ indicating a stronger uniformity. Thus, the aim in the construction of $(t, s)$-sequences in base $b$ is to make the value of the quality parameter $t$ as small as possible for given $b$ and $s$.

We briefly review earlier constructions of $(t, s)$-sequences. Sobol' [23] constructed $(t, s)$-sequences in base 2 for any dimension $s$. The value of $t$ depends on $s$ and is of the order of magnitude $s \log s$. The coefficient $A_s$ of the leading term $N^{-1}(\log N)^s$ in the discrepancy bound is then smaller than for the Halton sequence for sufficiently large $s$, but $A_s$ is still tending to $\infty$ at a superexponential rate as $s \to \infty$. Faure [2] obtained $(0, s)$-sequences in any prime base $b \geq s$, and if $b$ is chosen to be minimal, then we get $\lim_{s \to \infty} A_s = 0$. This construction was generalized by Niederreiter [8] who obtained $(0, s)$-sequences in any prime-power base $b \geq s$. The first construction yielding $(t, s)$-sequences for all dimensions $s$ and all bases $b$ is that of Niederreiter [9]. This construction includes all earlier constructions as special cases, and if $b$ is fixed, then $t$ is of the order of magnitude $s \log s$. A different approach to this construction which yields the same values of the quality parameter $t$ was discussed by Tezuka [24] and Tezuka and Tokuyama [26] (see also Tezuka [25, Chapter 6]).

The idea of using algebraic curves over finite fields (or, equivalently, global function fields) for the construction of $(t, s)$-sequences was first sketched in Niederreiter [11], [12]. In a detailed and improved form, this idea was presented in Niederreiter and Xing [13]. Further constructions of $(t, s)$-sequences using algebraic curves over finite fields were given in Niederreiter and Xing [15], [16] and Xing and Niederreiter [27]. To illustrate the progress achieved over the years, we list in Table 1 the values of $t$ for $(t, s)$-sequences in base 2 with $1 \leq s \leq 20$ achieved by four typical constructions: the construction of Sobol' [23] labeled "Sobol' 1967", that of Niederreiter [9] labeled "Nied 1988", that of Niederreiter and Xing [13] labeled "NX1 1995", and that of Niederreiter and Xing [15] labeled "NX2 1996". Note that in this table the values of $t$ from [9] improve on those from [23] for $s \geq 8$, the values from [13] improve on those from the previous constructions for $s \geq 16$, and the values from [15] substantially improve on those from all previous constructions for $s \geq 4$.

In Section 2 we recall the digital method, which is the standard construction principle for $(t, s)$-sequences, and in Section 3 we review the relevant background on algebraic curves over finite fields and global function fields. Four constructions of $(t, s)$-sequences using algebraic geometry are described, in the chronological order in which they were developed, in Sections 4 to 7. The final Section 8 contains a discussion of the implications of these constructions for the theory of low-discrepancy sequences, a table of current $t$-values, and information on the algebraic curves over finite fields that are required for the constructions.

Table 1

Values of $t$ in base 2 for $1 \leq s \leq 20$

| $s$ | Sobol' 1967 | Nied 1988 | NX1 1995 | NX2 1996 |
|---|---|---|---|---|
| 1 | 0 | 0 | 3 | 0 |
| 2 | 0 | 0 | 4 | 0 |
| 3 | 1 | 1 | 6 | 1 |
| 4 | 3 | 3 | 8 | 1 |
| 5 | 5 | 5 | 10 | 2 |
| 6 | 8 | 8 | 12 | 3 |
| 7 | 11 | 11 | 15 | 4 |
| 8 | 15 | 14 | 18 | 5 |
| 9 | 19 | 18 | 21 | 6 |
| 10 | 23 | 22 | 24 | 8 |
| 11 | 27 | 26 | 27 | 9 |
| 12 | 31 | 30 | 31 | 10 |
| 13 | 35 | 34 | 35 | 11 |
| 14 | 40 | 38 | 39 | 13 |
| 15 | 45 | 43 | 43 | 15 |
| 16 | 50 | 48 | 47 | 15 |
| 17 | 55 | 53 | 51 | 18 |
| 18 | 60 | 58 | 55 | 19 |
| 19 | 65 | 63 | 59 | 19 |
| 20 | 71 | 68 | 64 | 21 |

## 2. The Digital Method for the Construction of $(t, s)$-Sequences

The standard procedure for the construction of $(t, s)$-sequences uses the digital method introduced in [8, Section 6]; compare also with [10, Section 4.3]. To obtain $(t, s)$-sequences in a base $b \geq 2$, the digital method works with a commutative ring with identity and of finite order $b$. We concentrate here on the special case where this ring is a finite field $\mathbf{F}_q$ of prime-power order $q$. Information on the general case can be found in [10, Chapter 4] and in the recent paper of Larcher et al. [5]. The results of the latter paper suggest that the digital method is most powerful in the case where the underlying ring is chosen to be a finite field.

As before we write $Z_q = \{0, 1, \ldots, q-1\}$ for the set of digits in base $q$. We fix a dimension $s \geq 1$ and choose the following:

(S1) bijections $\psi_r : Z_q \to \mathbf{F}_q$ for $r \geq 0$ with $\psi_r(0) = 0$ for all sufficiently large $r$;

(S2) bijections $\eta_j^i : \mathbf{F}_q \to Z_q$ for $1 \leq i \leq s$ and $j \geq 1$;

(S3) elements $c_{j,r}^{(i)} \in \mathbf{F}_q$ for $1 \leq i \leq s$, $j \geq 1$, and $r \geq 0$.

For $n = 0, 1, \ldots$ let

$$n = \sum_{r=0}^{\infty} a_r(n) q^r$$

be the digit expansion of $n$ in base $q$, where $a_r(n) \in Z_q$ for $r \geq 0$ and $a_r(n) = 0$ for all sufficiently large $r$. We put

$$x_n^{(i)} = \sum_{j=1}^{\infty} y_{n,j}^{(i)} q^{-j} \quad \text{for } n \geq 0 \text{ and } 1 \leq i \leq s, \tag{4}$$

with

$$y_{n,j}^{(i)} = \eta_j^{(i)} \left( \sum_{r=0}^{\infty} c_{j,r}^{(i)} \psi_r(a_r(n)) \right) \in Z_q \quad \text{for } n \geq 0, 1 \leq i \leq s, \text{ and } j \geq 1.$$

Note that the sum over $r$ is always a finite sum. Now we define the sequence

$$\mathbf{x}_n = \left( x_n^{(1)}, \ldots, x_n^{(s)} \right) \in I^s \quad \text{for } n = 0, 1, \ldots. \tag{5}$$

**Definition 3.** If the sequence in (5) is a $(t, s)$-sequence in base $q$, then it is called a *digital $(t, s)$-sequence constructed over* $\mathbf{F}_q$. Here the truncations are required to operate on the expansions in (4).

The determination of the quality parameter $t$ for sequences constructed by the digital method is based on the following general notion. We consider a rectangular array $A$ of vectors from an arbitrary vector space $V$:

$$
\begin{array}{cccc}
\mathbf{a}_1^{(1)} & \mathbf{a}_2^{(1)} & \ldots & \mathbf{a}_m^{(1)} \\
\mathbf{a}_1^{(2)} & \mathbf{a}_2^{(2)} & \ldots & \mathbf{a}_m^{(2)} \\
\vdots & \vdots & & \vdots \\
\mathbf{a}_1^{(s)} & \mathbf{a}_2^{(s)} & \ldots & \mathbf{a}_m^{(s)}
\end{array}
$$

Such an $s \times m$ array is abbreviated as a two-parameter system $A = \{\mathbf{a}_j^{(i)} \in V : 1 \leq i \leq s, 1 \leq j \leq m\}$.

**Definition 4.** For $A = \{\mathbf{a}_j^{(i)} \in V : 1 \leq i \leq s, 1 \leq j \leq m\}$ we define $\varrho(A)$ to be the largest integer $d$ such that any system $\{\mathbf{a}_j^{(i)} : 1 \leq j \leq d_i, 1 \leq i \leq s\}$ with $0 \leq d_i \leq m$ for $1 \leq i \leq s$ and $\sum_{i=1}^{s} d_i = d$ is linearly independent in $V$ (here the empty system is viewed as linearly independent).

Now we consider the sequence in (5) obtained by the digital method with the elements $c_{j,r}^{(i)} \in \mathbf{F}_q$ in (S3). If $\mathbf{F}_q^{\infty}$ is the sequence space over $\mathbf{F}_q$, then we use these elements to set up the sequences

$$\mathbf{c}_j^{(i)} = \left( c_{j,0}^{(i)}, c_{j,1}^{(i)}, \ldots \right) \in \mathbf{F}_q^{\infty} \quad \text{for } 1 \leq i \leq s \text{ and } j \geq 1,$$

which are collected into the two-parameter system

$$C^{(\infty)} = \left\{ c_j^{(i)} \in \mathbf{F}_q^{\infty} : 1 \leq i \leq s \text{ and } j \geq 1 \right\}.$$

For $m \geq 1$ we define the projection

$$\pi_m : (c_0, c_1, \ldots) \in \mathbf{F}_q^{\infty} \mapsto (c_0, \ldots, c_{m-1}) \in \mathbf{F}_q^m,$$

and we put

$$C^{(m)} = \left\{ \pi_m(c_j^{(i)}) \in \mathbf{F}_q^m : 1 \leq i \leq s, 1 \leq j \leq m \right\}.$$

Finally, we set

$$\tau(C^{(\infty)}) = \sup_{m \geq 1} \left( m - \varrho(C^{(m)}) \right). \tag{6}$$

Then the following result was noted in [15, Lemma 2].

**Lemma 1.** *If the elements $c_{j,r}^{(i)} \in \mathbf{F}_q$ in (S3) are such that $\tau(C^{(\infty)}) < \infty$, then the sequence in (5) is a digital $(t, s)$-sequence constructed over $\mathbf{F}_q$ with $t = \tau(C^{(\infty)})$.*

The crucial issue in the application of the digital method is how to obtain elements $c_{j,r}^{(i)} \in \mathbf{F}_q$ for which $\tau(C^{(\infty)})$ is small. This will be the main point in the constructions to be described in Sections 4 to 7.

We remark that a construction of a $(t, s)$-sequence automatically yields infinitely many nets, in view of the following general principle: if there exists a $(t, s)$-sequence in base $b$, then for every integer $m \geq t$ there exists a $(t, m, s + 1)$-net in base $b$ (see [10, Lemma 4.22] and [15, Section 6]). There are constructions of nets, especially of a combinatorial nature, that are not based on this principle. For general surveys of constructions of nets we refer to Clayman *et al.* [1] and Mullen *et al.* [7].

## 3. Algebraic Curves over Finite Fields and Global Function Fields

Let $C$ denote a smooth, projective, absolutely irreducible algebraic curve defined over $\mathbf{F}_q$. Then $C$ has a finite number of $\mathbf{F}_q$-rational points, and for this number $N(C)$ we have the Weil-Serre bound

$$N(C) \leq q + 1 + g(C) \left\lfloor 2q^{1/2} \right\rfloor,$$

where $g(C)$ is the genus of $C$. In particular, the following definition makes sense.

**Definition 5.** For any prime power $q$ and any integer $g \geq 0$ let

$$N_q(g) = \max N(C),$$

where the maximum is extended over all curves $C$ with $g(C) = g$.

Information on $N_q(g)$ can be found in the survey article of Garcia and Stichtenoth [3] and also in [16, Section 9]. We mention here only the result of Serre [22] that

$$\overline{\lim}_{g\to\infty} \frac{N_q(g)}{g} > 0. \tag{7}$$

A unifying theme of the various constructions of digital $(t, s)$-sequences by means of algebraic curves is to take a curve $C$ with $N(C) \geq 1$ and to determine the elements $c_{j,r}^{(i)} \in F_q$ that are needed in (S3) in Section 2 from the coefficients in the expansions of certain points of $C$ in local coordinates at a fixed $F_q$-rational point of $C$. For the purpose of determining the $c_{j,r}^{(i)}$, it is often more convenient to work with the function field $K/F_q$ of $C$. The notation $K/F_q$ indicates that $F_q$ is the full constant field of the global function field $K$, i.e., that $F_q$ is algebraically closed in $K$. The genus $g(K/F_q)$ of $K/F_q$ is equal to $g(C)$ and the number $N(K/F_q)$ of rational places of $K/F_q$, i.e., of places of $K/F_q$ of degree 1, is equal to $N(C)$. It follows that the quantity $N_q(g)$ introduced in Definition 5 is also given by

$$N_q(g) = \max N(K/F_q),$$

where the maximum is extended over all global function fields $K/F_q$ with $g(K/F_q) = g$.

For a global function field $K/F_q$, we write $\nu_P$ for the normalized discrete valuation corresponding to the place $P$ of $K/F_q$. For any nonzero $k \in K$ let $Z(k)$, respectively $Q(k)$, be the set of zeros, respectively poles, of $k$. Then we define the *zero divisor* of $k$ by

$$(k)_0 = \sum_{P \in Z(k)} \nu_P(k)P,$$

the *pole divisor* of $k$ by

$$(k)_\infty = \sum_{P \in Q(k)} (-\nu_P(k))P,$$

and the *principal divisor* of $k$ by

$$(k) = (k)_0 - (k)_\infty.$$

For an arbitrary divisor $D$ of $K/F_q$, the $F_q$-vector space

$$\mathcal{L}(D) = \{k \in K\backslash\{0\} : (k) + D \geq 0\} \cup \{0\}$$

has a finite dimension which we denote by $l(D)$. We note also that if $P$ is a rational place of $K/F_q$ and $z$ is a local uniformizing parameter at $P$, then every $k \in K$ has an expansion

$$k = \sum_{r=v}^{\infty} a_r z^r,$$

where $\nu_P(k) \geq v$ and all $a_r \in \mathbf{F}_q$.

## 4. The First Construction

The first construction of digital $(t, s)$-sequences using algebraic curves over $\mathbf{F}_q$ is due to Niederreiter and Xing [13]. This construction can be viewed as a direct extension of the method of Niederreiter [9], where the latter method corresponds to the special case in which the algebraic curve is the projective line over $\mathbf{F}_q$ or, equivalently, the global function field is the rational function field over $\mathbf{F}_q$.

Let $K/\mathbf{F}_q$ be a global function field with $N(K/\mathbf{F}_q) \geq 1$ and let $P_\infty$ be a fixed rational place of $K/\mathbf{F}_q$. Let $n_1 < n_2 < \ldots$ be all pole numbers of $P_\infty$, i.e., the positive integers $n$ for which there exists a $k \in K$ with $(k)_\infty = nP_\infty$. By the Weierstrass gap theorem, there are exactly $g = g(K/\mathbf{F}_q)$ positive integers that are not pole numbers of $P_\infty$. Let $R$ be the ring of elements of $K$ that have no pole outside $P_\infty$. Given an integer $s \geq 1$, we choose $k_1, k_2, \ldots, k_s \in R$ satisfying the following two conditions:
(i) the zero sets $Z(k_1), Z(k_2), \ldots, Z(k_s)$ are pairwise disjoint;
(ii) $n_{e_i} - e_i < n_1$ for $1 \leq i \leq s$, where $e_i := -\nu_{P_\infty}(k_i) \geq 1$ for $1 \leq i \leq s$.

Since $n_r$ is a pole number of $P_\infty$, we can find $w_r \in R$ such that $(w_r)_\infty = n_r P_\infty$ for $r \geq 1$. Note that each $e_i, 1 \leq i \leq s$, is a pole number of $P_\infty$ since $(k_i)_\infty = e_i P_\infty$. Thus, for each $1 \leq i \leq s$ there exists a uniquely determined positive integer $f_i$ with $n_{f_i} = e_i$, and it is trivial that $f_i \leq e_i$. For each $1 \leq i \leq s$ we define the set

$$\{w_{i,0}, w_{i,1}, \ldots, w_{i,e_i-1}\} := \{1, w_1, w_2, \ldots, w_{e_i}\} \setminus \{w_{f_i}\}.$$

Furthermore, for $j \geq 1$ we write

$$j - 1 = Q(i,j)e_i + u(i,j)$$

with integers $Q(i,j)$ and $u(i,j)$, where $0 \leq u(i,j) < e_i$. It is easily seen that we always have

$$\nu_{P_\infty}\left(w_{i,u(i,j)}k_i^{-Q(i,j)-1}\right) \geq -g.$$

Hence we have the following expansions at $P_\infty$:

$$w_{i,u(i,j)}k_i^{-Q(i,j)-1} = z^{-g}\sum_{r=0}^{\infty} c_{j,r}^{(i)}z^r$$

with all $c_{j,r}^{(i)} \in \mathbf{F}_q$, where $z$ is a local uniformizing parameter at $P_\infty$.

These coefficients $c_{j,r}^{(i)}$ now serve as the elements in (S3) in the digital method for the construction of sequences described in Section 2. The bijections $\psi_r$ are chosen as in (S1) and the bijections $\eta_j^{(i)}$ are chosen as in (S2), but with the additional condition that $\eta_j^{(i)}(0) = 0$ for $1 \leq i \leq s$ and all sufficiently large $j$. For fixed $i$ it is clear that $Q(i,j) \to \infty$ as $j \to \infty$, and so for fixed $i$ and $r$ we have $c_{j,r}^{(i)} = 0$ for all sufficiently large $j$. If $\tau(C^{(\infty)})$ is as in (6), then the key result in [13] shows that

$$\tau(C^{(\infty)}) \le g + 1 + \sum_{i=1}^{s}(e_i - 1).$$

In view of Lemma 1, we are thus led to the following theorem.

**Theorem 1.** *The first construction yields a digital $(t,s)$-sequence constructed over $\mathbf{F}_q$ with*

$$t = g(K/\mathbf{F}_q) + 1 + \sum_{i=1}^{s}(e_i - 1),$$

*where $e_i = -\nu_{P_\infty}(k_i)$ for $1 \le i \le s$.*

An interesting special case of the first construction is discussed in [13] and [16, Section 5], namely the case where the global function field $K/\mathbf{F}_q$ has class number 1, or equivalently, where the Jacobian of the corresponding curve has only one $\mathbf{F}_q$-rational point. We assume again that $N(K/\mathbf{F}_q) \ge 1$ and let $P_\infty$ be a fixed rational place of $K/\mathbf{F}_q$. We list all places $\ne P_\infty$ of $K/\mathbf{F}_q$ by nondecreasing degrees in a sequence $P_1, P_2, \ldots$. Since $K/\mathbf{F}_q$ has class number 1, the divisor

$$P_i - (\deg(P_i))P_\infty$$

is principal for $i = 1, 2, \ldots$. Given $s \ge 1$, we choose $k_1, k_2, \ldots, k_s \in K$ such that

$$(k_i) = P_i - (\deg(P_i))P_\infty \quad \text{for } 1 \le i \le s.$$

Each $k_i$ is in $R$, and the condition (i) above is clearly satisfied. Furthermore, we have $e_i = -\nu_{P_\infty}(k_i) = \deg(P_i)$ for $1 \le i \le s$.

In [13] the following concrete example of a global function field $K/\mathbf{F}_q$ of genus 1 is presented: let $K = \mathbf{F}_2(x,y)$ with

$$y^2 + y = x^3 + x + 1.$$

The elliptic function field $K/\mathbf{F}_2$ has class number 1 and a unique rational place $P_\infty$. Furthermore, as for any elliptic function field, the pole numbers of $P_\infty$ are given by $n_r = r+1$ for all $r \ge 1$, and so the condition (ii) above is satisfied. It follows then from Theorem 1 that the first construction yields a digital $(E_2(s), s)$-sequence constructed over $\mathbf{F}_2$ with

$$E_2(s) = 2 + \sum_{i=1}^{s}(\deg(P_i) - 1).$$

The values of $E_2(s)$ for $1 \leq s \leq 20$ can be found in Table 1 in the column labeled "NX1 1995". It was noted in [16, Section 5] that $E_2(s)$ is of the order of magnitude $s \log s$. Thus, with regard to the order of magnitude of the $t$-values as a function of $s$, we do not get an improvement on the earlier constructions of Sobol' [23] and Niederreiter [9], but many individual values of $t = E_2(s)$ are smaller than in these previous constructions.

## 5. The Second Construction

The second construction of digital $(t, s)$-sequences using algebraic geometry was introduced by Niederreiter and Xing [15]. This construction yields substantial improvements on the construction in Niederreiter [9] and on the construction described in Section 4. The key idea of the second construction is to work with algebraic curves over $\mathbf{F}_q$ with many $\mathbf{F}_q$-rational points or, equivalently, with global function fields containing many rational places.

Let the dimension $s \geq 1$ and the prime power $q$ be given, and suppose that the global function field $K/\mathbf{F}_q$ satisfies $N(K/\mathbf{F}_q) \geq s + 1$. Let $P_\infty, P_1, P_2, \ldots, P_s$ be $s + 1$ distinct rational places of $K/\mathbf{F}_q$ and choose a positive divisor $D$ of $K/\mathbf{F}_q$ with $\deg(D) = g(K/\mathbf{F}_q)$ and $l(D) = 1$ (the existence of such a divisor $D$ will be discussed later). By the Riemann-Roch theorem it is easily seen that for $1 \leq i \leq s$ and $j \geq 1$ there exist elements $k_j^{(i)} \in \mathcal{L}(D + jP_i)$ with

$$\nu_{P_i}(k_j^{(i)}) = -\nu_{P_i}(D) - j.$$

Since $\nu_{P_\infty}(k_j^{(i)}) \geq -\nu_{P_\infty}(D) =: -v$, we have expansions at $P_\infty$ of the form

$$k_j^{(i)} = z^{-v} \sum_{r=0}^{\infty} b_{j,r}^{(i)} z^r \quad \text{for } 1 \leq i \leq s \text{ and } j \geq 1,$$

where $z$ is a local uniformizing parameter at $P_\infty$ and all $b_{j,r}^{(i)} \in \mathbf{F}_q$. For $1 \leq i \leq s$ and $j \geq 1$ we now define

$$c_{j,r}^{(i)} = \begin{cases} b_{j,r}^{(i)} & \text{for } 0 \leq r \leq v - 1, \\ \\ b_{j,r+1}^{(i)} & \text{for } r \geq v. \end{cases}$$

These $c_{j,r}^{(i)} \in \mathbf{F}_q$ are now used as the elements in (S3) in the digital method for the construction of sequences. The bijections $\psi_r$ and $\eta_j^{(i)}$ are chosen as in (S1) and (S2), respectively. If $\tau(C^{(\infty)})$ is as in (6), then it was shown in [15] that

$$\tau(C^{(\infty)}) \leq g(K/\mathbf{F}_q).$$

Taking into account Lemma 1, we arrive at the following result.

**Theorem 2.** *Under the conditions above, the second construction yields a digital $(t,s)$-sequence constructed over $\mathbf{F}_q$ with $t = g(K/\mathbf{F}_q)$.*

The existence of a positive divisor $D$ of $K/\mathbf{F}_q$ with $\deg(D) = g(K/\mathbf{F}_q)$ and $l(D) = 1$ is an interesting question in the theory of global function fields which has apparently not been considered before. In [15] the following result was proved by using properties of the zeta function of $K/\mathbf{F}_q$.

**Proposition 1.** *A positive divisor $D$ of $K/\mathbf{F}_q$ with $\deg(D) = g(K/\mathbf{F}_q)$ and $l(D) = 1$ exists if either* (i) $q = 2$ and $N(K/\mathbf{F}_q) \geq 4$; *or* (ii) $q \geq 3$ and $N(K/\mathbf{F}_q) \geq 2$.

It follows e.g. from the result in (7) that for every $q$ and $s$ there exists a genus $g$ with $N_q(g) \geq s+1$. Therefore, the following definition is meaningful.

**Definition 6.** For any prime power $q$ and any dimension $s \geq 1$ let

$$V_q(s) = \min\left\{g \geq 0 : N_q(g) \geq s + 1\right\}.$$

If $s \leq q$, then in the second construction we can choose $K/\mathbf{F}_q$ to be the rational function field over $\mathbf{F}_q$ and $D$ the zero divisor. By Theorem 2 we then get a digital $(t,s)$-sequence constructed over $\mathbf{F}_q$ with $t = 0 = V_q(s)$. If $s \geq q + 1$, then we choose $K/\mathbf{F}_q$ to be a global function field with $N(K/\mathbf{F}_q) \geq s + 1$ and $g(K/\mathbf{F}_q) = V_q(s)$. It follows that $N(K/\mathbf{F}_q) \geq 4$, and so by Proposition 1 there exists a positive divisor $D$ of $K/\mathbf{F}_q$ with $\deg(D) = g(K/\mathbf{F}_q)$ and $l(D) = 1$. Thus, the second construction can be applied and yields a digital $(t,s)$-sequence constructed over $\mathbf{F}_q$ with $t = g(K/\mathbf{F}_q) = V_q(s)$. We can summarize this as follows.

**Theorem 3.** *For every prime power $q$ and every dimension $s \geq 1$ there exists a digital $(V_q(s), s)$-sequence constructed over $\mathbf{F}_q$.*

The values of $V_2(s)$ for $1 \leq s \leq 20$ can be found in Table 1 in the column labeled "NX2 1996". A further discussion of values of $V_q(s)$ will be presented in Section 8. As a general result, it was shown in [15, Theorem 4] by using the class field theory of global function fields that $V_q(s) = O(s)$ with an absolute implied constant. In combination with Theorem 3 this shows that the second construction yields digital $(t,s)$-sequences constructed over $\mathbf{F}_q$ for which, for fixed $\overset{\cdot}{q}$, the values of $t$ grow at most linearly as a function of $s$. The fact that this growth rate of $t$ is indeed best possible will be discussed in Section 8.

## 6. The Third Construction

This construction due to Xing and Niederreiter [27] is more flexible than the second construction described in Section 5, in that it permits also the use of places (or closed points in the geometric language) of larger degree. If only rational places are employed in the third construction, then it is of the same quality as the second construction. Another aspect is that, in cases of practical interest, it is trivial to find the auxiliary divisor $D$ that is needed in the third construction, whereas the divisor $D$ for the second construction is sometimes harder to find (note that the proof of Proposition 1 in Section 5 is nonconstructive). The third construction is again based on the principle of the digital method in Section 2.

For a given dimension $s \geq 1$ and a prime power $q$, let $K/\mathbf{F}_q$ be a global function field containing at least one rational place $P_\infty$, and let $D$ be a positive divisor of $K/\mathbf{F}_q$ with $\deg(D) = 2g(K/\mathbf{F}_q)$ and $P_\infty \notin \operatorname{supp}(D)$. We choose $s$ distinct places $P_1, \ldots, P_s$ of $K/\mathbf{F}_q$ with $P_i \neq P_\infty$ for $1 \leq i \leq s$, and we put $e_i = \deg(P_i)$ for $1 \leq i \leq s$. With the abbreviation $g = g(K/\mathbf{F}_q)$ we have $l(D) = g + 1$ by the Riemann-Roch theorem. We choose a basis of $\mathcal{L}(D)$ in the following way. Note that $l(D - P_\infty) = g$ by the Riemann-Roch theorem and $l(D - (2g+1)P_\infty) = 0$, hence there exist integers $0 = n_0 < n_1 < \ldots < n_g \leq 2g$ such that

$$l(D - n_f P_\infty) = l(D - (n_f + 1)P_\infty) + 1 \quad \text{for } 0 \leq f \leq g.$$

Choose $w_f \in \mathcal{L}(D - n_f P_\infty) \backslash \mathcal{L}(D - (n_f + 1)P_\infty)$, then it is easily seen that $\{w_0, w_1, \ldots, w_g\}$ is a basis of $\mathcal{L}(D)$.

For each $1 \leq i \leq s$ we consider the chain

$$\mathcal{L}(D) \subset \mathcal{L}(D + P_i) \subset \mathcal{L}(D + 2P_i) \subset \ldots$$

of $\mathbf{F}_q$-vector spaces. By starting from the basis $\{w_0, w_1, \ldots, w_g\}$ of $\mathcal{L}(D)$ and successively adding basis vectors at each step of the chain, we obtain for each $n \geq 1$ a basis

$$\left\{ w_0, w_1, \ldots, w_g, k_1^{(i)}, k_2^{(i)}, \ldots, k_{ne_i}^{(i)} \right\}$$

of $\mathcal{L}(D + nP_i)$. Now let $z$ be a local uniformizing parameter at $P_\infty$. For $r = 0, 1, \ldots$ we put

$$z_r = \begin{cases} z^r & \text{if } r \notin \{n_0, n_1, \ldots, n_g\}, \\ w_f & \text{if } r = n_f \text{ for some } f \in \{0, 1, \ldots, g\}. \end{cases}$$

Note that then $\nu_{P_\infty}(z_r) = r$ for all $r \geq 0$. For $1 \leq i \leq s$ and $j \geq 1$ we have $k_j^{(i)} \in \mathcal{L}(D + mP_i)$ for some $m \geq 1$ and also $P_\infty \notin \operatorname{supp}(D + mP_i)$, hence $\nu_{P_\infty}(k_j^{(i)}) \geq 0$. Thus we have expansions at $P_\infty$ of the form

$$k_j^{(i)} = \sum_{r=0}^\infty a_{j,r}^{(i)} z_r \quad \text{for } 1 \leq i \leq s \text{ and } j \geq 1,$$

where all coefficients $a_{j,r}^{(i)} \in \mathbf{F}_q$. For $1 \leq i \leq s$ and $j \geq 1$ we now define the sequences

$$\mathbf{c}_j^{(i)} = \left( \widehat{a_{j,n_0}^{(i)}}, a_{j,1}^{(i)}, \ldots, \widehat{a_{j,n_1}^{(i)}}, a_{j,n_1+1}^{(i)}, \ldots, \widehat{a_{j,n_g}^{(i)}}, a_{j,n_g+1}^{(i)}, \ldots \right) \in \mathbf{F}_q^\infty,$$

where the hat indicates that the corresponding term is deleted. If we then write

$$\mathbf{c}_j^{(i)} = \left( c_{j,0}^{(i)}, c_{j,1}^{(i)}, \ldots \right),$$

then the terms $c_{j,r}^{(i)} \in \mathbf{F}_q$ serve as the elements in (S3) in the digital method described in Section 2. The bijections $\psi_r$ and $\eta_j^{(i)}$ are chosen as in (S1) and (S2), respectively. If $\tau(C^{(\infty)})$ is as in (6), then it was proved in [27] that

$$\tau(C^{(\infty)}) \leq g + \sum_{i=1}^{s}(e_i - 1).$$

In view of Lemma 1, this leads to the following consequence.

**Theorem 4.** *The third construction yields a digital $(t, s)$-sequence constructed over $\mathbf{F}_q$ with*

$$t = g(K/\mathbf{F}_q) + \sum_{i=1}^{s}(\deg(P_i) - 1).$$

Let $K/\mathbf{F}_q$ be a global function field with $N(K/\mathbf{F}_q) \geq s + 1$ and let $P_\infty, P_1, P_2, \ldots, P_s$ be $s + 1$ distinct rational places of $K/\mathbf{F}_q$. Then we can apply the third construction with $D = 2g(K/\mathbf{F}_q)P_1$, for instance, and this yields a digital $(t, s)$-sequence constructed over $\mathbf{F}_q$ with $t = g(K/\mathbf{F}_q)$, which is the same value of the quality parameter as that obtained from the second construction. In particular, for every $q$ and $s$ the third construction yields again a digital $(V_q(s), s)$-sequence constructed over $\mathbf{F}_q$. Some further examples for the use of the third construction are given in Section 8.

## 7. The Fourth Construction

This construction of digital $(t, s)$-sequences was presented in Niederreiter and Xing [16, Section 8]. It is not as powerful as the second and the third construction, but quite a bit simpler. As in the second construction, algebraic curves over $\mathbf{F}_q$ with many $\mathbf{F}_q$-rational points or, equivalently, global function fields containing many rational places are used. First, the following auxiliary result is needed.

**Lemma 2.** *Let $K/\mathbf{F}_q$ be a global function field of genus $g = g(K/\mathbf{F}_q) > 0$.*
*(i) If $P$ is a rational place of $K/\mathbf{F}_q$, then for every integer $j \geq 2g$ there exists an element $h_j \in K$ such that $(h_j)_\infty = jP$.*
*(ii) If $P$ and $Q$ are two different rational places of $K/\mathbf{F}_q$, then for every integer $j$ with $1 \leq j \leq 2g - 1$ there exists an element $h_j \in K$ such that*

$$(h_j)_\infty = (2g - j)P + jQ.$$

Let the dimension $s \geq 1$ and the prime power $q$ be given, and suppose that the global function field $K/\mathbf{F}_q$ satisfies $N(K/\mathbf{F}_q) \geq s + 1$.

Let $P_\infty, P_1, P_2, \ldots, P_s$ be $s+1$ distinct rational places of $K/\mathbf{F}_q$. Now we determine elements $h_j^{(i)} \in K$ for $1 \le i \le s$ and $j \ge 1$ as follows. If $g = g(K/\mathbf{F}_q) > 0$, then by Lemma 2 we can find $h_j^{(i)} \in K$ such that for $i = 1$ we have

$$(h_j^{(1)})_\infty = (2g + j - 1)P_1 \quad \text{for all } j \ge 1,$$

and for $2 \le i \le s$ we have

$$(h_j^{(i)})_\infty = \begin{cases} (2g - j)P_1 + jP_i & \text{if } 1 \le j \le 2g - 1, \\ jP_i & \text{if } j \ge 2g. \end{cases}$$

If $g = 0$, then we can find $h_j^{(i)} \in K$ such that

$$(h_j^{(i)})_\infty = jP_i \quad \text{for } 1 \le i \le s \text{ and } j \ge 1.$$

Given these elements $h_j^{(i)} \in K$ (in both cases $g > 0$ and $g = 0$), we now choose a local uniformizing parameter $z$ at $P_\infty$, and then we have the expansions

$$h_j^{(i)} = \sum_{r=0}^{\infty} b_{j,r}^{(i)} z^r \quad \text{for } 1 \le i \le s \text{ and } j \ge 1,$$

with all $b_{j,r}^{(i)} \in \mathbf{F}_q$. We define

$$c_{j,r}^{(i)} = b_{j,r+1}^{(i)} \quad \text{for } 1 \le i \le s, j \ge 1, \text{ and } r \ge 0.$$

The $c_{j,r}^{(i)} \in \mathbf{F}_q$ are now used as the elements in (S3) in the digital method described in Section 2. The bijections $\psi_r$ and $\eta_j^{(i)}$ are chosen as in (S1) and (S2), respectively. If $\tau(C^{(\infty)})$ is as in (6), then it was shown in [16] that

$$\tau(C^{(\infty)}) = 0 \quad \text{if } g(K/\mathbf{F}_q) = 0,$$
$$\tau(C^{(\infty)}) \le 2g(K/\mathbf{F}_q) - 1 \quad \text{if } g(K/\mathbf{F}_q) > 0.$$

By Lemma 1 we then get the following result.

**Theorem 5.** *Under the condition $N(K/\mathbf{F}_q) \ge s+1$, the fourth construction yields a digital $(t, s)$-sequence constructed over $\mathbf{F}_q$ with*

$$t = \begin{cases} 0 & \text{if } g(K/\mathbf{F}_q) = 0, \\ 2g(K/\mathbf{F}_q) - 1 & \text{if } g(K/\mathbf{F}_q) > 0. \end{cases}$$

If we optimize the choice of $K/\mathbf{F}_q$ as in the second construction (see Section 5), then we obtain that the fourth construction yields a digital $(W_q(s), s)$-sequence constructed over $\mathbf{F}_q$, where

$$W_q(s) = \begin{cases} 0 & \text{if } 1 \le s \le q, \\ 2V_q(s) - 1 & \text{if } s \ge q + 1, \end{cases}$$

and where $V_q(s)$ is as in Definition 6.

## 8. Consequences for $(t, s)$-Sequences

The constructions described in the previous sections yield $(t, s)$-sequences in a prime-power base $q$ obtained from the digital method with underlying ring $\mathbf{F}_q$. A general technique in which the digital method is used with a direct product of finite fields as the underlying ring allows the extension to arbitrary bases $b$ (see [15, Lemma 8]). Together with the construction in Section 5 this yields, for instance, the following result (see also [15, Corollary 2]).

**Theorem 6.** *Let* $b = \prod_{v=1}^{h} q_v$ *be the canonical factorization of the integer* $b \geq 2$ *into pairwise coprime prime powers* $q_1 < \ldots < q_h$. *Then for every dimension* $s \geq 1$ *there exists a* $(t, s)$*-sequence in base* $b$ *with*

$$t \leq \frac{c}{\log q_1} s + 1,$$

*where* $c > 0$ *is an absolute constant.*

It is an important fact that the order of magnitude of the bound on $t$ (as a function of $s$) in Theorem 6 is best possible. The following general lower bound, which was derived in [16, Theorem 8] from combinatorial results of Lawrence [6], shows that $t$ must grow at least linearly in $s$ for a fixed base $b$, no matter which construction of a $(t, s)$-sequence in base $b$ is used.

**Theorem 7.** *For any given base* $b \geq 2$ *and any given dimension* $s \geq 1$, *a* $(t, s)$*-sequence in base* $b$ *can exist only if*

$$t \geq \frac{s}{b} - \log_b \frac{(b-1)s + b + 1}{2},$$

*where* $\log_b$ *denotes the logarithm to the base* $b$.

We consider now the coefficient $C_b(s, t)$ of the leading term in the upper bound (2) for the star discrepancy of a $(t, s)$-sequence in base $b$. According to Theorem 6, for any fixed base $b \geq 2$ we can always achieve a value $t = t(b, s)$ of the quality parameter with $t(b, s) = O(s)$. In view of the formula (3) for $C_b(s, t)$, we then obtain

$$\log C_b(s, t(b, s)) \leq -s \log s + O(s), \tag{8}$$

where the implied constant depends only on $b$. In particular, $C_b(s, t(b, s))$ tends to 0 at a superexponential rate as $s \to \infty$. In the earlier constructions of $(t, s)$-sequences in base $b$ by Sobol' [23] (with $b = 2$) and Niederreiter [9] (with arbitrary $b$), the coefficient $C_b(s, t)$ tends to $\infty$ as $s \to \infty$ for fixed $b$.

In the constructions of $(t, s)$-sequences in base $b$ by Faure [2] (with $b$ prime) and Niederreiter [9] (with arbitrary $b$) the best results are obtained if the base $b$ is made to depend on the value of $s$. With such varying bases, the constructions in [2] and [9] yield a coefficient $C'(s)$ of the leading term $N^{-1}(\log N)^s$ in the discrepancy bound (2) which satisfies

$$\log C'(s) \leq -s \log \log s + O(s).$$

A comparison with (8) shows that our constructions yielding $t(b, s) = O(s)$ perform better with regard to the bound on the coefficient $C_b(s, t)$, and they have the additional practical advantage that we can work with a fixed base $b$ such as $b = 2$.

Theorems 6 and 7 solve the problem of the construction of $(t, s)$-sequences in base $b$ as far as the order of magnitude of $t$ in terms of $s$ is concerned. But for the numerical practice we require good explicit values of $t$ for a wide range of dimensions $s$ and for typical bases, such as the base $b = 2$. If we work with the two most powerful constructions of sequences presented here, namely those in Sections 5 and 6, then we need more information on the quantity $N_q(g)$ introduced in Definition 5. For the concrete implementation of these constructions of sequences, we require also an explicit description of the underlying algebraic curves. This leads to the problem of finding explicit constructions of algebraic curves over $\mathbf{F}_q$ with many $\mathbf{F}_q$-rational points, where the case of small $q$ is particularly interesting. Until recently, the aspect of explicit constructions was neglected in the literature on algebraic curves over finite fields.

The authors have embarked on a project with the aim of finding explicit algebraic curves over $\mathbf{F}_q$ having many $\mathbf{F}_q$-rational points for very small values of $q$ (like $q = 2, 3, 4, 5$) and for small values of the genus. Results are already available in the papers Niederreiter and Xing [14], [16, Section 9], [17]. The methods are based on the correspondence between algebraic curves over finite fields and global function fields (see Section 3) and on explicit constructions of global function fields with many rational places. The latter constructions use cyclotomic function fields as well as Artin-Schreier extensions and Kummer extensions of global function fields.

The authors have also obtained new theoretical results on the existence of algebraic curves over $\mathbf{F}_q$ with many $\mathbf{F}_q$-rational points, thus getting new values or lower bounds for $N_q(g)$. These results have been described in the papers Niederreiter and Xing [17], [18] and Xing and Niederreiter [28], [29]. The methods are based on Hilbert class fields and on the theory of Drinfeld modules.

In order to tabulate in a convenient manner the values of the quality parameter $t$ obtained from our constructions of digital $(t, s)$-sequences and from the results above on algebraic curves, we use the following number $d_q(s)$ from [16, Definition 8].

**Definition 7.** For any prime power $q$ and any dimension $s \geq 1$ let $d_q(s)$ be the least value of $t$ such that there exists a digital $(t, s)$-sequence constructed over $\mathbf{F}_q$.

We tabulate upper bounds for $d_q(s)$ for $q = 2, 3, 5$ and various ranges for the dimension $s$ in Table 2. Most of these bounds are obtained from the second construction, the resulting trivial inequality $d_q(s) \leq V_q(s)$, and upper bounds for $V_q(s)$, where $V_q(s)$ is as in Definition 6. Those few bounds for $d_q(s)$ that arise in a different manner are explained in Examples 1 to 5 below. The upper bounds for $V_q(s)$ are obtained from lower bounds for $N_q(g)$ in the papers on algebraic curves quoted above, where the most extensive set of data that is available is contained in a table of bounds for $N_2(g)$ in Xing and Niederreiter [29]. Entries in Table 2 that are marked with an asterisk represent the exact value of $d_q(s)$, and these values are in fact such that there cannot exist a $(t, s)$-sequence in base $q$ with a smaller value of $t$ by whatever construction.

### Table 2
#### Upper bounds for $d_q(s)$ for $q = 2, 3, 5$

| $s$ | 1 | 2 | 3 | 4 | 5 | 6 | 7 | 8 | 9 | 10 | 11 | 12 | 13 |
|---|---|---|---|---|---|---|---|---|---|---|---|---|---|
| $q = 2$ | 0* | 0* | 1* | 1* | 2* | 3 | 4 | 5 | 6 | 8 | 9 | 10 | 11 |
| $q = 3$ | 0* | 0* | 0* | 1* | 1* | 1* | 2 | 3 | 3 | 4 | 4 | 5 | 6 |
| $q = 5$ | 0* | 0* | 0* | 0* | 0* | 1* | 1* | 1* | 1* | 2 | 2 | 3 | 3 |

| $s$ | 14 | 15 | 16 | 17 | 18 | 19 | 20 | 21 | 22 | 23 | 24 | 25 | 26 |
|---|---|---|---|---|---|---|---|---|---|---|---|---|---|
| $q = 2$ | 13 | 15 | 15 | 18 | 19 | 19 | 21 | 23 | 25 | 25 | 29 | 31 | 31 |
| $q = 3$ | 7 | 7 | 9 | 9 | 9 | 11 | 12 | 12 | 13 | 13 | 15 | 15 | 15 |
| $q = 5$ | 3 | 3 | 4 | 4 | 5 | 5 | 6 | 8 | 9 | 9 | 10 | 10 | 12 |

| $s$ | 27 | 28 | 29 | 30 | 31 | 32 | 33 | 34 | 35 | 36 | 37 | 38 |
|---|---|---|---|---|---|---|---|---|---|---|---|---|
| $q = 2$ | 33 | 36 | 36 | 39 | 39 | 39 | 45 | 47 | 47 | 50 | 50 | 50 |
| $q = 3$ | 15 | | | | | | | | | | | |
| $q = 5$ | 12 | 12 | 12 | 13 | | | | | | | | |

| $s$ | 39 | 40 | 41 | 42 | 43 | 44 | 45 | 46 | 47 | 48 | 49 | 50 |
|---|---|---|---|---|---|---|---|---|---|---|---|---|
| $q = 2$ | 50 | 54 | 54 | 59 | 62 | 65 | 65 | 65 | 65 | 69 | 75 | 77 |

**Example 1.** The bound $d_2(33) \leq 45$ in Table 2 arises from the third construction in the following way. We use the theory of cyclotomic function fields for which convenient summaries can be found in [14] and [17]. Let $K/\mathbf{F}_2$ be the subfield of the cyclotomic function field $E$ with modulus $(x^2 + x + 1)^4 \in \mathbf{F}_2[x]$ fixed by the cyclic subgroup $< x >$ of $\mathrm{Gal}(E/\mathbf{F}_2(x))$. Since this cyclic subgroup has order 12, we get $[E : K] = 12$ and $[K : \mathbf{F}_2(x)] = 16$. The only ramified place in the extension $E/K$ is the place $P$ of $K/\mathbf{F}_2$ of degree 2 lying over $x^2 + x + 1$ and $P$ is totally ramified in $E/K$. If $Q$ is the unique place of $E$ lying over $P$, then by a similar calculation as in the proof of [14, Theorem 2] we get the different exponent $d(Q|P) = 32$. Hence the Hurwitz genus formula yields $12(2g(K/\mathbf{F}_2) - 2) + 2 \cdot 32 = 2g(E/\mathbf{F}_2) - 2$. Now $2g(E/\mathbf{F}_2) - 2 = 1024$ by [14, Proposition 1], and so $g(K/\mathbf{F}_2) = 41$.

The pole of $x$ and the zero of $x$ split completely in the extension $K/\mathbf{F}_2(x)$. Since $(x+1)^4 \equiv x^8 \bmod(x^2+x+1)^4$, the place $x+1$ of $\mathbf{F}_2(x)$ splits into four places of $K/\mathbf{F}_2$ of degree 4. Thus $N(K/\mathbf{F}_2) = 32$. Now let $P_\infty$ be one of the rational places of $K/\mathbf{F}_2$, let $P_1, \ldots, P_{31}$ be the remaining rational places of $K/\mathbf{F}_2$, put $P_{32} = P$, and let $P_{33}$ be one of the places of $K/\mathbf{F}_2$ of degree 4. Then Theorem 4 with $s = 33$ yields $t = 45$.

**Example 2.** Let $K/\mathbf{F}_2$ be the global function field in [29, Example 9], i.e., $K$ is the subfield of the cyclotomic function field $E$ with modulus $(x^2 + x + 1)(x^6 + x^3 + 1) \in \mathbf{F}_2[x]$ fixed by the cyclic subgroup $< x >$ of $\mathrm{Gal}(E/\mathbf{F}_2(x))$. Then it is shown in [29, Example 9] that $g(K/\mathbf{F}_2) = 54$ and $N(K/\mathbf{F}_2) = 42$. Furthermore, the place $x^6 + x^3 + 1$ of $\mathbf{F}_2(x)$ is totally ramified in the extension $K/\mathbf{F}_2(x)$, and so there exists a place $Q$ of $K/\mathbf{F}_2$ of degree 6. Let $P_\infty$ be one of the rational places of $K/\mathbf{F}_2$, let $P_1, \ldots, P_{41}$ be the remaining rational places of $K/\mathbf{F}_2$, and put $P_{42} = Q$. Then Theorem 4 yields $d_2(42) \leq 59$.

**Example 3.** Let $K/\mathbf{F}_2$ be the subfield of the cyclotomic function field $E$ with modulus $(x^2 + x + 1)^6 \in \mathbf{F}_2[x]$ fixed by the subgroup of $\mathrm{Gal}(E/\mathbf{F}_2(x))$ generated by $x$ and $x+1$. Since this subgroup has order 192, we get $[E : K] = 192$ and $[K : \mathbf{F}_2(x)] = 16$. The only ramified place in the extension $E/K$ is the place $P$ of $K/\mathbf{F}_2$ of degree 2 lying over $x^2 + x + 1$ and $P$ is totally ramified in $E/K$. If $Q$ is the unique place of $E$ lying over $P$, then we get the different exponent $d(Q|P) = 1280$. Hence the Hurwitz genus formula and [14, Proposition 1] yield $192(2g(K/\mathbf{F}_2) - 2) + 2 \cdot 1280 = 2g(E/\mathbf{F}_2) - 2 = 28 \cdot 2^{10}$, and so $g(K/\mathbf{F}_2) = 69$. All rational places of $\mathbf{F}_2(x)$ split completely in the extension $K/\mathbf{F}_2(x)$, thus $N(K/\mathbf{F}_2) = 48$. Since $(x^3 + x + 1)^2 \equiv x^{10}(x+1)^2 \bmod(x^2 + x + 1)^6$ and we cannot have $x^3 + x + 1 \equiv x^j(x+1)^m \bmod(x^2 + x + 1)^6$ with some $0 \leq j \leq 23$ and $0 \leq m \leq 7$, the place $x^3 + x + 1$ of $\mathbf{F}_2(x)$ splits into eight places of $K/\mathbf{F}_2$ of degree 6. Let $P_\infty$ be one of the rational places of $K/\mathbf{F}_2$, let $P_1, \ldots, P_{47}$ be the remaining rational places of $K/\mathbf{F}_2$, put $P_{48} = P$, and let $P_{49}$ be one of the places of $K/\mathbf{F}_2$ of degree 6. Then Theorem 4 yields $d_2(49) \leq 75$.

**Example 4.** Let $K = \mathbf{F}_3(x, y)$ be the Artin-Schreier extension of the rational function field $\mathbf{F}_3(x)$ with

$$y^3 - y = \frac{x^3 - x}{(x^2 + x - 1)^2}.$$

Then it is shown in [27, Example 1] that $g(K/\mathbf{F}_3) = 4$ and $N(K/\mathbf{F}_3) = 12$. Furthermore, the place $x^2 + x - 1$ of $\mathbf{F}_3(x)$ is totally ramified in the extension $K/\mathbf{F}_3(x)$, and so there exists a place $Q$ of $K/\mathbf{F}_3$ of degree 2. Let $P_\infty$ be one of the rational places of $K/\mathbf{F}_3$, let $P_1, \ldots, P_{11}$ be the remaining rational places of $K/\mathbf{F}_3$, and put $P_{12} = Q$. Then Theorem 4 yields $d_3(12) \leq 5$.

**Example 5.** Let $K = \mathbf{F}_5(x, y)$ be the Artin-Schreier extension of the

158

rational function field $\mathbf{F}_5(x)$ with

$$y^5 - y = \frac{x^5 - x}{(x^2 + 2)^3}.$$

Then it is shown in [17, Example 5.12A] that $g(K/\mathbf{F}_5) = 12$ and $N(K/\mathbf{F}_5) = 30$. Furthermore, the place $x^2 + 2$ of $\mathbf{F}_5(x)$ is totally ramified in the extension $K/\mathbf{F}_5(x)$, and so there exists a place $Q$ of $K/\mathbf{F}_5$ of degree 2. Let $P_\infty$ be one of the rational places of $K/\mathbf{F}_5$, let $P_1, \ldots, P_{29}$ be the remaining rational places of $K/\mathbf{F}_5$, and put $P_{30} = Q$. Then Theorem 4 yields $d_5(30) \leq 13$.

The third construction yields also a general upper bound for $d_q(s)$ which is completely explicit. In fact, it was shown in [27, Theorem 3] that for every prime power $q$ and every dimension $s \geq 1$ we have

$$d_q(s) \leq \frac{3q - 1}{q - 1}(s - 1) - \frac{(2q + 4)(s - 1)^{1/2}}{(q^2 - 1)^{1/2}} + 2.$$

In conclusion, we point out again that the algebraic-geometry approach to the construction of low-discrepancy sequences yields $(t, s)$-sequences in base $b$ where for fixed $b$ the quality parameter $t$ has the least possible order of magnitude when considered as a function of the dimension $s$. Table 1 documents the progress that has been achieved in terms of concrete values of $t$. Clearly, the algebraic-geometry approach yields the currently best construction method for low-discrepancy sequences. One of the practical implications of this work is that good low-discrepancy sequences are now available for much higher dimensions in quasi-Monte Carlo calculations.

## 9. References

[1] A.T. Clayman, K.M. Lawrence, G.L. Mullen, H. Niederreiter, and N.J.A. Sloane, Updated tables of parameters of $(t, m, s)$-nets, preprint, 1996.

[2] H. Faure, Discrépance de suites associées à un système de numération (en dimension $s$), *Acta Arith.* **41**, 337–351 (1982).

[3] A. Garcia and H. Stichtenoth, Algebraic function fields over finite fields with many rational places, *IEEE Trans. Information Theory* **41**, 1548–1563 (1995).

[4] J.H. Halton, On the efficiency of certain quasi-random sequences of points in evaluating multi-dimensional integrals, *Numer. Math.* **2**, 84–90 (1960); Berichtigung, *ibid.* **2**, 196 (1960).

[5] G. Larcher, H. Niederreiter, and W.Ch. Schmid, Digital nets and sequences constructed over finite rings and their application to quasi-Monte Carlo integration, *Monatsh. Math.* **121**, 231–253 (1996).

[6] K.M. Lawrence, Combinatorial bounds and constructions in the theory of uniform point distributions in unit cubes, connections with orthogonal arrays and a poset generalization of a related problem in coding theory, Ph.D. Dissertation, University of Wisconsin, Madison, 1995.

[7] G.L. Mullen, A. Mahalanabis, and H. Niederreiter, Tables of $(t, m, s)$-net and $(t, s)$-sequence parameters, *Monte Carlo and Quasi-Monte Carlo Methods in Scientific Computing* (H. Niederreiter and P.J.-S. Shiue, eds.), Lecture Notes in Statistics, Vol. 106, pp. 58–86, Springer, New York, 1995.

[8] H. Niederreiter, Point sets and sequences with small discrepancy, *Monatsh. Math.* **104**, 273–337 (1987).

[9] H. Niederreiter, Low-discrepancy and low-dispersion sequences, *J. Number Theory* **30**, 51–70 (1988).

[10] H. Niederreiter, *Random Number Generation and Quasi-Monte Carlo Methods*, SIAM, Philadelphia, 1992.

[11] H. Niederreiter, Pseudorandom numbers and quasirandom points, *Z. angew. Math. Mech.* **73**, T648–T652 (1993).

[12] H. Niederreiter, Factorization of polynomials and some linear-algebra problems over finite fields, *Linear Algebra Appl.* **192**, 301–328 (1993).

[13] H. Niederreiter and C.P. Xing, Low-discrepancy sequences obtained from algebraic function fields over finite fields, *Acta Arith.* **72**, 281–298 (1995).

[14] H. Niederreiter and C.P. Xing, Explicit global function fields over the binary field with many rational places, *Acta Arith.* **75**, 383–396 (1996).

[15] H. Niederreiter and C.P. Xing, Low-discrepancy sequences and global function fields with many rational places, *Finite Fields Appl.* **2**, 241–273 (1996).

[16] H. Niederreiter and C.P. Xing, Quasirandom points and global function fields, *Finite Fields and Applications* (S. Cohen and H. Niederreiter, eds.), London Math. Soc. Lecture Note Series, Vol. 233, pp. 269–296, Cambridge University Press, Cambridge, 1996.

[17] H. Niederreiter and C.P. Xing, Cyclotomic function fields, Hilbert class fields, and global function fields with many rational places, *Acta Arith.* **79**, 59–76 (1997).

[18] H. Niederreiter and C.P. Xing, Drinfeld modules of rank 1 and algebraic curves with many rational points. II, *Acta Arith.*, to appear.

[19] S. Ninomiya and S. Tezuka, Toward real-time pricing of complex financial derivatives, *Applied Math. Finance* **3**, 1–20 (1996).

[20] S.H. Paskov, Computing high dimensional integrals with applications to finance, Technical Report CUCS-023-94, Department of Computer Science, Columbia University, New York, 1994.

[21] S.H. Paskov and J.F. Traub, Faster valuation of financial derivatives, *J. Portfolio Management* **22**, 113–120 (1995).

[22] J.-P. Serre, Sur le nombre des points rationnels d'une courbe algébrique sur un corps fini, *C.R. Acad. Sci. Paris Sér. I Math.* **296**, 397–402 (1983).

[23] I.M. Sobol', The distribution of points in a cube and the approximate evaluation of integrals (Russian), *Zh. Vychisl. Mat. i Mat. Fiz.* **7**, 784–802 (1967).

[24] S. Tezuka, Polynomial arithmetic analogue of Halton sequences, *ACM Trans. Modeling and Computer Simulation* **3**, 99–107 (1993).

[25] S. Tezuka, *Uniform Random Numbers: Theory and Practice*, Kluwer Academic Publ., Norwell, MA, 1995.

[26] S. Tezuka and T. Tokuyama, A note on polynomial arithmetic analogue of Halton sequences, *ACM Trans. Modeling and Computer Simulation* **4**, 279–284 (1994).

[27] C.P. Xing and H. Niederreiter, A construction of low-discrepancy sequences using global function fields, *Acta Arith.* **73**, 87–102 (1995).

[28] C.P. Xing and H. Niederreiter, Modules de Drinfeld et courbes algébriques ayant beaucoup de points rationnels, *C.R. Acad. Sci. Paris Sér. I Math.* **322**, 651–654 (1996).

[29] C.P. Xing and H. Niederreiter, Drinfeld modules of rank 1 and algebraic curves with many rational points, preprint, 1996.

INSTITUTE OF INFORMATION PROCESSING, AUSTRIAN ACADEMY OF SCIENCES, SONNENFELSGASSE 19, A–1010 VIENNA, AUSTRIA
*E-mail:* niederreiter@oeaw.ac.at

DEPARTMENT OF MATHEMATICS, UNIVERSITY OF SCIENCE AND TECHNOLOGY OF CHINA, HEFEI, ANHUI 230026, P.R. CHINA

# A Monte Carlo Estimator Based on a State Space Decomposition Methodology for Flow Network Reliability

Stéphane Bulteau [1]
Mohamed El Khadiri [2]

ABSTRACT The exact evaluation of the probability that the maximum $st$-flow is greater than or equal to a fixed value $d$ in a stochastic flow network is an NP-hard problem. This limitation leads to consider Monte Carlo alternatives. In this paper, we show how to exploit the state space decomposition methodology of Doulliez and Jamoulle for deriving a Monte Carlo simulation algorithm. We show that the resulting Monte Carlo estimator belongs to the variance-reduction family and we give a worst-case bound on the variance-reduction ratio that can be expected when compared with the standard sampling. We illustrate by numerical comparisons that the proposed simulation algorithm allows substantial variance-reduction with respect to the standard one and it is competitive when compared to a previous work in this context.

## 1 Introduction

A basic mission of a flow network is the establishment of a flow between a source node $s$ and a sink node $t$ that exceeds a fixed demand $d$. In the stochastic case, the network components have random capacities and the probability that a maximum $st$-flow exceeds the demand is a performance measure, called *flow network reliability* measure. Several papers have been devoted to its evaluation when nodes do not limit flow transmission and arcs capacities are discrete, multi-valued and statistically independent random variables [1], [2], [3], [4]. In this case, one of the methods for computing exactly the parameter under consideration is based on the state space decomposition methodology of Doulliez&Jamoulle [1]. It starts by decomposing the state space into non-overlapping subsets: an operating set, failed

---

[1] Irisa, Campus de Beaulieu, 35042 Rennes Cédex, France. E-mail: sbulteau@irisa.fr
[2] C.E.R.L., I.U.T. Saint-Nazaire, Saint-Nazaire, France E-mail:elkhadiri@iut-saint-nazaire.univ-nantes.fr

sets, and undetermined sets. Each undetermined set is used as input of subsequent similar decomposition. The recursive process terminates when all generated sets are classed operating or failed. The probability that the random vector state belongs to any generated operating set is easy to compute and the flow network reliability parameter is the sum of these probabilities.

Unfortunately, the exact evaluation is an NP-hard problem. As a consequence, no algorithm is known to exist for solving exactly this problem in time bounded by a polynomial function of the network size. This limitation leads to consider Monte Carlo alternatives.

In [10], the authors exploit the decomposition methodology of Doulliez&Jamoulle for computing lower and upper bounds of the reliability parameter. These bounds have been exploited to construct an importance sampling estimator and an other one relying on stratified sampling to induce additional accuracy.

The aim of this paper is to show that the decomposition methodology of Doulliez&Jamoulle can be, as in exact context, recursively applied to construct a new estimator which belongs to the variance-reduction family.

This paper is organized as follows: The following section introduces some notations and the model definition. Section 3 briefly describes the principle of variance-reduction methods. In section 4, after recalling the state space decomposition methodology of Doulliez & Jamoulle, we show how to exploit it for constructing an estimator of the flow network reliability parameter. We show that this estimator belongs to the variance-reduction family and we give a lower bound on the variance-reduction ratio that can be expected when compared with the standard sampling. In Section 5, we illustrate by numerical comparisons the performance of the proposed estimator. Section 6 is devoted to some conclusions.

## 2   Notations and Model Definition

We resume in this section the model definition and some notations. For ease of explanation, additional definitions and notations will be given in adequate sections.

- $G = (V, A, \vec{C}, s, t, d)$ : the flow network

- $V$ : the set of nodes

- $A = \{e_1, \ldots, e_a\}$ : the set of arcs

- $s$ : the source node

- $t$ : the sink node

- $d$ : the demand required at node $t$

- for each arc $e_j \in A$,

    - $C_j$ : the random discrete capacity of arc $e_j$
    - $0 \leq c_{j1} < c_{j2} < \ldots < c_{jn_j} < +\infty$ : the $n_j$ possible values of the random variable $C_j$
    - $\Omega_j = \{c_{j1}, \ldots, c_{jn_j}\}$ : the state space of the random variable $C_j$
    - $p_{jn}$ : the probability that $C_j$ has capacity $c_{jn}$ in $\Omega_j$

- $\vec{C} = (C_1, \ldots, C_a)$ the random state vector of the network $G$

- $\Omega = \bigotimes_{j=1}^{a} \Omega_j$ : the network state space (the state space of random variable $\vec{C}$)

- $v(c_1, \ldots, c_a)$ : the value of maximum $st$-flow when the arc $e_j$ has capacity $c_j$, $e_j \in A$

- $\Phi$ the non-decreasing structure function of the network $G$ defined for all state vectors $\vec{C}$ by

$$\Phi(\vec{C}) = \begin{cases} 1 & \text{if } v(\vec{C}) \geq d \\ 0 & \text{otherwise} \end{cases}$$

- For any subset $R$ of $\Omega$, we associate the following definitions :

    - $\alpha(R) = (\alpha_1(R), \ldots, \alpha_a(R))$ : the lower limiting vector of $R$
    - $\beta(R) = (\beta_1(R), \ldots, \beta_a(R))$ : the upper limiting vector of $R$
    - $\Omega_j(R) = \{\alpha_j(R), \ldots, \beta_j(R)\}$, $e_j \in A$ : the set of possible capacities of arc $e_j$ given that $\vec{C} \in R$
    - for each $e_j \in A$,
        * $C_j[R]$ denotes the random variable having the same distribution as the conditional distribution of $C_j$ given that $C_j \in \Omega_j(R)$
        * $p_{jc}[R]$, denotes the probability that $C_j[R]$ is equal to $c$

$$p_{jc}[R] = \Pr\{C_j = c \mid C_j \in \Omega_j(R)\} = p_{jc} / \sum_{n \in \Omega_j(R)} p_{jn} \quad (1)$$

    - $\vec{C}[R] = (C_1[R], \ldots, C_a[R])$ : the random vector having the same distribution as the conditional distribution of $\vec{C}$ given that $\vec{C} \in R$
    - $G[R] = (V, A, \vec{C}[R], s, t, d)$ : the network that differs from $G$ only by its random state vector which is equal to $\vec{C}[R]$

- $\mathcal{B}[R] = \Phi(\vec{\mathcal{C}} \, [R])$ : the binary random state of $G[R]$ which takes value 1 if the maximal $st$-flow in $G[R]$ exceeds the demand $d$ and 0 otherwise

- $Rel(G[R]) = \mathrm{E}\{\mathcal{B}[R]\} = \mathrm{E}\left\{\Phi(\vec{\mathcal{C}} \, [R])\right\} = \mathrm{E}\left\{\Phi(\vec{\mathcal{C}}) \mid \vec{\mathcal{C}} \in R\right\}$ : the reliability of the flow network $G[R]$

- $Rel(G[\Omega]) = \mathrm{E}\{\mathcal{B}[\Omega]\} = \mathrm{E}\left\{\Phi(\vec{\mathcal{C}} \, [\Omega])\right\} = \mathrm{E}\left\{\Phi(\vec{\mathcal{C}})\right\}$ : the considered reliability parameter of the flow network under consideration

- A set $R \subseteq \Omega$ is *rectangular* if and only if there is a lower and an upper vector of capacities $\alpha(R) = (\alpha_1(R), \ldots, \alpha_a(R))$ and $\beta(R) = (\beta_1(R), \ldots, \beta_a(R))$, respectively, in $\Omega$ such that every vector of capacities $\vec{\mathcal{C}} = (c_1, \ldots, c_a)$ with $\alpha_j(R) \leq c_j \leq \beta_j(R)$ for all arcs $e_j$, belongs to $R$ and $R$ contains only those vectors [1].

## 3 The principle of the state space decomposition technique of Doulliez & Jamoulle

In this section, we recall how the decomposition methodology of Doulliez and Jamoulle [1] can be used in the exact evaluation of the reliability measure $Rel(G[R])$, for any rectangular subset of $\Omega$. This will be useful to understand how we design our simulation method in the next section.

If the lower vector $\alpha(R)$ is an operating state ($\Phi(\alpha(R)) = 1$), then all states in $R$ are operating. Consequently,

$$Rel(G[R]) = \mathrm{Pr}\left\{\vec{\mathcal{C}} \, [R] \in R\right\} = 1.$$

If the upper vector $\beta(R)$ is a failed state ($\Phi(\beta(R)) = 0$), then all states in $R$ are failed. Consequently,

$$Rel(G[R]) = \mathrm{Pr}\{\emptyset\} = 0.$$

In the non trivial case, the decomposition technique of Doulliez&Jamoulle produces a partition of the rectangular set $R \subseteq \Omega$ that contains operating and failed vectors into non-overlapping rectangular subsets :

- an operating rectangular subset $W(R)$ of operating vectors,

- failed rectangular subsets $F_j(R)$, $j \in \{1, \ldots, a\}$, of failed vectors,

- undetermined rectangular subsets $U_j(R)$, $j \in \{1, \ldots, a\}$, of vectors not yet classified as failed or operating.

The total expectation theorem is then applied to express the reliability measure of $G$ into the reliabilities of networks having same topology as $G$ and smaller state spaces :

$$\mathrm{E}\{\mathcal{B}[R]\} = \mathrm{E}\left\{\mathcal{B}[R] \mid \overrightarrow{C}[R] \in W(R)\right\} \mathrm{Pr}\left\{\overrightarrow{C}[R] \in W(R)\right\}$$
$$+ \sum_{j=1}^{a} \mathrm{E}\left\{\mathcal{B}[R] \mid \overrightarrow{C}[R] \in F_j(R)\right\} \mathrm{Pr}\left\{\overrightarrow{C}[R] \in F_j(R)\right\}$$
$$+ \sum_{j=1}^{a} \mathrm{E}\left\{\mathcal{B}[R] \mid \overrightarrow{C}[R] \in U_j(R)\right\} \mathrm{Pr}\left\{\overrightarrow{C}[R] \in U_j(R)\right\}.$$

Since $W(R)$ is an operating set and the $F_j(R)$'s are failed sets, we have $\mathrm{E}\left\{\mathcal{B}[R] \mid \overrightarrow{C}[R] \in W(R)\right\}$ is equal to 1 and $\mathrm{E}\left\{\mathcal{B}[R] \mid \overrightarrow{C}[R] \in F_j(R)\right\}$ is equal to 0. Moreover, the expectation of the variable $\mathcal{B}[R]$, given that $\overrightarrow{C}[R]$ belongs to $U_j(R)$ corresponds to the reliability measure of the network $G[U_j(R)]$. By these results, we obtain

$$\mathrm{E}\{\mathcal{B}[R]\} = \mathrm{Pr}\left\{\overrightarrow{C}[R] \in W(R)\right\} + \sum_{j=1}^{a} Rel(G[U_j(R)])\mathrm{Pr}\left\{\overrightarrow{C}[R] \in U_j(R)\right\}.$$

The same process is called to compute each parameter $Rel(G[U_j(R)])$ recursively until each leaf of the related tree consists of a rectangular set where all states are operating or failed. It results that the evaluation of the parameter $Rel(G[R])$ can be formalized by the following recursive formula:

$$Rel(G[R]) = \begin{cases} 0 \text{ if } \Phi(\beta(R)) = 0 \\ 1 \text{ if } \Phi(\alpha(R)) = 1 \\ \mathrm{Pr}\left\{\overrightarrow{C}[R] \in W(R)\right\} + \sum_{j=1}^{a} Rel(G[U_j(R)])\mathrm{Pr}\left\{\overrightarrow{C}[R] \in U_j(R)\right\} \\ \quad otherwise. \end{cases}$$

$$(2)$$

It suffices to replace $R$ by $\Omega$ in formula (2), for computing recursively the reliability parameter $Rel(G[\Omega])$.

**Remark 3.1** *The reader can see [1] for details about the computing process of the sets $W(R)$, $U_j(R)$, $F_j(R)$ and the probabilities that the random vector state $\overrightarrow{C}[R]$ belongs to them, even if this is not necessary to understand our work.*

When all arcs have only two possible capacities 1 or 0 and the demand is $d = 1$, the problem becomes the source-terminal reliability problem [8], [11] which is NP-hard [7]. Consequently, the general case considered here is also an NP-hard problem. This implies that the computational time will be prohibitive when the network size is large [9]. Therefore, we will employ the expression (2) later to deduce a recursive simulation process.

# 4 The state space decomposition technique of Doulliez & Jamoulle in the simulation context

## 4.1 The principle of variance-reduction methods

Suppose that $\mathcal{Z}$ is a random variable such that

$$\mathrm{E}\{\mathcal{Z}\} = Rel(G[\Omega]) \tag{3}$$

and let us denote by $\widehat{\mathcal{Z}}$,

$$\widehat{\mathcal{Z}} = \frac{1}{K}\sum_{i=1}^{K} \mathcal{Z}^{(i)}, \tag{4}$$

the unbiased sample mean based on $K$ independent and identically distributed random variables $\mathcal{Z}^{(1)}, \ldots, \mathcal{Z}^{(K)}$ with the same distribution function as the random variable $\mathcal{Z}$.

If $K$ is large enough, the central limit theorem can be applied to obtain a confidence interval [5]. In particular, the 99%–level confidence interval is such that

$$P\left(\mid Rel(G[\Omega]) - \widehat{\mathcal{Z}} \mid \leq \epsilon(\widehat{\mathcal{Z}})\right) \geq 99\% \tag{5}$$

where

$$\epsilon(\widehat{\mathcal{Z}}) = 2.575\sqrt{\mathrm{Var}\{\mathcal{Z}\}/K} \tag{6}$$

is the half length of the 99%–level confidence interval.

It results that the length of the confidence interval is proportional to the standard deviation of the variable $\mathcal{Z}$.

Since in general the variance of the random variable $\mathcal{Z}$ is unknown, its unbiased estimator

$$\widehat{V}_{\mathcal{Z}} = \frac{1}{K-1}\sum_{i=1}^{K}(\widehat{\mathcal{Z}} - \mathcal{Z}^{(i)})^2 \tag{7}$$

serves in formula (6) for computing the parameter $\epsilon(\widehat{\mathcal{Z}})$.

The unbiased standard Monte Carlo estimator of the parameter $Rel(G[\Omega])$ is a sample mean $\widehat{\mathcal{B}}[\Omega]$ based on the random variable $\mathcal{B}[\Omega]$ with variance equal to $\mathrm{E}\{\mathcal{B}[\Omega]\}(1 - \mathrm{E}\{\mathcal{B}[\Omega]\}) = Rel(G[\Omega])(1 - Rel(G[\Omega]))$ and the variance-reduction estimators are based on random variable having smaller variance than $\mathrm{Var}\{\mathcal{B}[\Omega]\}$ [6]. Then they offer more accurate estimates than the standard estimator, for a fixed sample size $K$.

## 4.2 The proposed estimator

### 4.2.1 Theoretical aspects

Based on the decomposition methodology of Doulliez and Jamoulle, we have shown in Section 3 that the recursive formula (2) can be used to recursively compute the reliability parameter $Rel(G[R])$. This formula can be rewritten as follows:

$$Rel(G[R]) = \begin{cases} 0 & \text{if } \Phi(\beta(R)) = 0 \\ 1 & \text{if } \Phi(\alpha(R)) = 1 \\ w(R) + u(R)\sum_{j=1}^{a} Rel(G[U_j(R)])\tilde{u}_j(R) & \text{otherwise.} \end{cases}$$

(8)

where

- $W(R)$, $F_j(R)$ and $U_j(R)$ denote the non-overlapping rectangular sets resulting from the decomposition of $R$ by the Doulliez and Jamoulle procedure

- $w(R) = \Pr\left\{\vec{C}\,[R] \in W(R)\right\}$

- $f_j(R) = \Pr\left\{\vec{C}\,[R] \in F_j(R)\right\}$, for $j \in \{1,\ldots,a\}$

- $u_j(R) = \Pr\left\{\vec{C}\,[R] \in U_j(R)\right\}$, for $j \in \{1,\ldots,a\}$

- $f(R) = \Pr\left\{\bigcup_{j=1}^{a} F_j(R)\right\} = \sum_{j=1}^{a} f_j(R)$

- $u(R) = \Pr\left\{\bigcup_{j=1}^{a} U_j(R)\right\} = \sum_{j=1}^{a} u_j(R) = 1 - w(R) - f(R)$

- $\tilde{u}_j(R) = \Pr\left\{\vec{C}\,[R] \in U_j(R) \mid (\vec{C}\,[R] \in \bigcup_{j=1}^{a} U_j(R))\right\} = u_j(R)/u(R)$,

  for $j \in \{1,\ldots,a\}$.

The recursive expression (8) leads us to propose the following result:

**Proposition 4.1** *For any rectangular set $R \subseteq \Omega$, let $\mathcal{Z}(R)$ be the random variable defined by the recursive formula*

$$\mathcal{Z}(R) = \begin{cases} 0 & \text{if } \Phi(\beta(R)) = 0 \\ 1 & \text{if } \Phi(\alpha(R)) = 1 \\ w(R) + u(R)\mathcal{Z}(\tilde{U}(R)) & \text{otherwise} \end{cases}$$

(9)

*where $\tilde{U}(R)$ is a random variable with values in the set $\{U_1(R),\ldots,U_a(R)\}$ of the undetermined rectangular sets resulting from the decomposition of the rectangle $R$ by the methodology of Doulliez and Jamoulle and defined by*

$$\Pr\left\{\tilde{U}(R) = U_j(R)\right\} = \tilde{u}_j(R), \text{ for } j \in \{1,\ldots,a\};$$

(10)

the remaining parameters are defined as used in the exact recursive formula (8). Then $\mathcal{Z}(R)$ verifies

$$E\{\mathcal{Z}(R)\} = E\{\mathcal{B}[R]\} = Rel(G[R]) \tag{11}$$

and

$$\text{Var}\{\mathcal{Z}(R)\} \le \text{Var}\{\mathcal{B}[R]\} = Rel(G[R])\,(1 - Rel(G[R])). \tag{12}$$

**Proof.** We will proceed by induction on the cardinality $m$ of the rectangle $R \subseteq \Omega$.

(a) Boundary condition:
Consider a rectangle $R$ with one element $C$ : then both $\mathcal{B}[R]$ and $\mathcal{Z}(R)$ have constant value equal to $Rel(G[R])$ $(Rel(G[R]) = \Phi(C) \in \{0, 1\})$. Consequently, they have the same expectation (equal to $\Phi(C)$) and the same variance (equal to 0).

(b) Inductive step:
Suppose that for all rectangles $R$ with cardinality smaller than $m$, $E\{\mathcal{Z}(R)\} = E\{\mathcal{B}[R]\}$ and $\text{Var}\{\mathcal{Z}(R)\} \le \text{Var}\{\mathcal{B}[R]\}$. We want to show that the same holds for all rectangles with cardinality $m$.

- If the lower vector $\alpha(R)$ is an operating state $(\Phi(\alpha(R)) = 1)$, then the network $G[R]$ is always operational. Consequently,

$$E\{\mathcal{B}[R]\} = E\{\mathcal{Z}(R)\} = 1$$

and

$$\text{Var}\{\mathcal{Z}(R)\} = \text{Var}\{\mathcal{B}[R]\} = 0.$$

- If the upper vector $\beta(R)$ is a failed state $(\Phi(\beta(R)) = 0)$, then the network $G[R]$ is always failed. Consequently,

$$E\{\mathcal{B}[R]\} = E\{\mathcal{Z}(R)\} = 0$$

and

$$\text{Var}\{\mathcal{Z}(R)\} = \text{Var}\{\mathcal{B}[R]\} = 0.$$

- In the non trivial case, we have

$$E\{\mathcal{Z}(R)\} = E\left\{w(R) + u(R)\mathcal{Z}(\tilde{U}(R))\right\} = w(R) + u(R)E\left\{\mathcal{Z}(\tilde{U}(R))\right\}.$$

Since the cardinality of each value of $\tilde{U}(R)$ is smaller than the cardinality of $R$, we apply the inductive hypothesis $(E\left\{\mathcal{Z}(\tilde{U}(R))\right\} = E\left\{\mathcal{B}[\tilde{U}(R)]\right\})$ to obtain

$$E\{\mathcal{Z}(R)\} = w(R) + u(R)E\left\{\mathcal{B}[\tilde{U}(R)]\right\}. \tag{13}$$

By the theorem of total expectation, we have

$$E\left\{B[\tilde{U}(R)]\right\} = \sum_{j=1}^{a} \Pr\left\{\tilde{U}(R) = U_j(R)\right\} E\left\{B[\tilde{U}(R)] \mid \tilde{U}(R) = U_j(R)\right\}.$$

Equation (13) becomes

$$
\begin{aligned}
E\left\{Z(R)\right\} &= w(R) + u(R)\sum_{j=1}^{a}\tilde{u}_j(R) E\left\{B[\tilde{U}(R)] \mid \tilde{U}(R) = U_j(R)\right\} \\
&= w(R) + u(R)\sum_{j=1}^{a}\tilde{u}_j(R) E\left\{B[U_j(R)]\right\} \\
&= w(R) + u(R)\sum_{j=1}^{a}\tilde{u}_j(R) Rel(G[U_j(R)]) \\
&= Rel(G[R]) = E\left\{B[R]\right\}.
\end{aligned}
$$

For $\mathrm{Var}\left\{Z(R)\right\}$, we have

$$
\begin{aligned}
\mathrm{Var}\left\{Z(R)\right\} &= \mathrm{Var}\left\{w(R) + u(R)Z(\tilde{U}(R))\right\} \\
&= (u(R))^2 \mathrm{Var}\left\{Z(\tilde{U}(R))\right\}.
\end{aligned}
$$

Since the cardinality of each value of $\tilde{U}(R)$ is smaller than the cardinality of $R$, we apply the inductive hypothesis to deduce that

$$\mathrm{Var}\left\{Z(R)\right\} \le (u(R))^2 \mathrm{Var}\left\{B[\tilde{U}(R)]\right\}. \tag{14}$$

Since $B[\tilde{U}(R)]$ is a binary random variable, we have

$$\mathrm{Var}\left\{B[\tilde{U}(R)]\right\} = E\left\{B[\tilde{U}(R)]\right\}\left(1 - E\left\{B[\tilde{U}(R)]\right\}\right) \tag{15}$$

By substituting in (13) $E\left\{Z(R)\right\}$ by its value $Rel(G[R])$, we deduce that

$$E\left\{B[\tilde{U}(R)]\right\} = (E\left\{Z(R)\right\} - w(R))/u(R) = (Rel(G[R]) - w(R))/u(R).$$

Consequently,

$$\mathrm{Var}\left\{B[\tilde{U}(R)]\right\} = \frac{(Rel(G[R]) - w(R))(u(R) + w(R) - Rel(G[R]))}{(u(R))^2}.$$

As

$$u(R) + w(R) + f(R) = 1$$

the last equality becomes

$$\text{Var}\left\{\mathcal{B}[\tilde{U}(R)]\right\} = \frac{(Rel(G[R]) - w(R))(1 - f(R) - Rel(G[R]))}{(u(R))^2}.$$

(16)

The substitution of $\text{Var}\left\{\mathcal{B}[\tilde{U}(R)]\right\}$ by its value in (14) gives

$$\text{Var}\left\{\mathcal{Z}(R)\right\} \leq (Rel(G[R]) - w(R))(1 - f(R) - Rel(G[R])). \quad (17)$$

From the decomposition process of Doulliez and Jamoulle, $W(R)$ is an operational set and $\bigcup_{j=1}^{a} F_j(R)$ is a failed set. Consequently, the values $w(R)$ and $1 - f(R)$ constitute a lower bound and an upper bound on $Rel(G[R])$, respectively. Then, $Rel(G[R]) \geq w(R) \geq 0$ and $1 - Rel(G[R]) \geq f(R) \geq 0$. We deduce

$$\text{Var}\left\{\mathcal{Z}(R)\right\} \leq Rel(G[R])(1 - Rel(G[R])).$$

From (a) and (b), we have the results (11) and (12) for every rectangular set $R \subseteq \Omega$.

The following proposition gives a worst-case bound on the variance-reduction ratio that can be achieved when the sampling is based on the random variable $\mathcal{Z}(R)$ defined by the recursive formula (9). Since this bound is always greater than 1, the sampling by $\mathcal{Z}(R)$ is always more accurate than the standard sampling.

**Proposition 4.2** *For any rectangular set $R \subseteq \Omega$ having both operational and failed vectors, a lower bound on the variance-reduction ratio $\frac{\text{Var}\{\mathcal{B}[R]\}}{\text{Var}\{\mathcal{Z}(R)\}}$ is given by the following inequality:*

$$\frac{\text{Var}\left\{\mathcal{B}[R]\right\}}{\text{Var}\left\{\mathcal{Z}(R)\right\}} \geq \frac{1}{1 - w(R)/Rel(G[R])} \times \frac{1}{1 - f(R)/(1 - Rel(G[R]))} > 1.$$

(18)

**Proof.** By using the result

$$\text{Var}\left\{\mathcal{B}[R]\right\} = Rel(G[R])(1 - Rel(G[R]))$$

and the inequality

$$\text{Var}\left\{\mathcal{Z}(R)\right\} \leq (Rel(G[R]) - w(R))(1 - f(R) - Rel(G[R])),$$

we deduce

$$\frac{\text{Var}\left\{\mathcal{B}[R]\right\}}{\text{Var}\left\{\mathcal{Z}(R)\right\}} \geq \frac{1}{1 - w(R)/Rel(G[R])} \times \frac{1}{1 - f(R)/(1 - Rel(G[R]))}. \quad (19)$$

Morevor, from the decomposition process of Doulliez and Jamoulle, we have $\bigcup_{j=1}^{a} U_j(R) \neq \emptyset$, and then $w(R) + f(R) \neq 0$. It results that we never have $w(R) = f(R) = 0$ and

$$\frac{1}{1 - w(R)/Rel(G[R])} \times \frac{1}{1 - f(R)/(1 - Rel(G[R]))} > 1. \qquad (20)$$

Result (18) follows from (19) and (20).

**Remark 4.3**
*For any rectangle $R \subseteq \Omega$, we can deduce from the formula (9) that the random variable $\mathcal{Z}(R)$ takes its values in the interval $[w(R), 1 - f(R)]$ and by the total probability theorem we can show that $\mathcal{Z}(R)$ is distributed as follows, for $z \in [w(R), 1 - f(R)]$:*

$$\Pr\{\mathcal{Z}(R) = z\} = \begin{cases} 1 & \text{if } u(R) = 0 \\ \sum_{j=1}^{a} \Pr\left\{\mathcal{Z}(U_j(R)) = \left(\frac{z - w(R)}{u(R)}\right)\right\} \Pr\left\{\tilde{U}(R) = U_j(R)\right\} \\ & \text{otherwise.} \end{cases}$$

*Then, the p.m.f. $f_{\mathcal{Z}(R)}$ can be deduced from the p.m.f. associated to the random variables $\mathcal{Z}(U_j(R))$ using the formula*

$$f_{\mathcal{Z}(R)}(z) = \begin{cases} 1 & \text{if } u(R) = 0 \\ \sum_{j=1}^{a} f_{\mathcal{Z}(U_j(R))}\left(\frac{z - w(R)}{u(R)}\right) \Pr\left\{\tilde{U}(R) = U_j(R)\right\} & \text{otherwise.} \end{cases}$$

*This function is not necessary for obtaining a trial of $\mathcal{Z}(R)$. We provide in the following section a procedure that exploits the formula (9) for generating recursively a trial of $\mathcal{Z}(R)$.*

## 4.2.2 The algorithm

For any rectangle $R \subseteq \Omega$, a procedure $\mathcal{Z}()$ that gives a trial of the random variable $\mathcal{Z}(R)$ defined by the recursive formula (9) can be summarized as follows:

Procedure $\mathcal{Z}(R)$

1. Check end recursion condition:
    1.1. Check if all vectors are operational: if $\Phi(\alpha(R)) = 1$ return 1.
    1.2. Check if all vectors are failed : if $\Phi(\beta(R)) = 0$ return 0.
2. Deduce from $G$ the network $G[R] = (V, A, \overrightarrow{C}[R], s, t, d)$ (see definition (1) for computing the distribution of $\overrightarrow{C}[R]$).

3. By the decomposition process of Doulliez and Jamoulle [1]:

    3.1. Find the upper and lower vectors of $W(R)$.

    3.2. For each $j \in \{1,\ldots,a\}$, find the upper and lower vector of $U_j(R)$.

    3.3. Compute the following parameters:

    3.3.1 $w(R) = \Pr\left\{ \vec{C}\,[R] \in W(R) \right\} = \prod\limits_{i=1}^{a} \sum\limits_{c \in \Omega_i(W(R))} p_{ic}[R].$

    3.3.2 $u_j(R) = \Pr\left\{ \vec{C}\,[R] \in U_j(R) \right\} = \prod\limits_{i=1}^{a} \sum\limits_{c \in \Omega_i(U_j(R))} p_{ic}[R], \ j \in \{1,\ldots,a\}.$

    3.3.3 $u(R) = \sum\limits_{j=1}^{a} u_j(R).$

4. Sample one undetermined rectangular set for the next recursive call:

    4.1. Compute the distribution of the discrete random variable $\widetilde{U}(R)$ (see Proposition 4.1).

    4.2. Accomplish a trial $U$ of $\widetilde{U}(R)$.

5. Recursive call: return$(w(R) + u(R) \times \mathcal{Z}(U))$.

**Remark 4.4**

*When the above procedure is called to obtain a trial of the random variable $\mathcal{Z}(R)$, at each step a recursive call is made for only one undetermined rectangular set whereas for the exact evaluation of $Rel(G[R])$ by formula (2), a recursive call is made for all undetermined rectangular sets $U_j(R)$, $j \in \{1,\ldots,a\}$. It results that a trial of $\mathcal{Z}(R)$ is obtained by considering only one branch of the tree related to the recursive exact evaluation.*

By calling $K$ times the procedure $\mathcal{Z}()$ with the parameter $R$ equal to $\Omega$ we collect $K$ trials of the random variable $\mathcal{Z}(\Omega)$ defined by (9) with $R$ equal to $\Omega$. These trials serve in formula (4) for estimating the measure $Rel(G[\Omega])$ and in formula (7) for estimating the parameter Var$\{\mathcal{Z}(\Omega)\}$ and the corresponding confidence interval (5) .

# 5   Numerical Results

In this section, we present numerical performances of the proposed Monte Carlo simulation algorithm. We use the network of 10 nodes and 25 arcs represented in Figure 1 [10]. For each arc $j$, the considered random discrete capacity $C_j$ is a Bernoulli random variable that has value 0 with probability $q$ and value $c_j$ with probability $p = 1 - q$. In Figure 1, the number on each arc $j$ indicates the capacity $c_j$. When the capacity of every arc $j$ is fixed at its upper value $c_j$, the maximum $st$-flow is equal to 71. Consequently, a demand greater than 71 at the sink node $t$ can not be satisfied.

Table 1 was used to validate the estimates given by the estimator $\widehat{\mathcal{Z}}$ based on $2^{16}$ trials of the random variable $\mathcal{Z}(\Omega)$ deduced from definition

| $d$ | $Rel(G)$ | $\widehat{Z}$ | $\delta(\widehat{Z})$ |
|---|---|---|---|
| 19 | 0.985506 | 0.985510 | $1.2 \times 10^{-4}$ |
| 36 | 0.909608 | 0.909392 | $5.7 \times 10^{-4}$ |
| 50 | 0.667960 | 0.667889 | $5.1 \times 10^{-4}$ |
| 60 | 0.456285 | 0.456406 | $3.9 \times 10^{-4}$ |
| 70 | 0.400488 | 0.400516 | $3.9 \times 10^{-4}$ |

TABLE 1. Simulation results for the network in figure 1 where $p = 0.9$, as a function of the demand $d$.

(9) by replacing $R$ by $\Omega$. We fixed $p$ at 0.9, and we tabulated for several demands the exact value of $Rel(G[\Omega])$ at column 2 and the estimate of $Rel(G[\Omega])$ by the estimator $\widehat{Z}$ at column 3. At column 4, we present the ratio $\delta(\widehat{Z}) = \epsilon(\widehat{Z})/Rel(G[\Omega])$ where $\epsilon(\widehat{Z})$ is defined by formula (6). Each ratio $\delta(\widehat{Z})$ is such that the probability of the event that the relative error overtakes the parameter $\delta(\widehat{Z})$ is smaller or equal to 1%. More formally, we have

$$\Pr\left\{\left(\frac{\mid Rel(G[\Omega]) - \widehat{Z} \mid}{Rel(G[\Omega])}\right) \leq \delta(\widehat{Z})\right\} \geq 99\%. \qquad (21)$$

The exact values of $Rel(G[\Omega])$ are derived by the exact algorithm of Doulliez and Jamoulle based on formula (2). Table 1 shows that estimates are close to exact values and the corresponding parameters $\delta(\widehat{Z})$ are small.

| $d$ | $\widehat{V}_Z$ | LBVR * | $\widehat{V}_B/\widehat{V}_Z$ ** | $\widehat{V}_B/\widehat{V}_{FS}[10]$ * ** |
|---|---|---|---|---|
| 19 | $1.413 \times 10^{-4}$ | 2.17 | 101.1 | 18.7 |
| 36 | $2.635 \times 10^{-3}$ | 1.74 | 31.2 | 3.6 |
| 50 | $1.120 \times 10^{-3}$ | 13 | 198.1 | 7.4 |
| 60 | $3.189 \times 10^{-4}$ | 105 | 777.8 | 575.0 |
| 70 | $2.466 \times 10^{-5}$ | 177 | 973.7 | 168.7 |

*LBVR is the lower bound on the variance ratio computed by formula (18).

**For $\widehat{V}_B$, the theoretical values $Rel(G[\Omega])(1 - Rel(G[\Omega]))$ are used.

***Results of this column concern the stratified sampling method proposed in [10].

TABLE 2. Evolution of the variance-reduction ratios, relative to the standard Monte Carlo estimator, of the estimator proposed in this paper and a previous one [10], as a function of the demand $d$.

In table 2, we provide for various demands an estimate of the variance of $Z$ (using formula (6) with $K = 2^{16}$) at column 2, a lower bound on the amount of variance-reduction with respect to the standard estimator (using the inequality (18) with $R = \Omega$) at column 3, the amount of variance-reduction with respect to the standard estimator at column 4. We also report the variance-reduction ratios relative to the standard Monte Carlo sampling of an other method [10]. This table shows for several values of the demand $d$ that the 2 estimators are more accurate than the standard one (the variance-reduction ratios are always greater than 1) and the proposed estimator is competitive when compared to the method in [10].

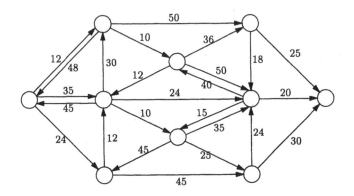

FIGURE 1. The network used for numerical illustrations [10].

# 6 Conclusions

The problem of the exact evaluation of the probability that the maximum $st$-flow exceeds a fixed value $d$ in a stochastic flow network is an NP-hard problem. Consequently, algorithms to compute it exactly have a high computational cost. When exact algorithms fail or when their computational time is prohibitive, Monte Carlo methods can supply an estimate in a reasonable time. In this paper, we propose to use the decomposition methodology of Doulliez & Jamoulle for constructing a Monte Carlo estimator. We show that the resulting estimator belongs to the variance-reduction family. We illustrate by numerical example that the proposed estimator offers substantial variance-reduction with respect to the standard Monte Carlo sampling. We also illustrate that the proposed estimator is competitive when compared to a previous Monte Carlo method.

# 7 REFERENCES

[1] P. Doulliez and E. Jamoulle. Transportation networks with random arc capacities. *R.A.I.R.O.*, nov. : 45–59, 1972.

[2] J. R. Evans. Maximal flow in probabilistic graphs - the discrete case. *Networks*, 6 : 161–183, 1976.

[3] J. V. Bukowski. On the determination of large-scale system reliability. *IEEE Transaction on Systems, Man and Cybernetics*, 12 : 538–548, 1982.

[4] S. H. Lee. Reliability in a flow network. *IEEE Transaction on Reliability*, 29 : 24–26, 1980.

[5] K.S. Trivedi. *Probability and Statistics with Reliability, Queuing, and Computer Science Applications*. Prentice-Hall, Inc., Englewood Cliffs, 1982.

[6] J.M. Hammersley and D.C. Handscomb. *Monte Carlo Methods*. Halsted Press, Wiley & Sons. Inc., New York, 1979.

[7] M.O. Ball. Computational complexity of network reliability analysis : an overview. *IEEE Transactions on Reliability*, 35 : 230–239, 1986.

[8] Cancela, H., and El Khadiri, M., "A recursive variance-reduction algorithm for estimating communication-network reliability," *IEEE Transaction on Reliability*, 44(4)599–602, 1995.

[9] El Khadiri, M., Direct evaluation and simulation of communication network reliability parameters Sequential and memory distributed parallel algorithms, PhD thesis, Rennes I, Campus de Beaulieu, 35042 Rennes, France, December 1992.

[10] Fishman, G.S. and Shaw, T.D., "Evaluating reliability of stochastic flow networks," *Probability in the Engineering and Informational Sciences*, 3:493–509, 1989.

[11] T. Elperin, I. Gertsbakh, and M. Lomonosov. Estimation of network reliability using graph evolution models. IEEE Trans. Reliab., 40(5)572–581, December 1991.

[12] Gondran, M. and Minoux, M., *Graphs and Algorithms*. Wiley-Interscience, Chichester, 1984.

# Monte Carlo and quasi-Monte Carlo algorithms for a linear integro-differential equation

## Ibrahim Coulibaly and Christian Lécot*

ABSTRACT We consider a linear Boltzmann-type transport equation in the $s$-dimensional unit cube $I^s = [0, 1)^s$. A quasi-Monte Carlo simulation algorithm is described. The equation is approximated by Euler's method and the simulation makes use of a $(0, 2s + 1)$-sequence. In addition, we use a technique involving renumbering the simulated particles at every time step. The convergence of the quasi-Monte Carlo simulation is studied. Experimental results are presented for a model problem whose solution can be found analytically. The results show that quasi-Monte Carlo algorithms can produce more efficient solutions than standard Monte Carlo algorithms, for $s \leq 3$. Of primary importance appears to be the way to do the renumbering.

## 1 Introduction

The basic idea of quasi-Monte Carlo algorithms is to replace random samples in Monte Carlo algorithms by evenly distributed points. In the case of numerical integration, quasi-Monte Carlo methods achieve a higher accuracy than Monte Carlo methods. A survey of numerous quasi-Monte Carlo methods is provided in the review article [12] and in the monograph [14] of Niederreiter. It seems to be promising to use deterministic instead of pseudorandom numbers in a simulation of the solution of an integral or a differential equation. Some quasi-Monte Carlo simulation schemes were recently proposed [1, 2, 4, 6, 9, 10, 15]. The results demonstrated that quasi-Monte Carlo algorithms could outperform standard Monte Carlo algorithms. In an earlier paper [6], a deterministic scheme was proposed for a nonlinear unidimensional Boltzmann equation. An error bound was provided, if the magnitude of the velocities of the simulated particles was used to reorder them in every time step.

In the present paper, we study a quasi-Monte Carlo simulation for a linear integro-differential multidimensional equation. The time is discretized by Euler's method. At each time step the solution is approximated by a sum

---

*Both authors are at: Laboratoire de Mathématiques, Université de Savoie, Campus scientifique, 73376 Le Bourget du Lac cedex, France

of $N$ Dirac measures, or particles. The particle movement is shown to be the quasi-Monte Carlo evaluation of an integral over the unit cube $I^{2s+1}$. The number of simulated particles is chosen to be $N = b^m$, with integers $b$ and $m$. We use a $(0, m, 2s + 1)$-net in base $b$ for the approximation. The integration error is reduced if the following relabeling technique is used at each time step. Let $m = d_1 + \ldots + d_s$, with integers $d_1, \ldots, d_s$. The particles are sorted in $b^{d_1}$ subsets of level 1, according to their coordinate $x_1$. Then the $b^{d_2} \cdots b^{d_s}$ particles of each subset of level 1 are sorted in $b^{d_2}$ subsets of level 2, according to their coordinate $x_2$, and so on. Finally the $b^{d_s}$ particles of each subset of level $s - 1$ are sorted according to their coordinate $x_s$. We derive the method as the natural generalization of a three-dimensional quasi-Monte Carlo method [8]. Here we prove that if the integers $d_1, \ldots, d_s$ are chosen so that $d_i \approx d_s/(s+2)$ for $i < s$, then a smaller error results. We also report a systematic computational study on a simple model problem which illustrates the theory. Moreover we show that by suitably coupling the timestep length to the number of particles used, the method performs uniformly well for all times.

The paper is organized as follows. In Section 1 a linear integro-differential Boltzmann-type equation is introduced. A quasi-Monte Carlo algorithm is described. In Section 2 a convergence proof is given. The error of the scheme is defined as the star discrepancy of the particles with respect to the exact solution. In Section 3 we consider a model problem where an exact solution is known, for $1 \leq s \leq 3$. Effective errors and experimental convergence rates are computed and compared with errors and convergence rates of a random simulation.

## 2 A quasi-Monte Carlo algorithm

We consider the integro-differential equation

$$\frac{\partial u}{\partial t}(x, t) =$$
$$\int_{I^s} (u(x', t)\gamma(x', x) - u(x, t)\gamma(x, x')) \, dx', \quad x \in I^s, \; t > 0, \quad (1)$$
$$u(x, 0) = u_0(x), \quad x \in I^s,$$

where

$$\int_{I^s} u_0(x)dx = 1.$$

In [8] a correspondence is established between equation (1), for $s = 3$, and the linear Boltzmann equation that describes the behavior of a mixture of two spatially homogeneous gases, when one of the components has a very small density and when the other species is in equilibrium. Our analysis is restricted to a bounded nonnegative kernel $\gamma$. We put

$$\|\gamma\|_\infty = \sup\{\gamma(x, x') \; : \; x, x' \in \overline{I}^s\}.$$

The *simple functions* are those whose range consists of only finitely many points in $[0, +\infty)$. The class of all simple measurable functions in $A$ will be denoted by $S(A)$. We use the following weak formulation

$$\frac{d}{dt}\int_{I^s}\sigma(x)u(x,t)dx =$$
$$\int_{I^{2s}}(\sigma(x') - \sigma(x))\gamma(x, x')u(x, t)dxdx', \quad \sigma \in S(\overline{I}^s), \qquad (2)$$

Denote the Dirac measure located at the point $y \in I^s$ by $\delta(x - y)$. We choose $N = b^m$ with integers $b \geq 2$ and $m \geq 0$. Let $d_1, \ldots, d_s$ be integers with $d_1 + \ldots + d_s = m$. We first sample $N$ points $x_j^{(0)}$ in $I^s$ from the initial distribution $u_0(x)dx$. We define $X^{(0)} = \{x_j^{(0)} \; : \; 0 \leq j < N\}$ and

$$u^{(0)}(x) = \frac{1}{N}\sum_{j=0}^{N-1}\delta(x - x_j^{(0)}).$$

A time step $\Delta t$ is chosen such that $\Delta t\|\gamma\|_\infty \leq 1$. We write $t_n = n\Delta t$ and $u_n(x) = u(x, t_n)$. We need a $(0, 2s + 1)$-sequence in base $b$: $y_0, y_1, \ldots$ for quasi-Monte Carlo approximation. Point sets $X^{(n)}$ of $N$ points $x_j^{(n)}$ in $I^s$ and measures

$$u^{(n)}(x) = \frac{1}{N}\sum_{j=0}^{N-1}\delta(x - x_j^{(n)})$$

are constructed by a step-by-step procedure.

(i) Reordering the particles: for integers $0 \leq a_i < b^{d_i}$, let $a = (a_1, \ldots, a_s)$. The $N$ points of $X^{(n)}$ are labeled $x_a^{(n)} = (x_{a,1}^{(n)}, \ldots, x_{a,s}^{(n)})$ so that

$$a_1 = b_1, \ldots, a_{i-1} = b_{i-1}, \; a_i < b_i \Rightarrow x_{a,i}^{(n)} \leq x_{b,i}^{(n)}.$$

(ii) Euler's method: define a measure $v^{(n)}$ by

$$\frac{1}{\Delta t}\int_{I^s}\sigma(x)\left(v^{(n)}(x) - u^{(n)}(x)\right) =$$
$$\int_{I^{2s}}(\sigma(x') - \sigma(x))\gamma(x, x')u^{(n)}(x)dx', \quad \sigma \in S(\overline{I}^s),$$

hence

$$\int_{I^s}\sigma(x)v^{(n)}(x) = \frac{1}{N}\sum_a\left(1 - \Delta t\int_{I^s}\gamma(x_a^{(n)}, x')dx'\right)\sigma(x_a^{(n)})$$
$$+ \frac{\Delta t}{N}\sum_a\int_{I^s}\sigma(x')\gamma(x_a^{(n)}, x')dx'. \qquad (3)$$

(iii) Quasi-Monte Carlo approximation: let $c_a$ be the characteristic function of

$$I_a = \prod_{i=1}^{s} \left[ \frac{a_i}{b^{d_i}}, \frac{a_i + 1}{b^{d_i}} \right),$$

and denote the characteristic function of

$$\{ x = (x', x_{s+1}) \in I^{s+1} \ : \ x_{s+1} < \Delta t \gamma(x_a^{(n)}, x') \}$$

by $\chi_a^{(n)}$. With each simple function $\sigma$ on $\overline{I}^s$ we associate a simple function $\Sigma^{(n)}$ on $\overline{I}^{2s+1}$ by

$$\Sigma^{(n)}(x) =$$
$$\sum_a c_a(x') \Big( \big(1 - \chi_a^{(n)}(x'', x_{2s+1})\big) \sigma(x_a^{(n)}) + \chi_a^{(n)}(x'', x_{2s+1}) \sigma(x'') \Big).$$

Then we have

$$\int_{I^s} \sigma(x) v^{(n)}(x) = \int_{I^{2s+1}} \Sigma^{(n)}(x) dx. \tag{4}$$

The measure $u^{(n+1)}$ is defined as follows.

$$\int_{I^s} \sigma(x) u^{(n+1)}(x) = \frac{1}{N} \sum_{j=0}^{N-1} \Sigma^{(n)}(y_{nN+j}), \quad \sigma \in S(\overline{I}^s).$$

**Remark 1** *For $0 \le j < N$, put*

$$a^{(n)}(j) = \big( \lfloor b^{d_1} y_{nN+j,1} \rfloor, \dots, \lfloor b^{d_s} y_{nN+j,s} \rfloor \big),$$

*where $\lfloor c \rfloor$ stands for the greatest integer $\le c$. We see that*

$$x_j^{(n+1)} = y''_{nN+j} \quad \text{if} \quad y_{nN+j,2s+1} < \Delta t \gamma\big( x_{a^{(n)}(j)}^{(n)}, y''_{nN+j} \big),$$
$$x_j^{(n+1)} = x_{a^{(n)}(j)}^{(n)},$$
$$\text{otherwise.}$$

## 3 Error estimations

We define the error of the scheme at time $t_n$ by

$$D_N^\star(X^{(n)}, u_n) = \sup_z |d_N^{(n)}(z)|,$$

where, for $z \in \overline{I}^s$,

$$d_N^{(n)}(z) = \frac{1}{N} \sum_a \sigma_z(x_a^{(n)}) - \int_{I^s} \sigma_z(x) u_n(x) dx,$$

and $\sigma_z$ is the characteristic function of $\prod_{i=1}^{s}[0, z_i)$. Then $D_N^{\star}(X^{(n)}, u_n)$ is the star $u_n$-discrepancy of the point set $X^{(n)}$ defined in the book of Hlawka, Firneis and Zinterhof [5, Chapter 1, Section 5]. We follow the error analysis of Euler's method. Consider the local discretization error

$$
\begin{aligned}
\varepsilon^{(n)}(z) \;=\; & \frac{1}{\Delta t} \int_{I^s} \sigma_z(x)\big(u_{n+1}(x) - u_n(x)\big)\,dx \\
& - \int_{I^{2s}} \big(\sigma_z(x') - \sigma_z(x)\big)\gamma(x, x')u_n(x)\,dx\,dx',
\end{aligned}
$$

the error term

$$
\begin{aligned}
e_N^{(n)}(z) = & \\
& \frac{1}{N}\sum_a \int_{I^s} \gamma(x_a^{(n)}, x')dx'\sigma_z(x_a^{(n)}) - \int_{I^{2s}} \gamma(x, x')\sigma_z(x)u_n(x)\,dx\,dx' - \\
& \frac{1}{N}\sum_a \int_{I^s} \gamma(x_a^{(n)}, x')\sigma_z(x')dx' + \int_{I^{2s}} \gamma(x, x')\sigma_z(x')u_n(x)\,dx\,dx',
\end{aligned}
$$

and the error of the quasi-Monte Carlo approximation

$$
\delta_N^{(n)}(z) = \frac{1}{N}\sum_{j=0}^{N-1} \Sigma_z^{(n)}(y_{nN+j}) - \int_{I^{2s+1}} \Sigma_z^{(n)}(x)\,dx,
$$

where $\Sigma_z^{(n)}$ is associated with $\sigma_z$. We have the recurrence formula

$$
d_N^{(n)}(z) = d_N^{(n-1)}(z) - \Delta t e_N^{(n-1)}(z) - \Delta t \varepsilon^{(n-1)}(z) + \delta_N^{(n-1)}(z). \tag{5}
$$

A bound for the error term $e_N^{(n)}(z)$ is obtained from the following lemmas.

**Lemma 1** *Suppose $\rho$ is a Riemann-integrable function on $\overline{I}^s$ such that $0 \le f$ and $\int_{I^s}\rho(x)dx = 1$. If $f$ is a function of bounded variation $V(f)$ on $\overline{I}^s$ in the sense of Hardy and Krause, and $x_0, \ldots, x_{N-1}$ are points in $I^s$, then*

$$
\left| \frac{1}{N}\sum_{j=0}^{N-1} f(x_j) - \int_{I^s} f(x)\rho(x)dx \right| \le V(f)\, D_N^{\star}(X, \rho).
$$

**Lemma 2** *Let $f$ be a function of bounded variation $V(f)$ on $\overline{I}^s$ in the sense of Hardy and Krause. If $z \in \overline{I}^s$, and if $\sigma_z$ is the characteristic function of $\prod_{i=1}^{s}[0, z_i)$, then $f\sigma_z$ has bounded variation on $\overline{I}^s$ in the sense of Hardy and Krause and*

$$
V(f\sigma_z) \le V(f) + |f(1, \ldots, 1)|.
$$

This leads to the following estimation.

**Proposition 1** *If $\gamma$ has bounded variation $V(\gamma)$ on $\overline{I}^{2s}$ in the sense of Hardy and Krause, then*

$$\sup_{z \in \overline{I}^s} |e_N^{(n)}(z)| \le (2V(\gamma) + \|\gamma\|_\infty) D_N^*(X^{(n)}, u_n).$$

We immediately obtain

$$|\varepsilon^{(n)}(z)| \le \int_{t_n}^{t_{n+1}} \int_{I^s} \left| \frac{\partial^2 u}{\partial t^2}(x, t) \right| dx\, dt. \tag{6}$$

Now we derive a bound for the error of the quasi-Monte Carlo approximation. The following inequality was established by Niederreiter [13, Lemma 3.4].

**Lemma 3** *Let $X$ be a $(t, m, s)$-net in base $b$. For any elementary interval $J' \subset I^{s-1}$ in base $b$ and for any $x_s \in \overline{I}$*

$$\left| \frac{A(J' \times [0, x_s), X)}{b^m} - \lambda_s(J' \times [0, x_s)) \right| \le b^{t-m}.$$

We can write

$$\delta_N^{(n)}(z) = \frac{A(F_z^{(n)}, Y^{(n)})}{N} - \lambda_{2s+1}(F_z^{(n)}) -$$

$$\frac{A(E_{z,1}^{(n)}, Y^{(n)})}{N} + \lambda_{2s+1}(E_{z,1}^{(n)}) + \frac{A(E_{z,2}^{(n)}, Y^{(n)})}{N} - \lambda_{2s+1}(E_{z,2}^{(n)}), \tag{7}$$

where

$$F_z^{(n)} = \{x = (x', x'', x_{2s+1}) \in I^{2s+1} : \sum_a c_a(x')\sigma_z(x_a^{(n)}) = 1\},$$

$$E_{z,k}^{(n)} = \{x = (x', x'', x_{2s+1}) \in I^{2s+1} : x_{2s+1} < \gamma_{z,k}^{(n)}(x', x'')\},$$

$$\text{for } k = 1, 2,$$

$$\gamma_{z,1}^{(n)}(x', x'') = \Delta t \sum_a c_a(x')\gamma(x_a^{(n)}, x'')\sigma_z(x_a^{(n)}), \quad (x', x'') \in \overline{I}^{2s},$$

$$\gamma_{z,2}^{(n)}(x', x'') = \Delta t \sum_a c_a(x')\gamma(x_a^{(n)}, x'')\sigma_z(x''), \quad (x', x'') \in \overline{I}^{2s}.$$

The point set $Y^{(n)}$ of the $y_{nN+j}$ with $0 \le j < N$ is a $(0, m, 2s + 1)$-net in base $b$. Since $F_z^{(n)}$ can be written as the disjoint union of elementary intervals in base $b$ of volume $b^{-m}$, we have

$$\frac{A(F_z^{(n)}, Y^{(n)})}{N} - \lambda_{2s+1}(F_z^{(n)}) = 0. \tag{8}$$

The following inequality is a generalization of an earlier result [7, Lemma 1].

**Lemma 4** *Let $f$ be a function of bounded variation $V(f)$ on $\overline{I}^s$ in the sense of Hardy and Krause. Let*

$$0 = x_{0,1} \leq x_{1,1} \leq \cdots \leq x_{n_1,1} = 1,$$
$$0 = x_{a_1,0,2} \leq x_{a_1,1,2} \leq \cdots \leq x_{a_1,n_2,2} = 1,$$
$$\text{for} \quad 0 \leq a_1 < n_1,$$

$$\cdots$$

$$0 = x_{a',0,s} \leq x_{a',1,s} \leq \cdots \leq x_{a',n_s,s} = 1,$$
$$\text{for} \quad a' = (a_1, \ldots, a_{s-1}), \quad 0 \leq a_i < n_i$$

*define partitions of $\overline{I}$ into subintervals. For $a = (a_1, \ldots, a_s)$ with integers $a_i$, $0 \leq a_i < n_i$ let*

$$\overline{I}_a = [x_{a_1,1}, x_{a_1+1,1}] \times [x_{a_1,a_2,2}, x_{a_1,a_2+1,2}] \times \cdots \times [x_{a',a_s,s}, x_{a',a_s+1,s}]$$

*and $y_a, z_a \in \overline{I}_a$. Then*

$$\sum_a |f(z_a) - f(y_a)| \leq V(f) \prod_{i=1}^{s} n_i \sum_{i=1}^{s} \frac{1}{n_i}.$$

**Proposition 2** *If $\gamma$ has bounded variation $V(\gamma)$ on $\overline{I}^{2s}$ in the sense of Hardy and Krause, then, for $k = 1, 2$,*

$$\left| \frac{A(E_{z,k}^{(n)}, Y^{(n)})}{N} - \lambda_{2s+1}(E_{z,k}^{(n)}) \right| \leq$$

$$\frac{1}{b^{d_s - (s+1)\lfloor \frac{d_s}{s+2} \rfloor}} + (V(\gamma) + \|\gamma\|_\infty)\Delta t \left( \sum_{i=1}^{s-1} \frac{1}{b^{d_i}} + \frac{s+1}{b^{\lfloor \frac{d_s}{s+2} \rfloor}} \right).$$

**Proof** Let $\delta_1, \ldots, \delta_{2s}$ be integers. For $\alpha = (\alpha', \alpha'') = (\alpha_1, \ldots, \alpha_{2s})$ with integers $0 \leq \alpha_i < b^{\delta_i}$ we put

$$I'_{\alpha'} = \prod_{i=1}^{s} \left[ \frac{\alpha_i}{b^{\delta_i}}, \frac{\alpha_i + 1}{b^{\delta_i}} \right), \quad I''_{\alpha''} = \prod_{i=s+1}^{2s} \left[ \frac{\alpha_i}{b^{\delta_i}}, \frac{\alpha_i + 1}{b^{\delta_i}} \right), \quad I_\alpha = I'_{\alpha'} \times I''_{\alpha''}$$

and

$$\underline{E}_{z,k}^{(n)} = \bigcup_\alpha I_\alpha \times \left[ 0, \inf_{I_\alpha} \gamma_{z,k}^{(n)} \right),$$

$$\overline{E}_{z,k}^{(n)} = \bigcup_\alpha I_\alpha \times \left[ 0, \sup_{I_\alpha} \gamma_{z,k}^{(n)} \right),$$

$$\partial E_{z,k}^{(n)} = \bigcup_\alpha I_\alpha \times \left[ \inf_{I_\alpha} \gamma_{z,k}^{(n)}, \sup_{I_\alpha} \gamma_{z,k}^{(n)} \right].$$

Since $\underline{E}_{z,k}^{(n)} \subset E_{z,k}^{(n)} \subset \overline{E}_{z,k}^{(n)}$, we have

$$\frac{A(\underline{E}_{z,k}^{(n)}, Y^{(n)})}{N} - \lambda_{2s+1}(\underline{E}_{z,k}^{(n)}) - \lambda_{2s+1}(\partial E_{z,k}^{(n)}) \le$$

$$\frac{A(E_{z,k}^{(n)}, Y^{(n)})}{N} - \lambda_{2s+1}(E_{z,k}^{(n)}) \le$$

$$\frac{A(\overline{E}_{z,k}^{(n)}, Y^{(n)})}{N} - \lambda_{2s+1}(\overline{E}_{z,k}^{(n)}) + \lambda_{2s+1}(\partial E_{z,k}^{(n)}). \tag{9}$$

Then it follows from Lemma 3 that

$$\left| \frac{A(E_{z,k}^{(n)}, Y^{(n)})}{N} - \lambda_{2s+1}(E_{z,k}^{(n)}) \right| \le$$

$$\frac{b^{\delta_1 + \dots + \delta_{2s}}}{b^m} + \frac{1}{b^{\delta_1 + \dots + \delta_{2s}}} \sum_\alpha \left( \sup_{I_\alpha} \gamma_{z,k}^{(n)} - \inf_{I_\alpha} \gamma_{z,k}^{(n)} \right). \tag{10}$$

Define

$$\sigma_{z,1}(\boldsymbol{x}', \boldsymbol{x}'') = \sigma_z(\boldsymbol{x}'), \quad \sigma_{z,2}(\boldsymbol{x}', \boldsymbol{x}'') = \sigma_z(\boldsymbol{x}''), \quad (\boldsymbol{x}', \boldsymbol{x}'') \in \overline{I}^{2s}.$$

If $\delta_1 = d_1, \dots, \delta_{s-1} = d_{s-1}$ and $\delta_s \le d_s$, we get

$$\sup_{I_\alpha} \gamma_{z,k}^{(n)} = \Delta t \sup \left\{ (\gamma \sigma_{z,k})(\boldsymbol{x}_{\alpha_1, \dots, \alpha_{s-1}, a_s}^{(n)}, \boldsymbol{x}'') : \right.$$

$$\left. \alpha_s b^{d_s - \delta_s} \le a_s < (\alpha_s + 1) b^{d_s - \delta_s}, \ \boldsymbol{x}'' \in I_{\alpha''}'' \right\}. \tag{11}$$

A similar statement holds for $\inf_{I_\alpha} \gamma_{z,k}^{(n)}$. Because the points $\boldsymbol{x}_\alpha^{(n)}$ are reordered, we can define partitions of $\overline{I}$:

$$0 = w_{0,1}^{(n)} \le w_{1,1}^{(n)} \le \dots \le w_{b^{d_1},1}^{(n)} = 1,$$

$$0 = w_{\alpha_1,0,2}^{(n)} \le w_{\alpha_1,1,2}^{(n)} \le \dots \le w_{\alpha_1,b^{d_2},2}^{(n)} = 1, \quad \text{for} \quad 0 \le \alpha_1 < b^{d_1},$$

$$\dots$$

$$0 = w_{\alpha_1,\dots,\alpha_{s-1},0,s}^{(n)} \le w_{\alpha_1,\dots,\alpha_{s-1},1,s}^{(n)} \le \dots \le w_{\alpha_1,\dots,\alpha_{s-1},b^{\delta_s},s}^{(n)} = 1,$$

$$\text{for} \quad 0 \le \alpha_1 < b^{d_1}, \dots, 0 \le \alpha_{s-1} < b^{d_{s-1}}$$

such that, for any integer $a_s$ with $\alpha_s b^{d_s - \delta_s} \le a_s < (\alpha_s + 1) b^{d_s - \delta_s}$, we have

$$\boldsymbol{x}_{\alpha_1,\dots,\alpha_{s-1},a_s}^{(n)} \in \left[ w_{\alpha_1,1}^{(n)}, w_{\alpha_1+1,1}^{(n)} \right] \times \dots \times \left[ w_{\alpha_1,\dots,\alpha_s,s}^{(n)}, w_{\alpha_1,\dots,\alpha_s+1,s}^{(n)} \right]. \tag{12}$$

By Lemmas 2 and 4 it follows that

$$\frac{1}{b^{\delta_1 + \dots + \delta_{2s}}} \sum_\alpha \left( \sup_{I_\alpha} \gamma_{z,k}^{(n)} - \inf_{I_\alpha} \gamma_{z,k}^{(n)} \right) \le (V(\gamma) + \|\gamma\|_\infty) \Delta t \sum_{i=1}^{2s} \frac{1}{b^{\delta_i}}. \tag{13}$$

If we choose $\delta_s = \ldots = \delta_{2s} = \left\lfloor \frac{d_s}{s+2} \right\rfloor$, we obtain the desired inequality. qed

We can combine Propositions 1 and 2 to derive an upper bound for the error of the quasi-Monte Carlo algorithm.

**Proposition 3** *Suppose $\gamma$ has bounded variation $V(\gamma)$ on $\bar{I}^{2s}$ in the sense of Hardy and Krause, and let $|\gamma| = 2V(\gamma) + \|\gamma\|_\infty$. Then*

$$D_N^\star(X^{(n)}, u_n) \le e^{|\gamma|t_n} D_N^\star(X^{(0)}, u_0) +$$

$$\Delta t \int_{(0,t_n)\times I^s} e^{|\gamma|(t_n - t)} \left| \frac{\partial^2 u}{\partial t^2}(x, t) \right| dx\, dt +$$

$$\frac{2e^{|\gamma|t_n}}{|\gamma|\Delta t} \left( \frac{1}{b^{d_s - (s+1)\lfloor \frac{d_s}{s+2} \rfloor}} + (V(\gamma) + \|\gamma\|_\infty)\Delta t \left( \sum_{i=1}^{s-1} \frac{1}{b^{d_i}} + \frac{s+1}{b^{\lfloor \frac{d_s}{s+2} \rfloor}} \right) \right).$$

## 4   Model problem results

We examine an example, in which the exact solution is known, to study the effectiveness of quasi-Monte Carlo (QMC) algorithm, when compared with a standard Monte Carlo (MC) algorithm devised by Nanbu [11]. If

$$\forall x \in \bar{I}^s \quad \int_{I^s} \gamma(x, x')dx' = \Gamma,$$

the Laplace transform of equation (1) is a Fredholm integral equation of the second kind whose solution is obtained by use of the method of successive approximation. With

$$\gamma(x, x') = \prod_{i=1}^s \left( 1 - \frac{1}{4}\left( 1 - 12\left( x_i - \frac{1}{2} \right)^2 \right)\left( 1 - 12\left( x_i' - \frac{1}{2} \right)^2 \right) \right),$$

and

$$u_0(x) = \prod_{i=1}^s \frac{1}{\pi\sqrt{x_i(1 - x_i)}},$$

we obtain

$$u(x, t) = e^{-t}u_0(x) + \sum_{r=0}^s (-5)^r \left( e^{-(1-(-1/5)^r)t} - e^{-t} \right) \cdot$$

$$\sigma_r\left( \frac{1}{8} - \frac{3}{2}\left( x_1 - \frac{1}{2} \right)^2, \ldots, \frac{1}{8} - \frac{3}{2}\left( x_s - \frac{1}{2} \right)^2 \right), \tag{14}$$

where $\sigma_0, \ldots, \sigma_s$ are the elementary symmetric polynomials of $x_1, \ldots, x_s$.

Both simulations are conducted by first sampling $N$ particles from the initial distribution. Since

$$u_0(x) = \prod_{i=1}^{s} u_{0,i}(x_i),$$

this sampling is done by mapping a $(0, m, s)$-net in base $b$, $\Xi$, to $\overline{I}^s$ using the inverse function of

$$U_0(x) = \left( \int_0^{x_1} u_{0,1}(w)dw, \ldots, \int_0^{x_s} u_{0,s}(w)dw \right).$$

A straightforward argument shows that

$$D_N^{\star}(X^{(0)}, u_0) = D_N^{\star}(\Xi). \tag{15}$$

The QMC algorithm makes use of a $(0, 2s + 1)$-sequence in base $b$. The following result was shown by Niederreiter [13, Corollary 5.17].

**Lemma 5** *A $(0, s)$-sequence in base $b$ can only exist if $s \leq b$.*

The construction by Faure [3] yields $(0, s)$-sequences in prime bases $\geq s$. The calculation of the discrepancy $D_N^{\star}(X^{(n)}, u_n)$ is very expensive in dimension $s \geq 2$, so an estimate of the error is computed:

$$\tilde{D}_N^{\star}(X^{(n)}, u_n) = \max_{1 \leq i \leq s} \sup_{z \in \overline{I}} |d_N^{(n)}(z_{(i)})|,$$

where $z_{(i)i'} = z$, if $i' = i$; otherwise $z_{(i)i'} = 1$. A convergence rate $N^{-p}$ is estimated by plotting the error exponent

$$p_N^{(n)} = -\log\left(\tilde{D}_N^{\star}(X^{(n)}, u_n)\right) / \log N.$$

For the simulations, the time steps are of size $\Delta t = 0.01$. A sample size of $N = b^{d_1 + \cdots + d_s}$ particles is used. The errors $\tilde{D}_N^{\star}(X^{(n)}, u_n)$ and error exponents $p_N^{(n)}$ of MC and QMC algorithms in dimension $s = 1, 2, 3$ are shown in Figures 1-5. Dashed lines correspond to MC simulations and solid lines correspond to QMC simulations. Thin lines correspond to small $N$ and thick lines correspond to large $N$.

We see in Figures 1, 2 and 4 that the QMC algorithm outperforms the MC algorithm, if $d_i \approx d_s/(s + 2)$ for $i < s$. Another choice can result in significant loss of efficiency: see Figures 3 and 5. In contrast with our preceding computational study [8], the error does not grow when $t$ is small: by suitably coupling $\Delta t$ to $N$, the error can be made arbitrarily small.

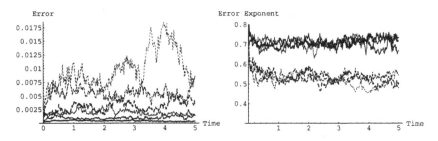

FIGURE 1. MC and QMC schemes, $s = 1$, $N = 3^8$, $3^9$, $3^{10}$.

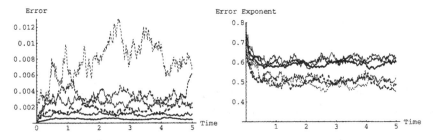

FIGURE 2. MC and QMC schemes, $s = 2$, $N = 5^{2+4}$, $5^{2+5}$, $5^{2+6}$.

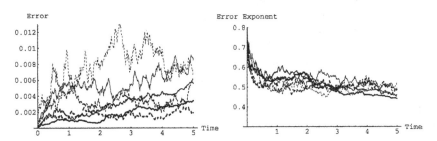

FIGURE 3. MC and QMC schemes, $s = 2$, $N = 5^{4+2}$, $5^{5+2}$, $5^{6+2}$.

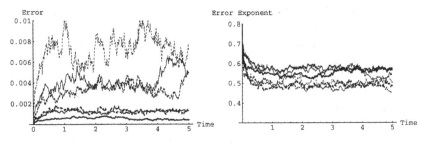

FIGURE 4. MC and QMC schemes, $s = 3$, $N = 7^{1+1+3}$, $7^{1+1+4}$, $7^{1+1+5}$.

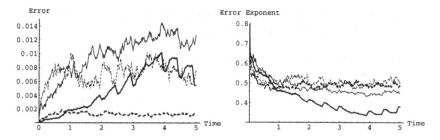

FIGURE 5. MC and QMC schemes, $s = 3$, $N = 7^{2+2+1}$, $7^{3+3+1}$.

# 5 Conclusion

A quasi-Monte Carlo algorithm for solving a linear integro-differential equation was analyzed in this paper. The time was discretized by Euler's method and the solution was approximated by a sum of Dirac measures (particles). An error bound was established when the particles are relabeled in every time step. The numerical results demonstrate that the new scheme can produce more accurate results than standard Monte Carlo simulation, when the correct renumbering technique is used. Other choices can fail to improve on Monte Carlo. The extension of the method to the spatially homogeneous nonlinear Boltzmann equation will be addressed in a forthcoming paper.

**Acknowledgements**
We would like to thank the Centre Grenoblois de Calcul Vectoriel du Commissariat à l'Énergie Atomique for providing access to the CRAY C90 computer.

# 6 REFERENCES

[1] R. E. Caflisch and B. Moskowitz, Modified Monte Carlo methods using quasi-random sequences, in: H. Niederreiter and P.J.-S. Shiue, Eds, *Monte Carlo and Quasi-Monte Carlo Methods in Scientific Computing* (Springer, New York, 1995) 1-16.

[2] C.S. Chen, The method of fundamental solutions and the quasi-Monte Carlo method for Poisson's equation, in: H. Niederreiter and P.J.-S. Shiue, Eds, *Monte Carlo and Quasi-Monte Carlo Methods in Scientific Computing* (Springer, New York, 1995) 158-167.

[3] H. Faure, Discrépance de suites associées à un système de numération (en dimension $s$), *Acta Arith.* **41** (1982) 337-351.

[4] A. Keller, A quasi-Monte Carlo algorithm for the global illumination problem in the radiosity setting, in: H. Niederreiter and P.J.-S. Shiue,

Eds, *Monte Carlo and Quasi-Monte Carlo Methods in Scientific Computing* (Springer, New York, 1995) 239-251.

[5] E. Hlawka, F. Firneis and P. Zinterhof *Zahlentheoretische Methoden in der Numerischen Mathematik* (R. Oldenburg Verlag, Wien, 1981).

[6] C. Lécot, A quasi-Monte Carlo method for the Boltzmann equation, *Math. Comp.* **56** (1991) 621-644.

[7] C. Lécot, Error bounds for quasi-Monte Carlo integration with nets, *Math. Comp.* **65** (1996) 179-187.

[8] C. Lécot and I. Coulibaly, A quasi-Monte Carlo scheme using nets for a linear Boltzmann equation, *SIAM J. Numer. Anal.*, to appear.

[9] W. J. Morokoff and R. E. Caflisch, A quasi-Monte Carlo approach to particle simulation of the heat equation, *SIAM J. Numer. Anal.* **30** (1993) 1558-1573.

[10] B. Moskowitz, Quasirandom diffusion Monte Carlo, in: H. Niederreiter and P.J.-S. Shiue, Eds, *Monte Carlo and Quasi-Monte Carlo Methods in Scientific Computing* (Springer, New York, 1995) 278-298.

[11] K.Nanbu, Stochastic solution method of the master equation and the model Boltzmann equation *J. Phys. Soc. Japan* **52** (1983) 2654-2658.

[12] H. Niederreiter, Quasi-Monte Carlo methods and pseudo-random numbers, *Bull. Amer. Math. Soc.* **84** (1978) 957-1041.

[13] H. Niederreiter, Point sets and sequences with small discrepancy, *Monatsh. Math.* **104** (1987) 273-337.

[14] H. Niederreiter, *Random Number Generation and Quasi-Monte Carlo Methods* (Society for Industrial and Applied Mathematics, Philadelphia, 1992).

[15] J. Spanier, Quasi-Monte Carlo methods for particle transport problems, in: H. Niederreiter and P.J.-S. Shiue, Eds, *Monte Carlo and Quasi-Monte Carlo Methods in Scientific Computing* (Springer, New York, 1995) 121-148.

# A Numerical Approach for Determination of Sources in Reactive Transport Equations

I. Dimov[1], U. Jaekel[2], H. Vereecken[2] and D. Wendt[3]

ABSTRACT  A new numerical approach for locating the source and quantifying its power is presented and studied. The proposed method may also be helpful when dealing with other transport problems.

The method uses a numerical integration technique to evaluate a linear functional arising from the solution of the adjoint problem. It is shown that using the solution of the adjoint equation an efficient numerical method can be constructed (a method, which is less time comsumpting than the usual schemes applied for solving the original problem). Our approach leads to a well-conditioned numerical problem.

A number of numerical tests are performed for a one-dimentional problem with a linear advection part. The results are compared with results obtained when solving the original problem. The code which realizes the presented method is written in FORTRAN 77.

## 1  Introduction

In this paper we discuss – for a simple but nontrivial example – how to obtain information on unknown sources in a transport equation from "observed" data. We assume that the data $J_i$ are of the form

$$J_i = \int dx \int dt\, p_i(x,t)u(x,t), \qquad (1)$$

where $x$ and $t$ are the space and time co-ordinates, respectively, $u(x,t)$ is the solution of the transport equation, and $p_i(x,t)$ is a function or distribution specified by the observation process. This covers a large number of important special cases. Some examples are:

---

[1]Central Laboratory for Parallel Computing, Bulgarian Academy of Sciences, Acad. G. Bonchev St.,bl. 25 A᛫, Sofia, 1113, Bulgaria, E-mail: dimov@amigo.acad.bg

[2]Research Centre Jülich, ICG-4, D-52428 Jülich, Germany, E-mail: u.jaekel@fz-juelich.de, h.vereecken@fz-juelich.de

[3]Institute of Theoretical Physics D, RWTH Aachen, D-52056 Aachen, Germany E-mail: dietmar@physik.rwth-aachen.de

1. $p_i(x,t) := \delta(x - x_i)\delta(t - t_i)$ : This coresponds to a measurement at time $t_i$ and position $x_i$.

2. $p_i(x,t) := \delta(x - x_i)H(t_i - t)$ : In this case $J_i$ is the total mass which has passed the position $x_i$ until a time $t_i$.

3. $p_i(x,t) := r(x_i - x)s(t_i - t)$ : This type occurs for measurements with only finite spatial and temporal resolution (characterized by the functions $r$ and $s$).

The information about the sources is reconstructed from the data using an adjoint formulation of the problem. We demonstrste its usefulness by solving two problems for a one-dimensional, not translation-invariant and hence nontrival, example:

1. We determine the set of points where a point source can be located such that the solution $u(x_1, t_1)$ at a fixed position and time is smaler or larger than a certain value $C$. As one of many possible applications one might think about the environmental problem of determining regions where possible pollutant sources must not be located, in order to keep pollutant concentrations below a prescribed level in a protected region.

2. We reconstruct the location and power of a steady state point source from two observations $u(x_1, t_1)$ and $u(x_2, t_2)$. A possible application would be the identification and characterization of a polutant source in an inaccessible area.

Our formulation leads to a well-conditioned problem and can be generalized to a large class of more complicated cases.

## 2  Formulation

Let us consider the following problem of $1D-$ transport:

$$\left(\frac{\partial}{\partial t} - L\right) u = \frac{\partial u}{\partial t} - \frac{\partial}{\partial x}(\alpha x u) - D\frac{\partial^2 u}{\partial x^2} = Q(t)\delta(x - x_0), \qquad (2)$$

$$u = u_0(x), \quad \text{for } t = 0. \qquad (3)$$

Assume that $u(x,t) \in W_1^2$ , where $W_1^2$ is a Sobolev space, $u(x,t) > 0$, $D > 0$, $-\infty < x < \infty$, $t \in [0,T]$ and the conditions providing the existence of the unique solution of the problem (2) - (3) in a weak formulation are fulfilled.

The problem, presented by (2) - (3) does not have an analytical solution, due to presence of the function $Q(t)$. In addition standard numerical methods used to solve the equation (2) are not efficient because there is a $\delta$ -function in (2).

The condition $u(x,t) \in W_1^2$ implies that

$$\int_0^T dt \int_{-\infty}^\infty u^{(2)}(x,t)dx < \infty, \tag{4}$$

which is a natural condition.

Now formulate the problems under consideration.

**Problem 1.** Find the subdomain $g$ of $G$ ($g \subset G$) such that if at any point of $g$ a source of power $Q(t)$ will be located the concentrations of the pollutants in a given point $x = x_1$ at a given time $t = t_1$ will be less than a given constant $C$,

that is to find the set of points $x_0$ for which

$$u(x_1, t_1) \le C. \tag{5}$$

**Problem 2.** Find the place of location $x_0$ and the power of the source $Q(t) = Q = const$ if two measurement data $u(x_1, t_1)$ and $u(x_2, t_2)$ are available. (It is assumed, that $x_1 \ne x_2$ and $t_1 \ne t_2$.)

The first problem (the problem of finding the subdomain $g$) is not so complicated if the solution of the adjoint formulation of the original problem (2) - (3) will be used.

## 3  Adjoint formulation

In fact, multiply the equation (2) by a function $u^*$, which will be defined letter, and integrate on time and space

$$\int_0^T dt \int_{-\infty}^\infty u^* \left( \frac{\partial u}{\partial t} - \frac{\partial}{\partial x}(\alpha x u) - D\frac{\partial^2 u}{\partial x^2} \right) dx$$

$$= \int_0^T Q(t)dt \int_{-\infty}^\infty u^* \delta(x - x_0)dx. \tag{6}$$

Applying the integration by part to (6) one can obtain (it is assumed that $u^*|_{t=T} = 0$):

$$\int_0^T dt \int_{-\infty}^\infty u^* \frac{\partial u}{\partial t} dx = \int_{-\infty}^\infty uu^* dx|_{t=0}^{t=T} - \int_0^T dt \int_{-\infty}^\infty u \frac{\partial u^*}{\partial t} dx$$

$$= -\int_{-\infty}^\infty u_0(x)u_0^*(x)dx - \int_0^T dt \int_{-\infty}^\infty u \frac{\partial u^*}{\partial t} dx, \tag{7}$$

$$\int_0^T dt \int_{-\infty}^{\infty} u^* \frac{\partial^2 u}{\partial x^2} dx$$

$$= \int_0^T \left( u^* \frac{\partial u}{\partial x} - u \frac{\partial u^*}{\partial x} \right) dx \Big|_{x=-\infty}^{\infty} + \int_0^T dt \int_{-\infty}^{\infty} u \frac{\partial^2 u^*}{\partial x^2} dx, \qquad (8)$$

and

$$\int_0^T dt \int_{-\infty}^{\infty} u^* \frac{\partial (xu)}{\partial x} dx = \int_{-\infty}^{\infty} xuu^* dx \Big|_{t=0}^{t=T} - \int_0^T dt \int_{-\infty}^{\infty} xu \frac{\partial u^*}{\partial x} dx$$

$$= -\int_{-\infty}^{\infty} xu_0(x)u_0^*(x) dx - \int_0^T dt \int_{-\infty}^{\infty} xu \frac{\partial u^*}{\partial x} dx. \qquad (9)$$

Let us assume that $u = u^* = 0$ when $x \to \infty$ or $x \to -\infty$.
Now from (7), (8) and (9) one can obtain:

$$\int_0^T dt \int_{-\infty}^{\infty} u \left( -\frac{\partial u^*}{\partial t} + \alpha x \frac{\partial u^*}{\partial x} - D \frac{\partial^2 u^*}{\partial x^2} \right) dx$$

$$- \int_{-\infty}^{\infty} (1 - x)u_0(x)u_0^*(x) dx = \int_0^T Q(t)u^*(x_0, t) dt \qquad (10)$$

Let the function $u^*$ satisfies the following adjoint equation:

$$-\frac{\partial u^*}{\partial t} + \alpha x \frac{\partial u^*}{\partial x^*} - D \frac{\partial^2 u}{\partial x^2} = p(x, t), \qquad (11)$$

with an initial condition

$$u^*(x, T) = 0, \qquad (12)$$

where the function $p(x, t)$ will be defined later.

It is possible to write (10) in the next form:

$$\int_0^T dt \int_{-\infty}^{\infty} p(x, t)u(x, t) dx = \int_0^T Q(t)u^*(x_0, t) dt + \int_{-\infty}^{\infty} (1-x)u_0(x)u_0^*(x) dx = J \qquad (13)$$

The equation (13) will be a basic equation for our numerical method. One can choose the function $p(x, t)$ in form of a product of two $\delta$-functions:

$$p(x, t) = \delta(x - x_1)\delta(t - t_1). \qquad (14)$$

Let us also assume that $Q(t)$ contains the Heaviside function $H(t)$, that is

$$Q(t) = q(t)H(t),$$

where $H(t) = 1$ for $t > 0$ and $H(t) = 0$ for $t \leq 0$. This means that the second term of the right-hand side of (13) vanishes.

Consider the following linear functional of the solution of the original problem:

$$J = \int_0^T dt \int_{-\infty}^{\infty} p(x,t)u(x,t)dx. \tag{15}$$

In this case we have

$$J = \int_0^T dt \int_{-\infty}^{\infty} u(x,t)\delta(x - x_1)\delta(t - t_1)dx = u(x_1, t_1). \tag{16}$$

Obviously, the value of the functional (15) is equal to the value of the concentration in point $x = x_1$ at the time moment $t = t_1$.

¿From the representation (13) it follows that instead of solving the original problem one can solve the adjoint problem for $u^*$:

$$-\frac{\partial u^*}{\partial t} + \alpha x \frac{\partial u^*}{\partial x} - D \frac{\partial^2 u^*}{\partial x^2} = \delta(x - x_1)\delta(t - t_1) \tag{17}$$

$$u^* = 0, \quad \text{for } t = T. \tag{18}$$

¿From the solution of the problem formulated by (17)–(18) one can estimate the functional (13) as a function depending of a parameter $x_0$ and solve the first problem under consideration. In fact, if the function $u^*(x, t, x_1, t_1)$ is the solution of the adjoint problem (17)– (18) (the solution $u^*(x, t, x_1, t_1)$ of the adjoint problem depends of parameters $x_1$ and $t_1$), then the presentation (13) defines a function of $x_0$ - $J(x_0, x_1, t_1)$. Now, solving the inequality

$$J(x_0, x_1, t_1) \leq C, \tag{19}$$

the set $g$ of points $x_0$ can be determined. That is the solution of the first problem under consideration.

Now, consider the second problem. Let the power of the source $q$ and its location $x_0$ be unknown. Two measurement data points - $u(x_1, t_1)$ and $u(x_2, t_2)$ are assumed to be available. One can then define two functions of $x_0$ - $F_1(x_0)$ and $F_2(x_0)$ using the following formal presentations:

$$\hat{J}(x_0, x_i, t_i) = \frac{J(x_0, x_i, t_i)}{q}, \quad i = 1, 2; \tag{20}$$

$$F_i(x_0) = \frac{\hat{J}(x_0, x_i, t_i)}{u(x_i, t_i)}, \quad i = 1, 2. \tag{21}$$

(It is possible to do this because $q$ is strongly positive.)

One can see, $x_0$ is the solution of equation

$$F_1(x) = F_2(x) \tag{22}$$

being the point of the location of the source of pollutants. The value of the power of the source $q$ can be determined using the expression:

$$q = \frac{1}{F_i(x_0)}. \qquad (23)$$

Now, the problem consists in finding an efficient numerical method for evaluating the functional

$$\int_0^{t_i} q(t)u^*(x_0, t)dt,$$

where $u^*(x_0, t)$ is the solution of the adjoint problem (17)–(18) and $x_0$ is a parameter.

## 4  The method

In a general case one needs an efficient method for evaluating the following linear functionals

$$\int_0^{t_i} q(t)u^*(x_0, t)dt, \qquad (24)$$

of the solution of the adjoint problem (17)–(18). It is known, that statistical numerical methods allow to find directly the unknown functional of the solution with a number of operations, necessary to solve the problem in one point of the domain [2], [2]. The statistical methods give statistical estimates for the functional of the solution by performing random sampling of a certain chance variable whose mathematical expectation is the desired functional. These methods have proved to be very efficient in solving multidimensional problems in composite domains [2], [2]. Moreover, it is shown that for some problems (including one dimensional ones) in the corresponding functional spaces the statistical methods have better convergence rate than the optimal deterministic methods in such functional spaces [2], [2], [2]. It is also very important that the statistical methods are very efficient when parallel or vector processors or computers are available, because above mentioned methods are inherently parallel and have loose dependencies. They are also well vectorizable. Using power parallel computers it is possible to apply Monte Carlo method or particle-tracking method for evaluating large-scale non-regular problems which sometimes are difficult to be solved by the well-known numerical methods.

The difficulties for finite difference and finite element methods appear from non-regularity of the right-hand side of the equation (17) and from so-called "artificial" (or "numerical scheme") diffusion. So, when $L$ is a general elliptic operator using a statistical numerical method , the functional (24) can be evaluated. In our case, the fundamental solution of the adjoint equation can be used.

In fact, introduce a new variable $t^* = T - t$, where $t^* \in [0, T]$ and present the adjoint problem in the next form:

$$\frac{\partial u^*}{\partial t^*} + \alpha x \frac{\partial u^*}{\partial x} - D \frac{\partial^2 u^*}{\partial x^2} = \delta(x - x_1)\delta(T - t^* - t_1) \qquad (25)$$

$$u^* = 0, \quad \text{for} \quad t^* = 0. \qquad (26)$$

The fundamental solution of the problem (25) can be found using the techniques [2], [2] and [2]:

$$\hat{u}^*(x, t) = \frac{1}{[\pi a(t)]^{1/2}} exp \left\{ -\frac{[x - b(t)]^2}{a(t)} \right\}, \qquad (27)$$

where the time-dependant functions $a(t)$ and $b(t)$ are defined below:

$$a(t) = \frac{2D}{\alpha} \left( 1 - e^{-2\alpha t} \right), \qquad (28)$$

$$b(t) = x_0 e^{-\alpha t}. \qquad (29)$$

The solution of the adjoint problem can be expressed now in the following form:

$$u^*(x, t^*) = \int_0^T dt^* \int_{-\infty}^\infty \hat{u}^*(x - \xi, t^* - \tau)\delta(\xi - x_1)\delta(T - \tau - t_1)d\xi d\tau. \quad (30)$$

The result of integration (30) will be a function defined by (27) with new arguments:

$$u^*(x - x_1, t_1 - t) = \frac{1}{[\pi a(t_1 - t)]^{1/2}} exp \left\{ -\frac{[x - x_1 - b(t_1 - t)]^2}{a(t_1 - t)} \right\}. \quad (31)$$

Substituting (31) into the functional (24) one can get:

$$\int_0^{t_1} q(t) \frac{1}{[\pi a(t_1 - t)]^{1/2}} exp \left\{ -\frac{[x_0 - x_1 - b(t_1 - t)]^2}{a(t_1 - t)} \right\} dt = J(x_0, x_1, t_1), \qquad (32)$$

The proposed numerical method is based on numerical integration scheme for evaluating the last parametrized functional. Using a simple discretization scheme one can get

$$J = \sum_{i=0}^{k-1} q_i \frac{1}{[\pi a(t_1 - t_i)]^{1/2}} exp \left\{ -\frac{[x_0 - x_1 - b(t_1 - t_i)]^2}{a(t_1 - t_i)} \right\} (t_{i+1} - t_i) + O(\tau), \qquad (33)$$

where $\tau$ is the maximal time-step

$$\tau = \max_i (t_{i+1} - t_i),$$

and

$$q_i = q(t_i).$$

The numerical scheme (33) has an error $O(\tau)$. It is possible to apply more complicated schemes of high order accuracy (say, $O(\tau^2)$, or $O(\tau^4)$). But from computational point of view, when the integrand is a highly-degree smooth function, which corresponds to the case under consideration it is better to use an adaptive numerical scheme. In our practical computations an adaptive numerical scheme is used. The integration starts with a regular coarse grid with a low number $k$ of grid-point. After evaluating the first coarse approximation to the functional (33) the total and local aposteriori variances are estimated. The obtained information is used for refinement of the mesh - in the subregions in which the local variance is too large the mesh is refined, such that after each step of refinement the number of nodes (mesh-points) increases 2 times. The refinement procedure is continued until a required accuracy of the integration is reached or until the maximum number of evaluations is executed.

## 5 The advection-diffusion-deposition (ADD) problem

The studied approach can be easely applied for transport problems with another elliptic operator $L$. In this section it will be shown how it can be done for an important from practical point of view problem of advection-diffusion-deposition transport. This problem is well-known and of great importance in air-pollution transport. Air pollutants emitted by different sources can be transported, by the wind, on long distances. Several physical processes (diffusion, deposition and chemical transformations) take place during the transport. Regions that are very far from the large emission sources may also be polluted. It is well-known that the atmosphere must be kept clean (or, at least, should not be polluted too much). It is also well-known that if the concentrations of some species exceed certain acceptable (or critical) levels, then they may become dangerous for plants, animals and humans. This is because it is very important to put into operator $L$ a term, which describes the depositions.

The problem can be formulated in the following form:

$$\left(\frac{\partial}{\partial t} - L\right) u = \frac{\partial u}{\partial t} - v\frac{\partial u}{\partial x} + \kappa u - D\frac{\partial^2 u}{\partial x^2} = Q(t)\delta(x - x_0), \qquad (34)$$

$$u = u_0(x), \quad \text{for} \quad t = 0, u = 0, \quad \text{for} \quad x \to \infty, \quad \text{and} \quad x \to -\infty. \qquad (35)$$

The different quantities that are involved in the mathematical model have the following meaning:

- the concentration is denoted by $u$;

- $v$ is the wind velocity;

- $D$ is the diffusion coefficient;

- the emission sources in the space domain are described by the function $Q\delta(x - x_0)$;

- $\kappa$ is the deposition coefficient.

Now we need to consider the following adjoint problem:

$$-\frac{\partial u^*}{\partial t} - \frac{\partial(vu^*)}{\partial x} + \kappa(x)u^* - D\frac{\partial^2 u^*}{\partial x^2} = p(x, t), \tag{36}$$

$$u^*(x, T) = 0, \tag{37}$$

The functional in this case has the same form as (24):

$$\int_0^{t_i} q(t)u^*(x_0, t)dt. \tag{38}$$

To compute the value of the functional we need to calculate

$$J(x_0, x_1, t_1)$$

$$= \frac{1}{2(D\pi)^{1/2}} \int_0^{t_1} \frac{q(t)}{(t_1 - t)^{1/2}} exp\left\{ -\left\{ \kappa(t_1 - t) + \frac{[x_0 - x_1 + v(t_1 - t)]^2}{4D(t_1 - t)} \right\} \right\} dt \tag{39}$$

¿From (39) one can obtain the following numerical scheme:

$$J(x_0, x_1, t_1) \approx \frac{1}{2(D\pi)^{1/2}} \times$$

$$\sum_{i=0}^{k-1} \frac{q_i}{(t_1 - t_i)^{1/2}} exp\left\{ -\left\{ \kappa(t_1 - t_i) + \frac{[x_0 - x_1 + v(t_1 - t_i)]^2}{4D(t_1 - t_i)} \right\} \right\} (t_{i+1} - t_i). \tag{40}$$

The numerical scheme (40) has an error $O(\tau)$. In this case, an adaptive numerical scheme presented in the previous section is also used.

# 6   Numerical examples and results

A number of numerical tests are performed for a one-dimentional problem with a linear advection part. The results are compared with results of solving the original problem. The code which realizes the presented method is written in FORTRAN 77.

**Test 1.** The first test consists in finding the nearest points $x_0$ (from both sides of $x_1$) of possible location of the source of given power $q$ for which the concentration of pollution $u(x_1, t_1)$ in a given point $x_1$ at a given time $t_1$ must be less than a given constant $C$.

In this case the functional of the solution of the adjoint problem is used and the numerical scheme (33) is applied. Some numerical results are shown on **fig. 1** for a source of power $q = 100$ and $D = 1.0$. The point $(x_1, t_1)$ is chosen as $(6.0, 10.0)$. The parameter of the linear advection in this example is $\alpha = 0.005$ and the constant $C$, which limits the level of the concentrations is $C = 100$. The computations are performed for different values of $x_0$ in the interval from 0 to 10.

The results show that there are two possible locations of the source with power $q$. One point is around $x_0 = 4$ and other one is around $x_0 = 7.8$. (The picture is almost symmetric, because the advection in this test is "small"). To check the results, the original problem

$$\left(\frac{\partial}{\partial t} - L\right) u = \frac{\partial u}{\partial t} - \frac{\partial}{\partial x}(\alpha x u) - D\frac{\partial^2 u}{\partial x^2} = Q(t)\delta(x - x_0)$$

$$u = u_0(x), \quad \text{for } t = 0$$

for the same values of parameters, but for $x_0 = 4.0$ is solved. The obtained solution of the original problem is also presented on **fig. 1**. One can see, that the value of the pollutant concentrations at the point $x = 6$ is really approximately equal to 100.

**Test 2.** In the second test different parameters $\alpha$ are considered (different levels of advection). Some numerical results are presented on **fig. 2** in the case when the value of the concentrations in the point $x = 5$ is limited.

**Test 3.** In the third test the problem of finding the location and the power of a source is considered. Some of the results of calculations of this type are presented on **fig. 3**. The first point for obtaining the measured data is $(x_1 = 6.0, t_1 = 10.0)$ and the second one is $(x_2 = 2.0, t_1 = 11.0)$. In fact, **fig. 3** presents the functions $F_1(x_0)$ and $F_2(x_0)$, i.e.

$$F_i(x_0) = \frac{\hat{J}(x_0, x_i, t_i)}{u(x_i, t_i)}, \quad i = 1, 2,$$

where the parametrized functional $\hat{J}(x_0, x_i, t_i)$ is calculated using the presented adaptive integration method. For the considered test-example, one can find the place of the location of the source: $x_0 \approx 4.3$, and also the power of the source, using the above presentation.

**Test 4.** In this test a number of numerical experiments for ADD problem are performed. The test in finding the nearest points $x_0$ (from both sides of $x_1$) of possible location of the source of given power $q$ for which the concentration of pollution $u(x_1, t_1)$ in a given point $x_1$ at a given time $t_1$ must be less than a given constant $C$. The functional of the solution of the

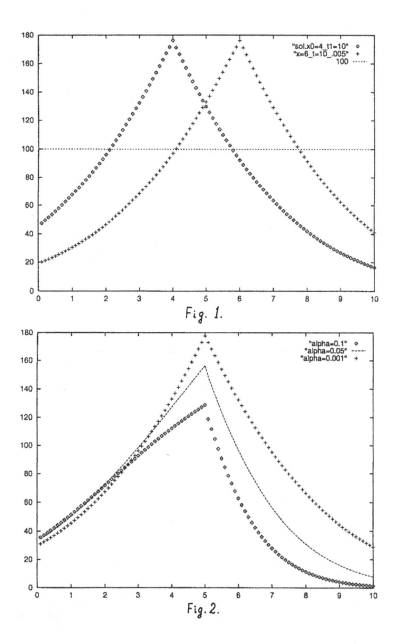

Fig. 1.

Fig. 2.

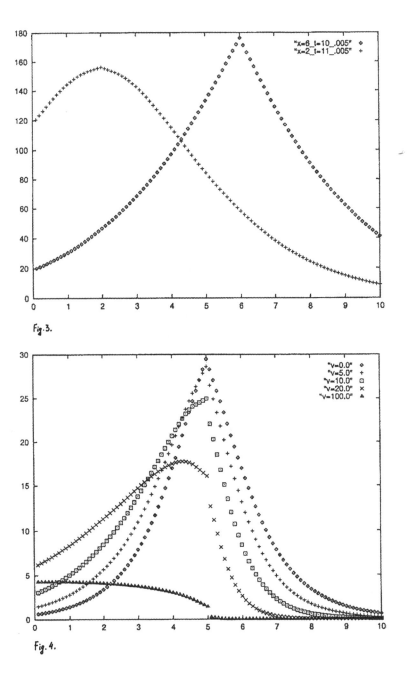

Fig. 3.

Fig. 4.

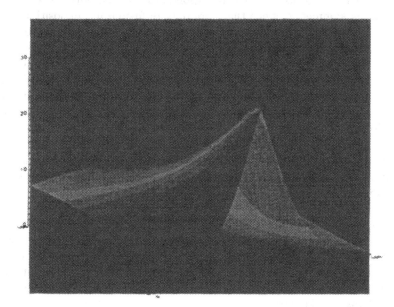

Fig. 5.

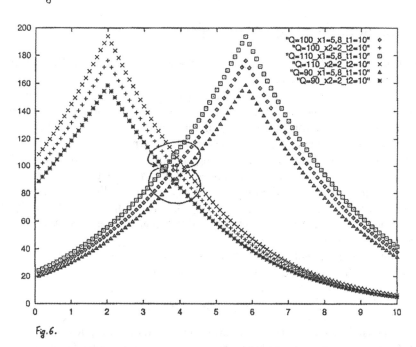

Fig. 6.

adjoint problem is used and the numerical scheme (40) is applied. The **fig. 4** contains results of calculations for different wind velocities. The first mesh function calculated for $v = 0.0$ corresponds to the case without wind - the solution is symmetric, since there are only diffusion and deposition. That means - for any critical level of concentrations $C$ the distance between $x_1 = 5.0$ and the location point $x_0$ from both sides of $x_0$ is the same. When the value of the wind velocity increases, then the solution becomes more and more unsymmetric. For high values of $v$ (when the advection is large) the maximum of the solution can be fare from the the point $x_1$. The computational results showed on **fig. 5** present the solution when both - location and advection are run. The symmetric case, which corresponds to the results on the front plane are for $v = 0.0$. Going to the larges value of $v$ , the solution decreases and becomes unsymmetric. The intersection of the solution with a plane parallel to the plane $(x, v)$ will defined the possible locations of sources for different values of the advection.

The performed numerical examples show that the problem of finding the source location is well-conditioned. This means that small perturbations of the initial data (measured concentrations) lead to small perturbations of the solution. A result of this type is expressed on **fig. 6**. It is because, the presented method does not use the solution of the inverse problem, which is usually ill-conditioned and needs some regularization techniques (like Tikhonov regularization).

## 7   Discussion and conclusion

In this work a new approach for recognizing the location and the power of a source have been presented.

The presented method uses a numerical integration method for evaluating a linear functional of the fundamental solution for the adjoint problem. It is shown that using the solution of the adjoint equation an efficient numerical method can be constructed (a method, which is less time comsumption then the method based on solving the original problem). The studied approach leads to a well-conditioned numerical problem.

A number of numerical tests are performed for one-dimentional problem with a linear advection part. The results are compared with results of solving the original problem. The code which realizes the presented method is written in FORTRAN 77.

The presented method is now applied for a case when the operator $L$ includes "diffusion" and "linear advection", i.e.

$$L = \frac{\partial}{\partial x}(\alpha x) + D\frac{\partial^2}{\partial x^2}. \tag{41}$$

It is possible to apply the same approach for more complicated operators $L$ (in fact, for any linear elliptic operator the same scheme is applicable).

The presented approach can be extended for two- and three–dimensional problems. In this case the advantage of the studied approach will consists in direct evaluating of the functional without solving the adjoint problem.

# 8  Acknowledgments

This work was done during the visit of the first author to the Research Centre Jülich, ICG-4, Jülich, Germany. It was also : partially supported by the Ministry of Science, Education and Technologies of Bulgaria under Grants # I 501/95 and # MM 449/94.

# 9  References

[1] N.S. Bahvalov, On the optimal estimations of convergence of the quadrature processes and integration methods, *Numerical methods for solving differential and integral equations* **Nauka**, Moscow, 1964, pp. 5–63.

[2] Bitzadze A.V., *Equations of the Mathematical Physics*, **Nauka**, 1982.

[3] I.Dimov, O.Tonev, Monte Carlo methods with overconvergent probable error, in *Numerical Methods and Applications*, **Proc. of Intern.Conf on Numerical Methods and Appl.,Sofia**, (Publ. House of Bulg. Acad. Sci**, Sofia, 1989), pp. 116–120.

[4] I. Dimov, O. Tonev, *Random walk on distant mesh points Monte Carlo methods*, **Journal of Statistical Physics**, vol. 70(5/6), 1993, pp. 1333 - 1342.

[5] I. Dimov, O. Tonev, *Monte Carlo algorithms: performance analysis for some computer architectures.* **Journal of Computational and Applied Mathematics, vol. 48** (1993), pp. 253–277.

[6] S.M. Ermakov, G.A. Mikhailov *Statistical Modeling*, **Nauka**, Moscow, 1982.

[7] H. Haken, *Synergetics*, **Springer– Verlag**, Berlin, Heidelberg, 1977.

[8] J.M. Hammersley, D.C. Handscomb, *Monte Carlo methods*, **John Wiley & Sons, inc.**, New York, London, Sydney, Methuen, 1964.

[9] L.V. Kantorovich, G.P. Akilov, *Functional analysis in normed spaces*, **Pergamon Press**, New York, 1964.

[10] G.Megson, V.Aleksandrov, I.Dimov, *Systolic Mat rix Inversion Using a Monte Carlo Method*, **Journal of Parallel Algorithms and Applications, vol. 3** (1994), pp. 311–330.

[11] G. A. Mikhailov *A new Monte Carlo algorithm for estimating the maximum eigenvalue of an integral operator*, **Docl. Acad. Nauk SSSR, 191**, No 5 (1970), pp. 993–996.

[12] G.A. Mikhailov, Optimization of the "weight" Monte Carlo methods **Nauka**, Moscow, 1987.

[13] I.M. Sobol *Monte Carlo numerical methods*, **Nauka**, Moscow, 1973.

[14] Tikhonov A.N., Samarski A.A., *Equations of the Mathematical Physics*. Moskow, **Nauka**, 1977.

# Monte Carlo Algorithms for Calculating Eigenvalues

## I.T. Dimov, A.N. Karaivanova, P.I. Yordanova

ABSTRACT  A new Monte Carlo approach for evaluating eigenvalues of real symmetric matrices is proposed and studied. Two Monte Carlo Almost Optimal (MAO) algorithms are presented. The first one is called *Resolvent Monte Carlo algorithm* (RMC) and uses Monte Carlo iterations by the resolvent matrix. The second one is called *Inverse Monte Carlo Iterative algorithm* (IMCI) and uses the presentation of the smallest eigenvalue by inverse Monte Carlo iterations. Estimators for speedup and for parallel efficiency are introduced. Estimates are obtained for both algorithms under consideration. Various typical models of computer architectures are considered.

Numerical tests are performed for a number of test matrices - general symmetric dense matrices, sparse symmetric matrices (including band symmetric matrices) with different behaviors on the vector machine CRAY Y–MP C92A.

Some information about the vectorization efficiency of the algorithms is obtained, showing that the studied algorithms are *well-vectorized*.

## 1  Introduction

In this work we deal with Monte Carlo algorithms for evaluating eigenvalues of symmetric matrices $A$, i.e. the values of $\lambda$ for which

$$Au = \lambda u \tag{1}$$

holds.

It is known that the problem of calculating the smallest eigenvalue of a matrix $A$ is more difficult from numerical point of view than the problem of evaluating the largest eigenvalue. Nevertheless, for many important applications in physics and engineering it is necessary to estimate the value of the smallest eigenvalue, because it usually defines the most stable state of the system which is described by the considered matrix.

There are , also, many problems in which it is important to have an efficient algorithm which is parallel and/or vectorizable. And for matrices with a large size which often appear in practice it is not easy to find efficient algorithms for evaluating the smallest eigenvalue when modern high–speed vector or parallel computers are used. For example, the problem of plotting the spectral portraits of matrices is one of the important problems where high-efficient vector and parallel algorithms are needed. The spectral portraits are used in stability analysis. The above mentioned problem leads to a large number of subproblems of evaluating the smallest eigenvalue of symmetric matrices.

Monte Carlo methods give statistical estimates for the functional of the solution by performing random sampling of a certain chance variable whose mathematical expectation is the desired functional. It is knowh that for some problems (including one dimensional ones) in the corresponding functional spaces Monte Carlo methods have better convergence rate than the optimal deterministic methods in such functional spaces [DT89], [DT93a]. An important advantage of these methods is that Monte Carlo ones allow to find directly the unknown functional of the problem solution (eigenvalues are also linear functionals of the eigenvectors) with a number of operations, necessary to solve the problem in one point of the domain [So73], [DT93].

Monte Carlo methods are very efficient when parallel processors or computers are available, because they are inherently parallel and have loose dependencies. In addition, they are well vectorizable by using powerful vector computers like CRAY Y–MP C92A.

The probable error for the usual Monte Carlo method (which does not use any additional apriori information about the regularity of the solution) [So73] is defined as:

$$r_N = c_{0.5} s(\theta) N^{-1/2}, \tag{2}$$

where $c_{0.5} \approx 0.6745$; $s(\theta)$ is the standard deviation.

An *superconvergent* Monte Carlo method is a method for which

$$r_N = c N^{-1/2 - \varepsilon(d)}, \tag{3}$$

where $c$ is a constant and $\varepsilon(d) > 0$ [So73], [DT89]. A method of this type is proposed by Dupach [Du56].

Another aspect of improvement of the Monte Carlo method is the reduction of the dispersion of the random variable $\theta$, whose mathematical expectation is equal to the value of a linear functional of the solution $J$. Let $\theta$ and $\hat{\theta}$ be two random variables such that $J = E\theta$, and $J = E\hat{\theta}$. Let, in addition, the conditions providing the existence and the finiteness of the dispersion $D\hat{\theta} = E\hat{\theta}^2 - (E\hat{\theta})^2$ be satisfied. The method for which $D\hat{\theta} < D\theta$ is called *efficient Monte Carlo method* [Ka50], [Mi87], [Di91], [DT93]. A method of this type is proposed by Kahn [Ka50] (for evaluation integrals) and by Mikhailov and Dimov (for evaluating of integral equations) [Mi87], [Di91]). The methods [Di91], [DT93] with the smallest $D\hat{\theta}_{MO}$ is called *Monte Carlo Optimal method* (MO). It is shown in [Di91] and [MAD94] that methods of this type are time-consuming. From an algorithmistic point of view it is more efficient to use Monte Carlo Almost Optimal method (MAO) with a dispersion $D\hat{\theta}_{MAO}$ for which $D\hat{\theta}_{MAO} \geq D\hat{\theta}_{MO}$, but

$|D\hat{\theta}_{MAO} - D\hat{\theta}_{MO}| \leq \varepsilon$, where $\varepsilon$ is a *small* parameter and the random variable $\hat{\theta}_{MAO}$ is easily constructed and the simulation of the corresponding random process for it is less time-consuming [MAD94].

The algorithms under consideration are algorithms of the second type (i.e. MAO algorithms). The value of the dispersion is reduced by means of a special kind of transition-density matrices.

# 2 Iterative Monte Carlo algorithms

Monte Carlo numerical algorithms may be divided in two classes - direct algorithms and iterative algorithms. Here we present *stationary linear iterative Monte Carlo algorithms* for evaluating eigenvalues of matrices.

## 2.1 Almost optimal Markov chains

The presented algorithms contain iterations with the original matrix, as well as Monte Carlo iterations with a resolvent matrix (used as iterative operator) of a given matrix.

Consider a matrix A:

$$A = \{a_{ij}\}_{i,j=1}^{n}, \quad A \in I\!R^{n \times n} \tag{4}$$

and a vector

$$f = (f_1, \ldots, f_n)^t \in I\!R^n \tag{5}$$

The matrix A can be considered as a linear operator $A[I\!R^n \to I\!R^n]$, so that the linear transformation

$$Af \in I\!R^n \tag{6}$$

defines a new vector in $I\!R^n$.

Since iterative Monte Carlo algorithms using the transformation (6) will be considered, the linear transformation (6) will be called "iteration". The algebraic transformation (6) plays a fundamental role in iterative Monte Carlo algorithms.

Now consider the following problem **P** for the matrix $A$:

**Problem P.** Evaluating of eigenvalues $\lambda(A)$:

$$Au = \lambda(A)u. \tag{7}$$

It is assumed that

(i)

$$\begin{cases} 1. & A \text{ is a symmetric matrix, i.e. } a_{ij} = a_{ji} \text{ for all } i, j = 1, \ldots, n; \\ 2. & \lambda_{min} = \lambda_n < \lambda_{n-1} \leq \lambda_{n-2} \leq \ldots \leq \lambda_2 < \lambda_1 = \lambda_{max}. \end{cases}$$

For the **problem P** under conditions (i) an iterative process of the type (6) can be used for calculating the largest eigenvalue:

$$\lambda_{max}(A) = \lim_{i \to \infty} \frac{(h, A^i f)}{(h, A^{i-1} f)}, \tag{8}$$

since for symmetric matrices $\lambda_{max}(A)$ is a real number.

We will be interested in evaluating the smallest eigenvalue $\lambda_{min}(A)$ using an iterative process of the type (6). It will be done by introducing a new matrix for realizing Monte Carlo iterations.

Let $A = \{a_{ij}\}_{i,j=1}^n$ be a given matrix and $f = \{f_i\}_{i=1}^n$ and $h = \{h_i\}_{i=1}^n$ are vectors. For the problems $P$ we create a stochastic process using the matrix A and vectors $f$ and $h$. Consider an initial density vector $p = \{p_i\}_{i=1}^n \in I\!R^n$, such that $p_i \geq 0, i = 1, \ldots, n$ and $\sum_{i=1}^n p_i = 1$. Consider also a transition density matrix $P = \{p_{ij}\}_{i,j=1}^n \in I\!R^{n \times n}$, such that $p_{ij} \geq 0$, $i, j = 1, \ldots, n$ and $\sum_{j=1}^n p_{ij} = 1$, for any $i = 1, \ldots, n$.

Define sets of permissible densities $P_h$ and $P_A$. The initial density vector $p = \{p_i\}_{i=1}^n$ is called *permissible* to the vector $h = \{h_i\}_{i=1}^n \in I\!R^n$, i.e. $p \in P_h$, if

$$p_i > 0, \quad \text{when } h_i \neq 0 \quad \text{and} \quad p_i = 0, \quad \text{when } h_i = 0 \quad \text{for} \quad i = 1, \ldots, n. \tag{9}$$

The transition density matrix $P = \{p_{ij}\}_{i,j=1}^n$ is called *permissible* to the matrix $A = \{a_{ij}\}_{i,j=1}^n$, i.e. $P \in P_A$, if

$$p_{ij} > 0, \text{when } a_{ij} \neq 0 \quad \text{and} \quad p_{ij} = 0, \text{when } a_{ij} = 0 \quad \text{for } i, j = 1, \ldots, n. \tag{10}$$

Consider the following Markov chain:

$$k_0 \to k_1 \to \ldots \to k_i. \tag{11}$$

where $k_j = 1, 2, \ldots, n$ for $j = 1, \ldots, i$ are natural random numbers.

The rules for constructing the chain (11) are:

$$Pr(k_0 = \alpha) = p_\alpha, \quad Pr(k_j = \beta | k_{j-1} = \alpha) = p_{\alpha\beta}. \tag{12}$$

Assume that

$$p = \{p_\alpha\}_{\alpha=1}^n \in P_h, \quad P = \{p_{\alpha\beta}\}_{\alpha,\beta=1}^n \in P_A. \tag{13}$$

Now define the random variables $W_j$ using the following recursion formula:

$$W_0 = \frac{h_{k_0}}{p_{k_0}}, \quad W_j = W_{j-1} \frac{a_{k_{j-1}k_j}}{p_{k_{j-1}k_j}}, \quad j = 1, \ldots, i. \tag{14}$$

The random variables $W_j$, $j = 1, \ldots, i$ can also be considered as weights on the Markov chain (12).

From all possible permissible densities we choose the following

$$p = \{p_\alpha\}_{\alpha=1}^n \in P_h, \ p_\alpha = \frac{|h_\alpha|}{\sum_{\alpha=1}^n |h_\alpha|}; \tag{15}$$

$$P = \{p_{\alpha\beta}\}_{\alpha,\beta=1}^n \in P_A, p_{\alpha\beta} = \frac{|a_{\alpha\beta}|}{\sum_{\beta=1}^n |a_{\alpha\beta}|}, \alpha = 1, \ldots, n. \tag{16}$$

Such a choice of the initial density vector and the transition density matrix leads to MAO algorithm. The initial density vector $p = \{p_\alpha\}_{\alpha=1}^n$ is called *almost optimal initial density vector* and the transition density matrix $P = \{p_{\alpha\beta}\}_{\alpha,\beta=1}^n$ is called *almost optimal density matrix*.

It is easy to show [So73], that under the conditions (i), the following equalities are fulfilled:

$$E\{W_i f_{k_i}\} = (h, A^i f), \quad i = 1, 2, \ldots; \tag{17}$$

$$\frac{E\{W_i f_{k_i}\}}{E\{W_{i-1} f_{k_{i-1}}\}} \approx \lambda_{max}(A), \quad \text{for sufficiently large "i"} \quad (P). \tag{18}$$

## 2.2 The Resolvent Monte Carlo algorithm (RMC)

Now consider an algorithm based on Monte Carlo iterations by the resolvent operator $R_q = [I - qA]^{-1} \in \mathbb{R}^{n \times n}$.

The following presentation

$$[I - qA]^{-m} = \sum_{i=0}^\infty q^i C_{m+i-1}^i A^i, \ |q|\lambda < 1 \tag{19}$$

is valid because of behaviours of binomial expansion and the spectral theory of linear operators (the matrix $A$ is a linear operator [KA64]). The eigenvalues of the matrices $R_q$ and $A$ are connected with the equality $\mu = \frac{1}{(1-q\lambda)}$, and the eigenfunctions coincide. According to (8), the following expression

$$\mu^{(m)} = \frac{([I - qA]^{-m} f, h)}{([I - qA]^{-(m-1)} f, h)} \xrightarrow{m \to \infty} \mu = \frac{1}{1 - q\lambda}, \quad f \in R^n, h \in R^n. \tag{20}$$

is valid. For a negative value of $q$, the largest eigenvalue $\mu_{max}$ of $R_q$ corresponds to the smallest eigenvalue $\lambda_{min}$ of the matrix $A$. Now, for constructing the method it is sufficient to prove the following theorem.

**Theorem 1.** *Let $\lambda'_{max}$ be the largest eigenvalue of the matrix $A' = \{|a_{ij}|\}_{i,j=1}^n$. If $q$ is chosen such that $|\lambda'_{max} q| < 1$, then*

$$([I - qA]^{-m} f, h) = E\{\sum_{i=0}^\infty q^i C_{m+i-1}^i W_i f(x_i)\}. \tag{21}$$

**Proof.** Since the expansion (19) converges in uniform operator topology [KA64] it converges for any vector $f \in \mathbb{R}^n$:

$$([I - qA]^{-m} f, h) = \sum_{i=0}^{\infty} q^i C^i_{m+i-1} (A^i f, h). \qquad (22)$$

For obtaining (22) from (21) one needs to apply (17) and to average every term of the presentation (22). Such averaging will be correct if $A$, $f$, $h$ and $q$ in (21) are replaced by their absolute values. If it is done the sum (21) will be finite since the condition $|\lambda'_{max} q| < 1$ is fulfilled. Thus, for a finite sum (21) there is a finite majorant summed over all terms and the expansion can be average over all terms. The theorem is proved.

After some calculations one can obtain

$$\lambda \approx \frac{1}{q}(1 - \frac{1}{\mu^{(m)}}) = \frac{(A[I - qA]^{-m} f, h)}{([I - qA]^{-m} f, h)} =$$

$$\frac{E \sum_{i=1}^{\infty} q^{i-1} C^{i-1}_{i+m-2} W_i f(x_i)}{E \sum_{i=0}^{\infty} q^i C^i_{i+m-1} W_i f(x_i)}. \qquad (23)$$

The coefficients $C^n_{n+m}$ are calculated using the presentation

$$C^i_{i+m} = C^i_{i+m-1} + C^{i-1}_{i+m-1}.$$

¿From the representation

$$\mu^{(m)} = \frac{1}{1 - |q|\lambda^{(m)}} \approx \frac{(h, [I - qA]^{-m} f)}{(h, [I - qA]^{-(m-1)} f)}, \qquad (24)$$

we obtain the following Resolvent Monte Carlo (RMC) algorithm for evaluating the smallest eigenvalue:

$$\lambda \approx \frac{1}{q}\left(1 - \frac{1}{\mu^{(m)}}\right) \approx \frac{E \sum_{i=0}^{l} q^i C^i_{i+m-1} W_{i+1} f(x_i)}{E \sum_{n=0}^{l} q^i C^i_{i+m-1} W_i f(x_i)}, \qquad (25)$$

where $W_0 = \frac{h_{k_0}}{p_{k_0}}$ and $W_i$ are defined by (14).

The parameter $q < 0$ has to be chosen so that to minimize the following expression

$$J(q, A) = \frac{1 + |q|\lambda_1}{1 + |q|\lambda_2}, \qquad (26)$$

or if $\lambda_1 = \alpha\lambda_2$, $(0 < \alpha < 1)$,

$$J(q, A) = 1 - \frac{|q|\lambda_2(1 - \alpha)}{1 + |q|\lambda_2}. \qquad (27)$$

We choose

$$q = -\frac{1}{2\|A\|} \qquad (28)$$

but sometimes a slightly different value of $q$ might give better results when a number of realizations of the algorithm is considered.

## 2.3   The Inverse Monte Carlo Iterative algorithm (IMCI)

Here an *Inverse Monte Carlo Iterative algorithm (IMCI)* is considered.

This algorithm can be applied when $A$ is a non-singular matrix. The algorithm has a high efficiency when the smallest by modulus eigenvalue of $A$ is much smaller then other eigenvalues. This algorithm can be realized as follow:

**1.** Calculate the inversion of the matrix $A$.

**2.** Starting from the initial vector $f_0 \in R^n$ calculate the sequence of Monte Carlo iterations:

$$f_1 = A^{-1}f_0, \ f_2 = A^{-1}f_1 \ldots, \ f_i = A^{-1}f_{i-1}, \ldots$$

The vectors $f_i \in R^n$ converge to the eigenvector corresponding to the smallest by modulus eigenvalue of $A$. In fact, we calculate the functionals

$$\frac{(Af_i, h_i)}{(f_i, h_i)} = \frac{(f_{i-1}, h_i)}{(f_i, h_i)}.$$

It is not necessary to calculate $A^{-1}$ because the vectors $f_k$ can be evaluated by solving the following system of equations:

$$Af_j = f_{j-1}, \ j = 1, \ldots, i.$$

When using Monte Carlo methods it is more efficient first to evaluate the inverse matrix using the algorithm proposed in [MAD94] and after that to apply the Monte Carlo iterations.

# 3   Time, Speedup and Parallel Efficiency Estimations

In this section the estimations for the mathematical expectation of the time, speedup and parallel efficiency will be presented. All three parameters define the quality of the parallel and/or vectorizable algorithms.

In order to estimate how the Monte Carlo algorithms depend on different computer architectures, we have considered the following models.

(1) A serial model with time $\tau$ required to complete a suboperation (for real computers this time is usually the clock period). The important feature of this model is that each operation has to be performed sequentially and one at a time.

(2) A pipeline model with startup time $s\tau$. Pipelining means an application of assembly-line techniques to improve the performance of the arithmetic operations.

(3) A multiprocessor configuration consisting of $p$ processors. Every processor of the multiprocessor system performs its own instructions on the data in its own memory.

The inherent parallelism of the Monte Carlo methods lies in the possibility of calculating each realization of the random variable $\theta$ on a different processor (or computer). There is no need for communication between the processors during the time of calculating the realizations - the only need for communication occurs at the end when the averaged value is to be calculated.

To estimate the performance of the Monte Carlo algorithms, we need to modify of the usual criterion.

Let a parallel algorithm $X$ for solving the given problem $Y$ be given. If $X$ is a deterministic algorithm then $T_p(X)$ is the time (or the number of suboperations) needed for realizing the algorithm $X$ on a system of $p$ processors.

If $X$ is a Monte Carlo algorithm it is impossible to obtain an exact theoretical estimation for $T_p(X)$ since every realization of the algorithm yields different values for $T_p(X)$. Theoretical estimations are possible only for the mathematical expectation, i.e. for $ET_p(X)$. Therefore the following estimator for the *speedup* of the Monte Carlo algorithm

$$S_p(X) = \frac{ET_1(X)}{ET_p(X)} \tag{29}$$

has to be considered. We shall call the algorithm $B$ *p-the-best* if

$$ET_p(X) \geq ET_p(B). \tag{30}$$

(Obviously, if an algorithm $D$ is a deterministic algorithm, then $ET_p(D) = T_p(D)$, for any $p = 1, 2, \ldots$.)

The *parallel efficiency* is defined as

$$E_p(X) = S_p(X)/p. \tag{31}$$

For many existing deterministic algorithms the parallel efficiency goes down rapidly when $p \geq 6$. In the general case the parallel efficiency of the deterministic algorithms strongly depends on the number of processors $p$. For Monte Carlo algorithms the situation is different. The parallel efficiency does not depend on $p$ and may be very close to 1 for a large number of processors $p$.

Further, we will consider two Monte Carlo algorithms for calculating the smallest eigenvalue of symmetric matrices and will get estimations for speedup and parallel efficiency. The both algorithms are iterative Monte Carlo algorithms. The only difference is that the matrices are used as linear operators in the Monte Carlo iterations. In fact, in both algorithms (RMC and IMCI) there is a random process which consists in jumping on element of the matrix.

## 3.1  Estimations for RMC algorithm

Every move in a Markov chain is done according to the following algorithm:

(i) generation of a random number (it is usually done in $k$ arithmetic operations where $k = 2$ or $3$);

(ii) determination of the next element of the matrix : this step includes a random number of logical operations [1];

(iii) calculating the corresponding random variable.

Since Monte Carlo Almost Optimal (MAO) algorithm is used, the random process never visits the zero-elements of the matrix $A$. (This is one of the reasons why MAO algorithm has high algorithmic efficiency for sparse matrices.)

Let $d_i$ be the number of non-zero elements of $i$-th row of the matrix $A$. Obviously, the number of logical operations $\gamma_L$ in every move of the Markov chain can be estimated using the following expression

$$E\gamma_L \approx \frac{1}{2}\frac{1}{n}\sum_{i=1}^{n} d_i = \frac{1}{2}d. \tag{32}$$

Since $\gamma_L$ is a random variable we need an estimation of the probable error of (32). It depends on the balance of the matrix. For matrices which are not very disbalanced and of not vary small-size ($n = 2, 3$), the probable error of (32) is negligible small in comparison with $\gamma_L$.

The number of arithmetic operations, excluding the number of arithmetic operations $k$ for generating the random number is $\gamma_A$.

The mathematical expectation of the operations needed for each move of any Markov chain is

$$E\delta = \tau \left[ (k + \gamma_A)l_A + \frac{1}{2}dl_L \right], \tag{33}$$

where $l_A$ and $l_L$ are the numbers of suboperations of the arithmetic and logical operations, respectively.

In order to obtain the initial density vector and the transition density matrix, the algorithm needs $d_i$ multiplications for obtaining the $i$-th row of the transition density matrix and $2dn$ arithmetic operations for constructing $\{p_{\alpha\beta}\}_{\alpha,\beta=1}^{n}$, where $d$ is determined by (32).

Thus, the mathematical expectation of the total number of operations becomes

$$ET_1(RMC) \approx 2\tau \left[ (k + \gamma_A)l_A + \frac{1}{2}dl_L \right] lN + 2\tau n(1 + d), \tag{34}$$

where $l$ is the numbers of moves in every realization of the Markov chain, and $N$ is the number of realizations.

It is worth noting that the main term of (34) does not depend on the size $n$ of the matrix. This result means that the time required for calculating

---

[1] Here the logical operation deals with testing the inequality "$a < b$".

the eigenvalue by RMC practically does not depend $n$. The parameters $l$ and $N$ depend on the spectrum of the matrix, but not depend on the size $n$. The above mentioned result was confirmed for a wide range of matrices during the realized numerical experiments.

One can get for the speedup of the pipeline model

$$S_{pipe}(RMC) \approx \frac{2\left[(k+\gamma_A)l_A + \frac{1}{2}dl_L\right]lN + 2n(1+d)}{4+3s+c+k+\gamma_A+l_A+2n+\left(\frac{1}{2}l_L+1\right)d+l+N}. \tag{35}$$

For the parallel efficiency, we have

$$E_{pipe}(RMC) \geq \frac{2l}{\frac{1}{N}+\frac{1}{B}}, \tag{36}$$

where

$$B = (k+\gamma_A)l_A + \frac{1}{2}dl_L.$$

For a rough estimation (supposing $N \approx B$) one can use the following simple expression for the speedup

$$S_{pipe}(RMC) \approx lN. \tag{37}$$

For the multiprocessor model with $p$ processors we have

$$S_p(RMC) \approx \frac{\left[(k+\gamma_A)l_A + \frac{1}{2}dl_L\right]lN + n(1+d)}{\left[(k+\gamma_A)l_A + \frac{1}{2}dl_L\right]l\frac{N}{p} + n\left[1+\frac{d}{p}\right]}. \tag{38}$$

Suppose that

$$\left[(k+\gamma_A)l_A + \frac{1}{2}dl_L\right]lN = \frac{1}{\varepsilon}n(p+d),$$

where $\varepsilon$ is a small positive number. Then for every $p \geq 1$

$$S_p(RMC) \geq \frac{p}{1+\varepsilon} \geq 1. \tag{39}$$

For the parallel efficiency we have

$$\frac{1}{1+\varepsilon} \leq E_p(RMC) \leq 1. \tag{40}$$

The last inequality shows that the parallel efficiency of RMC algorithm can be really very close to 1.

Now consider IMCI algorithm. In case of the scalar model of computers the algorithm requires: $[(k+\gamma'_A)l_A + \frac{1}{2}d'l_L]l'N'n$ operations for inverting the matrix $A$; $[(k+\gamma_A)l_A + \frac{1}{2}dl_L]lN$ operations for estimating the smallest eigenvalue iterating by the inverse matrix; $4n(1+d)$ operations for finding two initial density vectors and two transition density matrices (for inverting and for finding the smallest eigenvalue).

Thus, we have

$$S_{pipe}(IMCI) \approx \frac{\left[(k + \gamma_A)l_A + \frac{1}{2}dl_L\right]lNn + 4n(1 + d)}{8 + 5s + k + \gamma_A + l_A + 5n + \left(2 + \frac{1}{2}l_L\right)d + l + N}. \quad (41)$$

$$S_p(IMCI) \approx \frac{\left[(k + \gamma_A)l_A + \frac{1}{2}dl_L\right]lNn + 4n(1 + d)}{\left[(k + \gamma_A)l_A + \frac{1}{2}dl_L\right]l\frac{N}{p} + 4n\left[1 + \frac{n}{p}\right]}. \quad (42)$$

It is possible to use the following expression

$$E_p(IMCI) \approx \frac{1}{1 - \varepsilon}, \quad (43)$$

as a simple estimation of the parallel efficiency for the multiprocessor model, where

$$\varepsilon = \frac{4n(p + d)}{\left[(k + \gamma_A)l_A + \frac{1}{2}dl_L\right]lN}. \quad (44)$$

Hence it follows that the IMCI algorithm is usually more time-consumable. Nevertheles, it may happen even in practical calculations that IMCI algorithm is less time-consuming than RMC. This is due to the fact that the parameters $l$ and $N$ depend on the distribution of the eigenvalues of the matrix and in order to reach the given accuracy of calculations the value of $N'$ and $l'$ maybe much smaller than $N$ and $l$ correspondingly.

# 4  Numerical tests

In this section numerical results obtained by means of both RMC and IMCI algorithms will be presented. The code is written in FORTRAN 77 and $C^{++}$ and is performed on supercomputer CRAY Y–MP C92A.

Numerical tests are performed for a number of test matrices - general symmetric dense matrices, sparse symmetric matrices (including band sparse symmetric matrices) with different behaviors.

The results for matrices of size up to $n = 1024$ are shown in tables $1 - 4$.

The experimental results show that both IMCI and RMC algorithms give good results even in the case of small values of the parameters $m$ and $N$.

An information about the efficiency of the vectorization of the algorithms is received.

The Two-steps Power Monte Carlo algorithm has also been applied. It has a good efficiency when the ratio between the largest eigenvalue and the next eigenvalue is not "too close" to 1 for both matrices $A$ and $B$.

Table 1. Resolvent Monte Carlo algorithm (RMC) for MS4.R ($\lambda_{min} = -4.978$).

a) Number of Markov chains $N = 100$.

| $m$ | 2 | 2 | 3 | 3 | 3 |
|---|---|---|---|---|---|
| $l$ | 4 | 5 | 5 | 6 | 7 |
| $q$ | $-0.2$ | $-0.2$ | $-0.2$ | $-0.2$ | $-0.2$ |
| Calculated $\lambda_{min}$ | $-4.898$ | $-4.945$ | $-4.953$ | $-4.990$ | $-4.965$ |

RMC: CP time - 0.01s; HWM memory - 156 515;

b) Sensitivity of calculated $\lambda_{min}$ to the parameter $q$ (All calculations here are for $N = 100, m = 3$ and $l = 7$.)

| $q$ | $-0.01$ | $-0.02$ | $-0.08$ | $-0.10$ | $-0.12$ | $-0.15$ | $-0.20$ |
|---|---|---|---|---|---|---|---|
| Calculated $\lambda_{min}$ | $-4.900$ | $-4.910$ | $-4.954$ | $-4.976$ | $-4.990$ | $-5.00$ | $-4.965$ |

Table 2. Inverse Monte Carlo Iterative algorithm (IMCM) for MS512.2 ($\lambda_{min} = 0.2736$). (A general symmetric matrix of size 512.)

a) The number $N$ of Markov chains is fixed: $N = 80$.

| $m$ | Calculated $\lambda_{min}$ | Rel. error |
|---|---|---|
| 2 | 0.2736 | 0.0000 |
| 3 | 0.2733 | 0.0011 |
| 4 | 0.2739 | 0.0011 |
| 5 | 0.2740 | 0.0015 |
| 10 | 0.2732 | 0.0015 |
| 50 | 0.2738 | 0.0007 |
| 100 | 0.2757 | 0.0076 |

b) The number of iterations $m$ (number of moves in every Markov chain) is fixed: $m = 50$.

| $N$ | $Calculated$ $\lambda_{min}$ | $Rel.$ $error$ | $CP - time,$ $s$ | $HWM-$ $memory$ |
|---|---|---|---|---|
| 20 | 0.2729 | 0.0026 | 5.356 | 1137378 |
| 40 | 0.2742 | 0.0022 | 5.396 | 1137378 |
| 60 | 0.2748 | 0.0044 | 5.468 | 1137378 |
| 80 | 0.2739 | 0.0011 | 5.524 | 1137378 |
| 100 | 0.2736 | 0.0000 | 5.573 | 1137378 |
| 500 | 0.2737 | 0.0004 | 6.666 | 1137378 |
| 1000 | 0.2739 | 0.0011 | 8.032 | 1137378 |

**Remark:** The values for CP-time and HWM-memory are for CRAY Y-MP C92A.

Table 3. The number of iterations $m$ (number of moves in every Markov chain) is small and fixed - $m = 4$.

| $N$ | $Calculated$ $\lambda_{min}$ | $Rel.$ $error$ | $CP - time,$ $s$ | $HWM-$ $memory$ |
|---|---|---|---|---|
| 20 | 0.2737 | 0.0004 | 5.296 | 1137378 |
| 40 | 0.2749 | 0.0058 | $\star$ | 1137378 |
| 60 | 0.2754 | 0.0066 | $\star$ | 1137378 |
| 80 | 0.2739 | 0.0011 | $\star$ | 1137378 |
| 100 | 0.2736 | 0.0000 | $\star$ | 1137378 |
| 500 | 0.2737 | 0.0004 | $\star$ | 1137378 |
| 1000 | 0.2738 | 0.0007 | 5.514 | 1137378 |

**Remarks:**

- The values of CP-time and HWM-memory are for CRAY Y–MP C92A.

- "$\star$" - no estimated CP-time; the values of CP-time are between 5.296 s and 5.514 s.

Table 4. Inverse Monte Carlo Iterations algorithm (IMCM) for a
symmetric band matrix of size 1024 ($\lambda_{min} = -0.0001376$).

| $N$ | $m$ | $Calculated$ $\lambda_{min}$ | $Rel.$ $error$ | $CP - time,$ $s$ | $HWM-$ $memory$ |
|---|---|---|---|---|---|
| 5 | 10 | $-2.5178.10^{-4}$ | 0.342 | $\star$ | 1137378 |
| 10 | 10 | $-2.2573.10^{-4}$ | 0.203 | $\star$ | 1137378 |
| 20 | 10 | $-1.9562.10^{-4}$ | 0.043 | 23.30 | 1137378 |
| 30 | 10 | $-1.9307.10^{-4}$ | 0.029 | 23.34 | 1137378 |
| 40 | 10 | $-1.9165.10^{-4}$ | 0.022 | $\star$ | 1137378 |
| 50 | 10 | $-1.8942.10^{-4}$ | 0.010 | $\star$ | 1137378 |
| 80 | 10 | $-1.901.10^{-4}$ | 0.013 | 23.36 | 1137378 |

**Remarks:**

- "$\star$" - no estimations for CP-time.

- In this case the results for a small number of moves $m$ in every Markov
chain are not as good as those in the case of general symmetric matrix
of size 512.

# 5 Conclusion

Two different parallel and vectorizable Monte Carlo algorithms - RMC and
IMCI for calculating eigenvalues of symmetric matrices have been presented
and studied.

Estimators for speed-up as well as for parallel efficiency are introduced.
For both algorithms under consideration estimates are obtained. Different
typical models of computer architectures are considered. The above mentioned estimators use apriori information about considered matrix.

The both studied algorithms are "almost optimal" from statistical point
of view, i.e. the dispersion of the probable error of the random variable
which is equal to the value of $\lambda_{min}$ is "almost minimal" in the meaning of
the definition given in [Di91].

The studied algorithms are well-vectorized. For matrices $A$ with "well
distributed" eigenvalues, when the parameter $q$ is "not too large" the RMC
algorithm works well. When the largest eigenvalue of the resolvent matrix
$(I - qA)^{-1}$ is "well isolated" a small number of the power "$m$" is needed
($m = 3, \ldots, 10$). In this case the number of moves in every realization of
the Markov chain can also be small ($l = 5, \ldots, 10$) and the accuracy of
calculations is high. In this case the results are not very sensitive to the
value of the parameter $q$, which controls the speed of convergence of the
Monte Carlo iterative procedure. When $m > 35$ the results are sensitive to

$q$. This case takes place when $\lambda_1((I-qA)^{-1}/\lambda_2((I-qA)^{-1}$ is approximately equal to 1. In this case it is better to apply the IMCI algorithm.

IMCI algorithm is very efficient when the smallest eigenvalue is "well isolated". In this case the accuracy is high for arbitrary small power $m$ and number of moves in every Markov chain $l$. A good accuracy can by reached for a number of random trajectories $N \in [20,80]$. In same cases (for example, for the general symmetric matrix $MS512.2$ of size 512) the CP-time is less than the corresponding CP-time for the $NAG$-routine when CRAY Y–MP C92A machine is used for calculations.

*Acknowledgments:* This work was partially supported by the Ministry of Science, Education and Technologies of Bulgaria under Grants # I 501/95 and # MM 449/94.

# 6 References

[Di91] Dimov I. *Minimization of the Probable Error for Some Monte Carlo methods.* **Proc. Int. Conf. on Mathematical Modeling and Scientific Computation**, Varna, 1991

[DT89] I.Dimov, O.Tonev, Monte Carlo methods with overconvergent probable error, in *Numerical Methods and Applications, Proc. of Intern. Conf on Numerical Methods and Appl.,Sofia*, (Publ. **House of Bulg. Acad. Sci**, Sofia, 1989), pp. 116–120.

[DT93] I. Dimov, O. Tonev, *Random walk on distant mesh points Monte Carlo methods*, **Journal of Statistical Physics**, vol. 70(5/6), 1993, pp. 1333 - 1342.

[DT93a] I. Dimov, O. Tonev, *Monte Carlo algorithms: performance analysis for some computer architectures.* **Journal of Computational and Applied Mathematics, vol. 48** (1993), pp. 253–277.

[Du56] V. Dupach, Stochasticke pocetni metody, **Cas. pro pest. mat. 81**, No 1 (1956), pp. 55–68.

[Ka50] H. Kahn, Random sampling (Monte Carlo) techniques in neutron attenuation problems, **Nucleonics** , **6** No 5 (1950), pp. 27–33 ; **6**, No 6 (1950), pp. 60–65.

[KA64] L.V. Kantorovich, G.P. Akilov, *Functional analysis*, **Nauka**, Moscow, 1977.

[MAD94] G.Megson, V.Aleksandrov, **I.Dimov**, *Systolic Matrix Inversion Using a Monte Carlo Method*, **Journal of Parallel Algorithms and Applications, vol.4**, No 1 (1994).

220

[Mi70] G. A. Mikhailov *A new Monte Carlo algorithm for estimating the maximum eigenvalue of an integral operator*, **Docl. Acad. Nauk SSSR, 191,** No 5 (1970), pp. 993–996.

[Mi87] G.A. Mikhailov, Optimization of the "weight" Monte Carlo methods **Nauka**, Moscow, 1987.

[So73] I.M. Sobol, *Monte Carlo numerical methods*, **Nauka**, Moscow, 1973.

I.T. Dimov, A.N. Karaivanova, and P.I. Yordanova
Central Laboratory for Parallel Computing
Bulgarian Academy of Sciences
Acad. G. Bonchev St.,bl. 25 A
Sofia, 1113, Bulgaria
e-mail addresses:
dimov@amigo.acad.bg
anet@amigo.acad.bg
polina@copern.acad.bg

# Construction of digital nets from $BCH$-codes

Yves Edel
Jürgen Bierbrauer

ABSTRACT We establish a link between the theory of error-correcting codes and the theory of $(t, m, s)$-nets. This leads to the fundamental problem of net embeddings of linear codes. Our main result is the construction of four infinite families of digital $(t, m, s)$-nets based on $BCH$-codes.

## 1 Introduction

The needs of quasi-Monte Carlo methods of numerical integration have led Niederreiter [7] to the definition of $(t, m, s)$-nets. These are finite point sets in the $s$-dimensional Euclidean unit cube satisfying certain uniformity conditions. It has emerged from the work of Mullen/Schmid and Lawrence [6, 5] that $(t, m, s)$-nets are equivalent to certain finite combinatorial structures, which are closely related to **orthogonal arrays**, short $OA$. We will term these structures **ordered orthogonal arrays** or $OOA$. In this paper we consider linear $OOA$. These are $OOA$, which are vector spaces over an underlying finite field. The corresponding nets are special cases of what has been termed **digital nets.** For a definition of digital nets and a discussion of their primary applications we refer to [3]. Detailed discussions of the general theory of low-discrepancy point sets and sequences as well as their applications in numerical integration and the generation of pseudorandom numbers are to be found in Niederreiter's book [8].

We will describe a close relationship between linear $OOA$ and linear codes, and formulate the important problem to decide if a given code possesses a **net-embedding.** We then prove some systematic positive results in this direction by constructing certain infinite families of digital $(t, t + 4, s)$-nets, which are based on binary and ternary $BCH$-codes. Our main results are given in Theorem 1. The state of the art concerning net parameters is documented in the tables of [4] and [2]. It turns out that the parameters of our nets constitute considerable progress over what had been known before, the reason being that we succeed in making use of the theory of linear codes in a nontrivial way.

# 2 Basic definitions and statement of results

**Definition 1** *An* **ordered orthogonal array** *of* **depth** $l$ *denoted by*

$$OOA_\lambda(k, s, l, b)$$

*is an array with $b^k \lambda$ rows, $sl$ columns and entries from a $b$-set, where the columns occur in $s$ ordered* **blocks** $B_j, j = 1, 2, \ldots, s$ *of $l$ each, subject to the following condition:*
*Whenever $k = \sum_{j=1}^{s} k_j$, where $k_j \leq l$ for each $j$, then the set of $k$ columns consisting of the first $k_j$ columns from each block $B_j$ is independent.*

Here we call a set of columns **independent** if in the projection onto these columns each tuple of entries occurs the same number of times. This can be interpreted probabilistically as follows: identify the rows of the array with the points of a probability space, with uniform distribution. Interpret each column as a random variable with values in the $b$-set of entries. Assume each entry occurs with the same frequency in each column. Then a family of columns is independent if and only if the corresponding random variables are statistically independent. An $OOA$ is **linear** if $b = q$ is a prime-power, the set of entries is the field $\mathbb{F}_q$ and the rows form a linear subspace. It is then natural to consider generator matrices of linear $OOA$. These are matrices whose rows form a basis of the $OOA$. It is clear that in the linear case independence of columns is equivalent to linear independence of the corresponding columns in the generator matrix. We are led to the following Definition:

**Definition 2** *Let $q$ be a prime-power. An $M_q(s, l, m, k)$ is an $(m, sl)$-matrix with entries in $\mathbb{F}_q$, where the columns are divided into $s$* **blocks** $B_j, j = 1, 2, \ldots, s$ *of $l$ columns each, such that the following conditions are satisfied: whenever $k = \sum_{j=1}^{s} k_j$, where $k_j \leq l$ for all $j$, then the set of $k$ columns consisting of the first $k_j$ columns from each $B_j$ is linearly independent.*

Essentially equivalent concepts have been considered in [9] and [1]. Observe that the columns of each block are linear ordered: there is a first column, a second column, .... This is the reason why the name of ordered $OA$ has been chosen. Denote the sets of columns as considered in this Definition as **qualifying collections**. We call $s$ the **length**, $l$ the **depth**, $m$ the **dimension** and $k$ the **strength**. Denote by $(k_1, k_2, \ldots, k_s)$ the **type** of the qualifying collection in question (terms $k_j = 0$ are omitted, the order is immaterial). Values $l > k$ are not interesting as the $(k+1)$-st, $(k+2)$-nd ... columns of each block do not have to satisfy any condition. We will therefore restrict to $l \leq k$. It is clear that the row space of an $M_q(s, l, m, k)$ is a linear $OOA_{q^{m-k}}(k, s, l, q)$. If $l = k$ then a digital $(m - k, m, s)$-net in base $q$ can be constructed. For details we refer to Mullen/Schmid and Lawrence [6, 5].

Let $l' < l$. If we use only the $l'$ first rows per block we see that we get an $M_q(s, l', m, k)$ out of an $M_q(s, l, m, k)$. It has been observed by Mullen/Schmid and Lawrence [6, 5] in a slightly more general context that an $M_q(s, k - 1, m, k)$ yields an $M_q(s, k, m, k)$ in various ways: as $k$-th column of block $B_j$ we may choose the first column of some block $B_{j'}, j' \neq j$.

It is natural to start from depth 1. By definition an $M_q(s, 1, m, k)$ is an $(m, s)$–matrix each $k$ columns of which are linearly independent. Its rowspace is known as a **linear** $OA$ of strength $k$. Assume $s > m$. Its dual with respect to the usual dot-product is then an $(s - m)$-dimensional code $C$ of length $s$ and minimum distance $> k$. The parameters of such codes are often written in the form $[s, s - m, > k]_q$. Matrix $M_q(s, 1, m, k)$ is known as a **check matrix** of $C$ ( an $s$-tuple belongs to $C$ if and only if its scalar product with each row of the matrix vanishes). We collect this information in the following Lemma:

**Lemma 1** *Let $s > m$. If an $M_q(s, l, m, k)$ exists for some $l \geq 1$ then the family of first columns per block is a check matrix of a code $[s, s - m, > k]_q$.*

In particular the bounds on codes imply bounds on $M_q(s, l, m, k)$. We raise the question when this relationship between codes and digital nets can be inverted:

**Definition 3** *A $q$-ary linear code $C$ with parameters $[s, s - m, > k]$ possesses an $(m - k, m, s)$-net embedding if there is an $M_q(s, k, m, k)$ whose first columns per block form a check matrix of $C$.*

In the final section we will describe certain $BCH$-codes and construct net-embeddings. The results are as follows:

**Theorem 1** *The following digital nets exist and can be effectively constructed:*

- $M_2(2^{2r} + 1, 4, 4r, 4)$, *thus digital binary* $(4r - 4, 4r, 2^{2r} + 1)-nets$ $(r \geq 2)$.

- $M_2(2^r + 1, 4, 2r + 1, 4)$, *thus digital binary* $(2r - 3, 2r + 1, 2^r + 1)-nets$ $(r \geq 3)$.

- $M_2(2^r - 2, 4, 2r, 4)$, *thus digital binary* $(2r - 4, 2r, 2^r - 2)$-nets $(r$ *odd*, $r \geq 3)$.

- $M_3(3^r - 1, 4, 2r + 1, 4)$, *thus digital ternary* $(2r - 3, 2r + 1, 3^r - 1)-nets$ $(r \geq 2)$.

# 3 Constructions from $BCH$-codes

## 3.1 The first binary family

We consider the tower of finite fields

$$\mathbb{F}_2 \subset \mathbb{F}_{2^r} \subset \mathbb{F}_{2^{2r}} \subset \mathbb{F}_{2^{4r}} = F, \text{ where } r \geq 2.$$

Put $s = 2^{2r} + 1$ and let $W \subset F$ be the multiplicative subgroup of order $s$. Choose a basis of $F \mid \mathbb{F}_2$, define the binary $(4r, s)$–matrix $M$ whose columns are indexed by the $a \in W$, column $a$ being the $4r$-tuple of coefficients obtained when $a$ is developed with respect to the basis. We will have opportunity repeatedly to use the fact that $W \cap \mathbb{F}_{2^{2r}} = \{1\}$. This follows simply from $gcd(2^{2r} + 1, 2^{2r} - 1) = 1$.

Our first observation is that $M$ has rank $4r$. This is equivalent with the statement that the $\mathbb{F}_2$-vector space $< W >$ generated by $W$ is $< W > = F$. In fact, it is obvious that $< W >$ is closed under multiplication, so is a subfield. As $2^{2r} + 1$ divides the order of its multiplicative group, we obtain $< W > = F$.

Next we show that any four columns of $M$ are linearly independent. First of all, there is no 0-column and no two columns are identical. Assume three columns have vanishing sum. This amounts to $a = b + c$, where $a, b, c$ are pairwise different elements of $W$. Raising this equation to the $s$-th power we get

$$1 = a^s = (b+c)(b+c)^{2^{2r}} = (b+c)(b^{2^{2r}} + c^{2^{2r}}) = (b+c)\left(\frac{1}{b} + \frac{1}{c}\right).$$

Here we have used that the mapping $x \longrightarrow x^{2^{2r}}$ is a field-automorphism and that $b^s = c^s = 1$. Multiplying out we get $1 = x + \frac{1}{x}$, where $1 \neq x = b/c \in W$. Equivalently $x^2 + x + 1 = 0$. It follows that $x \in \mathbb{F}_4 \subset \mathbb{F}_{2^{2r}}$. We obtain the contradiction $1 \neq x \in W \cap \mathbb{F}_{2^{2r}}$.

Assume $a + b + c + d = 0$, where $a, b, c, d \in W$ are pairwise different. Write this in the form $a + b = c + d$, raise to power $s$ as before. We obtain $x + \frac{1}{x} = y + \frac{1}{y}$, where $x = a/b, y = c/d$. Removing the denominator and simplifying we obtain $0 = (x + y)(1 + xy)$. As a field has no divisors of zero we conclude that either $x = y$ or $x = 1/y$. This means in clear that either $ad = bc$ or $ac = bd$. This is our result when we use the partition $\{a, b, c, d\} = \{a, b\} \cup \{c, d\}$. We see that we can assume without restriction $ac = bd$. Use the partition $\{a, b, c, d\} = \{a, c\} \cup \{b, d\}$. Then either $ab = cd$ or $ad = bc$. None of these equations is compatible with the former equation ( in the first case we obtain by division $b/c = c/b$, hence $b = c$. In the second case a similar contradiction is obtained).

We have shown that matrix $M$ is an $M_2(s, 1, 4r, 4)$, equivalently the check matrix of a code $C$ with parameters $[s, s-4r, \geq 5]$. In fact, $C$ is a $BCH$-code and the parameters can also be obtained by invoking results from the theory of cyclic codes ( cyclotomic cosets for the dimension, the Roos bound for the minimum distance), but we preferred to give a direct treatment. We proceed to the construction of an $M_2(s, 3, 4r, 4)$ ( remember that this is equivalent with an $M_2(s, 4, 4r, 4)$). The blocks are indexed by the $a \in W$. Choose $\alpha \in \mathbb{F}_{2^r} \setminus \mathbb{F}_2, \beta \in \mathbb{F}_{2^{2r}} \setminus \mathbb{F}_{2^r}$. Define block $B_a$ as $B_a = (a, \alpha a, \beta a)$. We have to check that each qualifying collection of 4 columns is linearly independent. Type (1,1,1,1) has been checked already.

- type (2,1,1)

Assume $a, \alpha a, b, c$ are linearly dependent ($a, b, c \in W$, different). Clearly $\alpha a$ must be involved in the relation. It is impossible that $\rho a = b$ for some $\rho \in \mathbb{F}_{2^r}$ as otherwise $\rho = b/a \in \mathbb{F}_{2^r} \cap W = \{1\}$, contradiction. This shows that we must have $\rho a = b + c$, where $\rho = \alpha$ or $\rho = \alpha + 1$. Raising this to power $s$ we obtain $\rho = x + 1/x$, where $1 \neq x = b/c \in W$. Equivalently $x^2 + \rho x + 1 = 0$. This shows that $x$ must be in the quadratic extension $\mathbb{F}_{2^{2r}}$ of $\mathbb{F}_{2^r}$. This leads to our standard contradiction again.

- type (2,2)

Assume $a, \alpha a, b, \alpha b$ are linearly dependent ($a, b \in W, a \neq b$). Because of type (2,1,1) we know that $\alpha a$ and $\alpha b$ must be involved. $\alpha(a + b) = a + b$ is clearly impossible. We can assume $\alpha(a + b) = a$, hence $1/\alpha = 1 + x$, where $x = b/a$. We obtain the usual contradiction.

- type (3,1)

Assume $a, \alpha a, \beta a, b$ are linearly dependent ($a, b \in W, a \neq b$). We get $b = \rho a$ for some $\rho \in \mathbb{F}_{2^{2r}}$, leading to the usual contradiction.

## 3.2  The second binary family

Consider the finite fields

$$\mathbb{F}_2 \subset \mathbb{F}_{2^r} \subset \mathbb{F}_{2^{2r}} = F, \text{ where } r \geq 2.$$

Put $s = 2^r + 1$ and let $W \subset F$ be the multiplicative subgroup of order $s$. Choose a basis of $F \mid \mathbb{F}_2$, define the binary $(2r + 1, s)-$matrix $M$ such that the column corresponding to $a \in W$ is $(1, a)^t$, where the superscript $^t$ denotes transposition. As in the previous subsection it is clear that $W \cap \mathbb{F}_{2^r} = \{1\}$ and that $< W > = F$, where $< W >$ denotes the linear span. These facts will be freely used in the sequel. We begin by showing that the rows of $M$ are linearly independent, so that $M$ has full rank $2r + 1$. We have seen in the preceding subsection that the last $2r$ rows of $M$ are independent. It remains to show that the first row of $M$, which is constant $= 1$, is not contained in the linear span of the remaining rows. In order to see this it is handy to interpret the entries of $M$ in a different way: until now the choice of the basis of $F \mid \mathbb{F}_2$ had been irrelevant. Now we choose this basis as $z_i, i = 1, 2, \ldots, 2r$ such that $tr(z_i z_j) = \delta_{ij}$. Here we use the fact that every linear functional $: F \longrightarrow \mathbb{F}_2$ can be written as $x \longrightarrow tr(\alpha x)$ for some $\alpha \in F$. With this choice we can interpret the space generated by rows $2, 3, \ldots, 2r + 1$ as follows: the rows are indexed by $u \in F$, the columns by $a \in W$, with corresponding entry $tr(ua)$. We have to show that there is no $u \in F$ satisfying $tr(ua) = 1$ for all $a \in W$. Assume there is such an element $u$. It follows $tr(u^2 a^2) = 1$ for all $a \in W$. As $W$ has odd order $a^2$ varies over all elements of $W$, consequently $tr((u + u^2)W) = 0$. As $W$ generates $F$ as a vector space and the trace-form is a non-degenerate bilinear form it follows $u + u^2 = 0$, hence $u \in \mathbb{F}_2$. Certainly $u \neq 0$. It follows $u = 1$, thus $tr(a) = 1$ for every $a \in W$. Choosing $a = 1$ we obtain a contradiction.

Next we show that no five columns of $M$ are linearly dependent. As before, there is no 0-column, and no two columns are identical. The presence of the first row shows that the sum of an odd number of columns can never vanish. The remaining case of four columns of $M$ with vanishing sum is led to a contradiction exactly as in the previous subsection. For the present subsection it suffices to know that any four columns are linearly independent. The independence of any five columns of $M$ will be used in Subsection 3.3. We have shown that matrix $M$ is an $M_2(s, 1, 2r + 1, 5)$, equivalently the check matrix of a code $C$ with parameters $[s, s - 2r - 1, \geq 6]$. Again $C$ is a $BCH$-code, but we chose not to invoke the pertinent theory. We proceed to the construction of an $M_2(s, 3, 2r + 1, 4)$. Choose some $\alpha \in \mathbb{F}_{2^r} \setminus \mathbb{F}_2$. The choice of $\alpha$ will have to be restricted later. Define block $B_a$ as $B_a = ((1, a)^t, (1, \alpha a)^t, (0, a)^t)$. As before we have to check three types of quadruples of columns:

- type (2,1,1)

Assume $(1, a)^t, (1, \alpha a)^t, (1, b)^t, (1, c)^t$ are linearly dependent $(a, b, c \in W$, different). Certainly $(1, \alpha a)^t$ has to be involved in the linear relation, and the number of summands is 2 or 4. If the number of summands is 2 our standard contradiction is obtained. Assume the sum of all four columns vanishes: $a + \alpha a = b + c$. Raising this to power $s$ we obtain $(1 + \alpha)^s = x + \frac{1}{x}$, where $1 \neq x = b/c \in W$. As $(1 + \alpha)^s = (1 + \alpha)(1 + \alpha^{2^r}) = (1 + \alpha)(1 + \alpha) = 1 + \alpha^2$, this simplifies to

$$x^2 + (1 + \alpha^2)x + 1 = 0.$$

This is where we have to restrict the choice of $\alpha$. In fact, choose $\alpha \in \mathbb{F}_{2^r} \setminus \mathbb{F}_2$ such that $1 + \alpha^2 = \gamma + \frac{1}{\gamma}$ for some $\gamma \in \mathbb{F}_{2^r} \setminus \mathbb{F}_2$. Then the quadratic expression splits and we obtain $(x + \gamma)(x + \frac{1}{\gamma}) = 0$. It follows that $x \in \mathbb{F}_{2^r}$, and the usual contradiction is obtained. It remains to make sure that $\gamma$ can be chosen as required. The condition that $\alpha \notin \mathbb{F}_2$ is equivalent with $1 + \alpha^2 \notin \mathbb{F}_2$, hence with $\gamma + \frac{1}{\gamma} \notin \mathbb{F}_2$. This expression is certainly nonzero, and $\gamma + \frac{1}{\gamma} = 1$ is equivalent with $\gamma \in \mathbb{F}_4$. It follows that we need to assume $\gamma \in \mathbb{F}_{2^r} \setminus \mathbb{F}_4$ and $r > 2$.

- type (2,2)

This type presents no problems whatsoever.

- type (3,1)

Assume that $(1, a)^t$, $(1, \alpha a)^t$, $(0, a)^t$, and $(1, b)^t$ are linearly dependent with $a, b \in W, a \neq b$. As before $(0, a)^t$ has to be involved, and exactly two of the remaining three columns. The only critical case is $a + \alpha a = b$. This leads to $b/a = 1 + \alpha \in W \cap \mathbb{F}_{2^r}$, the usual contradiction.

## 3.3 The third binary family

Let $r \geq 3$ be an odd integer. Consider the fields

$$\mathbb{F}_2 \subset \mathbb{F}_{2^r} \subset \mathbb{F}_{2^{2r}} = F.$$

Put $s = 2^r + 1$. Observe that $F$ contains $\mathbb{F}_4 = \{0, 1, \omega, \omega^2\}$. Let $W \subset F$ be the multiplicative subgroup of order $s$, just as in the preceding subsection. $M$ is the binary $(2r + 1, s)$-matrix whose column corresponding to $a \in W$ is $(1, a)^t$, as before. We have shown that $M$ has maximal rank $2r + 1$ and that no 5 columns of $M$ are linearly dependent. This means that the $BCH$-code $B$ which has $M$ as a check matrix has minimum distance $\geq 6$. Let us consider the code $C$ which is obtained by truncating this $BCH$-code in the sense that the coordinate corresponding to $1 \in W$ is omitted. Then $C$ has length $s - 1 = 2^r$, minimum distance $\geq 5$ and a check matrix for $C$ is the $(2r, 2^r)$-matrix $M'$ with columns $(a + 1)^t$ corresponding to $a \in W \setminus \{1\}$. In fact, let $(\chi_a) \in C$, where $\chi_a \in \mathbb{F}_2, a \in W \setminus \{1\}$. Define $\chi_1 = \sum_{a \neq 1} \chi_a$. Then $(\chi_a)_{a \in W} \in B$ if and only if $0 = \sum_{a \in W} \chi_a a = \sum_{a \neq 1} \chi_a (a + 1)$. So far we have not been able to construct systematically an $M_2(2^r, 4, 2r, 4)$ whose set of first columns per block are the columns of $M'$. We are convinced that this ought to be possible as computer experiments led to positive results in the first cases $r \in \{3, 5, 7\}$. We proceed by considering matrix $M''$ obtained by restricting $M'$ to the columns $(a + 1)^t$, where $a \in W' = W \setminus \{1, \omega, \omega^2\}$. The block corresponding to $a \in W'$ is defined as $B_a = ((a + 1)^t, (\omega(a + 1))^t, 1^t)$. We claim that this defines an $M_2(2^r - 2, 3, 2r, 4)$. As before type (1,1,1,1) is all right by definition.

- type (2,1,1)

Assume at first $(a + 1), \omega(a + 1), (b + 1)$ are linearly dependent ($a, b \in W', a \neq b$). As the relation must involve $\omega(a + 1)$ and $1 + \omega = \omega^2$ we can assume without restriction $\omega(a + 1) = b + 1$. Raise this to power $s$

TABLE 1. An $M_3(8, 3, 5, 4)$

| 100 | 022 | 011 | 021 | 011 | 122 | 211 | 011 |
|-----|-----|-----|-----|-----|-----|-----|-----|
| 021 | 100 | 001 | 001 | 011 | 121 | 111 | 221 |
| 011 | 021 | 100 | 010 | 020 | 110 | 010 | 200 |
| 010 | 010 | 020 | 100 | 010 | 010 | 110 | 110 |
| 000 | 000 | 010 | 010 | 100 | 110 | 100 | 100 |

and observe that $s = 2^r + 1$ is a multiple of 3 as $r$ is odd. We obtain $a + 1/a = b + 1/b$, after simplification $(a+b)(a+1/b) = 0$. As $a \neq b$ we obtain $a = 1/b$. The original equation now reads $\omega(a + 1) = 1/a + 1 = (a + 1)/a$. As $a \neq 1$ we obtain $a = 1/\omega \in \mathbb{F}_4$, a contradiction.

Assume know that $a + 1, \omega(a + 1), b + 1, c + 1$ are linearly dependent. We have just shown that $b + 1$ and $c + 1$ must be involved in the relation. If $\omega(a + 1)$ was not involved this would contradict the fact that $M'$ has strength 4. We have without restriction $\omega(a + 1) = b + c$. Raising this to power $s$ we obtain $a + 1/a = b/c + c/b$. As all four summand are elements of $W$ we find, as in subsection 3.1, that we have without restriction $a = b/c$. The original relation now reads $\omega(a + 1) = b + c = c(a + 1)$, leading to the contradiction $c = \omega$.

- type (2,2)

Assume $a + 1, \omega(a+1), b+1, \omega(b+1)$ are linearly dependent. Case (2,1,1) shows that $\omega(a + 1)$ and $\omega(b + 1)$ must be involved in the relation. If the sum of all four terms vanishes, then clearly $a = b$, a contradiction. The only remaining case is without restriction $\omega(a + 1) = \omega^2(b + 1)$. After division by $\omega$ we are back to the case of type (2,1,1).

- type (3,1)

Assume $a + 1, \omega(a + 1), 1, b + 1$ are linearly dependent. We know that 1 is involved in the relation. If $b + 1$ was not involved we would obtain the contradiction $a \in \mathbb{F}_4$. Assume $b = a + 1$. Raising to power $s$ we obtain $1 = a + 1/a$, hence the contradiction $a \in \mathbb{F}_4$. It follows that the relation is without restriction $b = \omega(a + 1)$. Raising to power $s$ again we get the same contradiction $a \in \mathbb{F}_4$.

## 3.4 The ternary family

We constructed a ternary $(1, 5, 8)$−net by computer. An $M_3(8, 3, 5, 4)$ is given in Table 1.

Let $F = \mathbb{F}_{3^r}, s = 3^r - 1$, where $r \geq 3$. Define a ternary $(2r + 1, 3^r - 1)$-matrix $M$, whose columns are indexed by the nonzero elements $a \in F$.

Define this column as $u_a = (1, a, a^2)^t$. Here the elements of $F$ are represented by the coefficients when expressed in terms of a fixed basis of $F \mid \mathbb{F}_3$. In this case we do not find it rewarding to circumvent the theory of $BCH$-codes. Matrix $M$ is in fact a check matrix of a ternary primitive $BCH$-code with dimension $s - (2r + 1)$ and minimum distance $\geq 5$. In particular the rows of $M$ are independent and any four columns of $M$ are independent. We proceed to the construction of an $M_3(s, 3, 2r + 1, 4)$ : choose $u_a$ as first column in block $B_a$. The second column is $v_a = (0, \nu a, (\nu a)^2)^t$, the third column is $w_a = (0, 0, (\nu a)^2)^t$. We know that it suffices to check types $(2, 1, 1), (2, 2)$ and $(3, 1)$ to prove our claim. The crucial point is the choice of the element $\nu \in F$.

**Lemma 2** *For every $r > 2$ the field $F = \mathbb{F}_{3^r}$ contains an element $\nu$ such that $\nu$ generates the field $F$ over $\mathbb{F}_3, \nu$ is a nonsquare and $\nu - 1$ is a square.*

*Proof:* Let $\epsilon$ be a generator of the multiplicative group of $F$. In particular $\epsilon$ is a nonsquare. If $\epsilon - 1$ is a square we can use $\nu = \epsilon$ and are done. So assume $\epsilon - 1$ is a nonsquare. It follows that the quotient $1 - 1/\epsilon$ is a square. We distinguish the case when $r$ is even ( -1 is a square) and when $r$ is odd (equivalently -1 is a nonsquare). When $r$ is even we have that $1/\epsilon - 1$ is a square. Put $\nu = 1/\epsilon$. Assume now $r$ is odd. As $\epsilon - 1$ is a nonsquare we can repeat the argument above and assume that $\epsilon + 1$ is a nonsquare. It follows that $1 - 1/\epsilon$ and $1 + 1/\epsilon$ are squares, so that $1/\epsilon - 1$ is a nonsquare. Put $\nu = 1/\epsilon - 1$. ∎

Choose an element $\nu \in F$ satisfying the conditions of Lemma 2.

**Lemma 3** *Columns $u_a, u_b$, where $a, b$ are different nonzero elements of $F$, are always independent of any column $c = (0, \gamma, \gamma^2)$, where $\gamma \neq 0$.*

*Proof:* Assume there is a nontrivial linear combination of these columns. Observe that all our linear combinations are over $\mathbb{F}_3$. We can assume that the coefficient of $c$ is $= 1$. By changing the roles of $a, b$, if necessary, we get the equations $\gamma = a - b, \gamma^2 = a^2 - b^2$. As $\gamma^2 = a^2 + b^2 + ab$, we get $ab = b^2$, from which the contradiction $a = b$ is derived. ∎

- type $(2, 1, 1)$

Assume $v_a = xu_a + yu_b + zu_c$ with coefficients $x, y, z \in \mathbb{F}_3$. Lemma 3 shows that no coefficient vanishes. The first coordinate shows that $x + y + z = 0$. It follows that they must be equal, either $= 1$ or $= -1$. In the first case the remaining coordinates yield the equations $(\nu - 1)a = b + c, (\nu^2 - 1)a^2 = b^2 + c^2$. Adding the square of the first equation to the second yields after simplification $\nu(\nu - 1)a^2 = (b - c)^2$. This contradicts our assumption that $\nu(\nu - 1)$ is a nonsquare.
So assume $x = y = z = -1$. We get the equations $(\nu + 1)a = -b - c, (\nu^2 + 1)a^2 = -b^2 - c^2$. An analogous procedure yields after subtraction

and simplification $\nu a^2 = (b - c)^2$. This contradicts the assumption that $\nu$ is a nonsquare.

- type $(2, 2)$

Assume we have a nontrivial linear combination $wu_a + xv_a = yu_b + zv_b$ for some $a \neq b$. Observe at first that $v_a$ and $v_b$ are not linearly dependent. In fact, they are not equal as otherwise $a = b$, nor are they negatives of each other, or $(\nu a)^2 = -(\nu a)^2$, contradiction. It follows that $w, y$ cannot both vanish. The first coordinate shows $w = y$. We can assume $w = y = 1$. Lemma 3 shows that $xz \neq 0$. If $x = z = 1$, then the second coordinate shows $(\nu + 1)a = (\nu + 1)b$, hence $a = b$. The same contradiction is derived when $x = z = -1$. It follows that we can assume $x = 1, z = -1$. We obtain the equations $(\nu + 1)a = (1 - \nu)b$ and $(\nu^2 + 1)a^2 = (1 - \nu^2)b^2$. Subtracting the square of the first equation from the second we get after simplification $\nu - 1 = a^2/b^2$. Use this in the first equation, substituting for $(1 - \nu)$. We get $\nu + 1 = -a/b$. It follows $(\nu + 1)^2 = \nu - 1$. This yields $\nu^2 + \nu - 1 = 0$. We see that $\nu \in \mathbb{F}_9$, contradiction.

- type $(3, 1)$

Assume $wu_b = xu_a + yv_a + zw_a$. As the columns belonging to one block are independent, we may assume $w = 1$. The first coordinate shows $x = 1$. The second coordinate shows $y \neq 0$ as otherwise $a = b$. Lemma 3 shows $z \neq 0$. Assume $y + z = 0$. The last coordinate shows $b^2 = a^2$, hence $b = -a$. The second coordinate yields the contradiction $\nu \in \mathbb{F}_3$. Two cases are left: if $y = z = -1$, then $b = a(1 - \nu), b^2 = a^2(1 + \nu^2) = a^2(\nu^2 + \nu + 1)$. It follows $\nu = 0$. If finally $y = z = 1$, then the same procedure yields $\nu \in \mathbb{F}_3$ again. ∎

## 4 References

[1] M. J. Adams and B. L. Shader, *A construction for $(t, m, s)$-nets in base $q$*, SIAM Journal of Discrete Mathematics, to appear.

[2] A. T. Clayman, K. M. Lawrence, G. L. Mullen, H. Niederreiter, and N. J. A. Sloane, *Updated tables of parameters of $(t, m, s)$-nets*, manuscript.

[3] G. Larcher, H. Niederreiter, and W. Ch. Schmid, *Digital nets and sequences constructed over finite rings and their application to quasi-Monte Carlo integration*, Monatshefte Mathematik **121** (1996),231-253.

[4] K. M. Lawrence, A. Mahalanabis, G. L. Mullen, and W. Ch. Schmid, *Construction of digital $(t, m, s)$-nets from linear codes*, in S. Cohen and H. Niederreiter, editors, *Finite Fields and Applications (Glasgow, 1995)*, volume 233 of *Lecture Notes Series of the London Mathematical Society*, pages 189-208. Cambridge University Press, Cambridge, 1996.

[5] K. M. Lawrence, *A combinatorial characterization of* $(t, m, s)$-*nets in base b, J. Comb. Designs* **4** (1996),275-293.

[6] G. L. Mullen and W. Ch. Schmid, *An equivalence between* $(t, m, s)$-*nets and strongly orthogonal hypercubes,* Journal of Combinatorial Theory A **76** (1996), 164-174.

[7] H. Niederreiter, *Point sets and sequences with small discrepancy, Mo-natshefte Mathematik* **104** (1987),273-337.

[8] H. Niederreiter, *Random Number Generation and Quasi-Monte Carlo Methods,* Number **63** in CBMS–NSF Series in Applied Mathematics. SIAM, Philadelphia, 1992.

[9] W. Ch. Schmid and R. Wolf, *Bounds for digital nets and sequences,* Acta Arithmetica **78**(1997),377-399.

Yves Edel
Mathematisches Institut der Universität
Im Neuenheimer Feld 288
69120 Heidelberg (Germany)

Jürgen Bierbrauer
Department of Mathematical Sciences
Michigan Technological University
Houghton, Michigan 49931 (USA)

# Discrepancy lower bounds for special quasi-random sequences

## Henri Faure

ABSTRACT  Lower bounds for the discrepancy of special quasi-random sequences are given, showing that these sequences have the exact order $(Log\ N)^2$ within a multiplicative constant factor.

## 1  Introduction

Star-discrepancy for an infinite sequence $X = (x_n)_{n \geq 1}$ in the s-dimensional unit cube $I^s = [0, 1[^s$: A star-subinterval $P$ of $I^s$ is the product of $s$ intervals $[0, b_k[$ of $[0, 1[$. Given a set of positive integers $T$, let $A(P; T; X) = \#\{n \in T;\ x_n \in P\}$ and let $E(P; T; X) = A(P; T; X) - |T|\ |P|$. Then the star-discrepancy of the sequence $X$ is defined by

$$D^*(N, X) = \sup_{P \in \mathcal{P}_s^*}\ |\ E(P; ]0, N]; X)\ |$$

in which $\mathcal{P}_s^*$ is the set of star subintervals of $I^s$.

In arbitrary dimension, the only lower bound is that of K.F. Roth [12] obtained in 1954 : there exists a constant $C_s > 0$ such that, for every infinite sequence $X$, one has

$$\limsup_{N \to \infty}\ (D^*(N, X)/(LogN)^{s/2}) \geq C_s.$$

Only two improvements have been given:

by W.M.Schmidt [13] in 1972 and G. Halász [6] in 1981 who, by different methods, have got the order

$$Log\ N \quad \text{instead of} \quad \sqrt{Log\ N}$$

in one dimension, and by J. Beck [1] in 1989 with

$$(Log\ N)(Log\ Log\ N)^{\frac{1}{8} - \epsilon}$$

instead of $Log\ N$ in two dimensions.

A lot of sequences with very small irregularities have been constructed for numerical integration purposes; their discrepancy is bounded by a constant time $(Log\ N)^s$ in dimension $s$.

It is conjectured that $(Log\ N)^s$ is the exact order, because the lower bound of Roth is a bound for the mean-square discrepancy.

Chronologically, the following constructions may be quoted:
J.H. Halton [7] in 1960, I.M. Sobol' [14] in 1967, S. Srinivasan [15] in 1978, H. Faure [2] in 1982, H. Niederreiter [8,9] in 1987-1988 and recently by H. Niederreiter and C.P. Xing [10,11,16].
Note that even for these special sequences, the only lower bound was until now that of Roth or Beck in two dimensions.

In this paper we announce we have got the exact order $(Log\ N)^2$ for the discrepancy of our family of special $(0,2)$-sequences in prime base $b$ (see II), denoted by $S_b^f$, with $1 \leq f \leq b - 1$. If we set

$$d_s(X) = \limsup_{N \to \infty}\ (D^*(N, X)/(Log\ N)^s))$$

we already have the upper bound [2]

$$d_2(S_b^f) \leq \frac{1}{8}\ \left(\frac{b-1}{Log\ b}\right)^2.$$

With the notation above, our result is the following:

$$d_2(S_b^1) > \frac{0.2}{(b^2\ Log\ b)^2}\ ,$$

announced in [3], and

$$d_2(S_b^f) \geq \frac{c}{(b^2\ Log\ b)^2}\ ,$$

at least for any given reasonable $b$ (see corollary and remark in the next paragraph). These results are a joint work with H.Chaix and they generalize previous one's in base 2 announced in [4] and published in [5].


## 2 Sequences $S_b^f$ and theorem

Van der Corput sequence in base b: given integers $b \geq 2$ and $n \geq 1$, let $n - 1 = \sum_{r=0}^{\infty} a_r(n)\ b^r$ be the expansion of $(n - 1)$ in base $b$ ; the van der Corput sequence in base $b$ is $\phi_b(n) = \sum_{r=0}^{\infty} a_r(n)\ b^{-r-1}$.

For $f$ integer, $1 \leq f \leq b - 1$, let $\binom{h}{l}_f = \binom{h}{l}f^{h-l}$ be the generalized "$f$-binomial" coefficient. We note $S_b^f$ the sequence $(\phi_b, C^f\phi_b)$ in which $C^f\phi_b(n) = \sum_{j=0}^{\infty} y_j^f(n)\ b^{-j-1}$ with $y_j^f(n) = \sum_{r \geq j} \binom{r}{j}_f\ a_r(n) \bmod b$.

**Theorem.** Let be given $Q = [0, (b-1)q[ \times [0, q[$, with $q = \dfrac{b^{b-1}}{b^b - 1}$, and the sequence of integers

$$N_\lambda = \frac{b^{b^2(\lambda+1)} - 1}{b^{b^2} - 1}.$$

Then $E(Q; N_{b^\tau - 1}; S_b^f) = (K_b^f - \dfrac{b^{b-1}}{2(b^b - 1)})b^{2\tau}$

where $\tau$ is an integer greater than or equal to 1.

From the definition of $D^*(N, X)$ and of $d_s(X)$ in the introduction, we deduce the following corollary:

**Corollary.**
- For every prime number $b$, we have

$$d_2(S_b^1) > \frac{0.2}{(b^2 \ Log \ b)^2} \quad with \quad K_b^1 > \frac{b}{4(b-1)}$$

(already announced in [3]);
- For all prime numbers $b \le 101$ and all $f \ge 2$, we have

$$d_2(S_b^f) > \frac{0.1}{(b^2 \ Log \ b)^2} \quad with \quad K_b^f > 0.15.$$

**Remark.** The second part is obtained by numerical computation; in fact, with a long enough computation, we can attain a similar result for any given $b$ not too large; but at the present, we have no general lower bound $> \dfrac{1}{2b}$ for $K_b^f$.

## 3 Indications on the proof

The usual notions of elementary intervals, $(t, m, s)$-nets and $(t, s)$- sequences are not reminded.

Set $X(l, m) = \{x_{lb^m + 1}, \ldots, x_{(l+1)b^m}\}$, $E(N_\lambda) = E(Q; ] 0, N_\lambda]; S_b^f)$ and $E(S(T)) = E(Q; T; S_b^f)$, so that $E(S(l, m)) = E(S(]lb^m, (l+1)b^m]) = E(Q; ]lb^m, (l+1)b^m]; S_b^f)$.

**Lemma 1** (Expression for $E(N_\lambda)$).

Let the sequence $(l_\mu)$ be defined by

$$l_0 = 0 \text{ and } l_\mu = \sum_{i=1}^{\mu} b^{b^2 i} \text{ for } \mu \ge 1; \text{ then we have } E(N_\lambda) = \sum_{\nu=0}^{\lambda} E\left(S(l_{\lambda-\nu}, b^2\nu)\right).$$

This lemma comes from the additivity of the remainder $E(N_\lambda)$ and from the equality $N_\lambda = \sum_{i=0}^{\lambda} b^{b^2 i}$.

**Lemma 2** (Computing $E(S(l_{\lambda-\nu}, b^2\nu))$).
Set successively :

$$\alpha_k = \sum_{i=0}^{b\nu-k} (b-1)b^{-bi-1} \quad , \quad \alpha'_k = \alpha_k + b^{-b^2\nu+bk-b+1} \text{ for } 1 \le k \le b\nu;$$

$$\beta_1 = 0 \text{ and } \beta_k = \sum_{i=0}^{k-2} b^{-bi-1} \quad , \quad \beta'_k = \beta_k + b^{-bk+b-1} \text{ for } 1 \le k \le b\nu + 1.$$

Besides set:

$$\pi_k = [\alpha_k, \; \alpha'_k[ \; \times \; [\beta_k, \; \beta'_k[ \text{ for } 1 \le k \le b\nu, \; \pi_{b\nu+1} = [0,1[ \; \times \; [\beta_{b\nu+1}, \; \beta'_{b\nu+1}[$$

$$\text{and } D_\nu = Q \cap \bigcup_{k=1}^{\nu+1} \pi_k.$$

Then

$$E(S(l_{\lambda-\nu}, b^2\nu)) = A(D_\nu; \; S(l_{\lambda-\nu}, b^2\nu)) - \tfrac{b^{b-1}}{b^b-1}(\nu + \tfrac{b^{b-1}}{b^b-1}).$$

**Lemma 3** (Necessary and sufficient condition for $\pi_k \cap Q \cap S(l_{\lambda-\nu}, b^2\nu) \ne \phi$). For $1 \le \nu \le \lambda$ and $1 \le k \le b\nu$, consider the following determinant $\Delta_f(\lambda, \nu, k)$ of order $b\nu + 1$:

Then $A(\pi_k \cap Q; \; S(l_{\lambda-\nu}, b^2\nu)) = 1$ if and only if $\Delta_f(\lambda, \nu, k) = 0 \bmod b$.

The main ingredient is that $\pi_k$ contains exactly one point of $S(l_{\lambda-\nu}, b^2\nu)$, which is expressed by a linear Cramer-system; moreover, the point belongs to $Q$ and this property implies a condition of compatibility leading to $\Delta_f(\lambda, \nu, k) = 0$.

236

**Lemma 4** (Shifting operation).
Given a $k$-tuple $(l_1, \cdots, l_k)$ we define the Shift to be the operation getting the $k$-tuple $(fl_1, \ fl_2 + l_1, \ \cdots, fl_{k-1} + l_{k-2}, \ fl_k + l_{k-1})$.
Then after d shifts, $d \geq k$, the last term of the $k$-tuple is $\displaystyle\sum_{h=0}^{k-1} \binom{d}{h}_f l_{k-h}$.
This lemma is a consequence of the properties of f-binomial coefficients.

**Lemma 5** (Computation of $\Delta_f(\lambda, \nu, k)$)
Set $\quad \sigma(c,l) = \sum_{h=0}^{l} \binom{c}{h}_f, \ \tau(c,l) = \sum_{r=c}^{b^\tau - 1} \binom{r}{l}_f, \ \mu(c,l) = \sum_{r=0}^{c-1} \binom{r}{l}_f,$
and

$$\Delta_f(\nu, k) = \Delta_f(b^\tau - 1, \nu, k).$$

Then we have the formula (mod b):

$$f^e \Delta_f(\nu, bp + q) = (1 + f)^{q-1} \sigma(b^\tau - \nu + p, p) + (1 - f)^{b-q} \tau(b^\tau - \nu +$$
$$p, p) - f\mu(b^\tau - \nu + p, p) + f - f^{b^\tau - \nu + p + q - 1},$$

for $1 \leq q \leq b-1$ and $0 \leq p \leq \nu - 1$ ($e = (b^{\tau+1} - b\nu + bp + q - 2)(bp + q) + 1$), and an analogous formula for $q = 0$.
For this lemma, first we perform $(b^{\tau+1} - b\nu + k - 1)$ shifts on the $k$ last rows of $\Delta_f$; then we substract the $b\nu - k + 1$ first columns to the last one in order to yield $\Delta_f$ as a product of diagonal terms equal to 1 except the last one; finally the formula we got is reduced by means of the properties of f-binomial coefficients, in particular the Vandermonde convolution.

After these lemmas, the proof of the theorem consists essentially in accounting or estimating the number of zeros in sub-matrices of order $b^\tau$; for a given b (not too large), we are able to get a lower bound with the help of a computer, but until now, we have not found a general lower bound when $f > 1$.

The detailed proofs will be published in a full paper to appear later.

## REFERENCES

[1] J. BECK, A two dimensional van Aardenne-Ehrenfest theorem in irregularities of distribution, *Compositio Math.*, n° 72, 1989, p.269-339.

[2] H. FAURE, Discrépance de suites associées à un système de numération (en dimension s), *Acta Arith.*, XLI, 1982, p.337-351.

[3] H. FAURE, Discrepancy lower bound for two dimensional quasi-random sequences, *Proceedings Workshop on Quasi-Monte Carlo Methods and Their Applications*, K.T.Fang and F.J.Hickernell Ed., 1995, p.173-178.

[4] H. FAURE ET H. CHAIX, Minoration de discrépance en dimension deux, *C.R.Acad.Paris, série 1*, t.319, 1994, p.1-4.

[5] H. FAURE ET H. CHAIX, Minoration de discrépance en dimension deux, *Acta Arith.*, LXXVI.2, 1996, p.149-164.

[6] G. HALÀSZ, On Roth's method in the theory of irregularities of point distribution, *Recent Progress in Analytic Number Theory, Academic Press*, n° 2, 1981, p.79-94.

[7] J.H. HALTON, On the efficiency of certain quasi-random points in evaluating multi-dimensional integrals, *Numer. Math.*, n° 2, 1960, p.84-90.

[8] H. NIEDERREITER, Point sets and sequences with small discrepancy, *Monatsh. Math.*, n° 104, 1987, p.273-337.

[9] H. NIEDERREITER, Low discrepancy and low dispersion sequences, *J. Number Theory*, n° 30, 1988, p.51-70.

[10] H. NIEDERREITER AND C.P.XING, Low-discrepancy sequences obtained from algebraic function fields over finite fields, *Acta Arith.*, LXXII,1995.3, p.281-298.

[11] H. NIEDERREITER AND C.P.XING, Low-discrepancy sequences and global function fields with many rational places, *Finite Fields and their Appl.*, 2, 1996, p.241-273.

[12] K.F. ROTH, On irregularities of distribution, *Mathematika*, n° 1, 1954, p.73-79.

[13] W.M. SCHMIDT, Irregularities of distribution, *Acta Arith.*, XXI, 1972, p.45-50.

[14] I.M. SOBOL', On the distribution of points in a cube and the approximate evaluation of integrals, *USSR Comp. math. and Math. Physics*, n° 7, 1967, p.86-112.

[15] S. SRINIVASAN, On two dimensional Hammersley sequences, *J. of Number Theory*, n° 10, 1978, p.421-429.

[16] C.P. XING AND H. NIEDERREITER, A construction of low-discrepancy sequences using global function fields, *Acta Arith.*, LXXIII.1, 1995, p.87-102.

Institut Mathématique de Luminy
CNRS U.P.R. 9016
163 avenue de Luminy, case 930
F-13288 Marseille Cedex 09 France

or C.M.I, Université de Provence
39, rue Joliot-Curie
F-13453 Marseille, Cedex 13 France

# Computing Discrepancies Related to Spaces of Smooth Periodic Functions

Karin Frank
Stefan Heinrich

ABSTRACT  A notion of discrepancy is introduced, which represents the integration error on spaces of $r$-smooth periodic functions. It generalizes the diaphony and constitutes a periodic counterpart to the classical $L_2$-discrepancy as well as $r$-smooth versions of it introduced recently by Paskov [Pas93]. Based on previous work [FH96], we develop an efficient algorithm for computing periodic discrepancies for quadrature formulas possessing certain tensor product structures, in particular, for Smolyak quadrature rules (also called sparse grid methods). Furthermore, fast algorithms of computing periodic discrepancies for lattice rules can easily be derived from well–known properties of lattices. On this basis we carry out numerical comparisons of discrepancies between Smolyak and lattice rules.

## 1   Introduction

Discrepancies are a quantitative measure of the precision of multivariate quadratures. Their computation, however, often is a very complex task. Therefore algorithms are of interest which reduce the cost of computing discrepancies either for general quadrature formulas or for special classes. The general case was treated in [War72], [Hei95], while in [FH96] the authors developed a technique of using tensor product structures of quadratures in order to speed up the computation of discrepancy. For the class of Smolyak quadratures, which was so far practically inaccessible to discrepancy computations, this supplied highly efficient algorithms. In [FH96] discrepancies and their behavior under tensor products were studied in a general setting involving arbitrary kernels on the unit cube. The resulting discrepancies turned out to be the worst case integration error over the unit ball of the corresponding reproducing kernel Hilbert spaces of functions. This general approach incorporates the classical $L_2$-discrepancy and its $r$-smooth generalizations given by Paskov [Pas93]. It enabled us to compute these discrepancies for Smolyak quadratures and to compare them with known low discrepancy sequences as well as with standard Monte Carlo. For non-zero smoothness, the Smolyak rules performed very well. We refer to [FH96] for details.

The whole analysis of [FH96] was concerned with the non-periodic case (i. e. the corresponding reproducing kernel Hilbert spaces consist of non-periodic functions). On the other hand, an important class of multivariate quadratures — the class of lattice rules — was designed particularly for smooth periodic functions. Hence it would be desirable to have an analogue of the $r$-smooth discrepancy for the periodic case in order to compare lattice rules to other quadratures as e. g. Smolyak rules. The present paper is devoted to this task.

By using an appropriate kernel based on Bernoulli polynomials, we introduce the $r$-smooth periodic discrepancy $\tilde{D}_r(Q)$ of a quadrature $Q$. It possesses natural interpretations — it is the worst case error of $Q$ over the space of functions with square summable dominating mixed derivative of order $r$ and the average error over a certain related Wiener measure. For smoothness $r = 1$, we recover the diaphony introduced in [Zin76], [ZS78]. Our approach provides an efficient algorithm for computing $\tilde{D}_r(Q)$ for Smolyak quadratures $Q$. On the other hand, the well-known behavior of lattice rules on Bernoulli polynomials provides also an efficient algorithm for computing $\tilde{D}_r$ for this type of quadrature.

As a consequence of this analysis, we are in a position to compare Smolyak and lattice rules numerically. We compare both of them with Monte Carlo integration (more precisely, with the easily explicitly computed expectation of $\tilde{D}_r$ of truly random points). Although Smolyak rules represent a very general approach which leads to optimal (up to logarithmic factors) rates of convergence on many classes of functions including those considered here — our experimental findings revealed a considerably better performance of lattice rules. Smolyak rules, in turn, are much better than Monte Carlo in the present situation.

Summarizing, we think that the periodic discrepancies are a further tool to work out the advantages or disadvantages of various classes of multivariate quadrature formulas.

The paper is organized as follows. In Section 2, we recall the Smolyak construction, Section 3 briefly explains the general approach to discrepancy developed in [FH96]. The periodic discrepancy is introduced in Section 4, while fast algorithms are given in Section 5. The final Section 6 contains the results of numerical experiments.

# 2  Smolyak quadratures

In 1963, Smolyak [Smo63] introduced a special tensor product technique which describes the construction of higher dimensional quadrature formulas and approximation operators on the basis of a sequence of the corresponding one-dimensional objects. If the one-dimensional methods involved possess some optimality features, this technique allows to achieve almost optimal error rates in higher dimensions, too. For an extensive list of references on theoretical investigations as well as on numerical experiments see [WW95] and [NR96].

Let $G = [0,1]^d$. In the following we consider tensor products of quadrature formulas. Let $d = d_1 + d_2$, $d_1, d_2 \geq 1$, and $G = G_1 \times G_2$, with $G_1 = [0,1]^{d_1}$, $G_2 = [0,1]^{d_2}$. Let further $Q', Q''$ be quadrature formulas on $G_1, G_2$, respectively,

$$Q' = ((x_{1j_1}, v_{1j_1}))_{j_1=1}^{M_1} \; , \quad Q'' = ((x_{2j_2}, v_{2j_2}))_{j_2=1}^{M_2} \; ,$$

and $Q = Q' \otimes Q'' = ((x_j, v_j) : j = (j_1, j_2), \, j_1 = 1, ..., M_1, \, j_2 = 1, ..., M_2)$ their standard tensor product $x_j = (x_{1j_1}, x_{2j_2})$, $v_j = v_{1j_1} \cdot v_{2j_2}$. We want to approximate the $d$-dimensional integral

$$If = \int_G f(x) \, dx \, ,$$

where $f$ is a continuous function on $G$. Given a sequence $(Q_n)_{n=0}^{\infty}$ of one-dimensional quadrature rules on $[0,1]$ for continuous functions $f \in C([0,1])$,

$$Q_n f = \sum_{j=1}^{P_n} w_j^n \cdot f(x_j^n) \, ,$$

with $w_j^n \in \mathbb{R}$, $x_j^n \in [0,1]$, we construct the standard tensor product quadrature for the approximate computation of $If$ as

$$U_n^{(d)} f = \left( Q_n \otimes U_n^{(d-1)} \right) f = \sum_{j=0}^{n} (Q_j - Q_{j-1}) \otimes U_n^{(d-1)} f \, ,$$

where $U_n^{(1)} = Q_n$, $Q_{-1} \equiv 0$. Smolyak's approach modifies this definition, setting $Q_n^{(1)} = Q_n$ and defining recursively (see [Smo63])

$$Q_n^{(d)} f = \sum_{j=0}^{n} (Q_j - Q_{j-1}) \otimes Q_{n-j}^{(d-1)} f \, , \tag{1.1}$$

where again $Q_{-1} \equiv 0$. The point set $\Gamma_n^{(d)}$ exploited by the quadrature $Q_n^{(d)}$ is a so-called sparse grid. As was derived in [FH96], its cardinality can be calculated recursively by the formula

$$|\Gamma_n^{(d)}| = \sum_{j=0}^{n} |\Gamma_j^{(1)} \setminus \Gamma_{j-1}^{(1)}| \cdot |\Gamma_{n-j}^{(d-1)}| \, , \tag{1.2}$$

where $\Gamma_{-1}^1 = \emptyset$, provided that the one-dimensional grids $\Gamma_j^{(1)}$ are nested, that means $\Gamma_0^{(1)} \subset \Gamma_1^{(1)} \subset \ldots \subset \Gamma_n^{(1)}$. This condition will be satisfied in all concrete realizations of Smolyak rules we consider in this paper. Note that in contrast to the total number of points in a regular tensor product grid $N = O(M_n^d)$, under some natural assumptions on the sequence $(M_n)$, the number of points in the sparse grid $\Gamma_n^{(d)}$ is reduced to $|\Gamma_n^{(d)}| = O(M_n(\log M_n)^{d-1})$.

# 3   A general approach to discrepancy

Here we recall the general definition of discrepancies as given in [FH96]. Let $G = [0,1]^d$, let $H = L_2(G)$, the space of square-integrable functions with respect to the Lebesgue measure, and let $C(G, H)$ denote the space of continuous functions from $G$ into $H$ (where $H$ is endowed with the norm topology). Fix a function $B(x)$ $(x \in G)$ with $B \in C(G, H)$. For this $B$ we define a discrepancy $D_B$ as follows. Let $Q = ((x_1, v_1), \ldots, (x_M, v_M))$ be a quadrature formula on $G$, i. e. $x_j \in G$, $v_j \in \mathbb{R}$ $(j = 1, \ldots, M)$. Put

$$IB = \int_G B(x)\,dx$$

(the Bochner integral) and

$$QB = \sum_{j=1}^{M} v_j B(x_j).$$

Then the discrepancy $D_B(Q)$ is defined as

$$D_B(Q) = \|IB - QB\|_H. \tag{1.3}$$

Hence we get

$$\begin{aligned} D_B(Q)^2 &= (IB - QB, IB - QB) \\ &= C_B - 2F_B(Q) + S_B(Q, Q), \end{aligned} \tag{1.4}$$

where $(\cdot, \cdot)$ is the scalar product in $H$ and we used the notation

$$\begin{aligned} C_B &= (IB, IB) & \tag{1.5} \\ F_B(Q) &= (IB, QB) & \tag{1.6} \\ S_B(Q, R) &= (QB, RB). & \tag{1.7} \end{aligned}$$

Here $R$ can be any quadrature on $G$, $R = ((y_1, w_1), \ldots, (y_N, w_N))$. It follows readily from (1.3) that $D_B(Q)$ is the worst-case integration error of the quadrature $Q$ over the function class

$$\begin{aligned} W_B &= \{g \in C(G) : \exists h \in H, \|h\|_H \le 1 : \\ &\qquad \forall x \in G : g(x) = (B(x), h)\}. \end{aligned} \tag{1.8}$$

Here $C(G)$ denotes the space of continuous real functions on $G$. Setting $K(x, y) = (B(x), B(y))$, $x, y \in G$, we get that the class $W_B$ is the unit ball of the reproducing kernel Hilbert space generated by $K$.

In [FH96], the special case $B(x,t) = \frac{1}{(r!)^d}(t-x)_+^r$ was studied, where $r$ is a non-negative integer and $a_+ = a$ if $a > 0$ and $a_+ = 0$ otherwise. For $r = 0$ this gives the classical $L_2$-discrepancy, while for $r > 0$ one gets the $r$-smooth versions introduced by Paskov [Pas93]. The resulting classes $W_B$ are unit balls of Sobolev spaces of functions with $L_2$-bounded dominating mixed derivative, which satisfy certain (non-periodic) boundary conditions (see [FH96], [Pas93]). In this paper we shall consider the periodic case, which will be introduced in the next section.

## 4 Discrepancies related to Sobolev spaces of periodic functions

Let $r > 0$ be a natural number. Define for $x \in \mathbb{R}$

$$p_r(x) = 1 - (-1)^{\lfloor \frac{r}{2} \rfloor} \frac{(2\pi)^r}{r!} b_r(x)$$

where $b_r(x)$ denotes the $r$-th Bernoulli polynomial, and $\lfloor a \rfloor$ is the largest integer not exceeding $a$. It is well-known (see e. g. [GR65, 9.622, 9.623.3]) that for $x \in \mathbb{R}$,

$$p_r(\{x\}) = 1 + 2 \sum_{n=1}^{\infty} n^{-r} \cos 2\pi n x$$

if $r$ is even, and

$$p_r(\{x\}) = 1 + 2 \sum_{n=1}^{\infty} n^{-r} \sin 2\pi n x$$

if $r$ is odd. Here $\{a\} = a - \lfloor a \rfloor$ is the fractional part of a number $a \in \mathbb{R}$. For $r \geq 2$ the series converge absolutely and uniformly, for $r = 1$ it converges in $L_2([0,1])$. Now define for $x = (\xi_1, \ldots, \xi_d) \in \mathbb{R}^d$

$$P_r(x) = \prod_{l=1}^{d} p_r(\xi_l)$$

and for $x, t \in G$, $x = (\xi_1, \ldots, \xi_d)$, $t = (\tau_1, \ldots, \tau_d)$,

$$B_r^{(d)}(x,t) = P_r(\{x - t\}),$$

where $\{x - t\} = (\{\xi_1 - \tau_1\}, \ldots, \{\xi_d - \tau_d\})$. Although we deal with real functions only, it is convenient for us to use the following representation

$$p_r(\{x\}) = \sum_{n \in \mathbb{Z}} \sigma_r(n) \bar{n}^{-r} e^{2\pi i n x}, \tag{1.9}$$

where $\bar{n} = |n|$ if $n \neq 0$, $\bar{n} = 1$ if $n = 0$, and

$$\sigma_r(n) = \begin{cases} -\text{sign}(n)\, i & \text{if } n \neq 0 \text{ and } r \text{ is odd,} \\ 1 & \text{otherwise.} \end{cases}$$

For $n = (n_1, \ldots, n_d)$ we set

$$\bar{n} = \prod_{l=1}^{d} \bar{n}_l, \quad \sigma_r(n) = \prod_{l=1}^{d} \sigma_r(n_l).$$

Hence we get for $x, t \in G$

$$B_r^{(d)}(x, t) = \sum_{n \in \mathbb{Z}^d} \sigma_r(n) \bar{n}^{-r} e^{2\pi i(n, x-t)}. \tag{1.10}$$

It is easily derived from these representations and from (1.8) that $W_{B_r^{(d)}}$ is the unit ball of the Sobolev space of real functions

$$\mathcal{H}^r(G) = \{g \in L_2(G) : \|g\|_r^2 = \sum_{n \in \mathbb{Z}^d} \bar{n}^{2r} |\hat{g}(n)|^2 < \infty\}$$

where

$$\hat{g}(n) = \int_G g(t) e^{-2\pi i n t} dt$$

is the $n$-th Fourier coefficient (see e. g. [Fra95], where this space is considered). $\mathcal{H}^r(G)$ consists of all periodic functions $g \in L_2(G)$ whose generalized derivatives

$$\frac{\partial^{\alpha_1 + \ldots + \alpha_d} g(\xi_1, \ldots, \xi_d)}{\partial \xi_1^{\alpha_1} \ldots \partial \xi_d^{\alpha_d}}$$

belong to $L_2(G)$ whenever $0 \leq \alpha_l \leq r$ ($l = 1, \ldots, d$). Let us now consider the discrepancy defined by (1.3) and (1.4) on the basis of the function $B_r^{(d)}(x, t)$. Let $Q = ((x_j, v_j))_{j=1}^{M}$, $R = ((y_j, w_j))_{j=1}^{N}$ and denote

$$\tilde{D}_r(Q) = D_{B_r^{(d)}}(Q);$$
$$\tilde{S}_r(Q, R) = S_{B_r^{(d)}}(Q, R), \quad \tilde{F}_r(Q) = F_{B_r^{(d)}}(Q), \quad \tilde{C}_r = C_{B_r^{(d)}}.$$

In these notations, for the sake of simplicity we drop the dimension parameter, which is indicated by the arguments $Q, R$. By the discussion above, $\tilde{D}_r(Q)$ is the worst case error of a quadrature $Q$ over the unit ball of the space $\mathcal{H}^r(G)$. We consider the representation (1.4) and compute each of the terms. Using (1.9) and (1.10) we easily obtain

$$\tilde{C}_r = 1, \quad \tilde{F}_r(Q) = \sum_{j=1}^{M} v_j, \tag{1.11}$$

and

$$\tilde{S}_r(Q, R) = \sum_{j=1}^{M} \sum_{k=1}^{N} v_j w_k \sum_{m \in \mathbb{Z}^d} \bar{m}^{-2r} e^{2\pi i(m, x_j - y_k)}$$

$$= \sum_{j=1}^{M} \sum_{k=1}^{N} v_j w_k P_{2r}(\{x_j - y_k\}) \tag{1.12}$$

It follows that $\tilde{D}_r(Q)$ can be computed in $O(M^2)$ operations. Below we analyze situations, in which this can be done in a faster way. Note that for $r = 1$ and uniform weights $v_j = 1/M$ $(j = 1, \ldots, M)$, $\tilde{D}_1(Q)$ is the diaphony of the point set $(x_j, j = 1, \ldots, M)$ introduced and studied in [Zin76] and [ZS78]. To see this, observe that (1.4), (1.11) and (1.12) give for the case of weights satisfying $\sum_{j=1}^M v_j = 1$,

$$\tilde{D}_r(Q)^2 = \sum_{j,k=1}^M v_j v_k \left( \prod_{l=1}^d p_{2r}(\{x_{jl} - x_{kl}\}) - 1 \right).$$

It remains to compare $p_{2r}(\{x\})$ with the function $g(x)$ of [ZS78].

## 5   Fast computation of $\tilde{D}_r(Q)$

The recursive approach of [Hei95] could be applied to develop an algorithm for the computation of $\tilde{D}_r(Q)$. Formally, this would lead to an $O(M(\log M)^d)$ algorithm as in [Hei95, Section 4], but with a heavy dependence of the constants on $r$ and $d$. So only small $r$ and $d$ can be handled. In this section we concentrate on special types of quadratures — good lattice points and Smolyak rules.

Let $L$ be a $d$-dimensional integration lattice (see [Nie92, Section 5.3]) and let $Q_L$ be the associated quadrature rule on $G$ with node set $L \cap G'$, $G' = [0,1)^d$, and uniform weights. The computation of $\tilde{D}^r(Q_L)$ is easily accomplished following the lines of the analysis for the Korobov classes $\mathcal{E}_r^d$ (see [Kor63], [Nie92, Th.5.23]). According to Lemma 5.21 of [Nie92] we have

$$Q_L e^{2\pi i(n, \cdot)} = \begin{cases} 1 & \text{if } n \in L^\perp \\ 0 & \text{if } n \notin L^\perp, \end{cases}$$

where $L^\perp = \{n \in \mathbb{Z}^d : (n, x) \in \mathbb{Z} \text{ for all } x \in L\}$ is the dual lattice. Using this, the definition (1.3) and relation (1.10), we get

$$\begin{aligned}
\tilde{D}_r(Q_L) &= \|I B_r^{(d)} - Q_L B_r^{(d)}\|_{L_2(G)} \\
&= \left\| \sum_{n \in \mathbb{Z}^d} \sigma_r(n) \, \bar{n}^{-r} \left( (I - Q_L) e^{2\pi i(n, \cdot)} \right) e^{-2\pi i(n,t)} \right\|_{L_2(G)} \\
&= \left( \sideset{}{'}\sum_{n \in L^\perp} \bar{n}^{-2r} \right)^{1/2} = \left| (I - Q_L) B_{2r}^{(d)}(\cdot, 0) \right|^{1/2}
\end{aligned} \qquad (1.13)$$

As usual, $\sum'$ means that the summand for $n = (0, \ldots, 0)$ is left out. Hence in order to compute $\tilde{D}_r(Q_L)$ we have to determine the integration error of $Q_L$ on the single function

$$B_{2r}^{(d)}(x, 0) = P_{2r}(x) = \prod_{l=1}^d p_{2r}(\xi_l), \quad x = (\xi_1, \ldots, \xi_d) \in G'.$$

This is just the number theorists criterion considered in various papers dedicated to the search of optimal parameters for good lattice points, e. g. [Hab83], [SW90]. The evaluation of the integration error on $P_{2r}(x)$ is an $O(M)$ procedure. Note that (1.13) and Theorem 5.23 of [Nie92] imply that for lattice rules $Q_L$

$$\sup_{f \in B_{\mathcal{H}^r}} |(I - Q_L)f| = \sup_{g \in B_{\mathcal{E}^d_{2r}}} |(I - Q_L)g|^{1/2}$$

where $B_X$ denotes the unit ball of the space $X$, and the Korobov space $\mathcal{E}^d_r$ consists of all functions $g \in L_2(G)$ with

$$\|g\|_{\mathcal{E}^d_r} = \sup_{n \in \mathbb{Z}^d} \bar{n}^{-r}|\hat{g}(n)| < \infty.$$

Next we consider Smolyak quadratures. In [FH96], we developed a recursive algorithm which allows to reduce the computation of $D_B$ to the one-dimensional case provided $B(x,t)$ has product structure. This result applies to the present situation. The algorithm is the following.

As introduced in Section 2, the Smolyak quadrature rule on $[0,1]^d$, $d \geq 2$, satisfies the recursion

$$Q_n^{(d)} = \sum_{i=0}^{n} R_i \otimes Q_{n-i}^{(d-1)},$$

where $R_i = Q_i - Q_{i-1}$, $Q_{-1} \equiv 0$ and $Q_n^{(1)} = Q_n$. We fix a maximal level $n_{max}$ and apply a recursion over $d$ to calculate all quantities $\tilde{F}_r(Q_n^{(d)})$ and $\tilde{S}_r(Q_m^{(d)}, Q_n^{(d)})$ for $m, n = 0, 1, \ldots, n_{max}$.

The recursion starts from the univariate case by computing the terms $\tilde{F}_r(Q_n)$, $\tilde{S}_r(Q_m, Q_n)$. From these terms we get using the behavior of $\tilde{F}_r$ and $\tilde{S}_r$ under sums and tensor products as established in [FH96]

$$\tilde{F}_r(R_n) = \tilde{F}_r(Q_n) - \tilde{F}_r(Q_{n-1})$$
$$\tilde{S}_r(R_m, R_n) = \tilde{S}_r(Q_m, Q_n) - \tilde{S}_r(Q_{m-1}, Q_n)$$
$$- \tilde{S}_r(Q_m, Q_{n-1}) + \tilde{S}_r(Q_{m-1}, Q_{n-1})$$
$$\tilde{F}_r(Q_n^{(d)}) = \sum_{i=0}^{n} \tilde{F}_r(R_i) \cdot \tilde{F}_r(Q_{n-i}^{(d-1)})$$
$$\tilde{S}_r(Q_m^{(d)}, Q_n^{(d)}) = \sum_{i=0}^{m} \sum_{j=0}^{n} \tilde{S}_r(R_i, R_j) \cdot \tilde{S}_r\left(Q_{m-i}^{(d-1)}, Q_{n-j}^{(d-1)}\right)$$

Finally, the discrepancy $\tilde{D}_r(Q_n^{(d)})$ is given by (1.4), where by (1.11) $\tilde{C}_r = 1$ independently of the dimension.

To handle the one-dimensional quantities, let $Q = ((x_j, v_j))_{j=1}^M$ and $R = ((y_k, w_k))_{k=1}^N$ be quadratures on $[0,1]$. It is sufficient to transform $\tilde{S}_r(Q, R)$ into an efficiently computable form, since $\tilde{F}_r(Q)$ can be calculated in $O(M)$

operations simply by (1.11). We assume that the node sets are ordered in such a way that $x_1 \leq x_2 \leq \ldots \leq x_M$ and $y_1 \leq y_2 \leq \ldots \leq y_N$. Let us mention that the node sets of many quadrature rules are ordered by their definition. Then we determine for each $j = 1, \ldots, M$ an index $\nu(j)$ such that $x_j \geq y_k$ for each $k \leq \nu(j)$ and $x_j < y_k$ for $k > \nu(j)$. Using this we can rewrite the direct formula (1.12) as follows

$$\tilde{S}_r(Q, R) = \sum_{j=1}^{M} \sum_{k=1}^{N} v_j w_k \, p_{2r}(\{x_j - y_k\})$$

$$= \sum_{j=1}^{M} v_j \sum_{k=1}^{N} w_k - (-1)^r \frac{(2\pi)^{2r}}{(2r)!} \varphi^{(r)}$$

where

$$\varphi^{(r)} = \sum_{j=1}^{M} v_j \left[ \sum_{k=1}^{\nu(j)} w_k b_{2r}(x_j - y_k) + \sum_{k=\nu(j)+1}^{N} w_k b_{2r}(1 + x_j - y_k) \right]$$

$$= \sum_{l=0}^{2r} \alpha_{l,2r} \sum_{q=0}^{l} \binom{l}{q} \sum_{j=1}^{M} v_j x_j^{l-q} \, *$$

$$* \left[ \sum_{k=1}^{\nu(j)} w_k (-y_k)^q + \sum_{k=\nu(j)+1}^{N} w_k (1 - y_k)^q \right] \qquad (1.14)$$

Here $\alpha_{l,2r} \in \mathbb{R}$ are the coefficients of the Bernoulli polynomial $b_{2r}$

$$b_{2r}(x) = \alpha_{0,2r} + \alpha_{1,2r} x + \ldots + \alpha_{2r,2r} x^{2r} \,.$$

Both sums in (1.14) can be calculated in $O(M + N)$, if the inner sums are added up successively. We fix $r \geq 0$. Assume that there are reals $p > 1$, $c_1, c_2 > 0$ such that the number of nodes $P_n$ in the one-dimensional quadratures $Q_n$ satisfies

$$c_1 p^n \leq P_n \leq c_2 p^n \,.$$

This is a natural assumption for Smolyak quadratures. Fix $n_{max}$ and denote $P = P_{n_{max}}$. Obviously, $n_{max} = O(\log P)$. As was pointed out in [FH96], the complexity of the whole recursion process is then $O(P \log P + d(\log P)^4)$.

## 6 Numerical results

We will compare the $L_2$-discrepancies $\tilde{D}_r$ of two different Smolyak quadrature rules with rank-1 lattices, rank-2 lattices and Monte Carlo integration.

The Smolyak rules are fully determined by the recursion (1.1) and the sequence $(Q_n)$ $(n = 0, \ldots, n_{max})$ of one-dimensional quadratures. We will test two sequences used already in [FH96]. In both quadratures $Q_0$ is chosen as the midpoint rule

$$Q_0 f = f(0.5),$$

because otherwise the number of grid points would increase exponentially in $d$. As sequence $(Q_n)$ $(n \geq 1)$ of one-dimensional rules the Smolyak rule TR uses the sequence of composite trapezoidal rules on $2^n$ subintervals, whereas CC takes the sequence of Clenshaw-Curtis rules using $2^n$ subintervals (see [Bra77]). Note that by (1.2) the exact number of sample points employed by $Q_n^{(d)}$ coincides in both cases.

To compare Monte Carlo integration, we do not use any concrete random number generator, but compute the square mean of the discrepancy $\tilde{D}_r$

$$\mathbb{E}\tilde{D}_r(Q)^2 = \mathbb{E} \int_G \left( IB_r^{(d)}(\cdot, t) - \frac{1}{M} \sum_{i=1}^M B_r^{(d)}(\zeta_i, t) \right)^2 dt$$

$$= \frac{1}{M} \int_G \mathbb{E} \left( IB_r^{(d)}(\cdot, t) - B_r^{(d)}(\zeta_1, t) \right)^2 dt$$

$$= \frac{1}{M} \int_G \left( \int_G B_r^{(d)}(x, t)^2 dx - \left( \int_G B_r^{(d)}(x, t) dx \right)^2 \right) dt$$

$$= \frac{1}{M} \left( B_{2r}^{(d)}(0, 0) - 1 \right).$$

Finally, we include lattice rules into our comparisons. This family of quadrature formulas was developed by Korobov [Kor63] and Hlawka [Hla62], and is designed especially for the integration of periodic functions of several variables. In [SL89] it was shown that any lattice rule can be represented as a nonrepetitive expression of the form

$$Q_L f = \frac{1}{n_1 \ldots n_m} \sum_{j_1=1}^{n_1} \cdots \sum_{j_m=1}^{n_m} f\left( \left\{ j_1 \frac{z_1}{n_1} + \cdots + j_m \frac{z_m}{n_m} \right\} \right),$$

where $z_1, \ldots, z_m$ are integer vectors, and $n_{i+1}$ divides $n_i$ for $i = 1, \ldots, m-1$, $n_m > 1$. The rank $m$ and the invariants $n_1, \ldots, n_m$ are uniquely determined (see [SW90]). We will consider lattice rules of rank 1 with parameters as indicated in [Hab83], and lattice rules of rank 2 with parameters from [SW90].

All computations were carried out on a workstation of the series HP-9000 in the language C++. Due to cancellation in (1.4) it was necessary to use quadruple precision.

Unfortunately, the parameters for the good-lattice points were given only for certain numbers of points in [Hab83] and [SW90]. Neither can the number of sample points in the Smolyak quadratures be chosen voluntarily but depends on the dimension and the maximal level $n_{max}$. We tried to select such parameter combinations that the quadrature rules compared in

| r | CC | TR | Rank 1 | Rank 2 | MC |
|---|------|------|---------|---------|------|
|   | 13953 | 13953 | 12288 | 10404 | 13953 |
| 1 | 3.39e-01 | 2.08e-01 | 1.14e-02 | 9.64e-03 | 7.47e-02 |
| 2 | 4.91e-03 | 1.09e-03 | 2.70e-05 | 1.03e-05 | 4.69e-02 |
| 3 | 1.44e-04 | 8.18e-06 | 8.86e-08 | 1.48e-08 | 4.40e-02 |
| 4 | 5.98e-06 | 6.34e-08 | 3.06e-10 | 2.28e-11 | 4.34e-02 |

| r | CC | TR | Rank 1 | Rank 2 | MC |
|---|------|------|---------|---------|------|
|   | 18945 | 18945 | 16384 | 10332 | 18945 |
| 1 | 2.51 | 1.88 | 3.97e-02 | 4.63e-02 | 1.34e-01 |
| 2 | 1.63e-01 | 3.21e-02 | 2.08e-04 | 2.08e-04 | 7.24e-02 |
| 3 | 3.48e-02 | 9.31e-04 | 1.59e-06 | 1.33e-06 | 6.65e-02 |
| 4 | 1.34e-02 | 2.87e-05 | 1.33e-08 | 9.40e-09 | 6.53e-02 |

TABLE 1.1. Discrepancies with $M \approx 10^4$ points in $d = 3$ (top) and $d = 4$ (bottom)

one table work with as similar numbers of function values as possible. To be just against the Monte Carlo quadrature we computed the square mean of its discrepancy always with the highest number of points in the respective table. However, its performance turned out to be poor in comparison with the other methods.

Tables 1.1-1.3 report some numerical experiments for $M \approx 10^4$ (Table 1.1) and $M \approx 10^5$ (Tables 1.2, 1.3). The results almost speak for themselves. In contrast to [FH96], the Smolyak quadratures do not perform very well, particularly with moderate numbers of grid points in higher dimensions. Among them only the Smolyak quadrature based on the trapezoidal rule can achieve satisfactory results. This is due to the fact, that the trapezoidal rule is optimal on the class of smooth periodic functions on $[0, 1]$ (see [TWW88]).

Since good–lattice points were developed especially for periodic functions of several variables, it was to be expected that their discrepancy would be smaller than the discrepancy of the tensor product methods. However, we were surprised by the extend of superiority of the number–theoretic

| r | CC | TR | Rank 1 | Rank 2 | MC |
|---|------|------|---------|---------|------|
|   | 127105 | 127105 | 131072 | 100044 | 131072 |
| 1 | 37.42 | 43.05 | 1.18e-01 | 1.03e-01 | 2.18e-01 |
| 2 | 3.38 | 1.47 | 1.67e-03 | 8.04e-04 | 8.75e-02 |
| 3 | 1.37 | 1.33e-01 | 4.01e-05 | 1.15e-05 | 7.71e-02 |
| 4 | 8.89e-01 | 1.51e-02 | 1.05e-06 | 1.85e-07 | 7.51e-02 |

TABLE 1.2. Discrepancies with $M \approx 10^5$ points in $d = 6$

| r | CC | TR | Rank 1 | Rank 2 | MC |
|---|----|----|--------|--------|-----|
|   | 163841 | 163841 | 131072 | 100044 | 131072 |
| 1 | 6.16e-02 | 3.51e-02 | 1.53e-03 | 1.54e-03 | 2.18e-02 |
| 2 | 1.12e-04 | 2.31e-05 | 6.08e-07 | 3.25e-07 | 1.37e-02 |
| 3 | 3.84e-07 | 2.16e-08 | 3.29e-10 | 9.24e-11 | 1.28e-02 |
| 4 | 1.78e-09 | 2.09e-11 | 1.84e-13 | 2.83e-14 | 1.26e-02 |

| r | CC | TR | Rank 1 | Rank 2 | MC |
|---|----|----|--------|--------|-----|
|   | 113409 | 113409 | 131072 | 100044 | 131072 |
| 1 | 1.01 | 6.46e-01 | 8.04e-03 | 7.86e-03 | 5.08e-02 |
| 2 | 1.82e-02 | 2.89e-03 | 1.15e-05 | 6.14e-06 | 2.75e-02 |
| 3 | 9.31e-04 | 2.12e-05 | 2.44e-08 | 6.90e-09 | 2.53e-02 |
| 4 | 8.54e-05 | 1.64e-07 | 5.48e-11 | 8.54e-12 | 2.48e-02 |

TABLE 1.3. Discrepancies with $M \approx 10^5$ points in $d = 3$ (top) and $d = 4$ (bottom)

methods. As is only natural, the more general rank-2 methods showed even a slightly better performance than the rank-1 rules.

# 7 REFERENCES

[Bra77]  H. Brass. *Quadraturverfahren*. Vandenhoeck & Rupprecht, Göttingen, 1977.

[FH96]  K. Frank and S. Heinrich. Computing discrepancies of Smolyak quadrature rules. *J. Complexity*, 12:287 – 314, 1996.

[Fra95]  K. Frank. Complexity of local solution of multivariate integral equations. *J. Complexity*, 11:416 – 434, 1995.

[GR65]  I.S. Gradshteyn and I.M. Ryshik. *Tables of integrals, series and products*. Academic Press, NY, London, 1965.

[Hab83]  S. Haber. Parameters for integrating periodic functions of several variables. *Math. Comp.*, 41:115 – 129, 1983.

[Hei95]  S. Heinrich. Efficient algorithms for computing the $L_2$ discrepancy. TR 267/95 Universität Kaiserslautern, 1995. (to appear in Math. Comp.).

[Hla62]  E. Hlawka. Zur angenäherten Berechnung mehrfacher Integrale. *Monatsh. Math.*, 66:140 – 151, 1962.

[Kor63]  N.M. Korobov. *Number-Theoretic Methods in Approximate Analysis*. Fizmatgiz, Moscow, 1963. (in Russian).

250

[Nie92]    H. Niederreiter. *Random Number Generation and Quasi–Monte Carlo Methods*. CBMS–NSF 63. SIAM, 1992.

[NR96]    E. Novak and K. Ritter. High dimensional integration of smooth functions over cubes. to appear in Num. Math., 1996.

[Pas93]    S.H. Paskov. Average case complexity of multivariate integration for smooth functions. *J. Complexity*, 9:291 – 312, 1993.

[SL89]    I.H. Sloan and J.N. Lyness.    The representation of lattice quadratures as multiple sums. *Math. Comp.*, 52:81 – 94, 1989.

[Smo63]    S.A. Smolyak. Quadrature and interpolation formulas for tensor products of certain classes of functions (in Russian). *Dokl. Akad. Nauk SSSR*, 148:1042 – 1045, 1963. English transl.: *Soviet. Mat. Dokl.* 4, 240 - 243.

[SW90]    I.H. Sloan and L. Walsh. A computer search of rank-2 lattice rules for multidimensional quadrature. *Math. Comp.*, 54:281 – 302, 1990.

[TWW88] J. F. Traub, G. W. Wasilkowski, and H. Woźniakowski. *Information-Based Complexity*. Academic Press, New York, 1988.

[War72]    T.T. Warnock. Computational investigations of low discrepancy point sets. In S.K. Zaremba, editor, *Applications of Number Theory to Numerical Analysis*. Academic Press, New York, 1972.

[WW95]    G.W. Wasilkowski and H. Woźniakowski. Explicit cost bounds of algorithms for multivariate tensor product problems. *J. Complexity*, 11:1 – 56, 1995.

[Zin76]    P. Zinterhof. Über einige Abschätzungen bei der Approximation von Funktionen mit Gleichverteilungsmethoden. *Österr. Akad. Wiss. Math. Nat. Kl., S.-Ber.*, Abt. II 185:121 – 132, 1976.

[ZS78]    P. Zinterhof and H. Stegbuchner. Trigonometrische Approximation mit Gleichverteilungsmethoden. *Studia Sci. Math. Hungar.*, 13:273 – 289, 1978.

Karin Frank
RUS, Universität Stuttgart, Allmandring 30a, D–70550 Stuttgart, Germany.
email: frank@rus.uni-stuttgart.de

Stefan Heinrich
FB Informatik, Universität Kaiserslautern, PF 3049, D–67653 Kaiserslautern, Germany.
email: heinrich@informatik.uni-kl.de

# On correlation analysis of pseudorandom numbers

## Peter Hellekalek[1]

ABSTRACT In this paper we discuss the theoretical analysis of correlations between pseudorandom numbers. We present a new concept that allows to relate the discrepancy approach to the spectral test. Up to now, those two figures of merit for pseudorandom number generators were viewed as widely different. We discuss the most important examples of our approach as well as the underlying technique.

## 1 Introduction

> *"Monte Carlo results are misleading when correlations hidden in the random numbers and in the simulated system interfere constructively."*
> ... A. Compagner [Com95]

The goal of constructing and testing pseudorandom number generators is to provide the basic prerequisites for successful stochastic simulation. As Compagner's statement shows, correlation analysis of pseudorandom numbers is an absolute necessity to attain this objective. For this purpose certain numerical quantities are used, so-called "figures of merit" of pseudorandom number generators.

In this context, the most important quantities are *discrepancy* and the *spectral test*. More than two decades of practical experience have shown that these two figures of merit allow remarkably accurate predictions of the performance of pseudorandom numbers in empirical tests. The latter are nothing less than *prototypes* of simulation problems. As a consequence, discrepancy and the spectral test are highly relevant for the practice of pseudorandom number generation.

In this paper, we will present a new viewpoint that relates these figures of merit via so-called *Weyl sums*, the *concept of the weighted spectral test*. Several realizations of our concept that lead to numerical quantities of practical and theoretical relevance will be discussed in Section 4. The underlying technique will be demonstrated in Section 5.

---

[1] Research supported by the Austrian Science Foundation, project no. P11143-MAT

# 2 Prerequisites

Throughout this paper, we shall identify the $s$-dimensional torus $\mathbf{R}^s/\mathbf{Z}^s$ with the $s$-dimensional unit cube $[0,1[^s$. For a nonnegative integer $k$, let $k = \sum_{j=0}^{\infty} k_j 2^j$, $k_j \in \{0,1\}$, be the unique binary expansion of $k$. Every number $x \in [0,1[$ has a unique binary expansion $x = \sum_{j=0}^{\infty} x_j 2^{-j-1}$, $x_j \in \{0,1\}$, under the condition that $x_j \neq 1$ for infinitely many $j$. In the following, this uniqueness condition will be assumed without further notice. By a *binary rational* $c$, we understand an element $c = 0.c_0 c_1 \ldots \in [0,1[$ such that only finitely many digits $c_j$ are different from zero. The binary logarithm of $x \in [0,1[$ will be denoted by $\log_2 x$. If $x = 2^{-g} + x_g 2^{-g-1} + \ldots$, then the integer part of $\log_2 x$ is given by $\lfloor \log_2 x \rfloor = -g$.

**Definition 2.1** *The $k$-th Walsh function $w_k$, $k \geq 0$, to the base 2 is defined as*

$$w_k(x) := \prod_{j=0}^{\infty} (-1)^{x_j \cdot k_j}, \tag{1.1}$$

*where $x = 0.x_0 x_1 \ldots$ is the unique binary expansion of $x \in [0,1[$ and $k = \sum_{j=0}^{\infty} k_j 2^j$ is the unique binary expansion of the nonnegative integer $k$. If $\mathbf{k} = (k_1, \ldots, k_s)$ is an integer vector with nonnegative coordinates, then the $\mathbf{k}$-th Walsh function $w_{\mathbf{k}}$ on $[0,1[^s$ is defined as*

$$w_{\mathbf{k}}(\mathbf{x}) := \prod_{i=1}^{s} w_{k_i}(x_i), \quad \mathbf{x} = (x_1, \ldots, x_s) \in [0,1[^s. \tag{1.2}$$

**Definition 2.2** *For two digits $d$ and $d'$ in $\{0,1\}$, let*

$$d \oplus d' := d + d' \pmod{2}.$$

*For two numbers $x, y \in [0,1[$ with binary expansions $x = \sum_{j=0}^{\infty} x_j 2^{-j-1}$ and $y = \sum_{j=0}^{\infty} y_j 2^{-j-1}$, let $x \dotplus y$ denote the binary sum of $x$ and $y$,*

$$x \dotplus y := \sum_{j=0}^{\infty} (x_j \oplus y_j) 2^{-j-1} \pmod{1}.$$

*If $\mathbf{x}, \mathbf{y} \in [0,1[^s$, $\mathbf{x} = (x_1, \ldots, x_s)$, $\mathbf{y} = (y_1, \ldots, y_s)$, then*

$$\mathbf{x} \dotplus \mathbf{y} := (x_1 \dotplus y_1, \ldots, x_s \dotplus y_s).$$

**Remark 2.1** (1) The digits $(x \dotplus y)_j$ of the binary expansion of the number $x \dotplus y \in [0,1[$ need not coincide with $x_j \oplus y_j$. Each of the following conditions,

(C1) $x \dotplus y$ *not a binary rational,*

(C2) $x$ or $y$ *binary rationals,*

implies the equality

$$(x \dotplus y)_j = x_j \oplus y_j, \quad \forall j \geq 0. \tag{1.3}$$

(2) If the above identity (1.3) holds for $x, y \in [0, 1[$, then

$$w_k(x \dotplus y) = w_k(x) \, w_k(y). \tag{1.4}$$

In this paper, $\mathcal{F} = \{\chi_k\}$ will denote either the trigonometric function system $\mathcal{T} = \{e_k\}$,

$$e_k(\mathbf{x}) := \prod_{i=1}^{s} e^{2\pi\sqrt{-1}k_i x_i},$$

where $\mathbf{k} = (k_1, \ldots, k_s) \in \mathbf{Z}^s$ and $\mathbf{x} = (x_1, \ldots, x_s) \in [0, 1[^s$, or the Walsh function system $\mathcal{W} = \{w_k\}$ to the base two on the $s$-dimensional unit cube.

If $\phi$ is an integrable function on $[0, 1[^s$ and if $\mathbf{k} = (k_1, \ldots, k_s)$ is an integer vector with nonnegative coordinates, then let $\hat{\phi}(\mathbf{k})$ denote the $\mathbf{k}$-th Fourier coefficient of $\phi$ with respect to the function $\chi_\mathbf{k} \in \mathcal{F}$,

$$\hat{\phi}(\mathbf{k}) := \int_{[0,1[^s} \phi(\mathbf{x})\overline{\chi_\mathbf{k}(\mathbf{x})} \, d\mathbf{x}.$$

# 3 The Concept

Correlation analysis of pseudorandom numbers by means of discrepancy or the spectral test is based upon the following approach.

Suppose we are given pseudorandom numbers $x_0, x_1, \ldots$ in $[0, 1[$. To check for correlations between consecutive numbers, we construct either *overlapping* $s$-tuples $\mathbf{x}_n := (x_n, x_{n+1}, \ldots, x_{n+s-1})$ or *nonoverlapping* $s$-tuples $\mathbf{x}_n := (x_{ns}, x_{ns+1}, \ldots, x_{ns+s-1})$ and assess the empirical distribution of finite sequences $\omega = (\mathbf{x}_n)_{n=0}^{N-1}$ in the $s$-dimensional unit cube $[0, 1[^s$. The task is to measure how "well" $\omega$ is uniformly distributed. Strong correlations between consecutive pseudorandom numbers will lead to significant deviations of the empirical distribution of $\omega$ from uniform distribution, for some dimensions $s$. Of course, the restricted type of $s$-tuples that we consider cannot ensure against *long-range* correlations among the numbers $x_n$ themselves. For this topic, we refer the reader to De Matteis and Pagnutti [DMP90]. In the case of the explicit-inversive congruential pseudorandom number generator of Eichenauer-Herrmann [EH93], more general types of $s$-tuples have been considered (see also Niederreiter [Nie94, Nie95]). Interestingly, it has turned out that the behaviour of *full-period* sequences $\omega$ with respect to discrepancy or the spectral test allows very reliable predictions of the performance of the pseudorandom numbers $x_n$ themselves in empirical tests. If the full-period point set $\omega$ has a good distribution with respect to discrepancy or the spectral test in various dimensions $s$, then excellent empirical performance of the samples is highly probable[2], see Niederreiter [Nie95] or L'Ecuyer [L'E94] for background information. Practical evidence is that many target distributions will be simulated very well, see, for example, the empirical results of Leeb and Wegenkittl [LW96].

*Weyl sums* are the main tool to carry out this kind of a-priori correlation analysis, as it becomes apparent from Compagner [Com95] (see the notion of a "scanning ensemble" and Equation (19) in [Com95, page 5636]), the well-known work of Coveyou and MacPherson [CM67], and the important contributions of Niederreiter [Nie78, Nie92, Nie95]. The latter author has proved the fundamental inequalities to estimate discrepancy in terms of Weyl sums, see [Nie92, Theorem 3.10 and 3.16] and the related results of Hellekalek [Hel94].

**Definition 3.1** *Let* $\omega = (\mathbf{x}_n)_{n=0}^{N-1}$ *be a finite sequence in* $[0,1[^s$ *and let* $\mathcal{F} = \{\chi_{\mathbf{k}}\}$ *be a function system on the s-dimensional unit cube. The sum*

$$S_N(\chi_{\mathbf{k}}, \omega) := \frac{1}{N} \sum_{n=0}^{N-1} \chi_{\mathbf{k}}(\mathbf{x}_n)$$

*is called a Weyl sum with respect to* $\chi_{\mathbf{k}}$ *and* $\omega$.

Using the descriptive terminology of Compagner [Com95], we may also call $S_N(\chi_{\mathbf{k}}, \omega)$ the *correlation coefficient* of $\omega$ relative to $\chi_{\mathbf{k}}$.

The *interpretation* of Weyl sums in the context of correlation analysis of pseudorandom numbers is the following. The function $\chi_{\mathbf{k}}$ is viewed as an $s$-dimensional wave with frequency $\sqrt{k_1^2 + \ldots + k_s^2}$. It is a well-known fact that the distribution properties of $\omega$ depend more on low-frequency Weyl sums than on those with high frequencies, see the fundamental inequality of Erdös-Turán-Koksma in the monograph Kuipers and Niederreiter [KN74] and its variants adapted to the special case of pseudorandom number generation in Niederreiter [Nie92, Chapter 3.2]. The spectral test of Coveyou and MacPherson [CM67] (see also the discussion in Knuth [Knu67, Knu81]) and the Walsh spectral test of Tezuka [Tez87] are based on the very same Weyl sums or correlation coefficients. In the classical spectral test, where $\mathcal{F} = \mathcal{T}$, the test quantity is

$$\sigma_N(\omega) := \frac{1}{min\{\sqrt{k_1^2 + \ldots + k_s^2} : \mathbf{k} \neq 0, \ S_N(e_{\mathbf{k}}, \omega) \neq 0\}}.$$

In his Walsh spectral test, Tezuka [Tez87] considers a similar quantity for the system $\mathcal{F} = \mathcal{W}$,

$$\max \{\alpha : S_N(w_{\mathbf{k}}, \omega) = 0 \quad \forall \, \mathbf{k} = (k_1, \ldots, k_s) \neq 0, \ 0 \leq k_i < 2^{\alpha}\}.$$

---

[2]This relation between properties of the full period sequences in higher dimensions and the behaviour of -comparatively small- samples in low dimensions has not yet been put into rigorous mathematical form.

The notions of "range" and "order" of the vectors $\mathbf{k}$ that have been introduced by Compagner [Com95] may be considered as an interesting quantitative version of the spectral test approach. For additional results on this topic we refer to Percus and Percus [PP92] and Yuen [Yue77].

Inspired by the interpretation above, we will now introduce the concept of the *weighted spectral test*. Our goal is to define numerical quantities that (1) allow the same number-theoretic analysis of pseudorandom number generators as discrepancy, (2) are related to discrepancy, (3) provide for efficient practical computation of their value even in higher dimensions $s$, (4) are not restricted to sequences $\omega$ with lattice structure, as it is the case with the classical spectral test, and (5) possess a statistical background, like a known asymptotic distribution.

We proceed as follows. From a given finite sequence $(x_n)_n$ of pseudorandom numbers in $[0,1[$ we construct $s$-tuples $\mathbf{x}_n$ in the $s$-dimensional unit cube $[0,1[^s$. This will yield a finite sequence $\omega = (\mathbf{x}_n)_{n=0}^{N-1}$. We then choose an appropriate function system $\mathcal{F} = \{\chi_{\mathbf{k}}\}$ on $[0,1[^s$ and decreasing weights $\rho(\mathbf{k}) \geq 0$ such that

$$\sum_{\mathbf{k}} \rho(\mathbf{k}) = 1, \quad \rho(\mathbf{k}) \to 0 \quad \text{if} \quad \sqrt{k_1^2 + \ldots + k_s^2} \to \infty.$$

The weights $\rho(\mathbf{k})$ will be assigned to the Weyl sums $S_N(\chi_{\mathbf{k}}, \omega)$ in the following manner. We define the quantity

$$F_N(\mathcal{F}, \rho; \omega) := \left( \sum_{\mathbf{k} \neq 0} \rho(\mathbf{k}) |S_N(\chi_{\mathbf{k}}, \omega)|^2 \right)^{1/2} \tag{1.5}$$

and call it a *weighted spectral test*. If the weights $\rho(\mathbf{k})$ have been chosen in an appropriate way, then we can find a function $f : [0,1[^s \to \mathbf{R}$ such that the Fourier series of $f$ relative to $\mathcal{F}$ has the following properties:

$$\hat{f}(\mathbf{k}) = \rho(\mathbf{k}) \quad \forall \mathbf{k},$$
$$f = \sum_{\mathbf{k}} \hat{f}(\mathbf{k}) \chi_{\mathbf{k}} \quad \text{(pointwise)},$$
$$F_N^2(\mathcal{F}, \rho; \omega) = \frac{1}{N^2} \sum_{n=0}^{N-1} \sum_{m=0}^{N-1} f(\mathbf{x}_n \ominus \mathbf{x}_m), \tag{1.6}$$

where $\ominus$ is a difference operation on $[0,1[^s$ compatible with the function system $\mathcal{F}$. For example, in the case $\mathcal{F} = \mathcal{T}$, it will be the inverse of addition modulo one. If $\mathcal{F} = \mathcal{W}$, then it will be binary addition (which is the same as binary subtraction or the XOR operation). The quantity in (1.5) will be called the *diaphony* of $\omega$ relative to $\mathcal{F}$ and $f$.

## 4 Realizations

We will now give two exemplary realizations of the weighted spectral test.

**Example 4.1 (Classical Diaphony)** (see Hellekalek [Hel95a], Hellekalek and Niederreiter [HN96]) We choose $\mathcal{F} = \mathcal{T}$ and quadratic decrease in the weights,

$$\rho(\mathbf{k}) = \frac{1}{(1 + \pi^2/3)^s - 1} \prod_{i=1}^s (\max\{1, |k_i|\})^{-2}, \quad \mathbf{k} = (k_1, \ldots, k_s) \in \mathbf{Z}.$$

$$(1.7)$$

Then the weighted spectral test $F_N(\mathcal{T}, \rho; \omega)$ of $\omega$ can be realized by the function

$$f(\mathbf{x}) = -1 + \prod_{i=1}^s g(x_i), \quad \mathbf{x} = (x_1, \ldots, x_s) \in [0, 1[^s, \quad \text{where}$$

$$g(x) = 1 - \frac{\pi^2}{6} + \frac{\pi^2}{2}(1 - 2x)^2, \quad x \in [0, 1[, \tag{1.8}$$

and subtraction modulo one on $[0, 1[^s$. Zinterhof [Zin76] has proved the identity

$$F_N^2(\mathcal{T}, \rho; \omega) = \frac{1}{(1 + \pi^2/3)^s - 1} \cdot \frac{1}{N^2} \sum_{n=0}^{N-1} \sum_{m=0}^{N-1} f(\mathbf{x}_n - \mathbf{x}_m \bmod 1). \tag{1.9}$$

**Remark 4.1** (1) The name "diaphony" for the quantity in (1.9) stems from Zinterhof [Zin76]. Diaphony was introduced to estimate the error term in a special number-theoretic integration method. It is closely related to discrepancy, as Zinterhof and Stegbuchner [ZS78] and Stegbuchner [Ste80] have shown.
(2) Bounds for discrepancy that have been obtained by estimating exponential sums translate easily into bounds for diaphony. The case of pseudorandom number generators is particularly simple, see Hellekalek and Niederreiter [HN96, Proposition 2.2].
(3) Empirical results on diaphony applied to the assessment of pseudorandom number generators can be found in Auer and Hellekalek [AH94], Hellekalek [Hel95a], and Hellekalek and Niederreiter [HN96].
(4) The function $g$ in (1.8) is equal to the second Bernoulli polynomial, up to a normalizing constant. For higher-order decrease in the weights, we may choose for $g$ the $2r$-th Bernoulli polynomial, $r = 2, 3, \ldots$ This choice will result in weights $\rho(\mathbf{k})$ equal to (modulo normalization)

$$\prod_{i=1}^s \frac{1}{\max\{1, |k_i|\}^{2r}}.$$

Hence the concept of the weighted spectral test allows to model a great number of orders of decrease of the weights $\rho(\mathbf{k})$. The proof of this result follows from the methods presented in Hellekalek and Niederreiter [HN96].

**Example 4.2 (Dyadic Diaphony)** (see Hellekalek and Leeb [HL96]) We choose $\mathcal{F} = \mathcal{W}$, and the weights with quadratic decrease

$$\rho(\mathbf{k}) = \frac{1}{3^s - 1} \prod_{i=1}^{s} \rho(k_i),$$

where

$$\rho(k) := \begin{cases} 2^{-2g} & \text{if } k: \ 2^g \le k < 2^{g+1}, \text{with } g \ge 0, \ g \in \mathbf{Z}, \\ 1 & \text{if } k = 0, \end{cases}$$

for an integer vector $\mathbf{k} = (k_1, \dots, k_s)$ with nonnegative coordinates $k_i$. Then Hellekalek and Leeb [HL96] have proved that the weighted spectral test $F_N(\mathcal{W}, \rho; \omega)$ of $\omega$ can be realized by the function

$$f(\mathbf{x}) = -1 + \prod_{i=1}^{s} g(x_i), \quad \mathbf{x} = (x_1, \dots, x_s),$$

$$g(x) = \begin{cases} 3 - 3 \cdot 2^{1 + \lfloor \log_2 x \rfloor} & \text{if } x \in ]0, 1[, \\ 3 & \text{if } x = 0, \end{cases} \tag{1.10}$$

and binary addition,

$$F_N^2(\mathcal{W}, \rho; \omega) = \frac{1}{3^s - 1} \frac{1}{N^2} \sum_{n=0}^{N-1} \sum_{m=0}^{N-1} f(\mathbf{x}_n \dotplus \mathbf{x}_m). \tag{1.11}$$

**Remark 4.2** (1) Dyadic diaphony may be viewed as a quantitative and continuous version of Tezuka's Walsh spectral test in [Tez87].
(2) Both examples above of a weighted spectral test define measures of uniform distribution of sequences. This means that the diaphony converges to zero for $N$ to infinity if and only if the sequence $\omega$ is uniformly distributed. This fact is easily derived from Weyl's Criterion, see Kuipers and Niederreiter [KN74] and Hellekalek and Leeb [HL96].
(3) Both versions of diaphony may be interpreted as *mean squared integration errors*, see James, Hoogland and Kleiss [JHK96].
(4) If we consider diaphony as a random variable with argument $\omega$, then Leeb [Lee96] has computed its expectation and variance and proved preliminary results on the asymptotic distribution.

# 5 Techniques

In this section, we will construct a realization of the weighted spectral test that measures the irregularities in the distribution of the leading bits of pseudorandom numbers $x_n$.

For the ease of notation, we discuss the one-dimensional case first. Let $\omega = (x_n)_{n \geq 0}$ be a sequence of uniform pseudorandom numbers in $[0, 1[$. For fixed $g$, $g = 0, 1, 2, \ldots$, we consider the $(g + 1)$-th digit of every pseudorandom number $x_n$. Now, the Walsh function $w_{2^g}$ takes the value 1 if the $(g + 1)$-th digit $x_g$ of its argument $x$ equals 0, and $w_{2^g}(x) = -1$ if $x_g = 1$. The value of the Weyl sum $|S_N(w_{2^g}, \omega)|$, respectively of $|S_N(w_{2^g}, \omega)|^2$ is a numerical quantity that measures the irregularity of the distribution of the $(g + 1)$-th digit in the first $N$ elements of the sequence $\omega$. We assign the weight $2^{-1}$ to the correlation coefficient $|S_N(w_1, \omega)|^2$ of the first bit, the weight $2^{-2}$ to the correlation coefficient $|S_N(w_2, \omega)|^2$ of the second bit, and so on. We have chosen the weights such that they sum up to one. From the definition of the weights, it is clear that leading bits are more important to us than trailing bits. This is the usual case in applications of pseudorandom number generators. Altogether, the numerical quantity

$$\sum_{g=0}^{\alpha-1} \frac{1}{2^{g+1}} |S_N(w_{2^g}, \omega)|^2 \tag{1.12}$$

is a figure of merit of the distribution of the first $\alpha$ bits in the first $N$ pseudorandom numbers of the sequence $\omega$.

The generalization of this approach to higher dimensions is obvious. In dimension $s$, we obtain the figure of merit

$$F_N(\mathcal{W}, \rho; \omega) := \left( \frac{1}{2^s - 1} \sum_{\mathbf{k} \neq 0} \rho(\mathbf{k}) |S_N(w_{\mathbf{k}}, \omega)|^2 \right)^{1/2}. \tag{1.13}$$

with $\rho(\mathbf{k}) := \prod_{i=1}^{s} \rho(k_i)$, $\mathbf{k} = (k_1, \ldots, k_s)$,

$$\rho(k) := \begin{cases} 2^{-g-1} & \text{if } k = 2^g, \text{ with } g \geq 0, \\ 1 & \text{if } k = 0, \\ 0 & \text{otherwise.} \end{cases} \tag{1.14}$$

The reader should note that this numerical quantity will *not* be a measure of uniform distribution. In the definition of $F_N(\mathcal{W}, \rho; \omega)$ infinitely many Weyl sums $|S_N(w_{\mathbf{k}}, \omega)|^2$ are missing. As a consequence, there will be infinite sequences $\omega$ such that $\lim_{N \to \infty} F_N(\mathcal{W}, \rho; \omega) = 0$, without $\omega$ being uniformly distributed (see Remark 4.2(2) for the argument).

We now exhibit an appropriate function $f$ to realize the weighted spectral test defined in (1.13).

**Proposition 5.1** *Let* $g : [0,1[\to \mathbf{R}$, $g(x) := 2(1-x)$, *and let* $f : [0,1[^s \to$
$\mathbf{R}$,

$$f(\mathbf{x}) := -1 + \prod_{i=1}^{s} g(x_i), \quad \mathbf{x} = (x_1, \ldots, x_s).$$

*Then*

$$\hat{g}(k) = \rho(k) \quad \forall k$$

*and*

$$\hat{f}(\mathbf{k}) = \begin{cases} \rho(\mathbf{k}) & if \quad \mathbf{k} \neq 0, \\ 0 & if \quad \mathbf{k} = 0. \end{cases} \tag{1.15}$$

**Proof** The result for $\hat{f}(\cdot)$ follows from the result for $\hat{g}(\cdot)$. Suppose that
$2^\alpha \leq k < 2^{\alpha+1}$, with some nonnegative integer $\alpha$. Let $b_0, b_1, \ldots, b_\alpha$ be
arbitrary digits in $\{0,1\}$. Let

$$I(b_0, \ldots, b_\alpha) := \{x \in [0,1[ : x_j = b_j, \quad \forall j : 0 \leq j \leq \alpha\}.$$

Then $I(b_0, \ldots, b_\alpha)$ is an elementary binary interval of lenght $2^{-\alpha-1}$. We
see that

$$\hat{g}(k) = \sum_{b_0=0}^{1} \cdots \sum_{b_\alpha=0}^{1} \int_{I(b_0,\ldots,b_\alpha)} g(x) w_k(x) \, dx. \tag{1.16}$$

The Walsh function $w_k$ is constant on the interval $I(b_0, \ldots, b_\alpha)$ with the
value

$$w_k(0.b_0 \cdots b_\alpha) = \prod_{j=0}^{\alpha} (-1)^{b_j k_j}.$$

Hence

$$\hat{g}(k) = \sum_{b_0=0}^{1} \cdots \sum_{b_\alpha=0}^{1} \prod_{j=0}^{\alpha} (-1)^{b_j k_j} \int_{I(b_0,\ldots,b_\alpha)} g(x) \, dx.$$

For the ease of notation, we shall now calculate the Walsh coefficient $\hat{h}(k)$,
where $h(x) := x$ on the interval $[0,1[$. Trivially,

$$\hat{g}(k) = -2\hat{h}(k) \quad \text{for all } k \neq 0. \tag{1.17}$$

We have

$$\int_{I(b_0,\ldots,b_\alpha)} x \, dx = 2^{-\alpha-1} 0.b_0 \cdots b_\alpha + 2^{-2\alpha-3}.$$

Let $\sum_{j_0}$ denote $\sum_{b_0=0}^{1} \cdots \sum_{b_\alpha=0}^{1}$, where the summation over $b_{j_0}$ has been
omitted; let

$$B(j_0) := \frac{b_0}{2} + \ldots + \frac{b_{j_0-1}}{2^{j_0}} + \frac{b_{j_0+1}}{2^{j_0+2}} + \ldots + \frac{b_\alpha}{2^{\alpha+1}},$$

i.e. $B(j_0) = 0.b_0 \cdots b_\alpha - b_{j_0} 2^{-j_0-1}$, and let

$$C(j_0) := \prod_{\substack{j=0 \\ j \neq j_0}}^{\alpha} (-1)^{b_j k_j}.$$

Then

$$\hat{h}(k) = \sum_\alpha C(\alpha) \left( \sum_{b_\alpha=0}^{1} 2^{-\alpha-1} (B(\alpha) + b_\alpha 2^{-\alpha-1})(-1)^{b_\alpha k_\alpha} \right). \qquad (1.18)$$

This implies

$$\hat{h}(k) = -\frac{1}{2^{2\alpha+2}} \sum_\alpha C(\alpha).$$

Now $\sum_\alpha C(\alpha) = 2^\alpha$ if $k_j = 0$ for all $j$, $0 \leq j < \alpha$, and $\sum_\alpha C(\alpha) = 0$ otherwise. This yields

$$\hat{h}(k) = \begin{cases} -2^{-\alpha-2} & \text{if } k = 2^\alpha, \text{ with } \alpha \geq 0, \\ 2^{-1} & \text{if } k = 0, \\ 0 & \text{otherwise.} \end{cases} \qquad (1.19)$$

The result for $g$ follows.  □

**Remark 5.2** The proof of Proposition 5.1 is written such that the reader can easily calculate the Walsh coefficients of the function $h(x) = x$ for the generalized Walsh system to the base $q$ (see Hellekalek [Hel94, Hel95b] for its definition). It will turn out that

$$\hat{h}(k) = \frac{1}{q^{\alpha+1}(e^{-2\pi i k_\alpha/q} - 1)},$$

for all $k$ of the form $k = k_\alpha q^\alpha$, with $\alpha \geq 0$, $k_\alpha \in \{1, \ldots, q-1\}$, and $\hat{h}(k) = 0$ for all other $k \neq 0$.

**Theorem 5.3** *Let the functions $g$ and $f$ be as in Proposition 5.1. Then, for every sequence $\omega = (\mathbf{x}_n)_{n \geq 0}$ in $[0, 1[^s$ such that the coordinates of all points $\mathbf{x}_n$ fulfill Condition (C1) or (C2) in Remark 2.1, in particular, if the coordinates of all points $\mathbf{x}_n$ are binary rationals, we have the identity*

$$F_N^2(\mathcal{W}, \rho; \omega) = \frac{1}{2^s - 1} \frac{1}{N^2} \sum_{n=0}^{N-1} \sum_{m=0}^{N-1} f(\mathbf{x}_n \dotplus \mathbf{x}_m). \qquad (1.20)$$

**Proof** From Proposition 5.1 it follows that

$$f = \sum_{\mathbf{k} \neq 0} \rho(\mathbf{k}) w_{\mathbf{k}}$$

on $[0, 1[^s$. It is well-known that $f$ is equal to its Walsh series in every point of $[0, 1[^s$. Further, for all $\mathbf{x}$ and $\mathbf{y}$ in $[0, 1[^s$ that fulfill conditions (C1) or (C2) in Remark 2.1,

$$w_\mathbf{k}(\mathbf{x}\dotplus\mathbf{y}) = w_\mathbf{k}(\mathbf{x})\, w_\mathbf{k}(\mathbf{y}),$$

for all Walsh functions $w_\mathbf{k}$. The result follows. $\square$

The following notion of the *discretization* of the figure of merit defined in (1.13) is adapted to practical applications. It allows to choose the number $\alpha$ of bits we want to consider. When comparing the discretized and the original figure of merit, we will obtain an inequality of the Erdös-Turán--Koksma type, see Theorem 5.5.

**Definition 5.1** *Let $\alpha \in \mathbf{N}$; let $\rho$ be as in Relation (1.13) and let $\Delta(2^\alpha)$ and $\Delta^*(2^\alpha)$ be defined as $\Delta(2^\alpha) := \{\mathbf{k} = (k_1, \ldots, k_s) : 0 \le k_i < 2^\alpha \; \forall i, \; 1 \le i \le s\}$, and $\Delta^*(2^\alpha) := \Delta(2^\alpha) \setminus \{0\}$.*

*The discrete version of the weighted spectral test quantity in (1.13) of a sequence $\omega = (\mathbf{x}_n)_{n \ge 0}$ in $[0, 1[^s$ with respect to the first $\alpha$ digits is defined by*

$$F_N(\mathcal{W}, \rho, \alpha; \omega) := \left( \frac{1}{(2 - 2^{-\alpha})^s - 1} \sum_{\mathbf{k} \in \Delta^*(2^\alpha)} \rho(\mathbf{k}) |S_N(w_\mathbf{k}, \omega)|^2 \right)^{1/2}.$$

$$(1.21)$$

**Remark 5.4** (1) The term $(2 - 2^{-\alpha})^s - 1$ equals $\sum_{\mathbf{k} \in \Delta^*(2^\alpha)} \rho(\mathbf{k})$. Hence $0 \le F_N(\mathcal{W}, \rho, \alpha; \omega) \le 1$, for any $\alpha \in \mathbf{N}$ and any sequence $\omega$ in $[0, 1[^s$.
(2) If $\mathbf{k} \in \Delta^*(2^\alpha)$, then the value of the Walsh function $w_\mathbf{k}$ depends only on the first $\alpha$ digits of every coordinate $x_i$ of $\mathbf{x}=(x_1, \ldots, x_s)$.
(3) Let $\alpha \in \mathbf{N}$ be fixed and consider the regular lattice

$$\mathcal{P} := \left\{ (\frac{a_1}{2^m}, \ldots, \frac{a_s}{2^m}) : a_i \in \{0, 1, \ldots, 2^m - 1\} \right\}.$$

Let $N = 2^{ms}$. We find by elementary calculations that

$$F_N^2(\mathcal{W}, \rho; \mathcal{P}) = \frac{1}{2^s - 1} \left( (1 + \frac{1}{2^m})^s - 1 \right),$$

and that, for $\alpha > m$,

$$F_N^2(\mathcal{W}, \rho, \alpha; \mathcal{P}) = \frac{1}{(2 - 2^{-\alpha})^s - 1} \left( (1 + \frac{1}{2^m} - \frac{1}{2^\alpha})^s - 1 \right).$$

Hence the order of magnitude of $F_N(\mathcal{W}, \rho; \mathcal{P})$ and $F_N(\mathcal{W}, \rho, \alpha; \mathcal{P})$ is $N^{-1/s}$, as in the case of the discrepancy of regular lattices of $N$ points in dimension $s$.

For a given number $\alpha \in \mathbf{N}$, we let $p_\alpha$ denote the discretization of the binary expansion to $\alpha$ digits. If $x \in [0, 1[$ has the expansion $x = \sum_{j=0}^{\infty} x_j 2^{-j-1}$, then $p_\alpha(x) := \sum_{j=0}^{\alpha-1} x_j 2^{-j-1}$. We have $p_\alpha(x) \in 1/2^\alpha \mathbf{Z}$ modulo 1. For $\mathbf{x} = (x_1, \ldots, x_s)$ in $[0, 1[^s$ we define $p_\alpha(\mathbf{x}) := (p_\alpha(x_1), \ldots, p_\alpha(x_s))$.

**Theorem 5.5** *Let* $\omega = (\mathbf{x}_n)_{n \geq 0}$ *be as in Theorem 5.3 and let* $\alpha \in \mathbf{N}$. *If we define the functions* $g : [0, 1[ \to \mathbf{R}$ *and* $f : [0, 1[^s \to \mathbf{R}$ *as*

$$g(x) := 2(1 - x) - 2^{-\alpha}, \quad f(\mathbf{x}) := -1 + \prod_{i=1}^{s} g(x_i),$$

*then*

*(1) (Explicit formula for the discrete figure of merit)*

$$F_N^2(\mathcal{W}, \rho, \alpha; \omega) = \frac{1}{(2 - 2^{-\alpha})^s - 1} \frac{1}{N^2} \sum_{n=0}^{N-1} \sum_{m=0}^{N-1} f \circ p_\alpha(\mathbf{x}_n \dotplus \mathbf{x}_m) \quad (1.22)$$

*(2) (Inequality between the figures of merit)*

$$F_N^2(\mathcal{W}, \rho; \omega) \leq \frac{s}{2^\alpha} \frac{2^s}{2^s - 1}(1 - 2^{-\alpha-1})^{s-1} + \frac{(2 - 2^{-\alpha})^s - 1}{2^s - 1} F_N^2(\mathcal{W}, \rho, \alpha; \omega) \tag{1.23}$$

**Proof** Let $g_\alpha := g \circ p_\alpha$ and $f_\alpha := f \circ p_\alpha$. To prove (1), we observe that the function $p_\alpha$ is constant on the elementary binary subintervals $I(b_0, \ldots, b_{\alpha-1})$ of $[0, 1[$ of length $2^{-\alpha}$. Hence the function $g_\alpha$ is a step function on $[0, 1[$. The same is true for the function $h := p_\alpha$. We have

$$h(x) = \sum_{b_0=0}^{1} \cdots \sum_{b_{\alpha-1}=0}^{1} 0.b_0 \cdots b_{\alpha-1} \, \mathbf{1}_{I(b_0, \ldots, b_{\alpha-1})}(x),$$

where $\mathbf{1}_{I(b_0, \ldots, b_{\alpha-1})}(x)$ denotes the characteristic function of the elementary binary interval $I(b_0, \ldots, b_{\alpha-1})$. From Lemma 2(i) in Hellekalek [Hel94] we deduce that

$$\hat{h}(k) = 0 \quad \forall k \geq 2^\alpha.$$

Now, let $k \in \mathbf{N}$, $2^g \leq k < 2^{g+1}$, $0 \leq g < \alpha$. Then

$$\hat{h}(k) = \sum_{b_0=0}^{1} \cdots \sum_{b_{\alpha-1}=0}^{1} 0.b_0 \cdots b_{\alpha-1} \, w_k(0.b_0 \cdots b_{\alpha-1}) \frac{1}{2^\alpha}.$$

We proceed as in Proposition 5.1 and find that

$$\hat{h}(k) = \begin{cases} -2^{-g-2} & \text{if } k = 2^g, \\ \frac{1}{2}(1 - 2^{-\alpha}) & \text{if } k = 0, \\ 0 & \text{otherwise.} \end{cases} \tag{1.24}$$

As a consequence,

$$\widehat{g_\alpha}(k) = \rho(k) \quad \forall k : 0 \le k < 2^\alpha.$$

This implies

$$\widehat{f_\alpha}(\mathbf{k}) = \rho(\mathbf{k}) \quad \forall \mathbf{k} \in \Delta^*(2^\alpha).$$

Trivially,

$$f_\alpha(\mathbf{x}) = \sum_{\mathbf{k} \in \Delta^*(2^\alpha)} \widehat{f_\alpha}(\mathbf{k}) w_{\mathbf{k}}(\mathbf{x})$$

on $[0,1[^s$, further $p_\alpha(\mathbf{x}\dot{+}\mathbf{y}) = p_\alpha(\mathbf{x})\dot{+}p_\alpha(\mathbf{y}) \ \forall \ \mathbf{x}, \mathbf{y} \in [0,1[^s$. Part (1) of the theorem follows.

To prove part (2), we observe that, with the convention $\|f\|_A := 2^s - 1$ and $\|f_\alpha\|_A := (2 - 2^{-\alpha})^s - 1$,

$$F_N^2(\mathcal{W}, \rho; \omega) \le \frac{1}{\|f\|_A} \sup_{\mathbf{x} \in [0,1[^s} |f(\mathbf{x}) - f_\alpha(\mathbf{x})| + \frac{\|f_\alpha\|_A}{\|f\|_A} F_N^2(\mathcal{W}, \rho, \alpha; \omega). \quad (1.25)$$

The Mean Value Theorem of Calculus implies that

$$\sup_{\mathbf{x} \in [0,1[^s} |f(\mathbf{x}) - f_\alpha(\mathbf{x})| \le \frac{s}{2^\alpha} 2^s (1 - 2^{-\alpha-1})^{s-1}.$$

Then part (2) follows. $\square$

## 6 REFERENCES

[AH94]   T. Auer and P. Hellekalek. Independence of uniform pseudorandom numbers, part II: empirical results. In M. Vajteršic and P. Zinterhof, editors, *Proceedings of the International Workshop Parallel Numerics '94, Smolenice*, pages 59–73. Slovak Academy of Sciences, Bratislava, 1994.

[CM67]   R.R. Coveyou and R.D. MacPherson. Fourier analysis of uniform random number generators. *J. Assoc. Comput. Mach.*, 14:100–119, 1967.

[Com95]  A. Compagner. Operational conditions for random-number generation. *Phys. Review E*, **52**:5634–5645, 1995.

[DMP90] A. De Matteis and S. Pagnutti. Long-range correlations in linear and non-linear random number generators. *Parallel Computing*, 14:207–210, 1990.

[EH93]   J. Eichenauer-Herrmann. Statistical independence of a new class of inversive congruential pseudorandom numbers. *Math. Comp.*, 60:375–384, 1993.

[Hel94]    P. Hellekalek. General discrepancy estimates: the Walsh function
           system. *Acta Arith.*, **67**:209–218, 1994.

[Hel95a]   P. Hellekalek.   Correlations between pseudorandom numbers:
           theory and numerical practice. In P. Hellekalek, G. Larcher, and
           P. Zinterhof, editors, *Proceedings of the 1st Salzburg Minisym-
           posium on Pseudorandom Number Generation and Quasi-Monte
           Carlo Methods, Salzburg, 1994*, volume ACPC/TR 95-4 of *Tech-
           nical Report Series*, pages 43–73. ACPC – Austrian Center for
           Parallel Computation, University of Vienna, 1995.

[Hel95b]   P. Hellekalek.   General discrepancy estimates III: the Erdös-
           Turán-Koksma inequality for the Haar function system. *Mon-
           atsh. Math.*, **120**:25–45, 1995.

[HL96]     P. Hellekalek and H. Leeb. Dyadic diaphony. *Acta Arith.*, 1996.
           To appear.

[HN96]     P. Hellekalek and H. Niederreiter. The weighted spectral test:
           diaphony. 1996. Submitted to ACM Trans. Modeling and Com-
           puter Simulation.

[JHK96]    F. James, J. Hoogland, and R. Kleiss.    Multidimensional
           sampling for simulation and integration: measures, discrepancies,
           and quasi-random numbers. Preprint submitted to Computer
           Physics Communications, 1996.

[KN74]     L. Kuipers and H. Niederreiter. *Uniform Distribution of Se-
           quences.* John Wiley, New York, 1974.

[Knu67]    D.E. Knuth.   *The Art of Computer Programming, Vol. 2.*
           Addison-Wesley, Reading, Mass., 1967.

[Knu81]    D.E. Knuth.   *The Art of Computer Programming, Vol. 2.*
           Addison-Wesley, Reading, Mass., second edition, 1981.

[L'E94]    P. L'Ecuyer. Uniform random number generation. *Ann. Oper.
           Res.*, **53**:77–120, 1994.

[Lee96]    H. Leeb. A weak law for diaphony. Rist++ 13, Research Insti-
           tute for Software Technology, University of Salzburg, 1996.

[LW96]     H. Leeb and S. Wegenkittl. Inversive and linear congruential
           pseudorandom number generators in empirical tests. To appear
           in ACM Trans. Modeling and Computer Simulation, 1996.

[Nie78]    H. Niederreiter.   Quasi-Monte Carlo methods and pseudo-
           random numbers. *Bull. Amer. Math. Soc.*, **84**:957–1041, 1978.

[Nie92]   H. Niederreiter. *Random Number Generation and Quasi-Monte Carlo Methods*. SIAM, Philadelphia, 1992.

[Nie94]   H. Niederreiter. On a new class of pseudorandom numbers for simulation methods. *J. Comp. Appl. Math.*, **56**:159–167, 1994.

[Nie95]   H. Niederreiter. New developments in uniform pseudorandom number and vector generation. In H. Niederreiter and P.J.-S. Shiue, editors, *Monte Carlo and Quasi-Monte Carlo Methods in Scientific Computing*, volume 106 of *Lecture Notes in Statistics*, pages 87–120. Springer-Verlag, New York, 1995.

[PP92]   O. E. Percus and J. K. Percus. An expanded set of correlation tests for linear congruential random number generators. *Combin. Prob. Comput.*, 1:161–168, 1992.

[Ste80]   H. Stegbuchner. Zur quantitativen Theorie der Gleichverteilung mod 1. Arbeitsberichte, Mathematisches Institut der Universität Salzburg, 1980.

[Tez87]   S. Tezuka. Walsh-spectral test for GFSR pseudorandom numbers. *Comm. ACM.*, **30**:731–735, 1987.

[Yue77]   C.-K. Yuen. Testing random number generators by Walsh transform. *IEEE Trans. Comput.*, 26:329–333, 1977.

[Zin76]   P. Zinterhof. Über einige Abschätzungen bei der Approximation von Funktionen mit Gleichverteilungsmethoden. *Sitzungsber. Österr. Akad. Wiss. Math.-Natur. Kl. II*, **185**:121–132, 1976.

[ZS78]   P. Zinterhof and H. Stegbuchner. Trigonometrische Approximation mit Gleichverteilungsmethoden. *Studia Sci. Math. Hungar.*, **13**:273–289, 1978.

**Author's address:**

Peter Hellekalek, Institut für Mathematik, Universität Salzburg, Hellbrunner Straße 34, A-5020 Salzburg, Austria

e-mail address : `peter.hellekalek@sbg.ac.at`
WWW address: `http://random.mat.sbg.ac.at/`

# Quasi-Monte Carlo, Discrepancies and Error Estimates

Jiri Hoogland[1]
NIKHEF, Amsterdam, The Netherlands

Fred James[2]
CERN, Geneva, Switzerland

Ronald Kleiss[3]
University of Nijmegen, Nijmegen, The Netherlands

## Abstract

We present a survey of our recent work on the problem of defining an estimate for the error in Quasi-Monte Carlo integration. The key issue turns out to be the definition of an ensemble of quasi-random point sets that, on the one hand, includes a sufficiency of equivalent point sets, and on the other hand uses information on the degree of uniformity of the point set actually used, in the form of a discrepancy or diaphony. A few examples of such discrepancies are given. We derive the distribution of our error estimate in the limit of large number of points. In many cases, Gaussian limits are obtained. We also present numerical results for the $L_2$-discrepancy for a number of quasi-random sequences.

## 1   The error problem

We discuss the problem of integration of a (Riemann) square-integrable function over the $s$-dimensional unit hypercube $K = [0, 1)^s$, using a set of $N$ points $x_k$, $k = 1, 2, \ldots, N$. The actual integral is $J = \int_K f(x)\ dx$, and its numerical estimate is given by

$$S = \frac{1}{N} \sum_{k=1}^{N} f(x_k) \ . \tag{1}$$

Here and in the following, the dependence on $N$ is always implied. Depending on the way in which the points $x_k$ are chosen, we distinguish different integration methods: if the points come from some predetermined, deterministic scheme we have a quadrature rule, if they are considered to be

---

[1] e-mail: t96@nikhefh.nikhef.nl, research supported by Stichting FOM.
[2] e-mail: F.James@cern.ch
[3] e-mail: kleiss@sci.kun.nl, research supported by Stichting FOM.

independent random vectors distributed uniformly over $K$, we have Monte Carlo. An intermediate case is that of Quasi-Monte Carlo, where the points are considered to form the initial segment of a low-discrepancy sequence, but share some relevant properties of a truly random sequence[4]. The integration error is defined as $\eta = S - J$. Good integration methods are characterized by the fact that they typically lead to a small value of $|\eta|$, but, more importantly, allow one to obtain a good estimate of $|\eta|$. In the case of Monte Carlo, $\eta$ is a stochastic variable, and hence has a probability density $P_N(\eta)$. For quasi-random point sets used in Quasi-Monte Carlo, we may (as we shall specify more precisely later on) also interpret $\eta$ as having such a probability density: its form is the subject of this contribution.

In true Monte Carlo, the error density $P_N(\eta)$ is obtained by viewing the point set $\{x_1, x_2, \ldots, x_N\}$ as a typical member of an ensemble of random point sets, governed by the obvious uniform joint probability density

$$Q_N(x_1, x_2, \ldots, x_N) = 1 \ , \quad x_i \in K \ , \tag{2}$$

so that the $x_k$ are indeed iid uniform random vectors. The error $\eta$ is then a random variable over this ensemble, with the following well-known results. In the first place, $S$ is an unbiased estimator of $J$ in the sense that $E(\eta) = 0$, where the average $E(\cdot)$ is over the ensemble of points sets. In the second place, for large $N$, $P_N(\eta)$ weakly approaches a normal density according to the Central Limit Theorem. Finally, the variance of this distribution is given by $E(\eta^2) = V(f)/N$, where $V$ denotes the variance. Note that, since we average over the integration points, the error distribution depends only on the integrand itself.

The conceptual problem in the use of quasi-random rather than truly random point sets is the following: a quasi-random point set is *not* a 'typical' set of random points! Indeed, quasi-random point sets are very special, with carefully built-in correlations between the various points so that each new point tends to 'fill a gap' left by the previous ones. The usual Monte Carlo error estimate is therefore not really justified in Quasi-Monte Carlo. On the other hand, many different error bounds assure us that small errors are possible, and indeed likely, when we apply low-discrepancy point sets [1]. In the following, we shall discuss two approaches to a solution of this conundrum. We can only summarize the main results here: technical details and pictures can be found in the references.

---

[4] We shall not discuss the case of point sets with fixed, *predetermined* $N$.

# 2 The Bayesian approach

The first way around the aforementioned conceptual problem is to inter-change the rôles of integrand and point set: we view the integrand $f(x)$ as a typical member of some underlying class of functions and average over this class, so that the error depends only on the point set. In practice, the choice of function class often entails a good deal of idealism or pot luck, as usual in a Bayesian approach to probability. We discuss several examples, in which we denote by $\langle\rangle_\mu$ an average over the probability measure $\mu$ governing the function class.

## 2.1 The Woźniakowski Lemma

Let the integrand $f$ be chosen according to the Wiener (sheet) measure in $s$ dimensions[5]. This measure is Gaussian, with

$$\langle f(x)\rangle_\mu = 0 \quad , \quad \langle f(x), f(y)\rangle_\mu = \prod_{\nu=1}^{s} \min(x^\nu, y^\nu) \quad , \tag{3}$$

where the index $\nu$ labels the coordinates. We may then quote the Lemma from [2]:

$$\langle \eta \rangle_\mu = 0 \quad , \quad \langle \eta^2 \rangle_\mu = T(x_1^*, x_2^*, \ldots, x_N^*) \quad , \tag{4}$$

where $T$ stands for the $L_2$-discrepancy [1], and the $x_k^*$ denotes the 'reflected' point, with $(x_k^*)^\nu = 1 - (x_k)^\nu$. In [3] it is shown, moreover, that the density $P_N(\eta)$ in this case is normal with zero mean and variance $T$.

We have here the interesting general fact that the choice of a particular function class induces its own particular discrepancy. On the other hand, in many cases (such as in particle physics phenomenology) the Wiener measure is certainly not appropriate since it is dominated by integrands that are not locally smooth. In [4], *folded* Wiener sheet measures are studied with analogous results, but then again these describe functions that are much too smooth for many practical applications.

## 2.2 Induced discrepancies

In [3], we established the following general result. Let the integrand be chosen from some function class endowed with a measure such that

$$\langle f(x_1)\rangle_\mu = 0 \quad , \quad \langle f(x_1)f(x_2)\rangle_\mu = \int_L h(x_1, y)h(x_2, y)\, dy \tag{5}$$

for all $x_1, x_2$ in $K$, and for some $h(x, y)$. The space $L$ over which the variable $y$ runs is not necessarily the same as $K$: it is just defined such that the

---

[5]This implies that $f$ is a continuous function.

above equation holds (indeed, in the next secion it is just the set of non-negative integers). We can then define an *induced* quadratic discrepancy as the expected mean squared error:

$$D_2^{ind} = \langle \eta^2 \rangle_\mu = \int_L g(y)^2 \, dy \ ,$$

$$g(y) = \frac{1}{N} \sum_{k=1}^{N} h(x_k, y) - \int_K h(x, y) \, dx \ . \tag{6}$$

Note that $h$ is not necessarily in the same function class as the $f$, and indeed $y$ may be defined in a space quite different from $K$. For example, while the Wiener sheet measure discussed above describes continuous functions, the corresponding $h$ in that case turns out to be the characteristic function, which is discontinuous. Note that, whenever the function class measure is Gaussian, then $P_N(\eta)$ will also be Gaussian. Generalizations to higher moments can be found in [3, 5].

## 2.3   Orthonormal function bases

As an example in $s = 1$, let $u_n(x)$ , $n = 0, 1, 2, 3, \ldots$ be an orthonormal set of functions on $K$, as follows:

$$u_0(x) = 1 \ , \quad \int_K u_m(x) u_n(x) \, dx = \delta_{m,n} \ , \tag{7}$$

with $m, n = 0, 1, 2, \ldots$, and $\delta$ is the Kronecker delta. Let $f(x)$ admit a decomposition

$$f(x) = \sum_{n \geq 0} v_n u_n(x) \ , \tag{8}$$

and choose the measure such that the $v_n$ are normally distributed with mean zero and variance $\sigma_n^2$. The additional requirement $\sum_n \sigma_n^2 < \infty$ ensures that, on average, the integrand will indeed be square integrable. The induced quadratic discrepancy $D_2^{orth} = N \times D_2^{ind}$ is then given[6] by

$$\langle \eta^2 \rangle_\mu = \frac{1}{N} D_2^{orth} \ , \quad D_2^{orth} = \frac{1}{N} \sum_{n>0} \sigma_n^2 \sum_{k,l=1}^{N} u_n(x_k) u_n(x_l) \ . \tag{9}$$

Note that here we have taken a factor $1/N$ out of the discrepancy. This ensures that the expected discrepancy for truly random vectors will be independent of $N$. A special case is that of the Fourier class, which has

$$u_{2n} = \sqrt{2} \cos(2\pi n x) \ , \quad u_{2n-1} = \sqrt{2} \sin(2\pi n x) \ , \quad n = 1, 2, 3, \ldots \tag{10}$$

---

[6]This measure of discrepancy is also called *diaphony*.

The physically reasonable requirement that the *phase* of each mode $n$ (made up from $u_{2n-1}$ and $u_{2n}$) is uniformly distributed forces us to have the Gaussian measure, with in addition $\sigma_{2n-1} = \sigma_{2n}$. This implies that $D_2^{orth}$ is invariant under a translation mod 1 of all vectors $x_k$ by a constant vector. We then have

$$D_2^{orth} = \frac{2}{N} \sum_{n>0} \sigma_{2n}^2 \left| \sum_{k=1}^{N} \exp(2i\pi n x_k) \right|^2 . \tag{11}$$

Obviously, other orthonormal function bases are also possible, such as the system of Walsh functions; a further discussion, including the straightforward generalization to higher dimension, can be found in [5, 6, 7, 8]. Note that all such quadratic discrepancies are nonnegative by construction.


# 3 The discrepancy-based approach

Another way of establishing integration error estimates, which in our opinion does more justice to the spirit of Monte Carlo, is the following. Instead of considering all point sets of $N$ truly random points, with the joint probability density (2), we restrict ourselves to those point sets that have a given value of discrepancy, for some predefined type of discrepancy. In this way, information on the discrepancy performance of one's favorite quasi-random number sequence can be incorporated in the error estimate.


## 3.1 Conditional distribution of points

We have then, instead of (2), a joint probability density $Q_N$ for the $N$ points as follows. Let $D_N(x_1, \ldots, x_N)$ be *some* discrepancy (nonnegative by construction), defined on sets of $N$ points in $K$, and suppose its value for the actual point set that is used in the integration is $w$. Then,

$$H(w) = \int_K \delta\left(D_N(x_1, \ldots, x_N) - w\right) dx_1 \cdots dx_N ,$$

$$Q_N(w; x_1, \ldots, x_N) = \frac{1}{H(w)} \delta\left(D_N(x_1, \ldots, x_N) - w\right)$$

$$= 1 - \frac{1}{N} F_N(w; x_1, \ldots, x_N) , \tag{12}$$

so that $F_N$ measures the deviation from the density (2) assumed for truly random points. The function $\delta(.)$ indicates the Dirac delta distribution. The quantity $H(w)$ is of course just the probability density for the discrepancy over the iid uniform random numbers, an object interesting in its own right.

Let us also define marginal deviations as

$$F_k(w; x_1, \ldots, x_k) = \int\limits_K F_N(w; x_1, \ldots, x_N) \, dx_{k+1} \cdots dx_N \ . \qquad (13)$$

Again, the various $F_k$ depend of course on $N$. We can then simply establish, for instance, that, provided $F_1(w; x)$ vanishes for all $x$, we have $E(\eta) = 0$ and

$$E(\eta^2) = \frac{1}{N} \left[ V(f) - \frac{N-1}{N} \int\limits_K f(x) f(y) F_2(w; x, y) \right] dx \, dy \ , \qquad (14)$$

where $E(\cdot) = E(\cdot \| D_N)$ now denotes averaging with respect to $Q_N(w; \cdot)$. It is intuitively clear that we may expect a reduced error if $F_2$ is positive[7] when $x$ and $y$ are 'close' together in some sense, i.e. if the points in the point set 'repel' each other. Note that only a small, $\mathcal{O}(1/N)$, deviation from uniformity in $Q_N(w; \cdot)$ is sufficient.

## 3.2 Error probability distribution

In many cases, it is actually possible to compute the $F_2$ mentioned above. In fact, especially in the case of discrepancies defined using orthonormal function bases, we can do much more. Using a Feynman-diagram technique described in detail in [7, 9], we can establish results for $P_N(\eta)$ as an asymptotic expansion in $1/N$. Note that $P_N(\eta)$ also depends on the particular form of the discrepancy used. To leading order, we have [9]

$$P_N(\eta) = \frac{\sqrt{N/2\pi}}{2\pi i H(w)} \int\limits_{-i\infty}^{+i\infty} \frac{1}{\sqrt{B(z)}} \exp\left[ A(z) - \frac{\eta^2 N}{2B(z)} \right] dz \ ,$$

$$A(z) = -wz - \frac{1}{2} \sum_{n>0} \log(1 - 2z\sigma_n^2) \ ,$$

$$B(z) = \sum_{n>0} \frac{v_n^2}{1 - 2z\sigma_n^2} \ , \qquad (15)$$

where the $z$ integral runs to the left of any singularities. This result holds, for $N$ asymptotically large, for any discrepancy measure based on orthonormal functions as discussed above. Note that the choice of a particular set of $\sigma_n$ in the definition of $D_N$ is in principle independent of the actual values of $v_n$ in the integrand: as we shall see, we may take $\sigma_n$ zero for some values of $n$ while $v_n$ is nonvanishing for those $n$. An example is given below.

---

[7] Note that by assumption $F_2(w; x, y)$ vanishes upon integration over $x$ or $y$, and therefore must attain both positive and negative values.

The $1/N$ corrections are fully calculable, although we have not done so yet. Two corollaries follow immediately. In the first place,

$$\int_0^\infty H(w)P_N(\eta)\,dw = \sqrt{N/2\pi V(f)}\,e^{-\eta^2 N/2V(f)} \quad , \quad V(f) \equiv \sum_{n>0} v_n^2 \quad , \quad (16)$$

which recovers the Central Limit Theorem valid over the whole ensemble of $N$-point point sets with any $w$. In the second place, we obtain an integral representation for $H(w)$ by insisting that $P_N(\eta)$ be normalized to unity:

$$H(w) = \frac{1}{2\pi i}\int_{-i\infty}^{+i\infty} \exp\left[-zw - \frac{1}{2}\sum_{n>0}\log(1 - 2z\sigma_n^2)\right]\,dz \quad . \quad (17)$$

Generalizations of these results to more dimensions only affect the sums over $n$. We shall now discuss some applications of the above result. For definiteness, we assume the Fourier class of orthonormal functions.

### 3.3  Application 1: equal strengths

A simple model for a discrepancy is obtained by taking $\sigma_n = 1/2M$ for $n = 1, 2, \ldots, 2M$, and zero otherwise (with trivial extension to more dimensions). Let us then decompose the variance of the integrand as follows:

$$V(f) = \sum_{n>0} v_n^2 = V_1 + V_2 \quad , \quad V_1 = \sum_{n=1}^{2M} v_n^2 \quad , \quad V_2 = \sum_{n>2M} v_n^2 \quad , \quad (18)$$

so that $V_1$ contains that part of the variance to which the discrepancy is sensitive (the 'covered' part) and $V_2$ the rest (the 'uncovered' part). We then have [9]

$$H(w) = \frac{M^M}{\Gamma(M)}w^{M-1}e^{-Mw} \sim \exp\left(-\frac{M}{2}(w-1)^2\right) \quad ,$$

$$P_N(\eta) \sim \left(\frac{N}{2\pi(wV_1 + V_2)}\right)^{1/2}\exp\left(-\frac{\eta^2 N}{2(wV_1 + V_2)}\right) \quad , \quad (19)$$

where the approximations are valid for large $M$. We see that a new central limit theorem holds, where the variance of $f$ has been modified so that its covered part is reduced by a factor $w$, according to intuition.

### 3.4  Application 2: harmonic model in one dimension

Let us concentrate on the case $s = 1$, and take $\sigma_{2n-1} = \sigma_{2n} = 1/n$, so that $f$ is, on the average, square integrable, but its derivative is not. In that case we have,

$$H(w) = \sum_{m>0} (-1)^{m-1} m^2 e^{-wm^2/2} \tag{20}$$

which is, apart from a trivial scaling, precisely the probability density of the Kolmogorov-Smirnov statistic. This is somewhat surprising since that statistic is based on the $L_\infty$-norm of the standard star-discrepancy, an object quite different from the quadratic discrepancy based on orthonormal functions. In addition, we conjecture, that for values of $w$ small compared to its expectation value $E(w) = \pi^2/3$, we shall have

$$P_N(\eta) \propto \exp\left[-\frac{\eta^2 N}{2C}\right] \quad, \quad C = \mathrm{Var}(f)\frac{w}{E(w)} \quad; \tag{21}$$

this would again indicate a reduction of the error estimate for small $w$. To date, we have not yet been able to prove this assertion.

## 3.5 The $L_2$-discrepancy

We have also obtained some results for the quadratic form of the standard star-discrepancy [5, 6]. Although this is not based on orthonormal functions and the analysis is hence more complicated, we have obtained the moment-generating function $G(z) = E(e^{zw})$ for asymptotically large $N$, where $w$ now stands for $N$ times the $L_2$-discrepancy. More precisely, we have (cf. ref.[6], Eq.105 and its multidimensional analogue, Eq.108)

$$
\begin{aligned}
G(z) &= \exp(\psi(z))/\sqrt{\chi(z)} \;, \\
\psi(z) &= -\frac{1}{2} \sum_{n>0} Q_s(2n-1) \log(1-z_n) \;, \\
\chi(z) &= \frac{2^s}{2z} \sum_{n>0} Q_s(2n-1) \frac{z_n}{1-z_n} \;, \\
z_n &= (4/\pi^2)^s \frac{2z}{(2n-1)^2} \;.
\end{aligned}
\tag{22}
$$

Here, $Q_s(m)$ is the number of ways in which an odd positive integer $m$ can be written as a product of $s$ positive integers, including 1's. The function $H(w)$ can now be computed numerically for different $s$ values. We have done so, and find that $H(w)$ very slowly approaches a Gaussian distribution as $s$ increases. Indeed, the skewness of $H(w)$ is, for large $s$, approximately given by $(216/225)^{s/2}$ so that the approach to normality is indeed slow.

# 4 Numerical results for the $L_2$-discrepancy

Since we now have $H(w)$ for the $L_2$-discrepancy, we can reliably judge how well quasi-random number generators perform, since we can compare the discrepancy of their output with the behaviour of truly random points. Space does not permit us to show pictures, which can be found in [5, 6]. Here, we just describe the results. We have computed the $L_2$-discrepancy for a good pseudorandom generator (RANLUX [10]). This generator has been used succesfully for some time in particle physics and can, for present purposes, be considered as a reasonable imitation of a truly random number source. We have also used the Richtmyer, Halton, Sobol', and Niederreiter sequences [11]. We did this both exactly, and by Monte Carlo. The latter method is actually faster for $N$ larger than about 50,000 if we ask for 5% accuracy (this is checked for those values of $N$ for which we computed both the exact discrepancy and its Monte Carlo estimate). We made runs of up to $N = 150,000$, and considered the lowest and highest discrepancy value in subsequent intervals of 1000. These we compared with the well-known mean value $(2^{-s} - 3^{-s})/N$ for truly random points, and also plotted the standardized[8] form $\xi(w) = (w - E(w))/\sqrt{V(w)}$, where the variance $V(w)$ is $V(w) = 2(6^{-s} - 2(15/2)^{-s} + 9^{-s})/N^2$, as derived for instance in [6]. We considered dimensions from 1 up to 20. In all dimensions, RANLUX appears to mimic a truly random sequence quite well. The quasi-random generators perform very well in low dimensions, and generally the discrepancy falls further and further below that of a random sequence as $N$ increases. There are exceptions, however: for instance, the Sobol' sequence for $s = 11$ degrades, and is actually worse than random for $N \sim 60,000$, again rapidly improving for larger $N$. Apart from taking this as a warning, we have not investigated the reason for such behaviour in detail. The biggest surprise was when we plotted the variable $\xi(w)$ (for instance, for $N = 150,000$) as a function of $s$. It appears that, as measured in this way, the performance of all quasi-random generators *improves* with increasing $s$! We feel that this phenomenon is not simply due to the fact that the discrepancy loses its stringency in higher dimensions, since the low-discrepancy sequences appear to become more and more significantly different from a truly random sequence (as measured in terms of the 'natural' unit, the standard deviation): with increasing dimension, it becomes less and less probable to arrive at such low discrepancy values 'by coincidence', using truly random points. For $s$ larger than 15 or so, all generators become rather bad, which is of course due to the fact that the onset of the asymptotic regime occurs for larger $N$ as $s$ increases. But what is more striking is the fact that the old-fashioned and simple Richtmyer generator performs as well as the modern, sophisticated Sobol' and Niederreiter sequences. We take this as an indication that the Richtmyer generator deserves more study, in particular since we have not attempted any optimization of its 'lattice point'. A number of interesting issues remain, such as the rate of approach to the asymptotic limit, and the behaviour of the alternative definitions of

---

[8] Again, here the $E()$ stands for the expectation for truly random points.

discrepancies discussed above; so far, however, we have not yet been able to address these.

# 5 Summary

The main results of our studies can be summarized as follows. In the first place, we have established relations between given classes of integrands (such as those based on the Wiener (sheet) measure, or on orthonormal functions bases), and appropriate measures of non-uniformity (generalized discrepancies) of point set employed in integrating these functions (*cf.* section 2). In the second place, we have obtained results for the probability density of the error in the numerical intergation of a given function, where the additional information is used about the non-uniformity of the point set in terms of the value this point set yields for a given generalized discrepancy. In several instances, we recover a normal density, with a variance smaller than that expected for a truly random point set (*cf* Eqs. 19 and 21). Owing to their simplicity, these results can be of immediate use in practical error estimates. As a by-product we obtain the probability density of a number of given generalized discrepancies for truly random point sets. These are of relevance for the assesment of low-discrepancy sequences, since it is only by assigning confidence levels to a particular sequence's discrepancy that one can argue that it is, indeed, small compared to that of a truly random seuquence. In section 4, we have presented a first attempt to judge various sequences along such lines.

# 6 REFERENCES

[1] H. Niederreiter, *Random number generation and Quasi-Monte Carlo methods*, (SIAM, Philadelphia, 1992).

[2] H. Woźniakoski, *Average-case complexity of multivariate integration*, Bull. AMS **24**(1991)185-194.

[3] R. Kleiss, *Average-case complexity distributions: a generalization of the Woźniakowski lemma for multidimensional numerical integration*, Comp. Phys. Comm. **71**(1992)39-53.

[4] S. Paskov, *Average-case complexity of multivariate integration for smooth functions*, J. Complexity **9**(1993)291-312.

[5] J.K. Hoogland, *Radiative corrections, Quasi-Monte Carlo and Discrepancy*, Ph.D. thesis, University of Amsterdam, 1996.

[6] F. James, J. Hoogland and R. Kleiss, *Multidimensional sampling for simulation and integration: measures, discrepancies and quasi-random numbers*, Comp. Phys. Comm. **99** (1997) 180-220.

[7] J. Hoogland and R. Kleiss, *Discrepancy-based error estimates for Quasi-Monte Carlo. I: General formalism*, Comp. Phys. Comm. **98** (1996) 111-127.

[8] J. Hoogland and R. Kleiss, *Discrepancy-based error estimates for Quasi-Monte Carlo. II: Results for one dimension*, Comp. Phys. Comm. **98**(1996)128-136.

[9] J. Hoogland and R. Kleiss, *Discrepancy-based error estimates for Quasi-Monte Carlo. III: Error distributions and central limits*, Comp. Phys. Comm. **101**(1997) (in print).

[10] M. Lüscher, *A portable high-quality random number generator for lattice field theory simulations*, Comp. Phys. Comm. **79** (1994) 100-110; F. James, *RANLUX: a FORTRAN implementation of the high-quality pseudo-random number generator of Lüscher*, Comp. Phys. Comm. **79** (1994) 111-118.

[11] P. Bratley, B.L Fox and H. Niederreiter, *Algorithm 738: Program to generate Niederreiter's low-discrepancy sequences*, ACM Trans.Math.Software **20** (1994) 494-495.

# The Quasi-Random Walk

## Alexander Keller[1]

ABSTRACT We present the method of the quasi-random walk for the approximation of functionals of the solution of second kind Fredholm integral equations. This deterministic approach efficiently uses low discrepancy sequences for the quasi-Monte Carlo integration of the Neumann series. The fast procedure is illustrated in the setting of computer graphics, where it is applied to several aspects of the global illumination problem.

## 1 Introduction

The solution of Fredholm integral equations of the second kind with complex kernels as used in computer graphics is a very complex task. Analytical solutions are not available for realistic scene models and the complexity of the kernel due to unknown discontinuities makes usual Newton-Côtes style quadrature inefficient. Using a Neumann series expansion in order to avoid kernel discretization, the integration problem becomes infinite dimensional, so that due to the curse of dimension only Monte Carlo methods remain for the numerical approximation.

Based on random decisions the probabilistic Monte Carlo quadrature is independent of dimension with an expected convergence rate of $\mathcal{O}(N^{-\frac{1}{2}})$. Since on a computer no real random numbers can be supplied, deterministic pseudo-random numbers (usually the linear congruential method) are used to mimic real random numbers by statistical properties. The number theoretic investigation of these deterministic sampling patterns shows that the most important property for the convergence is the uniformity of distribution of the pseudo-random numbers. Consequently deterministic point sets with an improved uniform distribution were developed, disregarding all other statistical properties of random numbers. Under some smoothness assumptions on the integrand these so-called low discrepancy points provide a deterministic convergence rate of roughly $\mathcal{O}(N^{-1})$, which is quadratically faster than random sampling.

Already in [Hla62] quasi-Monte Carlo integration, as the numerical integration using low discrepancy sequences is called, has been applied for the calculation of point solutions of Fredholm integral equations. Further experiments have been carried out in [SP87, KMS94, Spa95], testing different low discrepancy sequences on physical problems described by integral equations. Here the kernels fulfilled the smoothness assumptions, so that

---

[1]FB Informatik, Universität Kaiserslautern, D-67653 Kaiserslautern, Germany

a theoretical analysis could be carried out, validating the experimental results. As opposed to the previous examples, low discrepancy sampling in computer graphics has had a bad start, since due to the kernel complexity the theoretical error bounds of the quasi-Monte Carlo method are far too pessimistic. Experiments have been performed only for the low dimensional problems of pixel supersampling [Shi91, Mit92]. In [Kel95, Kel96a, Kel96c], however, the application of quasi-Monte Carlo integration to several other problems of computer graphics has been investigated, discovering serious improvements in quality and speed of the new deterministic algorithms as compared to random predecessors especially for high dimensional problems.

After a brief introduction to the method of quasi-Monte Carlo integration (see [Nie92]) and the physics of global illumination calculations (see [CW93]), we present the new principle of the *quasi-random walk* for the approximation of several functionals of the solution of the *radiance (Fredholm integral) equation*. By numerical experiments the superiority of the low discrepancy algorithm is impressively illustrated.

## 2   Quasi-Monte Carlo Integration

The Monte Carlo method (see [KW86]) approximates an integral of $f$ over the $s$-dimensional unit cube $I^s := [0, 1)^s$ by sampling $f$ at $N$ random positions $P_N = \{x_0, \ldots x_{N-1}\} \subset I^s$ and calculating the average. By probabilistic arguments, it can be shown that for large $N$

$$\text{Prob}\left(\left|\int_{I^s} f(x)dx - \frac{1}{N}\sum_{i=0}^{N-1} f(x_i)\right| < \frac{3\sigma(f)}{\sqrt{N}}\right) \approx 0.997. \qquad (1.1)$$

Thus, the expected convergence rate of the Monte Carlo method is of order $O(N^{-\frac{1}{2}})$, where $\sigma^2(f)$ is the variance of $f$. This rate can be improved by variance reduction techniques like for example importance sampling or domain stratification.

For importance sampling, sample points are generated according to a density function $p$:

$$\int_{I^s} f(x)dx = \int_{I^s} f(x)\frac{p(x)}{p(x)}dx = \int_{I^s} \frac{f(y)}{p(y)}d\mu(y) \approx \frac{1}{N}\sum_{i=0}^{N-1} \frac{f(y_i)}{p(y_i)}. \qquad (1.2)$$

Using the multidimensional inversion method [HM72], we can uniquely define a mapping $\mu^{-1}$ by which $y_i = \mu^{-1}(x_i)$ can be generated out of uniformly distributed points $x_i$. The points $C_N = \{y_0, \ldots, y_{N-1}\}$ then are distributed according to the density $p$. The variance now is $\sigma^2\left(\frac{f}{p}\right)$, so if $p$ is roughly proportional to $f$, $\frac{f}{p}$ is less oscillating than $f$, reducing the variance by placing more samples into the more important regions.

Stratification means to partition the integration domain $I^s = \cup_{k=1}^{K} A_k$. Then $N_k = \lfloor N\lambda_s(A_k) \rfloor$ random samples are generated uniformly distributed inside $A_k$, where $\lambda_s$ is the $s$-dimensional Lebesgue-measure. For each domain the Monte Carlo estimates are calculated. Summing up the $K$ averages weighted by their support sizes $\lambda_s(A_k)$ yields an estimate of the integral, with a variance less or equal to $\sigma(f)$. One special choice is to partition the integration domain into $N$ equally sized domains and to place one random sample inside each of the subdomains. The partition simply is done by dividing each axis into $\sqrt[s]{N}$ intervals, which of course limits the choice of the sampling rate to $N = k^s$, $k \in \mathbb{N}$. Since the number of subdomains now depends on the sampling rate, the variance also depends on $N$, yielding a rate of $\mathcal{O}(N^{-\frac{s+1}{2s}})$ (see [Mit96]), which for large dimensions converges to the rate of (1.1). Thus stratification increases the uniformity of the sampling pattern, improving the rate of convergence for moderate dimensions. We define

**Definition 1** *The (star-) discrepancy is a measure of the uniform distribution of a given point set $P_N = \{x_0, \ldots, x_{N-1}\}$. It is defined with respect to the family of Lebesgue-measurable subsets $\mathcal{J}^* := \left\{ A = \prod_{j=1}^{s} [0, a_j) \subset I^s \right\}$ by*

$$D^*(P_N) := \sup_{A \in \mathcal{J}^*} \left| \lambda_s(A) - \frac{1}{N} \sum_{i=0}^{N-1} \chi_A(x_i) \right| ,$$

*where $\lambda_s$ is the Lebesgue-measure on $I^s$, and $\chi_A$ is the characteristic function of the set $A$.*

$D^*(P_N)$ can be interpreted as the worst case error of the quadrature formula defined by $P_N$ (with equal weights $\frac{1}{N}$) when integrating characteristic functions of elements from $\mathcal{J}^*$. It is bounded from below by (see [Nie92])

$$D^*(P_N) \geq B_s \frac{\log^{\frac{s-1}{2}} N}{N}$$

for some positive constant $B_s$, whereas random numbers only have a discrepancy of order $\mathcal{O}\left(\sqrt{\frac{\log \log N}{N}}\right)$ almost surely (see [Nie92]). In fact, infinite deterministic sampling point sequences can be constructed with a discrepancy of order $\mathcal{O}\left(\frac{\log^s N}{N}\right)$, and restricting to finite point sets the rate can even be improved by one logarithmic term (see [Nie92]). Although the lower bound is not reached, it is widely believed that the rate obtained by such constructions is optimal. By discarding almost any statistical properties, these so-called low discrepancy points are more uniformly distributed than (pseudo-) random numbers.

## 2.1 Low Discrepancy Points

Almost all low discrepancy constructions are based on the radical inverse function

$$\Phi_b(i) := \sum_{j=0}^{\infty} a_j(i) \, b^{-j-1} \in I \Leftrightarrow i = \sum_{j=0}^{\infty} a_j(i) \, b^j,$$

where $\Phi_b$ is the representation $(a_j)_{j=0}^{\infty}$ of $i \in \mathbb{N}_0$ in the integer base $b$ mirrored at the decimal point. The most simple $s$-dimensional low discrepancy sequence is the infinite Halton sequence (see [Nie92]):

$$x_i = (\Phi_{b_1}(i), \ldots, \Phi_{b_s}(i)), i \in \mathbb{N}_0,$$

where the base $b_j$ is the $j$-th prime number. The discrepancy of the Halton sequence is bounded by

$$D^*(P_N^{\text{Halton}}) < \frac{s}{N} + \frac{1}{N} \prod_{j=1}^{s} \left( \frac{b_j - 1}{2 \log b_j} \log N + \frac{b_j + 1}{2} \right)$$

with a constant that grows superexponentially with the dimension $s$. The theory of $(t, m, s)$-nets (or point sets) and $(t, s)$-sequences (see [Nie92] and this volume for more details) provides more elaborate constructions, where this constant goes to zero with increasing dimension $s$. As opposed to the Halton sequence, these constructions however are designed for a fixed number of dimensions and may expose strong correlations when restricted to less than $s$ dimensions. Since the algorithm proposed in the sequel needs an infinite sequence where the number of dimensions will change during integration, the Halton sequence is chosen, which in addition is very fast to generate using ACM algorithm 247 (see [HW64] or [Kel96c]).

## 2.2 Convergence

Using the discrepancy and the principle of importance sampling, the following generalized version of the Koksma-Hlawka inequality[1] has been established in [SM94]:

**Theorem 1** *Let $P_N = \{x_0, \ldots, x_{N-1}\}$ be a point set in $I^s$. For the points $y_i = \mu^{-1}(x_i)$, where $\mu^{-1}$ is defined by the multidimensional inversion method [HM72] according to the density $p$, we have*

$$\left| \int_{I^s} f(x)dx - \frac{1}{N} \sum_{i=0}^{N-1} \frac{f(y_i)}{p(y_i)} \right| \leq V\left(\frac{f}{p}\right) D^*(P_N)$$

*if $\frac{f}{p}$ is of bounded variation in the sense of Hardy and Krause.*

---

[1] The original inequality is obtained by setting $p \equiv 1$, then $\mu^{-1}$ is the identity.

Similar to the Monte Carlo bound, the error is split into a property of the function, i.e. the variation $V(\frac{f}{p})$ in the sense of Hardy and Krause (see [Nie92] for details), and the discrepancy as property of the sampling points. If the variation can be bounded by a constant, roughly speaking, integration by low discrepancy points yields a deterministic convergence rate of $\mathcal{O}(N^{-1})$, which is quadratically faster than the probabilistic one of random sampling.

Since in computer graphics typically the variation is infinite due to unknown discontinuities of the integrand, this inequality does not apply for error estimation. Other error bounds using isotropic discrepancy (see [Nie92]) guarantee the convergence but at a far too pessimistic rate as compared to numerical evidence. Taking a closer look, the difference between Monte Carlo and quasi-Monte Carlo integration is the generation of the discrete sampling density. Often a density function $p$ can be split off the integrand $f$, and similar to the principle of importance sampling (1.2) we get

$$\int_{I^s} f(x)dx = \int_{I^s} g(x)p(x)dx = \int_{I^s} g(y)d\mu(y) \approx \frac{1}{N} \sum_{i=1}^{N-1} g(y_i)$$

where the sample positions $C_N = \{y_0, \ldots, y_{N-1}\}$ are generated with respect to $p$ using the multidimensional inversion method. This way we avoid expensive evaluations of $g$ to be weighted by small values of the density $p$. Seen from a physical point of view the modeling of the discrete density $C_N$ corresponds to particle emission on the lights or scattering on the surface. To measure the quality of such a discrete density approximation we define

**Definition 2** *The deviation of $C_N$ from $p$, i.e. the discrepancy with respect to a density $p$ is given by*

$$D^*(C_N, p) := \sup_{A \in \mathcal{J}^s} \left| \int_{I^s} \chi_A(x)\, p(x)\, dx - \frac{1}{N} \sum_{i=0}^{N-1} \chi_A(y_i) \right|.$$

Using this definition, we find the following theorem in [HM72]:

**Theorem 2** *If $p$ is separable, i.e. $p(x^{(1)}, \ldots, x^{(s)}) = \prod_{j=1}^{s} p^{(j)}(x^{(j)})$, we have*

$$D^*(C_N, p) = D^*(P_N),$$

*otherwise*

$$D^*(C_N, p) \leq c\, (D^*(P_N))^{\frac{1}{s}},$$

*where $c \in \mathbb{R}^+$ and $s$ is the dimension.*

The first part of the theorem states that the discrepancy remains unchanged for separable density functions, whereas the second part, like estimates using isotropic discrepancy, only yields a very weak upper bound. Nevertheless, generating discrete densities by low discrepancy points is superior to the generation using random points. The true convergence rate of low discrepancy integration when applied to discontinuous functions still is an open problem, and only arguments based on very weak assumptions are available in [PTVF92, MC95]. Numerical evidence however proves the rate to be superior to that of jittered sampling, which is $\mathcal{O}(N^{-\frac{s+1}{2s}})$ (see e.g. graphs in [Kel96c]).

# 3 Global Illumination

The global illumination problem in computer graphics consists of calculating various functionals of the solution of a second kind Fredholm integral equation. Given the surface $S = \cup_{k=1}^{K} A_k$ of the scene, usually modeled by $K$ geometric primitives $A_k$ (like triangles, polygons, spheres, etc.), the surface properties $f_r$, and the light sources $L_e$, we have the following integral equation, known as radiance equation (see [CW93]), describing the radiance transfer

$$
\begin{aligned}
L(y, \vec{\omega}_r) &= (L_e + T_{f_r} L)(y, \vec{\omega}_r) \qquad\qquad (1.3)\\
&:= L_e(y, \vec{\omega}_r) + \int_{\Omega} L(h(y, \vec{\omega}_i), -\vec{\omega}_i)\, f_r(\vec{\omega}_i, y, \vec{\omega}_r)\, \cos\theta_i\, d\omega_i,
\end{aligned}
$$

where $L(y, \vec{\omega}_r)$ is the radiance measured in $\frac{W}{m^2 sr}$ leaving $y \in S$ into direction of $\vec{\omega}_r \in \Omega$ being the set of all directions of the hemisphere in point $y$ aligned normal to the surface. This radiance is the sum of the source contribution $L_e$ and all incoming light integrated over the hemisphere in $y$ attenuated by the projection term $\cos\theta_i$ and the bidirectional reflectance distribution function $f_r$. $\theta_i$ is the angle between the direction of incidence $\vec{\omega}_i$ and the surface normal in $y$. $f_r$ characterizes the surface properties, i.e. reflectivity and color. The hitpoint function $h$ returns the first point hit when shooting a ray from $y$ into direction $\vec{\omega}_i$. Its evaluation is very costly in computer graphics and bounded from below by $\mathcal{O}(\log K)$, thus algorithms for the solution of the integral equation should minimize the calls to $h$ in order to be fast. The central task in computer graphics now is the calculation of functionals of the form

$$
\langle L, W_e \rangle := \int_S \int_{\Omega} L(x, \vec{\omega}) W_e(x, \vec{\omega}) \cos\theta d\omega dx,
$$

where $W_e$ is a kind of detector functional. Using this scalar product and the definition of the adjoint operator $\langle T_{f_r} L, W \rangle = \langle L, T_{f_r}^* W \rangle$, the adjoint transport equation for the adjoint unit $W$ of $L$ is

$$
\begin{aligned}
W(y, \vec{\omega}_i) &= (W_e + T_{f_r}^* W)(y, \vec{\omega}_i) \qquad\qquad (1.4)\\
&:= W_e(y, \vec{\omega}_i) + \int_{\Omega} W(h(y, \vec{\omega}_i), \vec{\omega}_r) f_r(\vec{\omega}_r, h(y, \vec{\omega}_i), \vec{\omega}_i)\\
&\qquad\qquad\qquad\qquad \cos\theta(h(y, \vec{\omega}_i), \vec{\omega}_r) d\omega_r,
\end{aligned}
$$

where $W_e$ is the source of $W$ indicating the contribution of the location $(y, \vec{\omega}_i)$ to $W_e$.

For the subject of this work we restrict to diffuse reflection, i.e. $f_r = f_r(y) = \sum_{k=1}^{K} \chi_{A_k}(y) \frac{\rho_{d,k}}{\pi} \tau_{d,k}(y)$ only depends on the surface location, where $\rho_{d,k}$ is the reflectivity and $\tau_{d,k}$ the texture, i.e. the color function of the surface element $A_k$. Consequently the radiance $L = L(y)$ becomes independent of direction, too, and only diffuse light sources $L_e = L_e(y)$ are allowed. This setting, also known as the radiosity setting [CW93], is appropriate for light simulation in building interiors, where the most important transport mechanism is diffuse reflection. The extension to general reflection functions is done similar to [Kel96b]. For the sequel the substitution

$$dF(u_1, u_2) := d \sin^2 \frac{\pi}{2} u_1 du_2, (u_1, u_2) \in I^2 \qquad (1.5)$$

where $\vec{\omega} = (\arcsin \sqrt{u_1}, 2\pi u_2)$ is applied to transform the integrals from $\Omega$ onto the unit cube $I^2$ using importance sampling with respect to the $\cos \theta$ distribution.

# 4 The Quasi-Random Walk Algorithm

In realistic scenes we have the reflectivity $\rho := \|T_{f_r}\| < 1$, meaning that there is always less reflected than incident radiance. So the Neumann series for the solution of (1.3) is convergent and we expand

$$\langle L, W_e \rangle = \sum_{j=0}^{\infty} \langle T_{f_r}^j L_e, W_e \rangle.$$

Previous work (see e.g. [KMS94, Kel96a, Spa95]) showed that applying quasi-Monte Carlo integration for the evaluation of the truncated Neumann series expansion performs superior to random sampling, although suffering from underestimation caused by the truncation.

In experiments similar to [Spa95], where the deterministic paths were completed by random walks when determining how many dimensions can be integrated using low discrepancy sampling in order to outperform the random walk, it turns out that only the scattering from light sources, i.e. the first four dimensions, can be chosen deterministically. So evaluating the reflected light by the same number $N$ of samples as the directly emitted light, the higher powers of $T_{f_r}$ become evaluated too exactly, wasting a lot of work. Taking a look at a random walk simulation (see e.g. [KW86]) with Russian Roulette absorption, the reflectivity plays a crucial role. For the moment let $f_r = \frac{\rho}{\pi}$ be constant throughout the scene. Then after starting $N$ particles from the source, only $\rho N$ particles are expected not to be absorbed by the first surface interaction. In the next step only $\rho^2 N$ particles will continue their paths and so on.

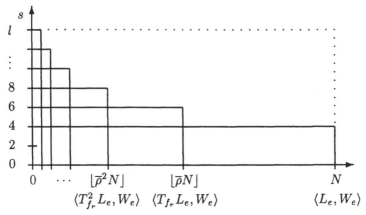

FIGURE 1. Usage of low discrepancy points in the quasi-random walk algorithm

## 4.1 Unidirectional Quasi-Random Walks

We now transfer this principle to quasi-Monte Carlo integration. As illustrated in figure 1, the term $\langle L_e, W_e \rangle$ representing the direct interaction of the light sources $L_e$ with the detector $W_e$ is evaluated using $N$ samples. The once reflected light $\langle T_{f_r} L_e, W_e \rangle$ is calculated using only $\lfloor \rho N \rfloor$ samples, and so on until $\lfloor \rho^{l+1} N \rfloor < 1$ which is guaranteed to happen since $\rho < 1$. As $\rho$ usually is not constant over the scene, we approximate

$$\rho \approx \overline{\rho} := \frac{\sum_{k=1}^{K} |A_k| \rho_{d,k}}{\sum_{k=1}^{K} |A_k|}.$$

This proves to be a good choice, since usually $\rho_{d,k}$ ranges only in a small interval of $[0.3, 0.7]$. Using the geometric series, the maximum number of rays shot is bounded by $\frac{1}{1-\overline{\rho}} N$. This deterministic upper bound uses less quasi-random numbers than the full set of $lN$ (see figure 1) as in e.g. [Kel96a] and overestimates the true number of rays by at most $l$. In order to approximate the diffuse radiance by

$$L(y) \approx L_e(y) + \sum_{k=1}^{K} \overline{L}_{r,k} W_{e,k}(y),$$

orthogonal detector functionals of the kind

$$W_e(y) = \sum_{k=1}^{K} W_{e,k}(y) = \sum_{k=1}^{K} \left( T_{f_r}^* \frac{\chi_{A_k}}{\sqrt{\pi |A_k|}} \right)(y)$$

are chosen, where the adjoint operator assures the light to be reflected at least once. Then all

$$\overline{L}_{r,k} := \langle L, W_{e,k} \rangle = \sum_{j=0}^{\infty} \langle T_{f_r}^j L_e, W_{e,k} \rangle = \sum_{j=0}^{\infty} \langle T_{f_r}^j L_e, T_{f_r}^* \frac{\chi_{A_k}}{\sqrt{\pi |A_k|}} \rangle$$

$$= \sum_{j=0}^{\infty} |S_0| \frac{\pi^{j+1}}{\sqrt{\pi |A_k|}} \int_{I^{2j+4}} L_e(y_0) \prod_{m=1}^{j+1} f_r(y_m) \chi_{A_k}(y_{j+1})$$

$$dy_0(u_0, u_1) dF(u_2, u_3) \cdots dF(u_{2j+2}, u_{2j+3})$$

are simultaneously calculated using $N$ deterministic paths. Following the previous principle, the quasi-Monte Carlo algorithm for path $i$ determines the number of reflections $j$ according to figure 1 (see also [Kel96b]). Using the vector $x_i = (u_{i,1}, \ldots, u_{i,2j+4})$ of the Halton sequence, the starting point $y_0 \in \text{supp}\, L_e$ is determined by $(u_{i,1}, u_{i,2})$. The initial direction of emission is selected by $\vec{\omega}_0 = (\arcsin \sqrt{u_{i,3}}, 2\pi u_{i,4})$ with respect to the density (1.5). Then the subsequent hitpoints are determined by $y_{m+1} = h(y_m, \vec{\omega}_m)$. On hitting a surface, the radiance of the particle is added to the corresponding $\overline{L}_{r,k}$. On each of the $j$ reflections scattering is performed by attenuating the particle by $f_r$ and then selecting the new direction by $(u_{i,2j'+3}, u_{i,2j'+4})$, $0 < j' \le j$, again with respect to the density $dF$. So $N - \lfloor \bar{\rho} N \rfloor$ paths of length 1 contribute to only $\langle L_e, W_e \rangle$, the subsequent $\lfloor \bar{\rho} N \rfloor - \lfloor \bar{\rho}^2 N \rfloor$ paths of length 2 contribute to $\langle L_e, W_e \rangle$ and $\langle T_{f_r} L_e, W_e \rangle$, and so on. This way Russian Roulette absorption is avoided, which considerably raises variance in random schemes. As the algorithm is deterministic, it is biased, but nevertheless consistent, since for $N \to \infty$ all terms of the Neumann series become included. By the evaluation of the higher operator powers with less particles, each particle (or ray shot) carries about the same amount of radiance. This way more particles are concentrated in the lower dimensions, exploiting the good discrepancy properties of the Halton sequence. Considering the higher dimensions, the behaviour of the particles approaches random behaviour, since due to the complicated scene geometry a small perturbation of the emission direction yields a completely different end point of the path.

In figure 2 we illustrate the numerical behaviour of the quasi-random walk algorithm as compared to a random walk for a test scene (a grey Cornell-Box with analytical solution, see [Kel95]). In the graphs the convergence, and the $L_2$ and $L_\infty$ error on the left, and the weighted counterparts (weighted by size of the surface elements $|A_k|$) on the right are displayed. For the random walks the error bars indicate the range covered by 20 independent random walks each using the same number of paths as the quasi-random walk. Obviously the better uniform distribution of the low discrepancy points yields a deterministic algorithm (of course without variance), which approaches the expectation faster and much smoother than the random algorithm. In [Kel96b] similar graphs can be found for other scenes where no analytical solution was accessible.

Instead of using the constant base functions on the surfaces, we can also use an adaptive wavelet representation of radiance. The results for the Haar wavelet can be seen in figure 3a. Using a tree wavelet (e.g. Haar- or Multi-wavelet) and non-standard decomposition, the decision of adding a

286

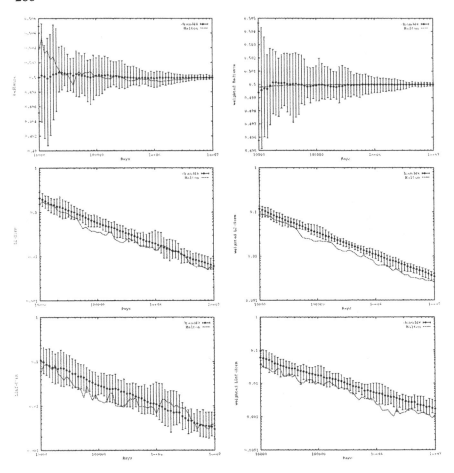

FIGURE 2. Measurements for the analytic Cornell-box

new level of detail is based on choosing the smaller error of either omitting the new level and having high accuracy due to higher number of particles per support, or adding the new level, with higher approximation accuracy, but decreased accuracy in the single wavelets, due to reduced number of particles per support. Compared to [GSCH93], a quasi-random walk calculating the wavelet coefficients works without kernel discretization, only the discretization error of the wavelet representation of radiance remains.

Although the results of the very fast quasi-random walk can be directly displayed using interpolating polygon hardware, it usually serves as a preprocess for photorealistic image synthesis in algorithms similar to [Kel96a]. A result of this postprocess with a quasi-random walk preprocessing can be seen in figure 4.

The quasi random walk profits from the coherence in the low dimensions, i.e. the coherent starting domain of the lightsources. The detector functional

$$W_{e,P}(x,\vec{\omega}) = \frac{\delta(\vec{\omega} - \vec{\omega}_{f_p})}{\cos\theta} \frac{1}{|P|} \chi_P(h(x,\vec{\omega})) \qquad (1.6)$$

describes the model of a pinhole camera with focal point $f_p$, where $P$ is the support of one pixel in the image matrix. When inserting $W_{e,P}$ in the adjoint transport equation (1.4), an even improved behaviour should be observed when calculating $W$ using a quasi-random walk, as $W$ originates from one point source, guaranteeing high coherence in the low dimensions. The *visual importance* $W$ can be used as an importance function to direct the radiance $L$ in a subsequent quasi-random walk to the areas of augmented interest [NNB+96]. This hints at iteration algorithms, first distributing $W_e$, used for the distribution of $L_e$, then again used for a more exact $W_e$ pass, until a suited stopping criterion is fulfilled.

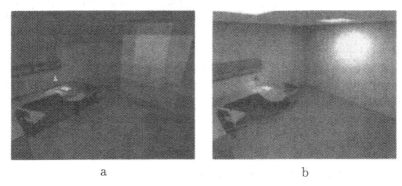

a                    b

FIGURE 3. Images generated using the quasi-random walk techniques for a) the Haar wavelet approximation of the radiance $L$ and b) bidirectional path tracing.

## 4.2  Bidirectional Quasi-Random Walks

Taking the $j$-th term of the Neumann expansion, we have several possibilities for its calculation using the adjoint operator

$$\langle T_{f_r}^j L_e, W_e \rangle = \sum_{n=0}^{j} w_{jn} \langle T_{f_r}^n L_e, T_{f_r}^{*(j-n)} W_e \rangle,$$

where obviously the weights must be chosen such that $\sum_{n=0}^{j} w_{jn} = 1$. The full series then is represented by the following triangular structure:

| $j$ | $N_j$ | | | |
|---|---|---|---|---|
| 0 | $N$ | | | $\langle L_e, W_e \rangle$ |
| 1 | $\rho N$ | | $\langle T_{f_r} L_e, W_e \rangle$ | $\langle L_e, T_{f_r}^* W_e \rangle$ |
| 2 | $\rho^2 N$ | $\langle T_{f_r}^2 L_e, W_e \rangle$ | $\langle T_{f_r} L_e, T_{f_r}^* W_e \rangle$ | $\langle L_e, T_{f_r}^{*2} W_e \rangle$ |
| $\vdots$ | $\vdots$ | | $\vdots$ | |

The left diagonal represents the unidirectional technique starting the paths from the light sources, whereas the right border describes the paths started from the detectors. The inner terms of the pyramid scheme are functionals linking segments independently started on the light sources and the detectors. The original stochastic algorithm for bidirectional path tracing has been introduced in [VG95, Laf96]. For the deterministic simulation, the Halton vector $x_i = (u_{i,1}, \ldots, u_{i,4j+2})$ is used in an interleaved way to exploit the good discrepancy in the lower dimensions: Choosing the pixel functional (1.6), $y_0^*$ is the first point hit when shooting a ray from the eye point through a point in the pixel specified by $(u_{i,1}, u_{i,2})$, $y_1^*$ is the next point hit after scattering according to $dF$ using $(u_{i,5}, u_{i,6})$. On the other hand, $y_0$ is the point of emission on a light source selected by $(u_{i,3}, u_{i,4})$, $y_1$ the first point hit after scattering with $(u_{i,7}, u_{i,8})$, and so on. According to the principle of the quasi-random walk, $N$ such pairs of paths are generated, where the number $j$ of reflections depends on the path number $i$ as in the unidirectional case (see figure 1). The evaluation of the terms up to level $j$ in the above table consists of two parts: $\langle L_e, T_{f_r}^{*j'} W_{e,P} \rangle$ is calculated by evaluating $L_e(y_{j'}^*)$ and multiplying by the accumulated bidirectional reflection distribution functions for $0 \leq j' \leq j$. The remaining terms for $\langle T_{f_r}^n L_e, T_{f_r}^{*(j'-n)} W_{e,P} \rangle$, $n > 0$, $j' + n \leq j$ are calculated by linking the corresponding points of both paths, using the visibility test

$$V(y_{j'-n}^*, y_n) := \begin{cases} 1 & \text{if } h(y_{j'-n}^*, y_n - y_{j'-n}^*) = y_n \\ 0 & \text{else} \end{cases}.$$

If both points are mutually visible, their contributions are multiplied and attenuated by the geometric term

$$G(y_{j'-n}^*, y_n) = \frac{\cos \theta_{y_{j'-n}^*} \cos \theta_{y_n}}{|y_{j'-n}^* - y_n|^2}$$

which results from a domain change from $\Omega$ to $S$ using $d\omega_x = \frac{\cos \theta_y}{|x-y|^2} dy$.

A sample image of the algorithm with the simple choice of $w_{jn} = \frac{1}{j+1}$ (although better choices for $w_{jn}$ exist, see [VG95]) is shown in figure 3b. Note, that this scheme is a single pass algorithm, working without discretization error. Due to the scene complexity, a very high sampling rate ($N \approx 1000$) is required for *each* pixel of the image, which can be reduced by techniques proposed in [VG95, Laf96]. Nevertheless this very expensive method is useful for reference calculations, when testing fast multi-pass algorithms as proposed in [Kel96a] in combination with unidirectional quasi-random walk preprocessing.

FIGURE 4. Photorealistic rendering of a conference room using the quasi-random walk preprocess. (The conference room and the office have been modeled by Anat Grynberg and Greg Ward and are available via http://radsite.lbl.gov/mgf/HOME.html in the *material and geometry format* MGF.)

# 5  Conclusion

We introduced the deterministic quasi-random walk technique for the approximation of functionals of the solution of a second kind Fredholm integral equation using low discrepancy sampling. The unidirectional variant of the new scheme is very fast, easy to apply and has only one free parameter, which is the number of simulation paths to use. Since the method is deterministic, the number of rays used can be calculated in advance, allowing precise simulation time and memory usage forecasts. For the radiosity problem in computer graphics the superiority of the new scheme is illustrated by numerical evidence, yielding that the quasi-random walk approaches the desired solution more smooth than random walks and of course without variance. In addition, the application to bidirectional path-tracing is indicated. Further research will concentrate on the development of an iteration scheme using both the transport equation and its dual, and the incorporation of further variance reduction techniques into quasi-Monte Carlo algorithms.

# 6 REFERENCES

[CW93]     M. Cohen and J. Wallace. *Radiosity and Realistic Image Synthesis*. Academic Press Professional, Cambridge, 1993.

[GSCH93]   S. Gortler, P. Schröder, M. Cohen, and P. Hanrahan. Wavelet Radiosity. In *Computer Graphics (ACM SIGGRAPH Annual Conference Series)*, pages 221 – 230, 1993.

[Hla62]    E. Hlawka. Lösung von Integralgleichungen mittels zahlentheoretischer Methoden I. *Sitzungsber., Abt. II, Österr. Akad. Wiss., Math.-Naturwiss. Kl.*, (171):103–123, 1962.

[HM72]     E. Hlawka and R. Mück. Über eine Transformation von gleichverteilten Folgen II. *Computing*, (9):127–138, 1972.

[HW64]     J. Halton and G. Weller. Algorithm 247: Radical-inverse quasirandom point sequence. *Comm. ACM*, 7(12):701–702, 1964.

[Kel95]    A. Keller. A Quasi-Monte Carlo Algorithm for the Global Illumination Problem in the Radiosity Setting. In H. Niederreiter and P. Shiue, editors, *Monte Carlo and Quasi-Monte Carlo Methods in Scientific Computing*, volume 106, pages 239–251. Springer, 1995.

[Kel96a]   A. Keller. Quasi-Monte Carlo Methods in Computer Graphics: The Global Illumination Problem. *Lectures in App. Math.*, 32:455–469, 1996.

[Kel96b]   A. Keller. Quasi-Monte Carlo Radiosity. In X. Pueyo and P. Schröder, editors, *Rendering Techniques '96 (Proc. 7th Eurographics Workshop on Rendering)*, pages 101–110. Springer, 1996.

[Kel96c]   A. Keller. The fast Calculation of Form Factors using Low Discrepancy Sequences. In *Proc. Spring Conference on Computer Graphics (SCCG '96)*, pages 195–204, Bratislava, Slovakia, 1996. Comenius University Press.

[KMS94]    A. Kersch, W. Morokoff, and A. Schuster. Radiative Heat Transfer with Quasi-Monte Carlo Methods. *Transport Theory and Statistical Physics*, 7(23):1001–1021, 1994.

[KW86]     M. Kalos and P. Whitlock. *Monte Carlo Methods, Volume I: Basics*. J. Wiley & Sons, 1986.

[Laf96]    E. Lafortune. *Mathematical Models and Monte Carlo Algorithms for Physically Based Rendering*. PhD thesis, Katholieke Universitiet Leuven, Belgium, 1996.

[MC95]     W. Morokoff and R. Caflisch. Quasi-Monte Carlo Integration. *J. Comp. Physics*, (122):218–230, 1995.

[Mit92]    D. Mitchell. Ray Tracing and Irregularities of Distribution. In *Proc. 3rd Eurographics Workshop on Rendering*, pages 61–69, Bristol, UK, 1992.

[Mit96]    D. Mitchell. Consequences of Stratified Sampling in Graphics. In *Computer Graphics (ACM SIGGRAPH Annual Conference Series)*, pages 277–280, 1996.

[Nie92]    H. Niederreiter. *Random Number Generation and Quasi-Monte Carlo Methods*. SIAM, Pennsylvania, 1992.

[NNB+96]  A. Neumann, L. Neumann, P. Bekaert, Y. Willem, and W. Purgathofer. Importance-Driven Stochastic Ray Radiosity. *Rendering Techniques '96 (Proc. 7th Eurographics Workshop on Rendering)*, pages 111–122, 1996.

[PTVF92]  H. Press, S. Teukolsky, T. Vetterling, and B. Flannery. *Numerical Recipes in C*. Cambridge University Press, 1992.

[Shi91]    P. Shirley. Discrepancy as a Quality Measure for Sampling Distributions. In *Eurographics '91*, pages 183–194, Amsterdam, North-Holland, 1991. Elsevier Science Publishers.

[SM94]     J. Spanier and E. Maize. Quasi-Random Methods for Estimating Integrals using relatively small Samples. *SIAM Review*, 36(1):18–44, March 1994.

[SP87]     P. Sarkar and M. Prasad. A comparative Study of Pseudo and Quasi Random Sequences for the Solution of Integral Equations. *J. Comp. Physics*, (68):66–88, March 1987.

[Spa95]    J. Spanier. Quasi-Monte Carlo Methods for Particle Transport Problems. In H. Niederreiter and P. Shiue, editors, *Monte Carlo and Quasi-Monte Carlo Methods in Scientific Computing*, pages 121–148. Springer, 1995.

[VG95]     E. Veach and L. Guibas. Optimally Combining Sampling Techniques for Monte Carlo Rendering. In *Computer Graphics (ACM SIGGRAPH Annual Conference Series)*, pages 419–428, 1995.

# Comparison of independent and stratified sampling schemes in problems of global optimization

Marina Kondratovich
Anatoly Zhigljavsky

ABSTRACT

Let $\mathcal{X}$ be a compact subset of $R^d$ and $\{x_1, \ldots, x_N\}$ be a sample of random points in $\mathcal{X}$. It is well–known that any properly organized stratified sampling procedure is superior to the independent sampling with respect to the variance of Monte Carlo estimates of integrals of functions in $L_2(\mathcal{X})$. We prove similar results for some perfomance characteristics important in global optimization. We also demonstrate that the stratified sample with the maximum stratification is optimum, in a suitable sense.

## 1 Introduction

Let $(\mathcal{X}, \mathcal{B}, P)$ be a measure space, $\mathcal{X}$ be a compact subset of $\mathbf{R}^d$, $\mathcal{B}$ be the $\sigma$-algebra of Borel subsets of $X$ and $P$ be a probability measure on $(\mathcal{X}, \mathcal{B})$, we call it *uniform distribution*. Let also $\mathcal{F} \subseteq C(\mathcal{X})$ be a functional class, $N$ be a fixed number, $\Xi = (x_1, x_2, ..., x_N) \in \mathcal{X}^N$, $M[f, \Xi] = \max_{x_i \in \Xi} f(x_i)$.

If $\Xi$ is a random vector in $\mathcal{X}^N$ with certain distribution $Q(d\Xi)$, then the ordered pair $\Pi = (M[f, \Xi], Q)$ is called *random search procedure* for the global maximum of $f \in F$.

Consider a partition $\mathcal{P}_m$ of $\mathcal{X}$ into $m$ disjoint connected subsets of positive measure:

$$\mathcal{X} = \bigcup_{i=1}^m \mathcal{X}_i, \ \mathcal{X}_i \in \mathcal{B}, \ q_i = P(\mathcal{X}_i) > 0 \text{ for } i = 1, \ldots, m, \ \mathcal{X}_i \cap \mathcal{X}_j = \emptyset \text{ for } i \neq j.$$

Since $P$ is a probability measure, $\sum_{i=1}^m q_i = 1$. Define the uniform probability measure $P_i$ on $\mathcal{X}_i$ by $P_i(A) = P(A \cap \mathcal{X}_i)/q_i$ for every $A \in \mathcal{B}$.

Given a partition $\mathcal{P}_m$ and a collection of integers $L = \{l_1, \ldots, l_m\}$ such that $\sum_{i=1}^m l_i = N$, the *stratified sample* $\Xi_{m,L}$ can be defined as

$$\Xi_{m,L} = (x_{1,1}, \ldots, x_{1,l_1}, \ldots, x_{m,1}, \ldots, x_{m,l_m}) \tag{1.1}$$

where for every $i = 1, \ldots, m$, $x_{i,1}, \ldots, x_{i,l_i}$ are independent random variables with the uniform distribution $P_i$ on $\mathcal{X}_i$. (In practice, an additional randomization on the order of generation of $x_{i,j}$ is sometimes useful as well.)

We shall call the stratified sample (1.1) *proper stratified sample* if the number of points in $\mathcal{X}_i$ is proportional to $q_i = P(\mathcal{X}_i)$, that is,

$$l_i = Nq_i \quad \text{for all } i = 1, \ldots, m. \tag{1.2}$$

The joint distribution of the random vector (1.1) is

$$Q_{m,L}(d\Xi_{m,L}) = \prod_{i_1=1}^{l_1} P_1(dx_{1,i_1}) \times \ldots \times \prod_{i_m=1}^{l_m} P_m(dx_{m,i_m}).$$

The random search procedure $\Pi_{m,L} = (M[f, \Xi_{m,L}], Q_{m,L})$ with $m > 1$ corresponds to the stratified sampling on $\mathcal{X}$, and $\Pi_1 = \Pi_{1,N} = (M[f, \Xi_{1,N}], Q_{1,N})$ corresponds to the independent sampling from the distribution $P$.

For a fixed $f \in \mathcal{F}$, let $\Psi_f(\Pi)$ be a criterion for comparison of procedures $\Pi$. In line with the general concept of domination, we say that $\Pi$ dominates $\Pi'$ in $\mathcal{F}$ if $\Psi_f(\Pi) \leq \Psi_f(\Pi')$ for every $f \in \mathcal{F}$ and there exists a function $f_* \in \mathcal{F}$ such that $\Psi_{f_*}(\Pi) < \Psi_{f_*}(\Pi')$.

Below we consider two related dominance criteria: (i) the cumulative distribution function (c.d.f.) of the record value $M[f, \Xi] = \max_{x_i \in \Xi} f(x_i)$ achieved at the sample points, and (ii) $k$-th moment of the difference $\max_{\mathcal{X}} f - M[f, \Xi]$, for every $k > 0$. Some other criteria are studied in [1]. The results of the paper confirm the assertion of [1] that the "reduction of randomness" and "increase of uniformity" improves the performance of Monte Carlo procedures for global optimization.

## 2 Stochastic dominance with respect to the record value

Let us consider the stochastic dominance when the criterion $\Psi_f(\Pi)$ is the c.d.f. of the record value $M[f, \Xi] = \max_{x_i \in \Xi} f(x_i)$:

$$F_{f,\Pi}(t) = P(M[f, \Xi] \leq t), \quad t \in (\min f, \max f). \tag{1.3}$$

In this case, the dominance of a procedure $\Pi$ over $\Pi'$ in $\mathcal{F}$ means that $F_{f,\Pi}(t) \leq F_{f,\Pi'}(t)$ for all real $t$ and $f \in \mathcal{F}$ and there exists $f_* \in \mathcal{F}$ such that $F_{f_*,\Pi}(t) < F_{f_*,\Pi'}(t)$ for all $t \in (\min f_*, \max f_*)$.

**Theorem 1.** Let $\mathcal{P}_m$ be a fixed partition of $\mathcal{X}$ into $m \leq N$ subsets, $\mathcal{F} = C^p(\mathcal{X})$ for some $0 \leq p \leq \infty$ and $\Pi_{m,L} = (M[f, \Xi_{m,L}], Q_{m,L})$ be a stratified sampling random search procedure such that $L = \{l_1, \ldots, l_m\}$, $l_i \geq 0$, $\sum_{i=1}^{m} l_i = N$. Then

(i) if the stratified sample $\Xi_{m,L}$ is proper, that is (1.2) holds, then the stratified sampling random search procedure $\Pi_{m,L}$ stochastically dominates the independent random sampling procedure $\Pi_1$ in $\mathcal{F}$, with respect to the criterion (1.3);

(ii) if (1.2) does not hold for at least one $i$, then $\Pi_{m,L}$ does not stochastically dominate $\Pi_1$: there exists $f^* \in \mathcal{F}$ such that for some $t$ $F_{m,L}(f_*, t) > F_1(f_*, t)$ where $F_{m,L}(f, t) = F_{f,\Pi_{m,L}}(t)$ and $F_1(f, t) = F_{f,\Pi_1}(t)$ are the c.d.f. (1.3) for the stratified and independent sampling procedures, respectively.

**Proof.** Let $f$ be an arbitrary function in $\mathcal{F}$. Then the c.d.f. $F_1(f, t)$ for the independent sampling procedure $\Pi_1 = (M[f, \Xi_{1,N}], Q_{1,N})$ can be rewritten as

$$F_1(f, t) = P(f(x_{1,j}) \leq t, j = 1, \ldots, N) = P^N(f(x_{1,j}) \leq t) = P^N(A_t)$$

where $A_t = f^{-1}((-\infty, t])$ is the inverse image of the set $(-\infty, t]$. Since $\{X_i\}_{i=1}^m$ is a complete system of events, we have

$$P(A_t) = \sum_{i=1}^m P(A_t \cap X_i) = \sum_{i=1}^m \beta_i$$

where

$$\beta_i = P(A_t \cap X_i), \quad i = 1, \ldots, m, \quad \sum_{i=1}^m \beta_i = P(A_t) \leq 1.$$

We thus have

$$F_1(f, t) = \left( \sum_{i=1}^m \beta_i \right)^N .$$

For the stratified sampling procedure $\Pi_{m,L} = (M[f, \Xi_{m,L}], Q_{m,L})$ the c.d.f. $F_{m,L}(f, t)$ can be analogously rewritten as

$$F_{m,L}(f, t) = P(f(x_{1,1}) \leq t, \ldots, f(x_{1,l_1}) \leq t, \ldots, f(x_{m,l_m}) \leq t) =$$

$$\prod_{i=1}^m P_i^{l_i}(f(x_{i,j}) \leq t) = \prod_{i=1}^m (P(\{f(x_{i,j}) \leq t\} \cap X_i)/q_i)^{l_i} = \prod_{i=1}^m \left( \frac{\beta_i}{q_i} \right)^{l_i} .$$

For every $i = 1, \ldots, m$, set

$$\gamma_i = \frac{l_i}{N}, \quad \alpha_i = \frac{\beta_i}{q_i} = P(A_t \cap X_i)/P(X_i).$$

Then

$$0 < \gamma_i < 1, \quad 0 \le \alpha_i \le 1 \quad \text{for } i = 1, \ldots, m, \quad \sum_{i=1}^{m} \gamma_i = 1$$

and the vector $\alpha = (\alpha_1, \ldots, \alpha_m)$ may get any value in the interior of the cube $[0, 1]^m$ depending on $f$ and $t$.

The representations for the c.d.f. $F_{m,L}(f, t)$ and $F_1(f, t)$ and (1.4) imply that the following two inequalities are equivalent:

$$F_{m,L}(f, t) \le F_1(f, t) \quad \Longleftrightarrow \quad \prod_{i=1}^{m} \alpha_i^{\gamma_i} \le \sum_{i=1}^{m} q_i \alpha_i$$

which we rewrite in a more convenient form

$$F_{m,L}(f, t) \le F_1(f, t) \quad \Longleftrightarrow \quad \sum_{i=1}^{m} \gamma_i \log \alpha_i \le \log \left( \sum_{i=1}^{m} q_i \alpha_i \right) \qquad (1.4)$$

Analogous equivalence takes place when the sign $\le$ in (1.4) is substituted for the strict inequality sign.

Let us now prove (i). If (1.2) holds then $\gamma_i = q_i$ for all $i = 1, \ldots, m$ and the validity of the second inequality in (1.4), for every $\alpha \in [0, 1]^m$ and thus for every $f \in \mathcal{F}$, follows from the concavity of the logarithm. Consider a function $f_* \in \mathcal{F}$ such that $0 \le f \le 1$, $f_*(x) = 0$ for all $x \in \mathcal{X}_1$ and $\max_{x \in \mathcal{X}_2} = 1$. Then

$$\alpha_1 = P(A_t \cap X_1)/P(X_1) = 1 \quad \text{and} \quad \alpha_2 = P(A_t \cap X_2)/P(X_2) < 1$$

for all $t \in (0, 1) = (\min f, \max f)$. Therefore the values $\alpha_i$ are not all equal each other and the strict concavity of the logarithm implies the strict inequality in (1.4).

Let us now turn to (ii). Assume that (1.2) does not hold. Then there exists $i_0 \le m$ such that $\gamma_{i_0} < q_{i_0}$. Consider a function $f^* \in \mathcal{F}$ such that $f^*(x) = 0$ for all $x \in \mathcal{X} \setminus \mathcal{X}_{i_0}$ and $\max_{x \in \mathcal{X}_{i_0}} = 1$. Then $\alpha_j = 1$ for all $j \neq i_0$ and $\alpha_{i_0}$ gets all values in $(0, 1)$ depending on $t$.

Let us show that for the function $f^*$ the inequality

$$\sum_{i=1}^{m} \gamma_i \log \alpha_i > \log \left( \sum_{i=1}^{m} q_i \alpha_i \right) \qquad (1.5)$$

holds for all sufficiently large $\alpha_{i_0} < 1$. Denote $\varepsilon = 1 - \alpha_{i_0} > 0$ and rewrite the inequality (1.5) as $h(\varepsilon) > 0$ where

$$h(\varepsilon) = \gamma_{i_0} \log(1 - \varepsilon) - \log(1 - q_{i_0} \varepsilon) .$$

At $\varepsilon = 0$ we have $h(0) = 0$ and $h'(0) = -\gamma_{i_0} + q_{i_0} > 0$, where we have used the fact that $\gamma_{i_0} < q_{i_0}$. This implies $h(\varepsilon) > 0$ for all sufficiently small $\varepsilon > 0$

and therefore the validity of (1.5). In its turn, it yields that the inequality $F_{m,L}(f^*, t) > F_1(f^*, t)$ holds for all $t$ sufficiently close to 1. ∎

**Corollary.** Analogously to (i) in Theorem 1 we can easily get that if $m' < m$, $\mathcal{P}_m$ is a subpartition of a partition $\mathcal{P}_{m'}$, $\Xi_{m,L}$ is a proper stratified sample and $\Pi_{m,L}$ and $\Pi_{m',L'}$ are the random search procedures, corresponding to the stratified samples $\Xi_{m,L}$ and $\Xi_{m',L'}$, then $\Pi_{m,L}$ stochastically dominates $\Pi_{m',L'}$ in $\mathcal{F} = C^p(\mathcal{X})$ for every $0 \leq p \leq \infty$. This particularly implies that the stratified sample $\Xi_{m,L}$ with the maximum stratification, that is, when $P(\mathcal{X}_i) = 1/m$ and $L = (1, \ldots, 1)$, generates the best possible random search procedure $\Pi_{m,L}$, with respect to the stochastic dominance based on the c.d.f. (1.3).

# 3  Asymptotic criteria

In the present section we only consider proper stratified sampling procedures $\Pi_{m,l} = \Pi_{m,L}$ where $P(\mathcal{X}_i) = 1/m$ and $l_i = l$ for all $i = 1, \ldots, m$. We also assume that $N = ml$, $l$=const, $m \to +\infty$, that is, the number of subsets in the partition $\mathcal{P}_m$ tends to infinity but the number of points in each subset stays constant.

As the criteria for comparison of procedures, consider now $k$-th moment of the random variable $(M[f] - M[f, \Xi_{m,l}])$ where $M[f] = \max_{x \in \mathcal{X}} f(x)$ and $\Xi_{m,l} = \Xi_{m,L}$ for $L = (l, \ldots, l)$:

$$\Psi_f(m, l) = \mathbf{E}(M[f] - M[f, \Xi_{m,l}])^k, \quad k > 0. \tag{1.6}$$

Theorem 1 implies that the stratified sampling procedure $\Pi_{m,l}$ is superior to the independent sampling procedure $\Pi_{1,N}$ with respect to the criteria (1.6), for every $k > 0$. Theorem 2 below establishes the qualitative result concerning this superiority.

Let the class of distributions $\{P\}$ and the functional class $\mathcal{F}_*$ of continuous functions $f(x) = \varphi(x - x_*)$ with a unique point of the global maximum $x_*(f)$, at this point $f(x_*) = M[f]$, satisfy the following two conditions.

**Condition A.** For the c.d.f. $F_f(t) = \int_{f(x) \leq t} P(dx)$, the function $V_f(v) = 1 - F_f(M[f] - 1/v)$, $v > 0$, regularly varies at infinity with some index $-\alpha < 0$, that is, $\lim_{v \to \infty} V_f(uv)/V_f(v) = u^{-\alpha}$ for all $u > 0$.

**Condition B.** The point $x_*$ has a certain distribution $R(dx)$ on $(\mathcal{X}, \mathcal{B})$ which is equivalent to the Lebesgue measure on $(\mathcal{X}, \mathcal{B})$.

The above conditions have been introduced and studied in [1,2]. It is shown in these works that conditions A and B hold in a rather general setup. Condition A is satisfied, for example, when $\mathcal{X}$ is a compact subset of $R^d$, the measure $P$ is equivalent to the Lebesgue measure in some neighbourhood of $x_*$, $\nabla\varphi(0) = 0$, and the matrix $\nabla^2\varphi(0)$ is non-singular, in this case $\alpha = d/2$.

The following theorem generalizes to arbitrary $k > 0$ the result of [1] for $k = 1, 2$.

**Theorem 2.** Assume that the conditions A and B are satisfied, $N = ml$, $l$=const, $m \to \infty$. Then for every $k > 0$ with $R$-probability 1

$$\frac{\Psi_f(m, l)}{\Psi_f(1, N)} = \frac{E(M[f] - M[f, \Xi_{m,l}])^k}{E(M[f] - M[f, \Xi_{1,N}])^k} = r(l, k, \alpha) + o(1), \quad N \to +\infty, \quad (1.7)$$

where

$$r(l, k, \alpha) = \frac{l^{k/\alpha}\Gamma(l+1)}{\Gamma(k/\alpha + l + 1)}, \quad (1.8)$$

and $\Gamma(\cdot)$ is the gamma function. Moreover, $r(l, k, \alpha) < 1$ for every $l, k, \alpha > 0$, function $r(l, k, \alpha)$ is strictly increasing as a function of $l$ and $\lim_{l \to \infty} r(l, k, \alpha) = 1$.

**Proof.** If the condition A holds, then

$$\lim_{N \to +\infty} F^N(M + (M - \theta_N)t) = \exp\{-(-t)^\alpha\}, \quad \forall t \leq 0,$$

where $F = F_f$, $M = M[f] = \max f$ and $\theta_N$ is the $(1 - 1/N)$-quantile of the c.d.f. $F$ $F(\theta_N) = 1-1/N$. This implies

$$M - F^{-1}(y) \sim (M - \theta_N)(-N \log y)^{1/\alpha}, \quad N \to \infty, \quad \forall y \in (0, 1]$$

For the independent sample

$$\Psi_f(1, N) = E(M[f] - M[f, \Xi_{1,N}])^k = N \int_{-\infty}^{+\infty} (M - x)^k F^{N-1}(x)dF(x) =$$

$$N \int_0^1 (M - F^{-1}(y))^k y^{N-1} dy \sim N \int_0^1 \left((M - \theta_N)(-N \log y)^{1/\alpha}\right)^k y^{N-1}dy =$$

$$N^{1+k/\alpha}(M - \theta_N)^k \int_0^1 (\log \frac{1}{y})^{k/\alpha} y^{N-1} dy =$$

$$N^{1+k/\alpha}(M - \theta_N)^k \int_0^{+\infty} z^{k/\alpha} e^{-Nz} dz = (M - \theta_N)^k \Gamma(k/\alpha + 1)$$

Therefore,

$$E(M[f] - M[f, \Xi_{1,N}])^k \sim (M - \theta_N)^k \Gamma(k/\alpha + 1), \quad N \to \infty \quad (1.9)$$

Consider the stratified sample. By the condition B, the probability that the point of the global maximum of $f$ is on the boundary of one of the sets $X_i$ is 0. Consider $l$ largest values from the collection $\{f(x_{i,j}); \ i = 1,\ldots,m; \ j = 1,\ldots,l\}$. Since f is a continuous function and the global maximum is attained at a single point $x_*$, then as $m \to +\infty$ all $l$ points with the largest function $f$ values are located in one set, denote it $X_{i_o}$. Therefore

$$F_o(t) = P_{i_o}(f(x) \le t) = m\left(P(f(x) \le t) - \frac{m-1}{m}\right) = mF(t) - m + 1.$$

Let $\theta_{ol}$ be the $(1-1/l)$-quantile of the c.d.f. $F_o(t)$. Since

$$1 - F_o(\theta_{ol}) = 1/l, \ 1 - mF(\theta_{ol}) + m - 1 = 1/l, \ 1 - F(\theta_{ol}) = \frac{1}{ml} = \frac{1}{n},$$

we get $\theta_N = \theta_{ol}$.

Since $F(t)$ satisfies condition A

$$F_o(M + (M - \theta_{ol})t) \sim m\exp\{-(-t)^{-\alpha}/(lm)\} - m + 1 \sim 1 - (-t)^{\alpha}/l.$$

Hence

$$M - F_o^{-1}(y) \sim (M - \theta_N)(l(1 - y))^{1/\alpha}, \ m \to +\infty.$$

When $m \to \infty$, we thus get for every $k > 0$

$$\Psi_f(m, l) = \mathbf{E}(M[f] - M[f, \Xi_{m,l}])^k = l\int_{-\infty}^{+\infty} (M - x)^k F_o^{l-1}(x)dF_o(x) =$$

$$l\int_0^1 (M - F_o^{-1}(y))^k y^{l-1}dy \sim l\int_0^1 \left((M - \theta_n)l^{1/\alpha}(1 - y)^{1/\alpha}\right)^k y^{l-1}dy =$$

$$(M - \theta_N)^k l^{1+k/\alpha} \int_0^1 y^{l-1}(1 - y)^{k/\alpha}dy =$$

$$(M - \theta_N)^k \frac{\Gamma(k/\alpha + 1)l^{k/\alpha}\Gamma(l + 1)}{\Gamma(k/\alpha + l + 1)}.$$

This and (1.9) yields (1.7).

Let us now fix $k, \alpha$ and study the function $r(l) = r(l, k, \alpha)$. Since $r(1) = 1/\Gamma(2 + k/\alpha) < 1$ and the application of the Stirling formula yields

$$\lim_{l\to\infty} r(l) = \lim_{l\to+\infty} \frac{l^{k/\alpha}(2\pi l)^{1/2} \exp(-l)l^l}{(2\pi(l + k/\alpha))^{1/2} \exp\{-(l + k/\alpha)\}(l + k/\alpha)^{l+k/\alpha}} = 1,$$

to complete the proof we only need to show that the function $r(l)$ is strictly increasing. Indeed,

$$\log r(l+1) - \log r(l) = \log\left(1 + \frac{1}{l}\right)^{k/\alpha} - \log\left(1 + \frac{k/\alpha}{l+1}\right)$$

This expression is positive for every $l, k, \alpha > 0$ since

$$\left(1 + \frac{1}{l}\right)^{k/\alpha} - \left(1 + \frac{k/\alpha}{l+1}\right) =$$

$$\left(\frac{k}{\alpha}\frac{1}{l} + \frac{\frac{k}{\alpha}(\frac{k}{\alpha}-1)}{2!}\frac{1}{l^2} + \ldots\right) - \left(1 + \frac{k}{l}\frac{1}{l} - \frac{k}{\alpha}\frac{1}{l^2} + \ldots\right) =$$

$$\frac{k}{\alpha}\left(\frac{k/\alpha - 1 + 2!}{2!l^2} + \frac{(k/\alpha - 1)(k/\alpha - 2) - 3!}{3!l^3} + \ldots\right)$$

and the positivity of the last expression follows from the fact that this series is alternating with rapid convergence and positive first term.  ∎

## 4   References

[1] A.A. Zhigljavsky *Theory of Global Random Search*, Kluwer Academic Publishers, Dordrecht, 1991.

Department of Mathematics
University of St.Petersburg
Bibliotechnaya sq. 2
198904 St.Petersburg
Russia

e-mail zh@stat.math.lgu.spb.su

# The rate of convergence to a stable law for the random sum of iid random variables

## Marcin Kotulski[1]

ABSTRACT We investigate, using Monte Carlo method, the rate of convergence to asymptotic distribution for the random sum of iid random variables, known in statistical physics as Continuous-Time Random Walk (CTRW). The rate of convergence, estimated from the simulation, is in agreement with theory. Simulated densities are compared with the limiting Lévy-stable densities.

## 1 Introduction

A Continuous-Time Random Walk (CTRW) introduced in ref. [1] is a walk with random waiting-times $T_i$ between successive random jumps $R_i$, i.e., CTRW is a random sum of iid random variables $R_i$. It proved to be a useful model of various aspects of anomalous transport, e.g. in turbulence, relaxation phenomena or intermittent chaotic systems. The signature of the anomalous diffusion is a non-Gaussian asymptotic distribution (propagator, diffusion front) of distance $R(t)$ reached up to time $t$ by a particle initially at the origin. The asymptotic distribution of $R(t)$ for CTRW is Lévy-stable distribution $s_{\alpha,\beta}(x)$ which has long power-law tails. We consider also the special case when the number of jumps $R_i$ in the sum $R(t)$ is non-random, which was called Lévy Flight by B. Mandelbrot.

While the limiting distribution of position $R(t)$ for large $t$ is known [2], it is of interest to estimate the rate of the convergence to the limiting law. We perform Monte-Carlo simulations in order to find the probability densities of $R(t)$ for small times $t$ and estimate how far these densities are from the limiting Lévy-stable density $s_{\alpha,\beta}(r)$. The estimated rate of convergence is

---

[1] Hugo Steinhaus Center for Stochastic Methods, Technical University of Wrocław, Wyspiańskiego 27, 50-370 Wrocław, Poland; kotulski@im.pwr.wroc.pl; http://www.im.pwr.wroc.pl/~kotulski

compared with theory. In order to obtain the reliable approximation of the densities for fixed times $t$, large samples of $10^8$ elements are generated.

## 2 Description of the model

Let us consider a Continuous-Time Random Walk described by means of two sequences of random variables $R_i$ and $T_i$, jump distances and waiting-time intervals, respectively, with a particle initially at the origin. For simplicity we restrict our attention to one-dimensional walks. The first instantaneous jump of random length $R_1$ takes place after a random waiting-time $T_1$, then the second instantaneous jump $R_2$ after time $T_2$, etc. In general, the $i$-th jump $R_i$ may depend on its waiting-time $T_i$ in an arbitrary way, but the pair $(R_i, T_i)$ is independent of the preceding and succeeding pairs of jumps and its waiting-times $(R_k, T_k)$. The special case when $R_i$ is independent of $T_i$ is called a decoupled memory CTRW as opposed to the coupled one with $R_i$ depending on $T_i$. It is evident that the model is entirely determined by $\psi(r, t)$, a two-dimensional joint probability density of the pair $(R_i, T_i)$, the same for each $i \geq 1$. The marginal densities $g(t)$ of $T_i$ and $f(r)$ of $R_i$ are

$$g(t) = \int_{-\infty}^{\infty} \psi(r, t) dr, \qquad f(r) = \int_{0}^{\infty} \psi(r, t) dt . \qquad (1)$$

We define a random variable $N_t$ as the number of jumps in the time interval $[0, t]$. Then, the position $R(t)$ of a walking particle at time $t$ is equal to a random sum of $N_t$ successive random jumps $R_i$, that is

$$R(t) = \sum_{i=0}^{N_t} R_i \quad \text{where} \quad N_t = \max\{k : \sum_{i=0}^{k} T_i \leq t\}, \ R_0 = 0, \ T_0 = 0. \ (2)$$

We search for the limiting distribution of position $R(t)$ for CTRW which, as we will see, depends on the tails of $T_i$ and $R_i$. Let us define parameters $\alpha$ and $\beta$ as follows. If $\langle R_i^2 \rangle < \infty$ we fix $\alpha = 2$ and $\beta = 0$, while if $\langle R_i^2 \rangle$ is infinite, the parameters $\alpha$ and $\beta$ describe the tails of the density $f(r)$ of the jump distances $R_i$,

$$f(-r) + f(r) \sim b r^{-\alpha - 1} \qquad \text{and} \qquad \frac{f(r)}{f(-r) + f(r)} \to \frac{1 + \beta}{2} \quad (3)$$

for $r \to \infty$, where $b$ is a constant, $0 < \alpha < 2$, $-1 \leq \beta \leq 1$.

We first consider the sum of non-random number of jumps $R_i$, i.e., variable $N_t$ is deterministic, $N_t = \lfloor t \rfloor$ and $T_i \equiv 1$. It is called Lévy Flight in physics. The Central Limit Theorem for *non-random sum* of jumps $R_i$ which are defined by Eqs. (1) and (3) reads

$$\Pr\left(\frac{\sum_{i=1}^{n} R_i - n\mu}{n^{1/\alpha}c} \leq r\right) \rightarrow S_{\alpha,\beta}(r) \equiv \int_{-\infty}^{r} s_{\alpha,\beta}(x)dx \ , \qquad (4)$$

where $\mu = \langle R_i \rangle$ if $1 < \alpha \leq 2$, $\mu = 0$ if $0 < \alpha < 1$, $c$ is a constant and $s_{\alpha,\beta}(x)$ is the standard Lévy-stable density. The standard Lévy-stable density $s_{\alpha,\beta}(x)$ with parameters $\alpha$, $\beta$ is defined by its characteristic function [3, 4]

$$\int_{-\infty}^{\infty} e^{i\omega x} s_{\alpha,\beta}(x)dx = \exp\{-|\omega|^{\alpha} \exp[-i(\pi/2)\beta K(\alpha)\text{sign}(\omega)]\} \ , \qquad (5)$$

where $K(\alpha) = \alpha - 1 + \text{sign}(1 - \alpha)$, $0 < \alpha \leq 2$, $\alpha \neq 1$ and $-1 \leq \beta \leq 1$. Note that density $s_{2,0}(x)$ is Gaussian and $s_{2,0}(x) = s_{2,\beta}(x)$ for any $\beta$.

We now turn to the study of asymptotic behavior of $R(t)$ for CTRW. Here we make use of Eq. (2), which states that $R(t)$ is a *random sum* of jumps $R_i$.

In the case of finite mean waiting-time $\tau = \langle T_i \rangle < \infty$ we have

$$\lim_{t \to \infty} N_t/t = 1/\tau \qquad \text{with probability 1} \ ,$$

thus the number $N_t$ of jumps up to the moment $t$ approximately equals $t/\tau$. In order to obtain the limiting distribution of $R(t)$ we replace $\sum_{i=1}^{n} R_i$ with $\sum_{i=0}^{N_t} R_i = R(t)$ and $n$ with $t/\tau$ in formula (4). As a result one finds that for $\mu = 0$ the limiting distribution is unchanged, while for $\mu \neq 0$ the limiting distribution may change [2].

For $\tau < \infty$, $\mu = 0$, we have as $t \to \infty$

$$\Pr\left(\frac{R(t)}{(t/\tau)^{1/\alpha}c} \leq r\right) \rightarrow S_{\alpha,\beta}(r) \equiv \int_{-\infty}^{r} s_{\alpha,\beta}(x)dx \ , \qquad (6)$$

where $c$ is a constant.

In this paper we make the following assumptions. Jumps $R_i$ have Mittag-Leffler (called also geometric stable) distribution [5, 6] defined by characteristic function

$$\varphi_{\alpha}(\omega) = \frac{1}{1 + (-i\omega)^{\alpha}}, \qquad 0 < \alpha < 1. \qquad (7)$$

We denote by $h(x,t)$ the probability density of the random variable on the left-hand side of Eqs. (4) or (6). Hence $h(x,t)$ is the *normalized* density of position $R(t)$ for the Lévy Flight or for the CTRW. For Mittag-Leffler distribution of $R_i$ the Local Limit Theorem is valid [7], thus we may express Eqs. (4) and (6) for large $t$ as

$$h(x,t) \approx s_{\alpha,\beta}(x). \qquad (8)$$

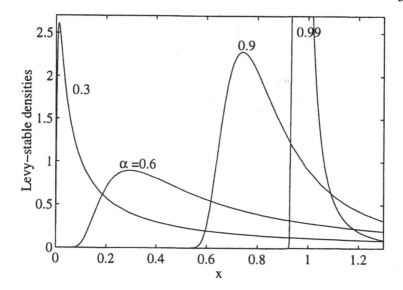

FIGURE 1. Lévy-stable densities $s_{\alpha,1}(x)$ for $\alpha = 0.3$, $0.6$, $0.9$, and $0.99$.

Note, that Mittag-Leffler random variable $R_i$ which is positive, has the L-shaped density and has the long tail,

$$\Pr(R_i > x) \sim \Gamma(1 - \alpha)^{-1} x^{-\alpha}.$$

It means that $R_i$ belongs to the domain of attraction of Lévy-stable law $s_{\alpha,1}(x)$. The general form of Lévy-stable characteristic function, defined by Eq. (5), can be simplified for $0 < \alpha < 1$, $\beta = 1$ and one obtains that the characteristic function of $s_{\alpha,1}(x)$ equals $\exp[-(-i\omega)^{\alpha}]$. Densities $s_{\alpha,1}(x)$ are plotted in Fig. 1. There is no closed form of $s_{\alpha,1}(x)$ except for $\alpha = 0.5$, that is $s_{0.5,1}(x) = (4\pi)^{-1/2} x^{-3/2} \exp[-1/(4x)]$.

The rate of convergence is difficult to find in general case, but for Mittag-Leffler distribution of jumps $R_i$ we have the following theorem which follows from Theorem 1.2 of [8] (see also [9, 10]). The behavior of characteristic functions, for $\omega \to 0$,

$$\varphi_\alpha(\omega) - \exp[-(-i\omega)^{\alpha}] = O(\omega^{2\alpha}), \qquad 0 < \alpha < 1,$$

leads to the rate of convergence $O(n^{-1})$,

$$\sup_{0 < x < \infty} |F_n(x) - S_{\alpha,1}(x)| = O(n^{-1}), \quad \text{as } n \to \infty,$$

where we denoted by $F_n(x)$ the distribution function of random variable $n^{-1/\alpha} \sum_{i=1}^{n} R_i$, which is the normalized sum of jumps $R_i$.

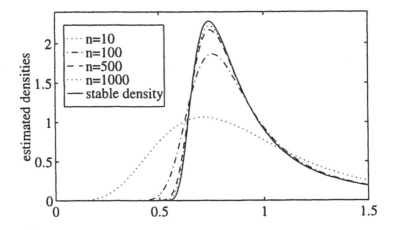

FIGURE 2. Densities $h(x, n)$ of Lévy Flight for various $n$ and $\alpha = 0.9$. The limiting Lévy-stable density $s_{0.9,1}(x)$ is plotted with solid line.

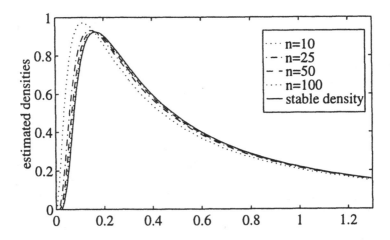

FIGURE 3. The same as in Fig. 2 for $\alpha = 0.5$.

# 3 Simulations

We present the results of Monte Carlo simulations of the model described in Section 2. Jumps $R_i$ have Mittag-Leffler distribution (7). Simulations are carried out for two values of parameter $\alpha$, namely $\alpha = 0.9$ and $\alpha = 0.5$. Random variables $R_i$ are generated along with formula

$$R_i = W^{1/\alpha} S_{\alpha,1}, \tag{9}$$

where $W$ is the exponential random variable with mean 1 and $S_{\alpha,1}$ is Lévy-stable random variable with density $s_{\alpha,1}(x)$. The algorithm for generation of random variable $S_{\alpha,1}$ is given in [3].

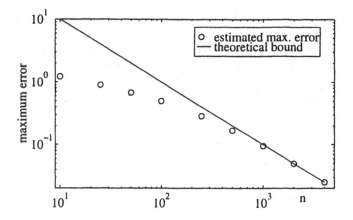

FIGURE 4. Simulated 'maximum error' $\sup_{0<x<\infty} |h(x,n) - s_{\alpha,1}(x)|$ for $\alpha = 0.9$ plotted for the chosen values of $n$ (circles). Solid line represents the theoretical bound.

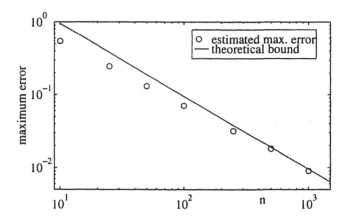

FIGURE 5. The same as in Fig. 4 for $\alpha = 0.5$.

First we estimate normalized densities $h(x,n)$ for Lévy Flight, Eq. (4). For the simulation process we choose times $n$ as follows: $n = 5, 10, 25, 50, 500$. The samples of $10^8$ elements are generated for each $n$. In Figs. 2 and 3 the estimated densities $h(x,n)$ and the limiting densities $s_{0.9,1}(x)$ or $s_{0.5,1}(x)$, (solid lines) are presented together. The Lévy-stable densities $s_{\alpha,1}(x)$ are calculated by numerical procedures.

In Figs. 4 and 5 simulated 'maximum error' $\sup_{0<x<\infty} |h(x,n) - s_{\alpha,1}(x)|$ is plotted as a function of $n$ in double logarithmic scale. Solid lines represent the theoretical bound. The numerical values obtained in the simulations are given in Table 1.

Next we estimate densities $h(x,t)$ for CTRW, Eq. (6). We assume that

TABLE 1. Numerical values of the 'maximum error' obtained in the simulations for $\alpha = 0.9$ and $\alpha = 0.5$.

| $n$ | 10 | 25 | 50 | 10 | 250 | 500 | 1000 | 2000 | 4000 |
|---|---|---|---|---|---|---|---|---|---|
| $\alpha = 0.9$ | 1.2 | .90 | .68 | .49 | .28 | .17 | .094 | .049 | .025 |
| $\alpha = 0.5$ | .54 | .24 | .13 | .070 | .031 | .018 | .0090 | | |

the jumps $R_i$ have Mittag-Leffler distribution with parameter $\alpha = 0.9$ and $\alpha = 0.5$ (see (7)), the waiting times $T_i$ are independent of jumps $R_i$ (decoupled case) and $T_i$ have exponential distribution with mean 1. Hence, the random index $N_t$ in (2) has Poisson distribution with mean $t$.

The results for CTRW are given in Figs. 6 and 7 for $t = 5, 10, 25$. The densities $h(x, t)$ for $t > 50$ are very close to the corresponding Lévy Flight densities. Only density $h(x, 5)$ in Fig. 7 departs significantly. It has a sharp peak at $x = 0$ because of positive probability $e^{-5}$ that no jump occurs up to time $t = 5$. This probability tends, as $t \to \infty$, exponentially to 0, so for larger times $t$ the peak disappears.

*Acknowledgments.*

I would like to thank the organizers of MC&QMC'96 in Salzburg, Austria, who partially supported my participation in the conference. The simulations were performed at Institute of Mathematics, Technical University of Wrocław and at Wrocław Centre of Networking and Supercomputing.

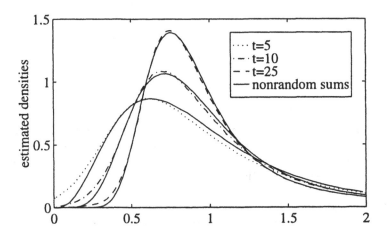

FIGURE 6. Densities $h(x, t)$ for CTRW with $\alpha = 0.9$ and $t = 5$ (dotted line), $t = 10$ (dashed-dotted line), $t = 25$ (dashed line). The corresponding Lévy Flight (i.e. non-random sum) densities are plotted with solid lines.

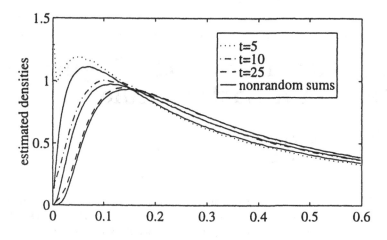

FIGURE 7. The same as in Fig. 6 for $\alpha = 0.5$.

# 4 References

[1] E. W. Montroll and G. H. Weiss, J. Math. Phys. **6**, 167 (1965).

[2] M. Kotulski, J. Stat. Phys. **81**, 777 (1995).

[3] A. Janicki and A. Weron, Simulation and Chaotic Behavior of $\alpha$-Stable Stochastic Processes, (Marcel Dekker, New York, 1994).

[4] V. M. Zolotarev, One-dimensional Stable Distributions, (American Mathematical Society, 1986).

[5] R. N. Pillai, Ann. Inst. Statist. Math. **42**, 157 (1990).

[6] K. Weron and M. Kotulski, Physica A (1996), in press.

[7] B. V. Gnedenko, Wiss. Z. Humboldt-Univ. Berlin, Math.-Nat. Reihe **3**, 287 (1954).

[8] A. Janssen and D. M. Mason, Probab. Th. Rel. Fields **86**, 253 (1990).

[9] V. Paulauskas, Lith. Math. J. **14**, 165 (1974).

[10] W. Macht and W. Wolf, Probab. Th. Rel. Fields **82**, 295 (1989).

# Some Bounds on the Figure of Merit of a Lattice Rule

## T. N. Langtry
## University of Technology, Sydney

ABSTRACT Lattice rules are quasi-Monte Carlo quadrature rules for multiple integrals defined on the $s$-dimensional cube $[0,1)^s$ that generalise the more widely-known method of good lattice points by allowing more than a single generating vector for the set of quadrature points. It is known that any $s$-dimensional lattice rule may be expressed as a multiple sum in terms of at most $s$ generating vectors. The minimal number of generating vectors required in this form for a given rule is called its *rank*. A rank 1 rule which has a generator with at least one component that is a unit is called *simple*. A key problem in the application of lattice rules is the identification of rules of various ranks that perform well with respect to certain standard criteria, one of these being the 'figure of merit' $\rho$. Various strategies have been used in conducting computer searches for rules that perform well with respect to this criterion, and some 'good' (with respect to $\rho$) rules have been found as a result. However, for a variety of reasons, constructive approaches are also of interest. In recent work the author has investigated the connections between the theories of Diophantine approximation and lattice rules. It was shown that bounds on the figure of merit of a rank 1 simple rule with a given generating vector could be expressed in terms of parameters determined by the best simultaneous Diophantine approximations to an $(s-1)$-dimensional projection of the generating vector. These bounds were interpreted as indicating that generators for good rules might be constructed from rational vectors that are poorly approximated, in a particular sense, by rational vectors with lower denominators. In this paper these results are extended to higher-rank rules.

## 1 Introduction

The aim of this work is to generalise to the case of higher-rank lattice rules the bounds, in terms of Diophantine approximations, on the figure of merit of rank 1 simple lattice rules that were established in [5]. Recent surveys of the theory and practice of lattice rules may be found in [6] and [7].

Lattice rules are quasi-Monte Carlo quadrature rules for multiple integrals defined on the $s$-dimensional cube $U^s = [0,1)^s$. It is known [8] that any $s$-dimensional lattice rule may be expressed as a multiple sum in terms of at most $s$ generating vectors of the form $\mathbf{g}_i/n_i$, where $\mathbf{g}_i \in \mathbb{Z}^s$ and $n_i \in \mathbb{Z}$:

$$Q_L(f) = \frac{1}{N} \sum_{j_1=0}^{n_1-1} \cdots \sum_{j_m=0}^{n_m-1} f\left(\left\{\sum_{i=1}^{m} j_i \mathbf{g}_i/n_i\right\}\right), \qquad (1.1)$$

where the braces indicate that addition is modulo $\mathbb{Z}^s$, each quadrature point occurs only once in the sum, $N = \prod_{i=1}^{m} n_i$ is the order of the rule, $1 \leq m \leq s$, and $n_{i+1} \mid n_i$, for $i = 1, \ldots, m - 1$, where $n_m > 1$. The values of $m$ and $n_1, \ldots, n_m$ in (1.1) are uniquely determined for a given lattice rule and are the *rank* and *invariants*, respectively, of the rule. The rank is the minimal number of generating vectors required for a given rule. A rank 1 rule which has a generator with at least one component that is ±1 is called *simple*. The *abscissa set* of a rule is the set of quadrature points. The *integration lattice* corresponding to the rule is the subset of $\mathbb{R}^s$ consisting of all integer linear combinations of the generating vectors for the rule and the standard Cartesian basis vectors. In the sequel, $L$ will denote an $s$-dimensional integration lattice and $Q_L$ the corresponding $s$-dimensional lattice rule. The *dual lattice*, denoted by $L^\perp$, is defined by

$$L^\perp = \{\mathbf{h} \in \mathbb{Z}^s : \mathbf{x} \cdot \mathbf{h} \in \mathbb{Z} \text{ for all } \mathbf{x} \in L\}.$$

The *figure of merit* $\rho(L)$ of an integration lattice is defined by

$$\rho(L) = \min\{r(\mathbf{h}) : \mathbf{h} \in L^\perp - \{\mathbf{0}\}\}, \qquad (1.2)$$

where

$$r(\mathbf{h}) = \prod_{i=1}^{s} \max(1, |h_i|).$$

Our approach will be based on some extensions of standard Diophantine approximation theory that are presented in §2. The main results are then established in §3. and some of their implications discussed in §4.

The results obtained in [5] generalised earlier results obtained in the two-dimensional case in [1] and [9], and rely upon the following definition and result—the interested reader may consult [3] and [4] for more detailed discussions of these.

**Definition 1 (Lagarias)** *Let* $\mathbf{v} \in \mathbb{R}^t$ *and for* $q \in \mathbb{N}$ *define*

$$\beta(q, \mathbf{v}) = \min\{\|q\mathbf{v} - \mathbf{p}\| : \mathbf{p} \in \mathbb{Z}^t\}, \qquad (1.3)$$

*where* $\|\cdot\|$ *is a given norm on* $\mathbb{R}^t$. *Where there is no ambiguity we shall usually suppress the second argument of* $\beta$. *The* best simultaneous approximation denominators (BSADs) $q_k$, *where* $k \in \mathbb{N}$, *for* $\mathbf{v}$ *with respect to* $\|\cdot\|$ *are defined by*

$$q_1 = 1, \quad q_k = \min\{q \in \mathbb{N} : q > q_{k-1}, \beta(q) < \beta(q_{k-1})\}.$$

*The* best simultaneous approximations (BSAs) *of* $\mathbf{v}$ *are the vectors* $\mathbf{v}_k = \mathbf{p}_k/q_k \in \mathbb{Q}^t$ *for which* $\mathbf{p}_k = \left(p_1^{(k)}, \ldots, p_t^{(k)}\right)$ *achieves the minimum* $\beta(q_k)$ *in* (1.3).

**Theorem 2 (Cassels [3, Ch. V, Theorem II])** *Let*

$$R_j(\mathbf{x}) = \sum_{i=1}^{m} \theta_{ji} x_i, \quad S_i(\mathbf{u}) = \sum_{j=1}^{n} \theta_{ji} u_j,$$

*for* $1 \le i \le m$ *and* $1 \le j \le n$, *where* $\theta_{ji} \in \mathbb{R}$. *Suppose that there is an integer vector* $\mathbf{x} \neq \mathbf{0}$ *in* $\mathbb{R}^m$ *and constants* $C$ *and* $X$ *such that* $0 < C < 1 \le X$, $\|\mathbf{x}\|_\infty \le X$ *and, for each* $j \in \{1, \ldots, n\}$,

$$\min_{z \in \mathbb{Z}} |R_j(\mathbf{x}) - z| \le C.$$

*Then there is an integer vector* $\mathbf{u} \neq \mathbf{0}$ *in* $\mathbb{R}^n$ *such that for each* $i \in \{1, \ldots, m\}$

$$\min_{z \in \mathbb{Z}} |S_i(\mathbf{u}) - z| \le D$$

*and* $\|\mathbf{u}\|_\infty \le U$, *where*

$$D = (\ell - 1) X^{(1-n)/(\ell-1)} C^{n/(\ell-1)},$$
$$U = (\ell - 1) X^{m/(\ell-1)} C^{(1-m)/(\ell-1)},$$
$$\ell = m + n.$$

Using Theorem 2 it was shown in [5] that for a rank 1 lattice rule $Q_L$ with generator $\mathbf{g}/N = (1, g_2, \ldots, g_s)/N$, the figure of merit $\rho = \rho(L(\mathbf{g}/N))$ defined by (1.2) satisfies

$$(s-1)^{-(s-1)} \max_{i \in \{1,\ldots,k-1\}} \left( \beta(q_i)^{s-2} \min(N\beta(q_i), q_i) \right)$$
$$\le \rho \le \min_{i \in \{1,\ldots,k-1\}} \left( \max \left( (s-1)^{s-1} q_i, (s-1)^s q_i^{1/(s-1)} \beta(q_i) \right) \right), \quad (1.4)$$

where $\beta(q) = \beta(q, \hat{\mathbf{g}}/N)$ and $\hat{\mathbf{g}} = (g_2, \ldots, g_s)$, and where the rational vectors $\mathbf{p}_1/q_1, \ldots, \mathbf{p}_{k-1}/q_{k-1}$ are the successive best approximations of $\hat{\mathbf{g}}/N$, excluding $\hat{\mathbf{g}}/N$ itself, that is, $\beta(q_i) = \|q_i \hat{\mathbf{g}}/N - \mathbf{p}_i\|_\infty$.

## 2  Diophantine approximations via a matrix

In attempting to generalise the bounds in (1.4) to the case of higher-rank lattice rules we must deal with the problem of finding best Diophantine approximations simultaneously, in some sense, to more than a single vector.

An examination of Theorem 2 suggests that we require a notion of approximation that indicates the proximity to an integer vector of the value of a linear transformation applied at a point with integral coordinates. Intuitively this gives an estimate of how close the transformation comes to having the property of mapping an element of the integer lattice into the integer lattice. Also problematic is the notion of a sequence of 'best approximation denominators', since the obvious measure of error for the notion we are describing would appear to be of the form

$$\beta(\mathbf{q}, V) = \min_{\mathbf{p} \in \mathbb{Z}^{t_1}} \|V\mathbf{q} - \mathbf{p}\|_\infty,$$

where $V$ is a $t_1 \times t_2$ matrix and $\mathbf{q} \in \mathbb{Z}^{t_2}$. The integer vector $\mathbf{q}$ in this expression that corresponds to the approximation denominators of Definition 1 can, however, be mapped naturally into $\mathbb{Z}^+$ via the $\ell_\infty$ norm, and more generally into $\mathbb{R}^+$ via other norms. These considerations motivate the following definition.

**Remark 3** *In the sequel, $M_{t_1, t_2}(\mathbb{R})$ will denote the set of $t_1 \times t_2$ matrices over $\mathbb{R}$.*

**Definition 4** *Let $V \in M_{t_1, t_2}(\mathbb{R})$. For $\mathbf{q} \in \mathbb{Z}^{t_2}$, define*

$$\beta(\mathbf{q}, V) := \min_{\mathbf{p} \in \mathbb{Z}^{t_1}} \|V\mathbf{q} - \mathbf{p}\|, \tag{1.5}$$

*where $\| \cdot \|$ is a given norm on $\mathbb{R}^{t_1}$ such that $\min\{\|\mathbf{z}\| : \mathbf{z} \in \mathbb{Z}^{t_1} - \{0\}\} = 1$. Such a norm will be called a* scaled norm. *We will use $\| \cdot \|$ to denote also the analogous norm on $\mathbb{R}^{t_2}$.*

(a) *The vector $\mathbf{q} \in \mathbb{Z}^{t_2}$ is a* best simultaneous approximation vector *(BSAV) for $V$ with respect to $\| \cdot \|$ if for $\mathbf{q}' \in \mathbb{Z}^{t_2}$,*

    (i)   *if $\|\mathbf{q}'\| < \|\mathbf{q}\|$ then $\beta(\mathbf{q}', V) > \beta(\mathbf{q}, V)$, and*

    (ii)  *if $\|\mathbf{q}'\| = \|\mathbf{q}\|$ then $\beta(\mathbf{q}', V) \geq \beta(\mathbf{q}, V)$.*

(b) *The* best simultaneous approximation pairs *(BSAPs) of $V$ with respect to $\| \cdot \|$ are the pairs $(\mathbf{p}, \mathbf{q}) \in \mathbb{Z}^{t_1} \times \mathbb{Z}^{t_2}$ such that*

    (i)   *$\mathbf{q}$ is a BSAV for $V$ with respect to $\| \cdot \|$, and*

    (ii)  *$\|V\mathbf{q} - \mathbf{p}\| = \beta(\mathbf{q}, V)$.*

(c) *The* best simultaneous approximation numbers *(BSANs) of $V$ with respect to $\| \cdot \|$ are defined by*

    (i)   *$Q_1 = 1$,*

    (ii)  *$Q_k = \min\{\|\mathbf{q}\| : \|\mathbf{q}\| > Q_{k-1}$ and $\mathbf{q}$ is a BSAV for $V$ with respect to $\| \cdot \|\}$, for $k > 1$.*

*We shall say that the BSAV $\mathbf{q}$ and the BSAP $(\mathbf{p}, \mathbf{q})$ are associated with the BSAN $Q$ if and only if $\|\mathbf{q}\| = Q$.*

The following proposition is an immediate consequence of the definition. It asserts that different BSAVs that produce the same minimal error in (1.5) are equivalent in the sense that they are associated with the same BSAN. In this sense, the BSANs of $V$ and their associated errors may be thought of as characterising the approximation properties of $V$ in which we are interested. The proof is straightforward and is omitted.

**Lemma 5** *Let* $V \in M_{t_1,t_2}(\mathbb{R})$ *and let* $\mathbf{q}$, $\mathbf{q}'$ *be BSAVs with respect to the scaled norm* $\|\cdot\|$ *for* $V$. *Then* $\|\mathbf{q}\| = \|\mathbf{q}'\| = Q$, *say, if and only if* $\beta(\mathbf{q}, V) = \beta(\mathbf{q}', V)$.

**Definition 6** *Let* $V \in M_{t_1,t_2}(\mathbb{R})$ *and let* $Q > 0 \in \mathbb{N}$ *be a BSAN of* $V$ *with respect to the scaled norm* $\|\cdot\|$. *The* best simultaneous approximation error (BSAE) *with respect to* $\|\cdot\|$ *at* $Q$ *for* $V$, *denoted by* $\beta(Q, V)$, *is*

$$\beta(Q, V) = \|V\mathbf{q} - \mathbf{p}\|, \tag{1.6}$$

*where* $(\mathbf{p}, \mathbf{q})$ *is a BSAP associated with* $Q$.

Once again, where there is no ambiguity we shall usually suppress the second argument of $\beta$. In this paper we shall be concerned with best approximations with respect to the $\ell_\infty$ norm. In extending the bounds in (1.4) to higher-rank lattice rules we shall require a bound on BSANs. Such a bound may be obtained as a corollary of the following result.

**Theorem 7** (Cassels [3, Ch. I, Theorem VI]) *Let* $L_j(\mathbf{x}) = \sum_{i=1}^m \theta_{ji} x_i$ *for* $j \in \{1, \ldots, n\}$ *be* $n$ *linear forms in* $m$ *variables and let* $V = (\theta_{ji})$. *To every real* $X > 1$ *there is an integral* $\mathbf{x} \neq 0$ *such that* $\|\mathbf{x}\|_\infty \leq X$ *and*

$$\min_{\mathbf{p} \in \mathbb{Z}^n} \|V\mathbf{x} - \mathbf{p}\|_\infty < X^{-m/n}.$$

**Corollary 8** *Let* $V \in M_{t_1,t_2}(\mathbb{R})$ *and let* $Q > 1$ *be a BSAN with respect to* $\|\cdot\|_\infty$ *of* $V$. *Then* $\beta(Q, V) < Q^{-t_2/t_1}$.

*Proof.* Since $Q > 1$ is a BSAN of $V$ with respect to $\|\cdot\|_\infty$ there is an associated BSAP $(\mathbf{p}, \mathbf{q})$ such that $Q = \|\mathbf{q}\|_\infty > 1$. Letting $(\theta_{ji}) = V$ and $X = Q$ in Theorem 7 we then have, for some integral vector $\mathbf{x} \neq 0$,

$$\min_{\mathbf{p}' \in \mathbb{Z}^{t_1}} \|V\mathbf{x} - \mathbf{p}'\|_\infty < Q^{-t_2/t_1},$$

with $\|\mathbf{x}\|_\infty \leq Q$. But since $Q$ is a BSAN and $Q = \|\mathbf{q}\|_\infty \geq \|\mathbf{x}\|_\infty$, we must have

$$\beta(Q, V) = \|V\mathbf{q} - \mathbf{p}\|_\infty \leq \min_{\mathbf{p}'} \|V\mathbf{x} - \mathbf{p}'\|_\infty < Q^{-t_2/t_1}. \qquad \square$$

# 3   Higher-rank lattice rules

In this section we generalise the bounds on the figure of merit from the case of rank 1 rules to that of higher-rank rules. In doing so we shall restrict our attention to rules that are properly $s$-dimensional in the sense that, intuitively, such a rule should not coincide with any of its lower-dimensional projections.

**Definition 9** *Let $x \in \mathbb{R}^s$ and for $j \in \{1, \ldots, s\}$ define*

$$\overline{\Pi}_j(\mathbf{x}) := (x_1, \ldots, x_{j-1}, 0, x_{j+1}, \ldots, x_s).$$

*Let $L$ be an integration lattice in $s$ dimensions and define*

$$L_j := \{\overline{\Pi}_j(\mathbf{x}) : \mathbf{x} \in L\} \cup \mathbb{Z}^s.$$

*The lattice rule $Q_L$ is called* proper *if for each $j \in \{1, \ldots, s\}$ we have $L \cap U^s \neq L_j \cap U^s$.*

We observe that if $L$ is not proper then, with some abuse of notation, $L \cap U^s = \overline{\Pi}_j(L \cap U^s)$, for some $j \in \{1, \ldots, s\}$. Hence, $\mathbf{h} = (0, \ldots, 0, 1, 0, \ldots, 0)$, with the unit in the $j$'th position, satisfies $\mathbf{x} \cdot \mathbf{h} \in \mathbb{Z}$ for all $\mathbf{x} \in L$, and thus $\rho(L) = r(\mathbf{h}) = 1$. Since we seek integration lattices that have as large a figure of merit as possible, we omit such lattices from further consideration in this paper.

**Lemma 10** *Let $Q_L$ be a proper $s$-dimensional lattice rule of order $N$ with abscissa set $L \cap U^s$ generated by $\{\mathbf{g}_1/N, \ldots, \mathbf{g}_t/N\}$, modulo the integer lattice $\mathbb{Z}^s$. Then, for each $e \in \{1, \ldots, s\}$ we have*

$$\max_{j \in \{1, \ldots, t\}} |g_{e,j}| > 0,$$

*where $g_{e,j}$ is the $e$'th component of $\mathbf{g}_j$.*

The proof of this result is straightforward, but it nevertheless allows us to simplify the statement of the main results, which follow. In the remainder of this section and in the following section we shall denote by $G/N$ an $s \times t$ rational matrix with column vectors $\mathbf{g}_1/N, \ldots, \mathbf{g}_t/N$ such that $\mathbf{g}_j \in \mathbb{Z}^s$ and $N \in \mathbb{Z}$. We shall denote by $g_{e,j}$ the element in the $e$'th row and $j$'th column of the matrix $G$.

**Lemma 11** *Let $Q_L$ be a proper $s$-dimensional lattice rule of order $N$ generated by the vectors $\mathbf{g}_1/N, \ldots, \mathbf{g}_t/N$, where $N, g_{i,j} \in \mathbb{Z}$ for $1 \leq i \leq s$ and $1 \leq j \leq t$. Let $G/N$ denote the $s \times t$ matrix whose $j$'th column is $\mathbf{g}_j/N$. For each $e \in \{1, \ldots, s\}$ define the $(s-1) \times t$ matrix $\widehat{G}^{(e)} = (g_{i,j})$, where $i \in \{1, \ldots, e-1, e+1, \ldots, s\}$ and $j \in \{1, \ldots, t\}$. Let $K_e$ denote the number of BSANs of $\widehat{G}^{(e)}/N$ with respect to $\|\cdot\|_\infty$ and let $Q_{e,1}, \ldots, Q_{e,K_e}$ denote* *these BSANs in ascending order, and let* $\beta\left(Q_{e,k}, \widehat{G}^{(e)}/N\right) = \beta(Q_{e,k})$. *Define*

$$b_{e,k}(G/N) := \tau^{-\tau} Q_{e,k}^{t-1} \beta\left(Q_{e,k}\right)^{s-2} \min\left(\beta\left(Q_{e,k}\right) \min_{\substack{j \in \{1,\ldots,t\} \\ g_{e,j} \neq 0}} \left|\frac{N}{g_{e,j}}\right|, Q_{e,k}\right),$$

$$(1.7)$$

*where* $\tau = s + t - 2$. *If* $\mathbf{h} \in L^{\perp}(G/N) - \{\mathbf{0}\}$ *then* $r(\mathbf{h}) \geq b(G/N)$, *where*

$$b(G/N) = \max_{e \in \{1,\ldots,s\}} \left(\max_{k \in \{1,\ldots,K_e-1\}} (b_{e,k}(G/N))\right). \qquad (1.8)$$

*Proof.* Let $\mathbf{h} \in L^{\perp} - \{\mathbf{0}\}$ and for each $e \in \{1,\ldots,s\}$ let

$$\widehat{\mathbf{h}}^{(e)} = (h_1,\ldots,h_{e-1},h_{e+1},\ldots,h_s)$$

and define $b_e = \max_{k \in \{1,\ldots,K_e-1\}} (b_{e,k}(G/N))$. For each $e \in \{1,\ldots,s\}$ we consider two cases. Firstly we use an argument by contradiction to show that, for $\widehat{\mathbf{h}}^{(e)} \neq \mathbf{0}$, we cannot have both

$$|h_e| < \tau^{-\tau} Q_{e,k}^{t-1} \beta\left(Q_{e,k}\right)^{s-1} \min_{\substack{j \in \{1,\ldots,t\} \\ g_{e,j} \neq 0}} \left|\frac{N}{g_{e,j}}\right|$$

and

$$\left\|\widehat{\mathbf{h}}^{(e)}\right\|_{\infty} < \tau^{-\tau} Q_{e,k}^t \beta\left(Q_{e,k}\right)^{s-2}$$

simultaneously, and hence that $r(\mathbf{h}) \geq b_e$. Secondly we show that if $\mathbf{h} \neq \mathbf{0}$ and $\widehat{\mathbf{h}}^{(e)} = \mathbf{0}$ then $r(\mathbf{h}) > b_e$. For notational convenience, we define

$$\overline{m}_{e,j} := \min_{\substack{j \in \{1,\ldots,t\} \\ g_{e,j} \neq 0}} \left|\frac{N}{g_{e,j}}\right| \qquad (1.9)$$

*Case* (i). Let $e \in \{1,\ldots,s\}$, $\mathbf{h} \in L^{\perp} - \{\mathbf{0}\}$ and $\widehat{\mathbf{h}}^{(e)} \neq \mathbf{0}$. Since $\mathbf{h} \in L^{\perp}$ so $(G^T/N)\mathbf{h} \in \mathbb{Z}^t$ and thus there exists a vector $\boldsymbol{\lambda} \in \mathbb{Z}^t$ such that $(G^T/N)\mathbf{h} = \boldsymbol{\lambda}$. Now

$$\boldsymbol{\lambda} = \frac{G^T}{N}\mathbf{h} = \frac{h_e}{N}\begin{pmatrix} g_{e,1} \\ \vdots \\ g_{e,t} \end{pmatrix} + \frac{\left(\widehat{G}^{(e)}\right)^T}{N}\widehat{\mathbf{h}}^{(e)}. \qquad (1.10)$$

Thus, rearranging (1.10),

$$h_e\left(g_{e,1}/N,\ldots,g_{e,t}/N\right)^T = \boldsymbol{\lambda} - \frac{\left(\widehat{G}^{(e)}\right)^T}{N}\widehat{\mathbf{h}}^{(e)},$$

from which it follows that

$$|h_e| \max_{j \in \{1,\ldots,t\}} \left| \frac{g_{e,j}}{N} \right| = \left\| \lambda - \frac{\left(\widehat{G}^{(e)}\right)^T}{N} \widehat{h}^{(e)} \right\|_\infty . \tag{1.11}$$

Next we assume that there is at least one $k \in \{1, \ldots, K_e - 1\}$ such that

$$\left\| \lambda - \frac{\left(\widehat{G}^{(e)}\right)^T}{N} \widehat{h}^{(e)} \right\|_\infty < \tau^{-\tau} Q_{e,k}^{t-1} \beta(Q_{e,k})^{s-1}$$

and

$$\left\| \widehat{h}^{(e)} \right\|_\infty < \tau^{-\tau} Q_{e,k}^t \beta(Q_{e,k})^{s-2},$$

simultaneously. Now if $\tau^{-\tau} Q_{e,k}^t \beta(Q_{e,k})^{s-2} \leq 1$ then $\left\| \widehat{h}^{(e)} \right\|_\infty = 0$, which contradicts $\widehat{h}^{(e)} \neq 0$. Thus we must have $\tau^{-\tau} Q_{e,k}^t \beta(Q_{e,k})^{s-2} > 1$. Also, by Corollary 8, $\beta(Q_{e,k}) < Q_{e,k}^{-t/(s-1)}$, that is, $\beta(Q_{e,k})^{s-1} < Q_{e,k}^{-t}$, and so

$$\tau^{-\tau} Q_{e,k}^{t-1} \beta(Q_{e,k})^{s-1} < \tau^{-\tau} Q_{e,k}^{t-1} Q_{e,k}^{-t} = \tau^{-\tau} Q_{e,k}^{-1} \leq 1, \tag{1.12}$$

since $s, t \geq 1$ and $Q_{e,k} \geq 1$. Hence, by Theorem 2 with

$$\ell = \tau - 1, \quad n = t, \quad m = s - 1, \quad (\theta_{ji}) = \frac{\left(\widehat{G}^{(e)}\right)^T}{N}, \quad x = \widehat{h}^{(e)},$$
$$C = \tau^{-\tau} Q_{e,k}^{t-1} \beta(Q_{e,k})^{s-1} < 1, \quad 1 \leq X < \tau^{-\tau} Q_{e,k}^t \beta(Q_{e,k})^{s-2},$$

there exists an integer vector $\mathbf{q} \in \mathbb{Z}^t - \{0\}$ such that

$$\beta\left(\mathbf{q}, \frac{\widehat{G}^{(e)}}{N}\right) \leq D$$
$$< \tau \left(\tau^{-\tau} Q_{e,k}^t \beta(Q_{e,k})^{s-2}\right)^{(1-t)/\tau} \left(\tau^{-\tau} Q_{e,k}^{t-1} \beta(Q_{e,k})^{s-1}\right)^{t/\tau}$$
$$= \beta(Q_{e,k}),$$

and

$$\|\mathbf{q}\|_\infty \leq U$$
$$< \tau \left(\tau^{-\tau} Q_{e,k}^t \beta(Q_{e,k})^{s-2}\right)^{(s-1)/\tau} \left(\tau^{-\tau} Q_{e,k}^{t-1} \beta(Q_{e,k})^{s-1}\right)^{(2-s)/\tau}$$
$$= Q_{e,k}.$$

However, this contradicts the statement that $Q_{e,k}$ is a BSAN of $\widehat{G}^{(e)}/N$. Recalling that $\max_{j \in \{1,\ldots,t\}} |g_{e,j}/N| > 0$ by Lemma 10, we therefore have, for each $k \in \{1,\ldots,K_e - 1\}$, either

$$|h_e| \max_{j \in \{1,\ldots,t\}} \left| \frac{g_{e,j}}{N} \right| \geq \tau^{-\tau} Q_{e,k}^{t-1} \beta(Q_{e,k})^{s-1},$$

by (1.11), that is, since $1/\max_{j \in \{1,\ldots,t\}} |g_{e,j}/N| = \overline{m}_{e,j}$,

$$|h_e| \geq \tau^{-\tau} Q_{e,k}^{t-1} \beta(Q_{e,k})^{s-1} \overline{m}_{e,j},$$

or

$$\left\| \widehat{\mathbf{h}}^{(e)} \right\|_\infty \geq \tau^{-\tau} Q_{e,k}^{t} \beta(Q_{e,k})^{s-2},$$

and hence

$$r(\mathbf{h}) \geq \max \left( |h_e|, \left\| \widehat{\mathbf{h}}^{(e)} \right\|_\infty \right)$$

$$\geq \tau^{-\tau} Q_{e,k}^{t-1} \beta(Q_{e,k})^{s-2} \min \left( \beta(Q_{e,k}) \overline{m}_{e,j}, Q_{e,k} \right).$$

*Case* (ii). Now let $\mathbf{h} \neq \mathbf{0}$ and $\widehat{\mathbf{h}}^{(e)} = \mathbf{0}$. Then $h_e \neq 0$ and, since $L$ is proper, $\max_{j \in \{1,\ldots,t\}} |g_{e,j}/N| > 0$. Thus

$$\max_{j \in \{1,\ldots,t\}} |g_j/N \cdot \mathbf{h}| = \max_{j \in \{1,\ldots,t\}} |g_{e,j} h_e/N| = \mu \geq 1$$

for some $\mu \in \mathbb{N}$ and so $|h_e| = |\mu| \overline{m}_{e,j} \geq \overline{m}_{e,j}$. Thus

$$r(\mathbf{h}) = |h_e| \geq \overline{m}_{e,j}.$$

Now by Corollary 8, as we saw in (1.12), $\tau^{-\tau} Q_{e,k}^{t-1} \beta(Q_{e,k})^{s-1} < 1$ for each $k \in \{1,\ldots,K_e - 1\}$ and so

$$r(\mathbf{h}) \geq \overline{m}_{e,j} > \max_{k \in \{1,\ldots,K_e-1\}} \left( \tau^{-\tau} Q_{e,k}^{t-1} \beta(Q_{e,k})^{s-1} \overline{m}_{e,j} \right) \geq b_e.$$

Combining both cases we have $r(\mathbf{h}) \geq b_e$ for all $\mathbf{h} \in L^\perp - \{\mathbf{0}\}$, for each $e \in \{1,\ldots,s\}$, as required. $\square$

We may also determine an upper bound on $\rho$.

**Lemma 12** *With the notation of Lemma 11, define*

$$B_{e,k}(G/N) := \tau^{s-1} Q_{e,k}^{-1+st/\tau} \beta(Q_{e,k})^{-(s-1)(t-1)/\tau} \times$$

$$\max \left( Q_{e,k}^{(s-2)/\tau}, \tau \beta(Q_{e,k})^{(s-1)/\tau} \min_{\substack{j \in \{1,\ldots,t\} \\ g_{e,j} \neq 0}} \left| \frac{N}{g_{e,j}} \right| \right), \quad (1.13)$$

*where $\tau = s + t - 2$. Under the conditions of Lemma 11 there exists $\mathbf{h} \in L^\perp(G/N) - \{\mathbf{0}\}$ such that $r(\mathbf{h}) < B(G/N)$, where $B$ is defined by*

$$B(G/N) = \min_{e \in \{1,\ldots,s\}} \left( \min_{k \in \{1,\ldots,K_e-1\}} (B_{e,k}(G/N)) \right). \quad (1.14)$$

*Proof.* Define

$$B_e\,(G/N) = \min_{k\in\{1,\dots,K_e-1\}}(B_{e,k}(G/N))\,.$$

For each $e \in \{1,\dots,s\}$ we have $1 > \beta(Q_{e,k}) = \left\|\left(\widehat{G}^{(e)}/N\right)\mathbf{q}_{e,k} - \mathbf{p}_k\right\|_\infty$, where $\mathbf{q}_{e,k}$ is a BSAV associated with $Q_{e,k}$, for each $k \in \{1,\dots,K_e-1\}$. Consequently, for each $k$,

$$\min_{z\in\mathbb{Z}}\left|\left(\frac{\widehat{G}^{(e)}}{N}\mathbf{q}_{e,k}\right)_j - z\right| \le \beta(Q_{e,k}) < 1$$

for each $j \in \{1,\dots,s-1\}$, and applying Theorem 2 with

$$m = t,\quad n = s-1,\quad (\theta_{ij}) = \frac{\widehat{G}^{(e)}}{N},\quad \mathbf{x} = \mathbf{q}_{e,k},$$
$$C = \beta(Q_{e,k}) < 1,\quad 1 \le X = Q_{e,k},$$

it follows that there exists an integer vector $\widehat{\mathbf{h}}^{(e)} \ne \mathbf{0}$ in $\mathbb{Z}^{s-1}$ such that

$$\min_{z\in\mathbb{Z}^t}\left\|\frac{\left(\widehat{G}^{(e)}\right)^T}{N}\widehat{\mathbf{h}}^{(e)} - \mathbf{z}\right\|_\infty \le D, \tag{1.15}$$

and $\left\|\widehat{\mathbf{h}}^{(e)}\right\|_\infty \le U$, where

$$D = \tau Q_{e,k}^{(2-s)/\tau}\beta(Q_{e,k})^{(s-1)/\tau},$$
$$U = \tau Q_{e,k}^{t/\tau}\beta(Q_{e,k})^{(1-t)/\tau}.$$

Hence

$$r(\widehat{\mathbf{h}}^{(e)}) \le \left\|\widehat{\mathbf{h}}^{(e)}\right\|_\infty^{s-1} \le \tau^{s-1}Q_{e,k}^{(s-1)t/\tau}\beta(Q_{e,k})^{(1-t)(s-1)/\tau}. \tag{1.16}$$

Also, choosing $\mathbf{z} = \lambda \in \mathbb{Z}^t$ as a vector which achieves the minimum in (1.15) and choosing $h_e$ such that

$$\frac{h_e}{N}(g_{1,e},\dots,g_{t,e})^T = \lambda - \frac{\left(\widehat{G}^{(e)}\right)^T}{N}\widehat{\mathbf{h}}^{(e)},$$

we have

$$\left\|\lambda - \frac{\left(\widehat{G}^{(e)}\right)^T}{N}\widehat{\mathbf{h}}^{(e)}\right\|_\infty = |h_e|\max_{j\in\{1,\dots,t\}}\left|\frac{g_{e,j}}{N}\right| \le D.$$

Now $L$ is proper and so $1/\overline{m}_{e,j} = \max|g_{e,j}/N| > 0$ by Lemma 10, where $\overline{m}_{e,j}$ is defined in (1.9), and hence

$$|h_e| = \overline{m}_{e,j}\left\|\lambda - \frac{\left(\widehat{G}^{(e)}\right)^T}{N}\widehat{\mathbf{h}}^{(e)}\right\|_{\infty} \tag{1.17}$$

$$\leq \overline{m}_{e,j}\tau Q_{e,k}^{(2-s)/\tau}\beta\left(Q_{e,k}\right)^{(s-1)/\tau}.$$

Computing the product of the bounds on $r\left(\widehat{\mathbf{h}}^{(e)}\right)$ and $|h_e|$ in (1.16) and (1.17) we obtain

$$r(\mathbf{h}) = r\left(\widehat{\mathbf{h}}^{(e)}\right)\max(1,|h_e|)$$

$$\leq \max\left(\tau^{s-1}Q_{e,k}^{(s-1)t/\tau}\beta\left(Q_{e,k}\right)^{(s-1)(1-t)/\tau},\right.$$

$$\left.\tau^s Q_{e,k}^{(2-s)/\tau+(s-1)t/\tau}\beta\left(Q_{e,k}\right)^{(s-1)/\tau+(s-1)(1-t)/\tau}\overline{m}_{e,j}\right)$$

$$= B_{e,k}(G/N).$$

Since this holds for each $k \in \{1,\ldots,K_e-1\}$ we have $r(\mathbf{h}) \leq B_e(G/N)$, for each $e \in \{1,\ldots,s\}$, and hence $r(\mathbf{h}) \leq B(G/N)$. $\square$

The next result establishes that matrices of the form considered in Lemmas 11 and 12 that differ by an integer matrix have equivalent Diophantine approximation properties. The proof is straightforward and is omitted.

**Lemma 13** *Let $G/N$ be a rational $s \times t$ matrix and let $Z$ be an $s \times t$ integer matrix. Then for $\mathbf{q} \in \mathbb{Z}^t$ and $\|\cdot\|$ a scaled norm on $\mathbb{R}^s$,*

$$\beta\left(\mathbf{q}, \frac{G}{N} + Z\right) = \beta\left(\mathbf{q}, \frac{G}{N}\right).$$

Lemma 13 implies that in Lemmas 11 and 12 we need only consider sets of generators such that each generating vector lies in $U^s$, since any matrix $G/N$ of generators with columns $\mathbf{g}_1/N,\ldots,\mathbf{g}_t/N$ lying outside $U^s$ differs by an integer matrix from one with column vectors all of which are contained in $U^s$. Of course bounds on $\rho(L)$ can be established from these results by considering just a single generator set for the quadrature points—however, the significance of Lemma 13 is that the best bound on $\rho$ that can be established from Lemmas 11 and 12 can be determined in a finite number of operations, since the result implies that we need consider only generator sets whose elements lie in $U^s$, and there are only finitely many of these. The following result is an immediate consequence of Lemmas 11, 12 and 13.

**Theorem 14** *Under the conditions of Lemmas 11 and 12,*

$$\max_{G/N \in \mathcal{G}} b(G/N) \leq \rho(L) \leq \min_{G/N \in \mathcal{G}} B(G/N),$$

*where $\mathcal{G}$ denotes the set of matrices whose column vectors lie in $U^s$ and form a generator set (modulo $\mathbb{Z}^s$) for the abscissa set of $Q_L$, that is, $L \cup U^s$.*

# 4 Relationships with other bounds

When the conditions of Lemmas 11 and 12 are restricted to the case $t = 1$, and when we fix $e = 1$ and consider just a single matrix, say $G/N = (\mathrm{g}/N)$, where $\mathrm{g} = (1, g_2, \ldots, g_s) \in \mathbb{Z}^s$, we recover the bounds in Theorems 3.5 and 3.6 of [5]. When we further restrict consideration to the case $s = 2$, $t = 1$, $\mathrm{g} = (1, g_2)$, where $\gcd(g_2, N) = 1$, we obtain the relations

$$\max_{k \in \{1, \cdots, K-1\}} \left(\min\left(q_k, |q_k g_2 - p_k N|\right)\right)$$
$$\leq \rho \leq \min_{k \in \{1, \ldots, K-1\}} \left(\max\left(q_k, q_k |q_k g_2 - p_k N|\right)\right)$$
$$= \min_{k \in \{1, \ldots, K-1\}} \left(q_k |q_k g_2 - p_k N|\right).$$

The final equality follows from the fact that $p_k/q_k$ is a convergent of the continued fraction for $g_2/N$, and hence $|q_k g_2 - p_k N| \geq 1$. These relations are consistent with the explicit formula for $\rho$ in this case, due to Borosh and Niederreiter [2]:

$$\rho(L) = \min_{k \in \{1, \ldots, K-1\}} \left(q_k |q_k g_2 - p_k N|\right),$$

and the earlier results of [1] and [9].

It should be noted that the lower bound, even in this case, is generally not tight. As demonstrated in the rank 1 case in [5], the lower and upper bounds obtained in Lemmas 11 and 12 may become increasingly loose as the dimension increases. Nevertheless, if $t$, $Q_{e,k}$ and $\min |N/g_{e,j}|$ are held constant in (1.7) and (1.13) and we consider the variations in $b(G/N)$ and $B(G/N)$ that are due to $\beta(Q_{e,k})$, we observe that, as the exponents of $\beta(Q_{e,k})$ in (1.7) are non-negative, $b$ is a non-decreasing function of $\beta(Q_{e,k})$ whereas, for $t \geq 2$, (1.13) indicates that $B$ is a non-increasing function of $\beta(Q_{e,k})$. The behaviour of the lower bound suggests that lattices with good values of $\rho$ may be found amongst those with a set of generators that induce a mapping, represented by the corresponding matrix $G/N$, that approximates an integer mapping poorly, in the sense that $\beta(Q_{e,k})$ is large for each $e$ and each $k$. The behaviour of the upper bound is more equivocal. For $t = 1$ we may draw the same conclusion as from the lower bound, but not for $t \geq 2$. Given the looseness of these bounds for even moderate dimensions, however, it is conceivable that a closer analysis may yield more information. Nevertheless, the present analysis suffices to demonstrate a link between the theory of Diophantine approximation and lattice rules of arbitrary rank.

## 5 REFERENCES

[1] N. S. Bakhvalov. Approximate computation of multiple integrals. *Vestnik Moskov. Univ. Ser. Mat. Meh. Astr. Fiz. Him.*, 4:3–18, 1959.

[2] I. Borosh and H. Niederreiter. Optimal multipliers for pseudo-random number generation by the linear congruential method. *BIT*, 23:65–74, 1983.

[3] J. W. S. Cassels. *An Introduction to Diophantine Approximation*. Cambridge University Press, Cambridge, UK, 1957.

[4] J. C. Lagarias. Some new results in simultaneous diophantine approximation. In P. Ribenboim, editor, *Proceedings of the Queen's University Number Theory Conference 1979*, volume 54 of *Queen's Papers in Pure and Applied Mathematics*, pages 453–474, 1980.

[5] T. N. Langtry. An application of Diophantine approximation to the construction of rank 1 lattice quadrature rules. *Math. Comp.*, 65:1635–1662, 1996.

[6] H. Niederreiter. *Random Number Generation and Quasi-Monte Carlo Methods*. SIAM (Society for Industrial and Applied Mathematics), Philadelphia, Pennsylvania, 1992.

[7] I. H. Sloan and S. Joe. *Lattice Methods for Multiple Integration*. Oxford University Press, Oxford, 1994.

[8] I. H. Sloan and J. N. Lyness. The representation of lattice quadrature rules as multiple sums. *Math. Comp.*, 52:81–94, 1989.

[9] S. K. Zaremba. Good lattice points, discrepancy and numerical integration. *Ann. Mat. Pura Appl.*, 73:293–317, 1966.

Author's address: School of Mathematical Sciences, University of Technology, Sydney, PO Box 123, Broadway NSW 2007, Australia.

# Quasi-Monte Carlo integration of digitally smooth functions by digital nets

## Gerhard Larcher, Gottlieb Pirsic* and Reinhard Wolf, Salzburg

ABSTRACT  In a series of papers by the first author and several co-authors, a "digital lattice rule" for the numerical integration of digitally smooth functions by digital nets was developed and investigated. In this paper we give the general concepts of the method and we prove an error estimate which in some sense summarizes and generalizes the currently known error estimates in this field.

## 1  Introduction

In a series of papers a method for the numerical integration of multivariate functions which are represented by rapidly converging multivariate Walsh series was developed. (See for example [2],[3] and [6].) Such functions may be called "functions which are smooth in a digital sense".

This is motivated by the analogy to multivariate Fourier series: as is well known, functions which are smooth in an ordinary sense are represented by rapidly converging multivariate Fourier series. Further there are various concepts of digital derivations of functions (see [1],[3] and [12]). The smoothness of functions with respect to these derivatives is in strong connection to the convergence speed of different Walsh series representations of the function. See [3] or [13].

In this paper we do not consider these connections. We introduce the method via a very general concept of Walsh series over a finite commutative group. This concept was introduced in [2]. (For a quite similar concept see also [12].)

Further we formulate and prove a generalization of the main error estimate in this theory as it was given in [2], Theorem 1 and - already in a more general form - in [14], Theorem 4.

In Section 2 we give the concept of Walsh functions over a finite commutative group and of rapidly converging Walsh series. In Section 3 we recall

*The first two authors were supported by the Austrian Research Foundation (FWF), Project P 11009 MAT

the notion of digital nets over a finite commutative ring with identity. In Section 4 we formulate and prove the error estimate.

## 2 Digitally smooth functions

Let $G$ be a finite commutative group of order $e \geq 2$, $Z_e := \{0, 1, \ldots, e-1\}$ and

$$\varphi : Z_e \longrightarrow G \text{ with } \varphi(0) = 0$$

be a bijection.

Let $\psi : G \to \widehat{G}$ be any isomorphism between $G$ and its dual group $\widehat{G}$.

For a non-negative integer $n$ with $e$-adic representation $n = \sum_{j=0}^{r} n_j \cdot e^j$, $n_r \neq 0$ if $n \neq 0$, and a real $x \in [0, 1)$ with $e$-adic representation $x = \sum_{j=1}^{\infty} x_j \cdot e^{-j}$ we define

$$\text{wal}_n^{(G,\varphi,\psi)}(x) := \prod_{j=0}^{r} (\psi(\varphi(n_j)))(\varphi(x_{j+1})).$$

For non-negative integers $n_1, \ldots, n_s$ and $\mathbf{x} = (x^{(1)}, \ldots, x^{(s)}) \in [0, 1)^s$ let

$$\text{wal}_{n_1,\ldots,n_s}^{(G,\varphi,\psi)}\left(x^{(1)}, \ldots, x^{(s)}\right) := \prod_{j=1}^{s} \text{wal}_{n_j}^{(G,\varphi,\psi)}\left(x^{(j)}\right).$$

For reals $\alpha > 1$ and $c > 0$ by $_{(G,\varphi,\psi)}E_s^\alpha$ we denote the class of all functions

$$f : [0, 1)^s \longrightarrow \mathbf{C}$$

which can be represented by a Walsh series of the form

$$f\left(x^{(1)}, \ldots, x^{(s)}\right) = \sum_{n_1,\ldots,n_s=0}^{\infty} c(n_1, \ldots, n_s) \cdot \text{wal}_{n_1,\ldots,n_s}^{(G,\varphi,\psi)}\left(x^{(1)}, \ldots, x^{(s)}\right)$$

with complex Walsh-Fourier coefficients satisfying the estimate

$$|c(n_1, \ldots, n_s)| \leq \frac{c}{(\bar{n}_1 \cdots \bar{n}_s)^\alpha}$$

for all $(n_1, \ldots, n_s) \neq (0, \ldots, 0)$. (By $\bar{n}$ we denote $\max(n, 1)$ .)

## 3 Digital $(t, m, s)$-nets

The concept of digital $(t, m, s)$-nets over a finite commutative ring with identity was given by Niederreiter (see [7],[8] or [9].)

**Definition 1** *Let* $b \geq 2, s \geq 1$, *and* $0 \leq t \leq m$ *be integers. Then a point set* $\{x_0, \ldots, x_{N-1}\}$ *consisting of* $N = b^m$ *points of* $[0,1)^s$ *forms a* $(t, m, s)$-*net in base* $b$, *if* $|\{n : 0 \leq n < N : x_n \in J\}| = b^t$ *for every subinterval* $J$ *of* $[0,1)^s$ *of the form*

$$J = \prod_{i=1}^{s} \left[ a_i b^{-d_i}, (a_i + 1) b^{-d_i} \right)$$

*with integers* $d_i \geq 0$ *and* $0 \leq a_i < b^{d_i}$ *for* $1 \leq i \leq s$ *and with* $s$-*dimensional volume* $b^{t-m}$.

**Definition 2** *Let* $b \geq 2, s \geq 1$, *and* $m \geq 1$ *be integers. Let* $R$ *be a finite commutative ring with identity and of order* $b$.

*Let* $\sigma : R \rightarrow Z_b$ *be a bijection with* $\sigma(0) = 0$.

*Let* $C_1, \ldots, C_s$ *with* $C_i = \left( c_{jk}^{(i)} \right)_{j,k=0}^{m-1}$ *be* $s$ $m \times m$ *matrices over* $R$.

*For* $n \in Z_{b^m}$ *let* $n = \sum_{j=0}^{m-1} n_j \cdot b^j$ *be the* $b$-*adic digit representation of* $n$. *We put*

$$x_n^{(i)} = \sum_{j=1}^{m} y_{nj}^{(i)} \cdot b^{-j} \text{ for } 0 \leq n < b^m$$

*and* $1 \leq i \leq s$, *with*

$$y_{nj}^{(i)} = \sigma \left( \sum_{k=0}^{m-1} c_{jk}^{(i)} \cdot \sigma^{-1}(n_k) \right) \in Z_b$$

*for* $n \in Z_{b^m}, 1 \leq i \leq s, 1 \leq j \leq m$.

*If for some integer* $t$ *with* $0 \leq t \leq m$ *the point set* $x_n = \left( x_n^{(1)}, \ldots, x_n^{(s)} \right) \in [0,1)^s$ *for* $n = 0, \ldots, b^m - 1$ *is a* $(t, m, s)$-*net in base* $b$, *then it is called a digital* $(t, m, s)$-*net constructed over* $R$ *with respect to* $\sigma$.

For highly efficient methods for the construction of digital $(t, m, s)$-nets of high quality (i.e. with small $t$), see for example [9],[10],[11] or the paper of Niederreiter and Xing in this volume.

## 4 The error estimate

For the formulation of the result we need one further quantity which in some sense represents a distance between finite commutative groups of "comparable order".

**Definition 3** *Let* $G_1, G_2$ *be finite commutative groups of order* $b$ *and* $e$.

*Let* $\sigma : G_1 \longrightarrow Z_b$ *and* $\varphi : Z_e \longrightarrow G_2$ *be bijections with* $\sigma(0) = 0$ *and* $\varphi(0) = 0$.

*Assume that there exist positive integers $M$ and $L$ with $b^M = e^L =: d$.*
*Define*

$$\tau : G_1^M \longrightarrow G_2^L$$

*in the following way:*

$$\tau : (g_1, \ldots, g_M) \in G_1^M \to \sigma(g_1) + \cdots + \sigma(g_M) b^{M-1} =:$$

$$=: h_1 + \cdots + h_L \cdot e^{L-1} \in Z_d \to (\varphi(h_L), \ldots, \varphi(h_1)) \in G_2^L .$$

*Then*

$$\beta_{G_1,G_2,\varphi,\sigma} := \log_d \left( \max_{\chi \in \widehat{G_1^M}} \sum_{\xi \in \widehat{G_2^L}} \left| \frac{1}{d} \cdot \sum_{x \in G_1^M} \xi(\tau(x)) \cdot \overline{\chi(x)} \right| \right)$$

*(Here by $\log_d$ we denote the logarithm to base $d$.)*

We have

**Theorem 1** *Let $R$ be a finite commutative ring with identity, with additive group $G_1$ and of order $b \geq 2$. Let $G_2$ be a finite commutative group of order $e \geq 2$.*

*Let $L$ and $M$ be positive integers with $b^M = e^L$ and let $\sigma : G_1 \to Z_b$ and $\varphi : Z_e \to G_2$ be bijections with $\sigma(0) = 0$ and $\varphi(0) = 0$. Let $\alpha > 1 + \beta_{G_1,G_2,\varphi,\sigma}$ and $c > 0$ be reals.*

*Let $\mathbf{x}_0, \ldots, \mathbf{x}_{N-1}$ be a digital $(t, mM, s)$-net over $R$ with respect to $\sigma$. Then*

$$\left| \int_{[0,1)^s} f(\mathbf{x}) \, d\mathbf{x} - \frac{1}{N} \sum_{k=0}^{N-1} f(\mathbf{x}_k) \right| \leq K \cdot b^{t\left(\alpha - \beta_{G_1,G_2,\varphi,\sigma}\right)} \cdot \frac{(\log N)^{s-1}}{N^{\alpha - \beta_{G_1,G_2,\varphi,\sigma}}}$$

*for all $f \in {}_{(G_2,\varphi,\psi)}E_s^\alpha$ with an arbitrary isomorphism $\psi : G_2 \to \widehat{G_2}$. ($K$ is a constant depending only on $b, e, c, \alpha$ and $s$.)*

**Remark 1** *We have $0 \leq \beta_{G_1,G_2,\varphi,\sigma} < 1/2$ (see Lemma 1 in [14]) and $\beta_{G_1,G_2,\varphi,\sigma} = 0$ if and only if $\tau : G_1^M \to G_2^L$ as defined in Definition 3 is an isomorphism (see Theorem 1 in [14]).*

**Remark 2** *The result of the Theorem was proved for $b = e$ in [14], Theorem 4 and for $G_1 = G_2$ and $\sigma = \varphi^{-1}$ in [2], Theorem 1.*

**Remark 3** *The quantity $\beta_{G_1,G_2,\varphi,\sigma}$ was explicitly calculated for various examples in [4] and [5].*

**Proof of the theorem**
First step:
let

$$\tilde{\varphi} : Z_{e^L} \to G_2^L : k_0 + \cdots + k_{L-1} e^{L-1} \mapsto (\varphi(k_{L-1}), \ldots, \varphi(k_0)).$$

For $\psi : G_2 \to \widehat{G_2}$ given let

$$\tilde{\psi} : G_2^L \to \widehat{G_2^L} : (g_0, \ldots, g_{L-1}) \mapsto \psi(g_0) \times \ldots \times \psi(g_{L-1})$$

where for $(h_0, \ldots, h_{L-1}) \in G^L$ we define

$$(\psi(g_0) \times \cdots \times \psi(g_{L-1})) ((h_0, \ldots, h_{L-1})) :=$$

$$(\psi(g_0))(h_0) \cdots (\psi(g_{L-1}))(h_{L-1}).$$

Then for all $k \in \mathbf{N}_0$ there exists an unique $\tilde{k} \in \mathbf{N}_0$, such that

$$\mathrm{wal}_k^{(G_2, \varphi, \psi)} = \mathrm{wal}_{\tilde{k}}^{(G_2^L, \tilde{\varphi}, \tilde{\psi})}.$$

This may be checked by the following:
let $k$ have $e$-adic representation

$$k = k_0 + \cdots + k_r e^r =$$

$$= (k_0 + \cdots + k_{L-1} e^{L-1}) + (k_L + \cdots + k_{2L-1} e^{L-1}) \cdot e^L + \cdots$$

$$\cdots + (k_{tL} + \cdots + k_r e^{r-tL} + 0 \cdot e^{r-tL+1} + \cdots + 0 \cdot e^{L-1}) \cdot e^{tL}.$$

Set

$$\tilde{k} = (k_{L-1} + \cdots + k_0 e^{L-1}) + (k_{2L-1} + \cdots + k_L e^{L-1}) \cdot e^L + \cdots$$

$$\cdots + (0 + \cdots + 0 \cdot e^{(t+1)L-r-1} + k_r e^{(t+1)L-r} + \cdots + k_{tL} e^{L-1}) \cdot e^{tL}.$$

Let $x \in [0, 1)$ have $e$-adic representation

$$x = \frac{x_1}{e} + \frac{x_2}{e^2} + \cdots + \frac{x_{r+1}}{e^{r+1}} + \cdots =$$

$$= (e^{L-1} x_1 + \cdots + x_L) \cdot \frac{1}{e^L} + (e^{L-1} x_{L+1} + \cdots + x_{2L}) \cdot \frac{1}{(e^L)^2} + \cdots$$

$$\cdots + \left( e^{L-1} x_{tL+1} + \cdots + e^{(t+1)L-r-1} x_{r+1} + \right.$$

$$\left. + e^{(t+1)L-r-2} x_{r+2} + \cdots + x_{(t+1)L} \right) \cdot \frac{1}{(e^L)^{t+1}} + \cdots.$$

Then (we write $\chi_k$ for $\psi(\varphi(k))$, and note that $\chi_0$ is the trivial character)

$$\operatorname{wal}_k^{(G_2,\varphi,\psi)}(x) = \chi_{k_0}(\varphi(x_1)) \cdots \chi_{k_{L-1}}(\varphi(x_L)) \cdots$$

$$\cdots \chi_{k_{tL}}(\varphi(x_{tL+1})) \cdots \chi_{k_r}(\varphi(x_{r+1})) \cdot \chi_0(\varphi(x_{r+2})) \cdots \chi_0(\varphi(x_{(t+1)L})) =$$

$$= \tilde{\psi}((\varphi(k_0),\ldots,\varphi(k_{L-1})))((\varphi(x_1),\ldots,\varphi(x_L))) \cdots$$

$$\cdots \tilde{\psi}((\varphi(k_{tL}),\ldots,\varphi(k_r),\varphi(0),\ldots,\varphi(0)))$$

$$((\varphi(x_{tL+1}),\ldots,\varphi(x_{r+1}),\varphi(x_{r+2}),\ldots,\varphi(x_{(t+1)L}))) =$$

$$= \tilde{\psi}(\tilde{\varphi}(k_{L-1}+\cdots+k_0 e^{L-1}))(\tilde{\varphi}(x_L+\cdots+x_1 e^{L-1})) \cdots$$

$$\cdots \tilde{\psi}\left(\tilde{\varphi}\left(0+\cdots+0\cdot e^{(t+1)L-r-1}+k_r e^{(t+1)L-r}+\cdots\right.\right.$$

$$\cdots\left.\left. +k_{tL}e^{tL-1}\right)\right)(\tilde{\varphi}(x_{(t+1)L}+\cdots+x_{tL+1}e^{tL-1})) =$$

$$= \operatorname{wal}_{\tilde{k}}^{(G_2^L,\tilde{\varphi},\tilde{\psi})}(x).$$

So $_{(G_2,\varphi,\psi)}E_s^\alpha(c) = {}_{(G_2^L,\tilde{\varphi},\tilde{\psi})}E_s^\alpha(c)$ and it suffices to consider the case $e = b^M$ (i.e. $L = 1$).

Second step:

By following the proof of Theorem 4 in [14] we see that there even the following slightly stronger result is shown:

Let $\Gamma$ be a finite commutative group of order $\gamma$. Let $\sigma' : \Gamma \to Z_\gamma$ be a bijection with $\sigma'(0) = 0$.

For $\mathbf{x} = (x^{(1)},\ldots,x^{(s)}) \in [0,1)^s$ with $x^{(i)} = \sum_{j=1}^\infty \frac{y_j^{(i)}}{\gamma^j}$ and non-negative integers $d_1,\ldots,d_s$ let

$$T_{d_1,\ldots,d_s}(\mathbf{x}) := \left((\sigma')^{-1}\left(y_1^{(1)}\right),\ldots,(\sigma')^{-1}\left(y_{d_1}^{(1)}\right),(\sigma')^{-1}\left(y_1^{(2)}\right),\ldots\right.$$

$$\left.\ldots,(\sigma')^{-1}\left(y_{d_2}^{(2)}\right),\ldots,(\sigma')^{-1}\left(y_1^{(s)}\right),\ldots,(\sigma')^{-1}\left(y_{d_s}^{(s)}\right)\right) \in \Gamma^{d_1+\cdots+d_s}.$$

Let $\Lambda$ be a finite commutative ring with additive group $\Gamma$ and let $\Omega$ be a finite commutative group also of order $\gamma$. Let $\varphi' : Z_\gamma \to \Omega$ be a bijection with $\varphi'(0) = 0$.

Let $\mathbf{x}_0,\ldots,\mathbf{x}_{N-1}$ with $N = \gamma^m$ be a $(t',m,s)$-net in base $\gamma$ with the property that for all non-negative integers $d_1,\ldots,d_s$ the set

$$T := T_{d_1,\ldots,d_s}(\{\mathbf{x}_0,\ldots,\mathbf{x}_{N-1}\})$$

is a subgroup of $\Gamma^{d_1+\cdots+d_s}$ with at least $\gamma^{\min(d_1+\cdots+d_s,m-t')}$ elements and such that for all $z \in T$ we have

$$|\{n : 0 \le n < N, T_{d_1,\ldots,d_s}(\mathbf{x}_n) = z\}| = \frac{\gamma^m}{|T|} \le \gamma^{\max(m-(d_1+\cdots+d_s),t')}.$$

Then for all $c > 0$ and $\alpha > 1 + \beta_{\Gamma,\Omega,\sigma',\varphi'}$ and all $f \in {}_{(\Omega,\varphi',\psi')}E_s^\alpha(c)$
($\psi' : \Omega \to \widehat{\Omega}$ arbitrary) we have

$$\left| \int\limits_{[0,1)^s} f(\mathbf{x})\, d\mathbf{x} - \frac{1}{N} \sum_{k=0}^{N-1} f(\mathbf{x}_k) \right| \le K \cdot \gamma^{t'\left(\alpha - \beta_{\Gamma,\Omega,\sigma',\varphi'}\right)} \cdot \frac{(\log N)^{s-1}}{N^{\alpha - \beta_{\Gamma,\Omega,\sigma',\varphi'}}}$$

($K$ is a constant depending only on $\gamma, s, \alpha$ and $c$.)

Third step:

Let $\mathbf{x}_0, \ldots, \mathbf{x}_{N-1}$ be a $(t, mM, s)$-net in base $b$. Let $(t'-1) \cdot M < t \le t'M$, then, by Lemma 2.9 in [7], $\mathbf{x}_0, \ldots, \mathbf{x}_{N-1}$ is a $(t', m, s)$-net in base $b^M$.

Fourth step:

So altogether it remains to show the following for the digital $(t, mM, s)$-net $\mathbf{x}_0, \ldots, \mathbf{x}_{N-1}$ over $R$ with respect to $\sigma$:

let

$$\tilde{\sigma} : G_1^M \to Z_{b^M} : (g_1, \ldots, g_M) \to \sigma(g_1) + \cdots + \sigma(g_M) \cdot b^{M-1},$$

and let $T^{(1)}_{d_1, \ldots, d_s}$ be defined like $T_{d_1, \ldots, d_s}$ in step two for $\Gamma = G_1^M$ and $\sigma' = \tilde{\sigma}$. Then $T^{(1)} := T^{(1)}_{d_1, \ldots, d_s}(\{\mathbf{x}_0, \ldots, \mathbf{x}_{N-1}\})$ is a subgroup of $G_1^{M(d_1 + \cdots + d_s)}$ with at least $b^{M \cdot \min(d_1 + \cdots + d_s, m - t')}$ elements ($t'$ like in step three) and such that for all $z \in T^{(1)}$ we have

$$\left| \{ n : 0 \le n < N, T_{d_1, \ldots, d_s}(\mathbf{x}_n) = z \} \right| = \frac{b^{mM}}{|T|} \le b^{M \cdot \max\left(m - (d_1 + \cdots + d_s), t'\right)}.$$

This may be checked by the following:

Let $T^{(2)}_{d_1, \ldots, d_s}$ be the function $T_{d_1, \ldots, d_s}$ for $\Gamma = G_1$ and $\sigma' = \sigma$. Then by Lemma 1b in [2] we have

$$T^{(2)} := T^{(2)}_{d_1 M, \ldots, d_s M}(\{\mathbf{x}_0, \ldots, \mathbf{x}_{N-1}\})$$

is a subgroup of $G_1^{d_1 M + \cdots + d_s M} = G_1^{M(d_1 + \cdots + d_s)}$ with at least $b^{\min(M(d_1 + \cdots + d_s), mM - t)}$ elements and such that for all $z \in T^{(2)}$ we have

$$\left| \left\{ n : 0 \le n < N, T^{(2)}_{d_1 M, \ldots, d_s M}(\mathbf{x}_n) = z \right\} \right| = \frac{b^{Mm}}{|T^{(2)}|} \le$$

$$\le b^{\max(Mm - M(d_1 + \cdots + d_s), t)} \le b^{M \cdot \max\left(m - (d_1 + \cdots + d_s), t'\right)}.$$

Obviously we also have $T^{(2)} = T^{(1)}$.

Applying now step two for $\Gamma = G_1^M$, $\Omega = G_2^L$, $\varphi' = \tilde{\varphi}$ from step one and $\sigma' = \tilde{\sigma}$ from above (note that $\tilde{\varphi} \circ \tilde{\sigma} = \tau : G_1^M \to G_2^L$ from Definition 3) yields the result.

$\square$

# 5 References

[1] P.L. Butzer and H.J. Wagner. Walsh–Fourier series and the concept of a derivative. *Applicable Analysis*, **3**:29–46, 1973.

[2] G. Larcher, H. Niederreiter, and W. Ch. Schmid. Digital nets and sequences constructed over finite rings and their application to quasi-Monte Carlo integration. *Monatsh. Math.*, **121**:231–253, 1996.

[3] G. Larcher and W. Ch. Schmid. Multivariate Walsh series, digital nets and quasi-Monte Carlo integration. In H. Niederreiter and P. J.-S. Shiue, editors, *Monte Carlo and Quasi-Monte Carlo Methods in Scientific Computing (Las Vegas, 1994)*, volume 106 of *Lecture Notes in Statistics*, pages 252–262. Springer, New York, 1995.

[4] G. Larcher, W. Ch. Schmid, and R. Wolf. Representation of functions as Walsh series to different bases and an application to the numerical integration of high-dimensional Walsh series. *Math. Comp.*, **63**:701–716, 1994.

[5] G. Larcher, W. Ch. Schmid, and R. Wolf. Quasi-Monte Carlo methods for the numerical integration of multivariate Walsh series. *Math. and Computer Modelling*, **23**(8/9):55–67, 1996.

[6] G. Larcher and C. Traunfellner. On the numerical integration of Walsh series by number-theoretic methods. *Math. Comp.*, **63**:277–291, 1994.

[7] H. Niederreiter. Point sets and sequences with small discrepancy. *Monatsh. Math.*, **104**:273–337, 1987.

[8] H. Niederreiter. Low-discrepancy point sets obtained by digital constructions over finite fields. *Czechoslovak Math. J.*, **42**:143–166, 1992.

[9] H. Niederreiter. *Random Number Generation and Quasi-Monte Carlo Methods*. Number **63** in CBMS–NSF Series in Applied Mathematics. SIAM, Philadelphia, 1992.

[10] H. Niederreiter and C. P. Xing. Low-discrepancy sequences and global function fields with many rational places. *Finite Fields Appl.*, **2**:241–273, 1996.

[11] H. Niederreiter and C. P. Xing. Quasirandom points and global function fields. In S. Cohen and H. Niederreiter, editors, *Finite Fields and Applications (Glasgow, 1995)*, volume 233 of *Lect. Note Series of the London Math. Soc.*, pages 269–296. Camb. Univ. Press, Cambridge, 1996.

[12] C.W. Onneweer. Differentiability for Rademacher series on groups. *Acta Sci. Math.*, **39**:121–128, 1977.

[13] G. R. Pirsic. Schnell konvergierende Walshreihen über Gruppen. Master's thesis, University of Salzburg, 1995.

[14] R. Wolf. A distance measure on finite abelian groups and an application to quasi-Monte Carlo integration. Preprint, University of Salzburg, 1996.

Gerhard Larcher, Gottlieb Pirsic and Reinhard Wolf
Institut für Mathematik
Universität Salzburg
Hellbrunnerstraße 34
A–5020 Salzburg
Austria
e-mail addresses:
GERHARD.LARCHER@SBG.AC.AT
PIRSIC@WST.EDVZ.SBG.AC.AT
REINHARD.WOLF@SBG.AC.AT

# Weak limits for the diaphony

## Hannes Leeb[1]

ABSTRACT In one dimension, we represent the diaphony as the $L_2$-norm of a random process which is found to converge weakly to a second order stationary Gaussian; up to scaling, this implies the asymptotic distributions of the diaphony and the ∗-discrepancy to coincide. Further, we show that properly normalized, the diaphony of $n$ points in dimension $d$ is asymptotically Gaussian if both $n$ and $d$ increase with a certain rate.

## 1 Introduction

For a sequence $\omega = (\omega_i)_{i \in \mathbf{N}}$ in $[0,1]^d$, the diaphony (Zinterhof [Zin76]) of its first $n$ elements is defined as

$$F_n(\omega) = \left( \sum_{k \in \mathbf{Z}^d} \rho_k^2 \left| \frac{1}{n} \sum_{l=1}^{n} e^{2\pi \sqrt{-1} k \cdot \omega_l} \right|^2 \right)^{1/2}, \tag{1}$$

where for $k = (k_1, \ldots, k_d) \in \mathbf{Z}^d \setminus \{0\}$, $\rho_k = \prod_{i=1}^{d} \rho_{k_i}$, $\rho_{k_i} = \max\{1, |k_i|\}^{-1}$, $\rho_{(0,\ldots,0)} = 0$, and $k \cdot \omega_l$ denotes the standard inner product in $\mathbf{R}^d$.

It is known that $\omega$ is equidistributed modulo 1 if and only if its diaphony tends to zero [Zin76] (which implies a strong law: $\lim_{n \to \infty} F_n(\omega) = 0$ for almost all $\omega$). The most prominent measure for equidistribution modulo 1, the ∗-discrepancy [Nie92], is also widely used in statistics as the two-sided Kolmogorov–Smirnov test. While the ∗-discrepancy of $n$ points in $d$ dimensions can, in general, be computed in $O(n^d)$, their diaphony is $O(dn^2)$. This renders it an interesting alternative to the ∗-discrepancy in higher dimensions.

Recently, the diaphony and similar quantities, and their stochastic behaviour have received the attention of physicists [JHK96]. In this paper, we study the asymptotic distribution of the diaphony in two different setups. In Section 2, we show that the diaphony of $n$ points in one dimension is equal to a continuous function of a random element in the space of cadlag functions on $[0,1]$ which converges weakly to a second order stationary Gaussian process. This gives an explicit representation of its asymptotic distribution, which in turn implies that in one dimension, up to a constant, both the diaphony and the ∗-discrepancy have the same asymptotic law.

[1] Research supported by the Austrian Science Foundation (FWF), project no. P11143-MAT

In Section 3, we show that properly normalized, the diaphony of $n$ points in $d$ dimensions converges weakly to a standard Gaussian if both $n$ and $d$ increase with a certain rate. Concluding comments are given in Section 4.

# 2 Weak convergence in dimension one

Let $D$ be the space of functions on $[0, 1]$ which are right-continuous and have left-hand limits and $\mathcal{D}$ the Borel $\sigma$-algebra with respect to the Skorohod topology on $D$. See [Bil68] for the theory of weak convergence of probability measures on $(D, \mathcal{D})$.

Throughout this section, let $d = 1$, let $(u_i)_{i \in \mathbb{N}}$ be a sequence of independent random variables, each uniformly distributed on $[0, 1]$, and let $F_n$ be the diaphony of its first $n$ elements; for $t \in \mathbb{R}$, let $\{t\} := t \bmod 1$,

$$f(t) := \pi - 2\pi\{t\}$$

and

$$g(t) \quad := \quad \int_0^1 f(s)f(s - t)ds \tag{2}$$

$$= \quad \sum_{k=-\infty}^{\infty} \rho_k^2 \exp(2\pi\sqrt{-1}kt) \tag{3}$$

$$= \quad -\frac{\pi^2}{6} + \frac{\pi^2}{2}(1 - 2t)^2 \qquad (0 \le t \le 1); \tag{4}$$

for $t \in [0, 1]$, let

$$X_n(t) := n^{-1/2} \sum_{i=1}^{n} f(t - u_i). \tag{5}$$

Note that $X_n$ is a random element in $(D, \mathcal{D})$.

With these definitions and (3), we can represent $F_n$ as a function of $X_n$.

**Lemma 1**
$$\sqrt{n}F_n = \|X_n\|_2. \tag{6}$$

To derive a weak limit for $X_n$, we need three short lemmata. The first follows from straightforward integration and the $d$-dimensional central limit theorem [Bil86, p.398].

**Lemma 2** *For any $t_1, \ldots, t_d \in [0, 1]$ with $t_i \ne t_j$, $(X_n(t_1), \ldots, X_n(t_d))$ converges weakly to a Gaussian in $R^d$ with expectation vector $(0, \ldots, 0)$ and covariance matrix $(g(t_i - t_j))_{i,j=1}^d$.*

**Lemma 3** *For $0 \le t \le t + h \le 1$ and $\epsilon > 0$,*

$$P\left(|X_n(t) - X_n(t + h)| \ge \epsilon\right) \le 2\epsilon^{-2}(g(0) - g(h)). \tag{7}$$

**Proof:** For $0 \leq t \leq t+h \leq 1$ and $\epsilon > 0$, expansion of terms and integration gives $E\left((X_n(t) - X_n(t+h))^2\right) = 2(g(0) - g(h))$. $\square$

**Lemma 4** *For $0 \leq t_1 \leq t \leq t_2 \leq 1$, $n \geq 1$ and $\epsilon > 0$,*

$$P\left(|X_n(t) - X_n(t_1)| \geq \epsilon, \, |X_n(t_2) - X_n(t)| \geq \epsilon\right) \leq 96\pi^4 \epsilon^{-2} |t_2 - t_1|^2. \quad (8)$$

**Proof:** Following [Bil68, p.106], for $i = 1, \ldots, n$, let $\alpha_i$ be $2\pi(t_1 - t + 1)$ or $2\pi(t_1 - t)$ according as $u_i$ lies in $(t_1, t]$ or not, and let $\beta_i$ be $2\pi(t - t_2 + 1)$ or $2\pi(t - t_2)$ according as $u_i$ lies in $(t, t_2]$ or not. The random vectors $(\alpha_i, \beta_i)$ are independent and take on the values $2\pi(t_1 - t, t - t_2)$, $2\pi(t_1 - t, t - t_2 + 1)$ and $2\pi(t_1 - t + 1, t - t_2)$ with respective probabilities $1 - t_2 + t_1$, $t_2 - t$ and $t - t_1$. Now $E(\alpha_i) = E(\beta_i) = 0$, so we get

$$E\left((X_n(t) - X_n(t_1))^2 (X_n(t_2) - X_n(t))^2\right)$$
$$= n^{-2} E\left(\left(\sum_{i=1}^{n} \alpha_i\right)^2 \left(\sum_{i=1}^{n} \beta_i\right)^2\right)$$
$$\leq 6(2\pi)^4 (t - t_1)(t_2 - t).$$

$\square$

**Theorem 1** *There exists a Gaussian process $X$ on $(D, \mathcal{D})$ with $E(X(t)) = 0$, $E(X(s)X(t)) = g(s - t)$ for $s, t \in [0, 1]$, such that*

$$X_n \xrightarrow{w} X.$$

**Proof:**

Using the Portmanteau Theorem, the existence of $X$ follows from Lemma 2, (7), (8), and Theorem 15.7 from [Bil68].

Again using the Portmanteau Theorem, weak convergence of $X_n$ to $X$ will follow from Lemma 2, (8) and Theorem 15.6 from [Bil68] if we can show that $X$ is almost certainly left-continuous at 1. But this is implied by (7) and equation (15.16) from [Bil68]. $\square$

From this, it follows that $\phi(X_n) \xrightarrow{w} X$ for any function $\phi$ on $D$ which is continuous except on a set $B$ with $P(X \in B) = 0$.

**Lemma 5** *For $1 \leq p \leq \infty$, the mapping $x \mapsto \|x\|_p$ is continuous on $D$.*

**Proof:** Let $x_n \to x$ in the Skorohod topology. Then there exist continuous, strictly increasing functions $\lambda_n$ on $[0, 1]$ with $\lambda_n(0) = 0$, $\lambda_n(1) = 1$, such that $\|\lambda_n - e\|_\infty \to 0$ and $\|x_n \circ \lambda_n - x\|_\infty \to 0$, where $e$ denotes the identity map.

For $p = \infty$, we have

$$\left|\|x_n\|_\infty - \|x\|_\infty\right| = \left|\|x_n \circ \lambda_n\|_\infty - \|x\|_\infty\right| \leq \|x_n \circ \lambda_n - x\|_\infty \to 0.$$

For $1 \leq p < \infty$, we have

$$| \, ||x_n||_p - ||x||_p| \leq ||x_n - x||_p \leq ||x_n - x \circ \lambda_n^{-1}||_p + ||x \circ \lambda_n^{-1} - x||_p.$$

The first integrand on the right-hand side is bounded by $||x_n - x \circ \lambda_n^{-1}||_\infty = ||x_n \circ \lambda_n - x||_\infty$, which goes to zero by choice of the $\lambda_n$. Again by choice of the $\lambda_n$, the second integrand goes to zero at every continuity point of $x$ which is almost everywhere. Since $x \in D$ is bounded, the Theorem of Dominated Convergence completes the proof. $\qquad \Box$

**Theorem 2** *If $W_0$ is a Brownian Bridge, independent of $(u_i)_{i \in N}$, then*

$$\frac{\sqrt{n}}{2} F_n \xrightarrow{w} ||W_0||_\infty.$$

A proof of this is given in [Lee96] and is also obtainable from [JHK96] using the results from [Wat61]; Theorem 1 provides a significantly shorter approach.

**Proof:** With Theorem 1, (6) and Lemma 5, it is sufficient to show that the distributions of $1/2||X||_2$ and $||W_0||_\infty$ coincide.

Let $(N_k)_{k \in N}$ be an independent sequence of standard Gaussian random variables, independent of $(u_i)_{i \in N}$. Following [KS47], a Gaussian process $Y$ with $E(Y(t)) = 0$ and $E(Y(s)Y(t)) = g(s - t)$, whose Fourier-expansion is given by (3), has the representation

$$Y(t) = \sqrt{2} \sum_{k=1}^{\infty} \rho_k (N_{2k-1} cos(2\pi kt) + N_{2k} sin(2\pi kt)).$$

Note that the distributions of $||X||_2$ and $||Y||_2$ coincide. Integration gives

$$||Y||_2^2 = \sum_{k=1}^{\infty} \rho_k^2 (N_{2k-1}^2 + N_{2k}^2), \qquad (9)$$

a series whose independent terms are exponentially distributed with respective parameter $2\rho_k^2$. A comparison of the characteristic function of (9) to that of the quantity $U^2$ considered in [Wat61] and the distribution of $U^2$ derived therein give

$$P(1/4||Y||_2^2 \leq t) = P(||W_0||_\infty^2 \leq t).$$

$\qquad \Box$

The characteristic function of (9), which coincides with that of $||X||_2^2$, provides us with another interesting relation.

**Corollary 1** *If $W_0$, $W_0'$ are independent Brownian Bridges, independent of the limit $X$ from Theorem 1, then, for real $t$,*

$$P(\pi^{-2}||X||_2^2 \leq t) = P(||W_0||_2^2 + ||W_0'||_2^2 \leq t).$$

**Proof:** The asymptotic distribution of the Cramér–von Mises criterion is that of $||W_0||_2^2$ described in [And93]. Comparison of the characteristic function of $\pi^{-2}||X||_2^2$ obtained from (9) to that of $||W_0||_2^2 + ||W_0'||_2^2$ (see [And93]) yields the result. □

## 3   Weak convergence in growing dimension

We are to apply the central limit theorem to a martingale closely related to $F_n$. Elementary properties of martingales and notations as in [Bil86] will be used.

Throughout this section, let $(u_i)_{i \in \mathbf{N}}$ and $g$ as before, and, for $d \in \mathbf{N}$, let $v_{d,i} := (u_{(i-1)d+1}, \ldots, u_{id})$; for reals $(x_i)_{i \in \mathbf{N}}$, let ·

$$h_d(x_1, \ldots, x_d) := \prod_{j=1}^{d}(g(x_i) + 1) - 1;$$

denote the coordinate-wise sum and difference of $d$-tuples by $+$ and $-$ and the convolution on the $d$-dimensional torus by $*$; for $d \in \mathbf{N}$ and $2 \leq k$, let

$$M_k := \sum_{l=1}^{k-1} h_d(v_{d,k} - v_{d,l}).$$

To ease up notation in expressions involving $M_k$, we state the dependence on $d$ explicitly and write $v_k$ for $v_{d,k}$ and $h$ for $h_d$.

**Lemma 6** *Let $d \in \mathbf{N}$, $x$ a real-valued function on $\mathbf{R}^d$ with period 1 in each coordinate, and let $\{k, l\}, \{m\}, \{i_1, \ldots, i_m\} \subset \mathbf{N}$. Then $x(v_{d,k} - v_{d,l})$ is distributed as $x(v_{d,1})$, and $x(v_{d,k} - v_{d,l})$ and $\{v_{d,i_1}, \ldots, v_{d,i_m}\}$ are independent if and only if $\{k, l\} \not\subseteq \{i_1, \ldots, i_m\}$.*

**Proof:** Using conditional expectations, the lemma follows from the translational invariance of the $d$-dimensional Lebesgue-measure. □

**Lemma 7** *If $d \in \mathbf{N}$ and $F_n$ is the diaphony of $v_{d,1}, \ldots, v_{d,n}$, then*

$$n^2 F_n^2 - nh_d(0) = 2 \sum_{k=2}^{n} M_k \qquad (10)$$

*is a martingale with zero expectation.*

**Proof:** Equality (10) is established by Fourier-expansion of $h_d$ which can be derived from (3). The martingale property is implied by Lemma 6 since $E(h_d(v_{d,1})) = 0$. □

**Lemma 8** *For fixed $d \in \mathbf{N}$ and any $k, l \geq 2$, we have*

$$
\begin{aligned}
E(M_k) &= 0, \\
E(M_k^2) &= (k-1)E(h^2), \\
E(M_k^4) &= 3(k-1)(k-2)E(h^2)^2 + (k-1)E(h^4), \qquad and \\
E(M_k^2 \cdot M_l^2) &= (k-1)(l-1)E(h^2)^2 + \\
&\quad 2(l-1)(l-2)E((h*h)^2) \qquad (l < k).
\end{aligned}
$$

**Proof:** We prove the last equality; the derivation of the others is similar. $M_k^2 \cdot M_l^2$ is a sum of terms of the form

$$
\alpha = h(v_k - v_{i_1}) \cdot h(v_k - v_{i_2}) \cdot h(v_l - v_{j_1}) \cdot h(v_l - v_{j_2})
$$

with $1 \leq i_1, i_2 < k$ and $1 \leq j_1, j_2 < l$. With Lemma 6, the expectation of $\alpha$ is non-zero only if either $i_1 = i_2$ and $j_1 = j_2$, or $i_1 \neq i_2$ and $\{i_1, i_2\} = \{j_1, j_2\}$.

If $i_1 = i_2$ and $j_1 = j_2$, Lemma 6 gives $E(\alpha) = E(h^2)^2$. This holds for $(k-1)(l-1)$ terms in $M_k^2 \cdot M_l^2$.

The other case, $i_1 \neq i_2$ and $\{i_1, i_2\} = \{j_1, j_2\}$, occurs $2(l-1)(l-2)$ times in $M_k^2 \cdot M_l^2$. If, say, $i_1 = j_1$ and $i_2 = j_2$, we have

$$
E\left(\alpha \| v_k, v_{i_1}, v_{i_2}\right) = h(v_k - v_{i_1}) \cdot h(v_k - v_{i_2}) \cdot h * h(v_{i_1} - v_{i_2})
$$

since $h(v_k) = h(-v_k)$. Conditioning this with respect to $v_{i_1}$ and $v_{i_2}$ gives $E(\alpha) = E((h*h)^2)$. $\qquad \square$

Let $(d_n)_{n \in \mathbf{N}}$ be a sequence of positive integers, and, for each $n$, let $F_{(n)}$ be the diaphony of $v_{d_n,1}, \ldots, v_{d_n,n}$. For $\mu_{(n)} := E(nF_{(n)}^2)$ and $\sigma_{(n)}^2 := V(nF_{(n)}^2)$, lemmata 6, 7 and 8 give

$$
\begin{aligned}
\mu_{(n)} &= (1 + g(0))^{d_n} - 1 \\
&= (1 + \pi^2/3)^{d_n} - 1, \tag{11}
\end{aligned}
$$

$$
\begin{aligned}
\sigma_{(n)}^2 &= 2\left(\left(1 + \int_0^1 g(x)^2 dx\right)^{d_n} - 1\right)\frac{n-1}{n} \\
&= 2((1 + \pi^4/45)^{d_n} - 1)\frac{n-1}{n}. \tag{12}
\end{aligned}
$$

Note that for the $*$-discrepancy, the first two moments in higher dimensions are not known.

**Theorem 3** *Let $c = \int_0^1 (g(x)+1)^4 dx / (\int_0^1 (g(x)+1)^2 dx)^2$ and $N$ be a real-valued standard Gaussian random variable independent of $(u_i)_{i \in \mathbf{N}}$. If the sequence $(d_n)_{n \in \mathbf{N}}$ is nondecreasing and goes to infinity such that $c^{d_n} = o(n^2)$, then*

$$
\frac{nF_{(n)}^2 - \mu_{(n)}}{\sigma_{(n)}} \xrightarrow{w} N.
$$

**Proof:** For $n \geq 2$ and $2 \leq k \leq n$, let $\mathcal{G}_{n,k} := \sigma(v_{d_n,1}, \ldots, v_{d_n,k})$; for $d = d_n$, let $M_{n,l} := 2(n\sigma_{(n)})^{-1} M_l$ and $S_{n,k} := \sum_{l=2}^{k} M_{n,l}$ (note the dependence of $S_{n,k}$ on the dimension $d_n$). Then $\{S_{n,k}, \mathcal{G}_{n,k} : 2 \leq k \leq n, \ n \geq 2\}$ is a zero-mean square-integrable martingale array with $\mathcal{G}_{n,k} \subseteq \mathcal{G}_{n+1,k}$. Moreover, $S_{n,n} = (nF_{(n)}^2 - \mu_{(n)})/\sigma_{(n)}$.

$S_{n,n}$ converges weakly to $N$ by Theorem 3.2 from [HH80] if

$$E\left(\max_{2 \leq k \leq n} M_{n,k}^2\right) \quad \text{is bounded in } n, \tag{13}$$

$$\max_{2 \leq k \leq n} |M_{n,k}| \xrightarrow{p} 0 \quad \text{and} \tag{14}$$

$$\sum_{k=2}^{n} M_{n,k}^2 \xrightarrow{L_2} 1, \tag{15}$$

where $\xrightarrow{p}$ and $\xrightarrow{L_2}$ respectively denote convergence in probability and in $L_2$.

For condition (13), note that the integrand is bounded by $\sum_{k=2}^{n} M_{n,k}^2$; Lemma 8 and (12) give

$$E\left(\sum_{k=2}^{n} M_{n,k}^2\right) = 1. \tag{16}$$

For condition (14), let $\epsilon > 0$; then

$$P\left(\max_{2 \leq k \leq n} |M_{n,k}| > \epsilon\right) = P\left(\cup_{k=2}^{n}\{|M_{n,k}| > \epsilon\}\right) \leq \epsilon^{-4} \sum_{k=2}^{n} E(M_{n,k}^4).$$

Lemma 8 gives

$$\limsup_n \sum_{k=2}^{n} E(M_{n,k}^4) \leq \limsup_n \frac{2E(h_{d_n}^4)}{n^2 E(h_{d_n}^2)^2}, \tag{17}$$

and integration shows that the right-hand side equals $2\limsup_n c^{d_n}/n^2 = 0$, by choice of $c$ and the sequence $(d_n)_{n \in \mathbf{N}}$.

Finally, for condition (15), note that (16) gives

$$E\left(\left(\sum_{k=2}^{n} M_{n,k}^2 - 1\right)^2\right) = \sum_{k=2}^{n} E(M_{n,k}^4) + 2\sum_{k=3}^{n} \sum_{l=2}^{k-1} E(M_{n,k}^2 \cdot M_{n,l}^2) - 1. \tag{18}$$

The first term on the right-hand side goes to zero with $n$ by (17). With Lemma 8 and Parseval's Theorem, we get a constant $K$, independent of $n$ and $d_n$, such that the limes superior of (18) is bounded by

$$K \limsup_n \frac{E((h_{d_n} * h_{d_n})^2)}{E(h_{d_n}^2)^2} = K \limsup_n \left(\frac{1 + \sum_{k=-\infty}^{\infty} \rho_k^8}{(1 + \sum_{k=-\infty}^{\infty} \rho_k^4)^2}\right)^{d_n},$$

which, of course, goes to zero. $\qquad \square$

# 4 Comments

The diaphony is used as a deterministic quantity in both Monte Carlo and quasi-Monte Carlo [CF93, Hel96, Zin76]. By giving the limiting distributions of three associated random quantities (5), (6) and (10), thereby relating the diaphony to Gaussian processes in Theorem 1, the ∗-discrepancy in Theorem 2, the Cramér–von Mises criterion in Corollary 1, and the normal distribution in Theorem 3, we have contributed a certain probabilistic background.

Our approach in Section 2 is based on a heuristic idea of Doob [Doo49] which was rigidly proven in the famous work of Donsker [Don52], combined with the explicit representation of the limiting process obtained by Kac and Siegert [KS47]. The same technique was also used in one dimension by Anderson and Darling [AD52], Bickel and Breiman [BB83], and Watson [Wat61].

However, the fairly simple structure of the diaphony and the limiting Gaussian process, and its strong relation to harmonic analysis might provide a generalisation of Theorem 1 for higher dimensions: the obvious candidate for the limiting process in dimension $d$ is a second order stationary Gaussian with a covariance function whose Fourier-coefficients are given by $\rho_k^2$ ($k \in \mathbf{Z}^d$).

Concerning the relationship between the limit laws of the diaphony and the ∗-discrepancy found in Theorem 2, we go with Watson [Wat61]: "It is very surprising".

Kleiss [Kle96] points out that the asymptotic distribution of the diaphony, properly scaled, is Gaussian if first the sample size goes to infinity and then the dimension. Theorem 3 shows how this limit can be obtained if dimension and sample size grow simultaneously.

Finally, note that the concept of diaphony is not restricted to the exponential function system: Hellekalek [Hel96] presents a general framework which relates the concept of equidistribution modulo 1 to correlation analysis and the spectral test. Hellekalek and the author give a variant of diaphony based on dyadic Walsh-functions in [HL96]; other variants defined via not necessarily complete function systems, which hence do not necessarily give measures for equidistribution, are studied by James, Hoogland and Kleiss in [JHK96].

*Acknowledgments:* The author would like to thank his thesis' supervisor, Peter Hellekalek, for all the discussions, support which made this present work possible, and Prof. Pierre L'Ecuyer, for valuable suggestions and for the invitations to Montreal where most of this paper emerged. Further credits go to Maximilian Thaler, for invaluable support and discussions on Theorem 3, and to Clemens Amstler, for providing us with the relation in Lemma 1.

338

# 5 References

[AD52]  T.W. Anderson and D.A. Darling. Asymptotic theory of certain "goodness of fit" criteria based on stochastic processes. *Ann. Math. Stat.*, **23**:193–212, 1952.

[And93]  T.W. Anderson. Goodness of fit tests for spectral distributions. *Ann. Statist.*, **21**:830–847, 1993.

[BB83]  P.J. Bickel and L. Breiman. Sums of functions of nearest neighbor distances, moment bounds, limit theorems and a goodness of fit test. *Ann. Probab*, **11**:185–214, 1983.

[Bil68]  P. Billingsley. *Convergence of Probability Measures*. John Wiley & Sons, Inc., New York, 1968.

[Bil86]  P. Billingsley. *Probability and Measure*. John Wiley & Sons, Inc., New York, 2nd edition, 1986.

[CF93]  H. Chaix and H. Faure. Discrepancy and diaphony in dimension one (French). *Acta Arith.*, **63**:103–141, 1993.

[Don52]  M.D. Donsker. Justification and extension of Doob's heuristic approach to the Komogorov-Smirnov theorems. *Ann. Math. Stat.*, **23**:277–281, 1952.

[Doo49]  J.L. Doob. Heuristic approach to the Kolmogorov-Smirnov theorems. *Ann. Math. Stat.*, **20**:393–403, 1949.

[Hel96]  P. Hellekalek. On correlation analysis of pseudorandom numbers. submitted to the Proceedings of the Second International Conference on Monte Carlo and Quasi-Monte Carlo Methods in Scientific Computing, 1996.

[HH80]  P. Hall and C.C. Heyde. *Martingale Limit Theory and its Application*. Probability and Mathematical Statistics. Academic Press, Inc., San Diego, California, 1980.

[HL96]  P. Hellekalek and H. Leeb. Dyadic diaphony. *Acta Arith.*, to appear, 1996.

[JHK96]  F. James, J. Hoogland, and R. Kleiss. Multidimensional sampling for simulation and integration: measures, discrepancies, and quasi-random numbers. to appear in Comp. Phys. Comm., 1996.

[Kle96]  R. Kleiss. private communications, 1996.

[KS47]  M. Kac and A.J.F. Siegert. An explicit representation of a stationary Gaussian process. *Ann. Math. Stat.*, **18**:438–442, 1947.

[Lee96]   H. Leeb. The asymptotic distribution of diaphony in one dimension. G-96-52, GERAD - École des Hautes Études Commerciales, Montréal, 1996.

[Nie92]   H. Niederreiter. *Random Number Generation and Quasi-Monte Carlo Methods.* SIAM, Philadelphia, 1992.

[Wat61]   G.S. Watson. Goodness–of–fit tests on a circle. *Biometrika,* **48**:109–114, 1961.

[Zin76]   P. Zinterhof. Über einige Abschätzungen bei der Approximation von Funktionen mit Gleichverteilungsmethoden. *Sitzungsber. Österr. Akad. Wiss. Math.-Natur. Kl. II,* **185**:121–132, 1976.

**Author's address :**
Hannes Leeb
Institut für Mathematik, Universität Salzburg
Hellbrunner Straße 34, A-5020 Salzburg
Austria

e-mail: `leeb@random.mat.sbg.ac.at`
WWW: `http://random.mat.sbg.ac.at/~leeb/`

# Quasi-Monte Carlo Simulation of Random Walks in Finance

William J. Morokoff     Russel E. Caflisch

ABSTRACT The need to numerically simulate stochastic processes arises in many fields. Frequently this is done by discretizing the process into small time steps and applying pseudo-random sequences to simulate the randomness. This paper address the question of how to use quasi-Monte Carlo methods to improve this simulation. Special techniques must be applied to avoid the problem of high dimensionality which arises when a large number of time steps are required. One such technique, the generalized Brownian bridge, is described here. The method is applied to a classical problem from finance, the valuation of a mortgage backed security portfolio. When expressed as an integral, this problem is nominally 360 dimensional. The analysis of the integrand presented here explains the effectiveness of the quasi-random sequences on this high dimensional problem.

## 1   Introduction

Monte Carlo is often the only effective numerical method for evaluating integrals and other functionals over paths generated by an underlying stochastic process. Applications range from determining levels of radiation emission to pricing financial derivatives. The standard Monte Carlo method using pseudo-random sequences can be quite slow, however, because its convergence rate is only $O(N^{-1/2})$ for $N$ sample paths. Quasi-Monte Carlo methods, using deterministic sequences that are more uniform than random, can be much faster with errors approaching size $O(N^{-1})$ in optimal cases.

This dramatic improvement in convergence rate has the potential for significant gains both in computational time and in range of application of Monte Carlo methods. The present work was spurred by improvements for quantitative finance problems observed in a number of earlier studies

[1, 7, 9]. These papers were all motivated by the results of Paskov [8] on mortgage backed securities.

The effectiveness of quasi-Monte Carlo methods does have some important limitations. First, quasi-Monte Carlo methods are valid for integration problems, but may not be directly applicable to simulations, due to the correlations between the points of a quasi-random sequence. This problem can be overcome in many cases by writing the desired result of a simulation as an integral, but the resulting integral is often of very high dimension (e.g. dimension 360 for a mortgage of length 30 years).

This leads to a second limitation: The improved accuracy of quasi-Monte Carlo methods is generally lost for problems of high dimension or problems in which the integrand is not smooth. This loss of effectiveness has been documented for a series of test problems in [3, 4, 5]. Several researchers in computational finance have recently reported great success with quasi-Monte Carlo computation of problems of very high dimension [1, 7, 9]. One purpose of this paper is to show that, at least for some of these results, the problems are actually of moderate dimension when cast in the proper form.

While quasi-Monte Carlo may be very successful on relatively simple problems, obtaining the same success on more complicated applications can be difficult due to these limitations. In this study we show how to overcome some of these limitations for the mortgage backed security problem by a reformulation of the Monte Carlo representation so that the effective dimension is of moderate size. One of our main conclusions is that the range of application of quasi-Monte Carlo methods can be significantly extended by modification of the standard Monte Carlo techniques.

The outline of this paper is the following: In Section 2 a test case from finance, the valuation of a mortgaged-backed security portfolio, is formulated. Our main technical tool for formulating the problem with reduced effective dimension is the generalized Brownian bridge representation of a random walk, which is described in Section 3. Computational results for the mortgage-backed security problem are presented in Section 4 along with an interpretation of these results in terms of a Taylor expansion of the integrand. Conclusions are discussed in Section 5.

## 2  Mortgage-Backed Securities

Consider a security backed by mortgages of length M months with fixed interest rate $i_0$, which is the current interest rate at the beginning of the mortgage. The present value of the security is then

$$PV = E(v)$$

$$= E(\sum_{k=1}^{M} u_k m_k) \tag{2.1}$$

in which $E$ is the expectation over the random variables involved in the interest rate fluctuations. The variables in the problem are the following:

$u_k$ = discount factor for month $k$

$m_k$ = cash flow for month $k$

$i_k$ = interest rate for month $k$

$w_k$ = fraction of remaining mortgages prepaying in month $k$

$r_k$ = fraction of remaining mortgages at month $k$

$c_k$ = (remaining annuity at month $k$)/$c$

$c$ = monthly payment

$\xi_k$ = an $N(0, \sigma)$ random variable.

This notations follows that of Paskov [8], except that our $c_k$ is denoted $a_{M-k+1}$ by Paskov.

Several of these variables are easily defined:

$$u_k = \prod_{j=0}^{k-1}(1+i_k)^{-1}$$

$$m_k = cr_k((1-w_k) + w_k c_k)$$

$$r_k = \prod_{j=1}^{k-1}(1-w_j)$$

$$c_k = \sum_{j=0}^{M-k}(1+i_0)^{-j}.$$

Following Paskov, we use a model for the interest rate fluctuations and the prepayment rate given by

$$i_k = K_0 e^{\xi_k} i_{k-1}$$

$$= K_0^k e^{\xi_1 + \cdots + \xi_k} i_0 \tag{2.2}$$

$$w_k = K_1 + K_2 \arctan(K_3 i_k + K_4)$$

in which $K_1, K_2, K_3, K_4$ are constants of the model. The constant $K_0 = e^{-\sigma^2/2}$ is chosen to normalize the log-normal distribution, i.e. so that $E(i_k) = i_0$. The initial interest rate $i_0$ is an additional constant that must be specified.

In this study we do not divide the cash flow of the security among a group of tranches, as in [8], but only consider the total cash flow. Nevertheless, the results should be indicative of a more general computation involving a number of tranches.

The expectation $PV$ can be written as in integral over $R^M$ with Gaussian weights

$$g(\xi) = (2\pi\sigma^2)^{-1/2}e^{-\xi^2/2\sigma^2}. \qquad (2.3)$$

This is transformed into an unweighted integral by a mapping $\xi = G(x)$ with $G'(x) = g(\xi)$, which takes a uniformly distributed variable $x$ to an $N(0,\sigma)$ variable $\xi$. The formula for $PV$ is

$$
\begin{aligned}
PV &= \int_{R^M} v(\xi_1,\ldots,\xi_M)g(\xi_1)\ldots g(\xi_1)d\xi_1\ldots d\xi_M \\
&= \int_{[0,1]^M} v(G(\xi_1),\ldots,G(\xi_M))dx_1\ldots dx_M. \qquad (2.4)
\end{aligned}
$$

Note that in quasi-Monte Carlo evaluation of an expectation involving a stochastic process with $M$ time steps, the resulting integral is $M$ dimensional.

In the numerical study below, we consider the parameters

$$(i_0, K_1, K_2, K_3, K_4, \sigma^2) = (.007, .01, -.005, 10, .5, .0004). \qquad (2.5)$$

The interest rate corresponds to a yearly rate of 8.4%. The variance in interest rate increments $\sigma^2$ leads to yearly fluctuations of size .5%. In this examples the length of the loans is taken to be 30 years ($M = 360$). The prepayment rate is nearly linear in the interest rate, in the range of interest; this results in the integrand being a nearly linear function of the Gaussian increment variables $\xi$. As discussed below, this provides the key insight into understanding the success of quasi-Monte Carlo methods for this problem. Finally, the monthly payment $c$, which only acts as a scaling factor, is set to unity.

## 3 The Generalized Brownian Bridge

The interest rate in the example above follows geometric Brownian motion, which by Equation (2.2) may be generated from standard Brownian motion. We now describe a quasi-Monte Carlo method for generating standard Brownian motion paths. Any stochastic process path which can be written as a function of a standard Brownian motion path can be generated through this procedure. This includes geometric Brownian motion, as well as mean reverting processes and processes with time dependent drifts and variances.

Since Brownian motion is a Markov process, it is most natural to generate a discrete time Brownian motion random walk $x_{n+1} = b(t_{n+1})$ at time $t_{n+1} = t_n + \Delta t$ as a random jump from its value $x_n = b(t_n)$ through the formula

$$x_{n+1} = x_n + \sqrt{\Delta t} \, z \tag{3.1}$$

in which $z$ is sampled from $N(0, \sigma)$. More generally, any future point $x_m$, $(m > n)$ may be generated by

$$x_m = x_n + \sqrt{(m-n)\Delta t} \, z \, . \tag{3.2}$$

Any point of the walk in the middle can then be generated from knowledge of the past, $x_n$, and the future $x_m$ according to the *Brownian bridge* formula [2]

$$x_k = (1 - \gamma)x_n + \gamma x_m + \sqrt{\gamma(1-\gamma)(m-n)\Delta t} \, z \tag{3.3}$$

where $n \leq k \leq m$ and

$$\gamma = \frac{k-n}{m-n} \, .$$

We remark that this formula is valid only for generating one step $k$ between steps $n$ and $m$, as any subsequently generated steps must be correlated with $x_k$. Equation (3.3) may continue to be used, however, by simply replacing one of the endpoints with the most recently generated point $x_k$. Note that variance of the random part of the Brownian bridge formula (3.3) for generating $x_k$ is reduced by a factor $1/(1-\gamma)$ compared with the variance for generating $x_k$ with formula (3.2).

The standard method of generating a random walk $x_k$ is based on the updating formula (3.1). The initial value is $x_0 = 0$. Each subsequent value $x_{k+1}$ is generated from the previous value $x_k$ using formula (3.1), with independent normal variables $z_k$. The variance associated with each $z_k$ in the standard method is constant at $\sigma^2$.

Another method, which we refer to as the *Brownian bridge discretization* can be based on (3.3). Suppose we wish to determine the path $x_0, x_1, \ldots, x_D$, and for convenience assume that $D$ is a power of 2. The initial value is $x_0 = 0$. The next value generated is $x_D = \sigma\sqrt{D\Delta t} \, z_1$. Then the value at the mid point $x_{D/2}$ is determined from the Brownian bridge formula (3.3) with $\gamma = 1/2$. Subsequent values are found at the successive mid-points; i.e. $x_{D/4}, x_{3D/4}, x_{D/8}, \ldots$, sweeping along the breadth of the domain at each level of refinement.

Although the total variance associated with each $x_k$ in this representation is the same as in the standard discretization, the variance associated with the $z_k$ is no longer constant. It has been redistributed so that much more of the variance is contained in the first few steps of the Brownian bridge discretization, while the later steps have significantly smaller variance due to the factor of 2 reduction in the variance arising in formula (3.3). This reduces the effective dimension of the random walk simulation, which increases the accuracy of quasi-Monte Carlo. Moskowitz and Caflisch [6] applied this method to the evaluation of Feynman-Kac integrals and showed the error to be substantially reduced when the number of time steps, which is equal to the dimension of the corresponding integral, is large. Since the mortgage-backed securities problem described above depends on a random walk, and can be written as a discretization of a Feynman-Kac integral, we were naturally led to apply this alternate discretization to the problem .

The Brownian bridge approach allows for a great deal of generalization. Another possibility involves a rearrangement of the breadth-first discretization described above in a depth-first fashion, such that the $x_k$ are generated in the following order:

$$x_0, \ x_D, \ x_{D/2}, \ x_{D/4}, \ x_{D/8}, \ \ldots, \ x_1, x_{3D/4}, \ x_{3D/8}, \ \ldots, \ x_3, \ \ldots x_{(D-1)}$$

In fact, formula (3.3) provides the means for generating the steps of the random walk in any order desired. Moreover, the number of terms in the walk $D$, representing the dimension of the problem, need not be a power of two.

To formalize these extensions, we introduce now the *generalized Brownian bridge discretization*. The path of the random walk may be expressed as a vector

$$\vec{x} = (x_1, \ldots, x_D)^T,$$

as may the independent random numbers

$$\vec{z} = (z_1, \ldots, z_D)^T .$$

The standard method of generating the random walk sets $x_1 = z_1$, $x_2 = z_1 + z_2$, etc. This may be written in matrix notations as

$$\vec{x} = A\vec{z} \tag{3.4}$$

where here and in the following sections, the matrix $A$ is defined as

$$A = \begin{pmatrix} 1 & 0 & 0 & 0 & \ldots & 0 \\ 1 & 1 & 0 & 0 & \ldots & 0 \\ 1 & 1 & 1 & 0 & \ldots & 0 \\ & & & \cdot & & \\ & & & \cdot & & \\ & & & \cdot & & \\ 1 & 1 & 1 & 1 & \ldots & 1 \end{pmatrix} . \tag{3.5}$$

The Brownian bridge discretization described above can also be seen as a linear combination of the $\vec{z}$, so that there exist a matrix $B$ such that

$\vec{x} = B\,\vec{z}$. We define now the generalized Brownian bridge discretization to be any matrix $B$ such that the paths $B\,\vec{z}$ correspond to the same stochastic process as the paths $A\,\vec{z}$. Because a Gaussian process is completely specified by it's covariance, if the paths $B\,\vec{z}$ and $A\,\vec{z}$ have the same covariance, they will necessarily be sampled from the same process. The covariance of the paths $\vec{x_A}$ and $\vec{x_B}$ and given by

$$E(\vec{x_A}\vec{x_A}^T) \quad = \sigma^2\,AA^T \tag{3.6}$$

$$E(\vec{x_B}\vec{x_B}^T) \quad = \sigma^2\,BB^T \tag{3.7}$$

Thus the matrix $B$ will correspond to a generalized Brownian bridge discretization if and only if

$$B\,B^T = A\,A^T. \tag{3.8}$$

It is important to remember that random Monte Carlo methods will not be affected by how the random walk path is generated. From the integration point of view, this follows from the fact that for any function $f(\vec{x})$, under the change of variables $\vec{z'} = B^{-1}A\vec{z}$, we have that

$$E(f(\vec{x})) = E(f(A\vec{z})) = E(f(B\vec{z'})) \tag{3.9}$$

for any $B$ satisfying (3.8). In particular, the variance of a given function of the path, which is expressed as an integral, is independent of the path generating matrix, so the error is also not affected.

As demonstrated in [6], the combination of quasi-random sequences and the Brownian bridge discretization can lead to significant error reduction. One can imagine that if the last step of the walk is more uniformly distributed, then the set of paths so generated is necessarily more uniformly chosen from the space of all possible paths, leading to smaller integration errors. It is also possible that knowledge of the integrand can be exploited to focus the first, most uniformly distributed dimensions of the sequence on the steps of the random walk (or subspace spanned by the steps of the walk) which contribute most to the error. For example in the mortgage-backed security problem, the later months, for which the majority of the loans have already been prepaid, contribute very little to the error, as do the initial months when the interest rate is tightly tied to its initial value. The greatest variance comes from the middle months, which suggests that a generalized Brownian bridge scheme which focuses on these months may lead to smaller errors.

The first Brownian bridge discretization described above is an example of an important sub-class of the generalized discretizations, namely those that concern generating the steps of the walk sequentially according to a specified permutation $\Pi = (\pi_1, \ldots, \pi_D)^T$ of the first $D$ integers. The unique Brownian bridge matrix $B_\Pi$ corresponding to this permutation may be generated as follows. Let $P$ be the permutation matrix defined by $\Pi = P \cdot (1, \ldots, D)^T$. Then

$$B_\Pi = P\,R_\Pi \tag{3.10}$$

where $R_\Pi$ is the unique lower triangular matrix obtained from the Cholesky decomposition

$$R_\Pi R_\Pi^T = P^T A A^T P \qquad (3.11)$$

where again $A$ is given by (3.5). It is easily checked that $B_\Pi$ satisfies (3.8).

The Brownian bridge formula (3.3) shows that each term of a permutation defined discretization may be expressed as a linear combination of exactly two previously determined steps of the path. Thus the path $\vec{x}$ may be generated recursively in $\mathcal{O}(D)$ steps. On the other hand, the matrix $R_\Pi$ will in general be a dense lower triangular matrix, so that generating $\vec{x} = B_\Pi \vec{z}$ will be an $\mathcal{O}(D^2)$ operation. This may lead to a significant increase in the computation time necessary to generate the paths $\vec{x}$. We remark here, however, that it is always possible to generate a permutation based discretization in $\mathcal{O}(D)$ steps.

A permutation based Brownian bridge discretization has the interpretation that the individual steps of the random walk $\vec{x}$ are generated in a specific order. Equation (3.8) allows, however, for a more general interpretation. Using the fact that $B$ satisfies (3.8) if and only if $B = AQ$ for some orthogonal matrix $Q$, we see that generating the random walk as $B\vec{z}$ is equivalent to applying $A$ to an orthogonal transformation of $\vec{z}$. Such a transformation may lead to a diagonalization of the integrand, concentrating much of the variance of the problem into a few principle directions, and thereby reducing the effective dimension. There is a computational price to be paid for this approach in that, for a general orthogonal matrix $Q$, it will not be possible to generate the corresponding random walk $B\vec{z}$ by recursion, but will require an $\mathcal{O}(D^2)$ procedure.

# 4  Numerical Results

The value $PV$ for the mortgage-backed security example was calculated to be 131.787. The variance in this value is 41.84 and the variance in the antithetic computation of this value (described below) is .014. The mean length of a mortgage in this case is 100.9 months and the median length is 93 months.

We now describe the accuracy of various integration methods for this problem as a function of $N$, the number of paths. For each of these results, we present the root-mean-square of the error over 25 independent computations. Moreover, the computations for different values of $N$ are also all independent. For random Monte Carlo, independence is defined in the probabilistic sense. This is approximated with a pseudo-random sequence by using non-overlapping subsequences for all runs and all $N$. This approach to generating independent samples through non-overlapping subsequences is extended to quasi-random sequences. Although the probabilistic meaning is lost for these highly correlated deterministic sequences, the average over 25 runs based on distinct, non-overlapping subsequences leads to a reasonable estimate of the likely error that will result from using $N$ quasi-random paths. The results are plotted in terms of error vs. $N$, both in log base 10.

First, we perform straightforward Monte Carlo evaluation, with results plotted in Figure 1. The top curve shows results from Monte Carlo using standard pseudo-random points, with the error decreasing at the expected rate of $N^{-1/2}$. The second curve is the result of using the Sobol' sequence for the first 50 dimensions (time steps), followed by pseudo-random for the remaining 310 time steps. This shows almost no gain through this limited use of quasi-Monte Carlo. The third graph shows a dramatic improvement using quasi-random for all 360 dimensions of the problem. The 360 dimensional Sobol' sequence was generated using a part of the code FINDER, written by Spassimir Paskov. We are grateful to Paskov and Joseph Traub for providing us with this code.

These results are consistent with the results of Paskov [8, 9] and Ninomiya and Tezuka [7]. They seem to contradict the observations above that the effectiveness of quasi-Monte Carlo is lost in high dimensions. This result is deceptive,however, because of the special nature of this problem in which the value is almost entirely linear in the integration variables. For such a linear problem, only the one-dimensional projections of the quasi-random sequence are significant. The high-dimensional sequences generated here have the property that all of the one-dimensional projections are equally well distributed. So they do well on linear functions of any dimensionality. We will show next that they do not necessarily do well on high dimensional nonlinear functions, even quadratic functions, presumably due to the poor quality of some of the two-dimensional projections.

The linear terms in the integrand can all be eliminated through the use of antithetic variables. This means that for every path $\{x_n\}$, we also use the path $\{-x_n\}$. The resulting computational results are plotted on Figure 2. The top curve of this figure shows the error due to standard Monte Carlo, without antithetic variables, for reference. When antithetic variables are used with standard Monte Carlo the error is reduced by more than a factor of 50, as shown in the curve labeled *MC-anti*. Straightforward use of 360 dimensional quasi-random sequences, labeled *QR-anti* does not improve this result. This shows that once the linear terms are removed, straightforward quasi-Monte Carlo is not effective, due to the high-dimension of the problem. The final result, labeled *QR-anti,BB*, shows the result of using the Brownian bridge representation with antithetic variables. This gives an additional error reduction over random antithetic variables by a factor of nearly 10 (for larger $N$) and increases the rate of error decay from $N^{-1/2}$ to approximately $N^{-3/4}$. This improvement is due to the decrease in effective dimension in the Brownian bridge representation.

As further evidence of the nearly linearity of this problem, we next con-

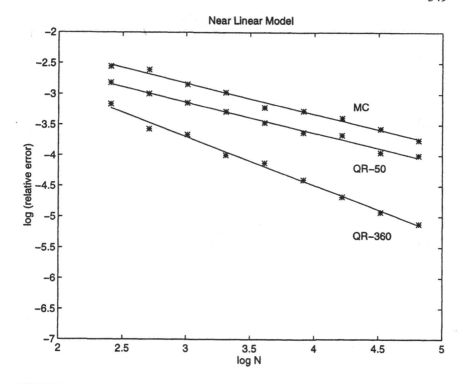

FIGURE 1. Error vs. N (log base 10) for the Nearly Linear Problem, using standard Monte Carlo (MC), quasi-random in the first 50 dimensions (QR-50), and quasi-random for all dimensions (QR-360).

sider the effect of the quadratic terms of the Taylor expansion of the integrand about the point $(\xi_1, \ldots, \xi_M) = (0, \ldots, 0)$. The quadratic terms can be calculated exactly by taking second derivatives of the integrand. We then use them as a control variate; i.e., we subtract the exact quadratic terms from the integrand for the Monte Carlo evaluation, and then add back the exact integral of the quadratic terms. The resulting error, for standard Monte Carlo with antithetic variables, is another factor of 5 to 10 smaller, as shown in Figure 3.

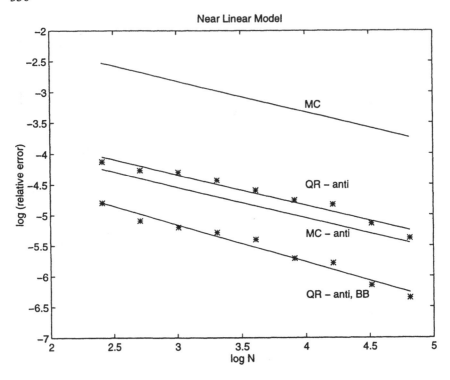

FIGURE 2. Error vs. N (log base 10) for the Nearly Linear Problem, using standard Monte Carlo with antithetic variables (MC-anti), quasi-random with antithetic variables (QR-anti), and quasi-random with antithetic variables and the Brownian bridge discretization (QR-anti,BB). The error for standard Monte Carlo is also plotted for reference.

# 5 Conclusions

Our main conclusions are as follows. While Quasi-Monte Carlo methods provide significant improvements in accuracy and computational speed for problems of moderate dimension, their effectiveness is lost on problems of high dimension, except in special cases, such as linear problems. However, some problems that have a large nominal dimension can be reformulated to have a moderate-sized effective dimension, so that the effectiveness of quasi-Monte Carlo is recovered. One means of achieving this reduction in dimensionality is the generalized Brownian bridge representation of random walks, which is effective on problems such as the valuation of mortgage-backed securities described here. We have also demonstrated

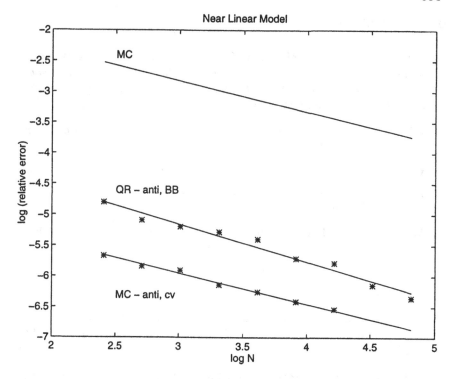

FIGURE 3. Error vs. N (log base 10) for the Nearly Linear Problem, using standard Monte Carlo with antithetic variables and with the exact quadratic terms as a control variate (MC-anti). The results from standard Monte Carlo and from quasi-random with antithetic variables and the Brownian bridge discretization are also plotted for reference.

that Quasi-Monte Carlo can be effectively combined with the variance reduction technique of antithetic variables. In general for the problems that we have considered, there is very little gained from using high-dimensional quasi-random sequences.

Instead of using high-dimensional quasi-random sequences, our recommendation is the following: First analyze the problem to determine the most important dimensions. The Brownian bridge representation, combined with antithetic variables, is one way of doing this. For the most important dimensions, using quasi-random variables, and for the remaining dimensions, quasi-random or pseudo-random variables can be used.

We believe that range of applicability for quasi-Monte Carlo methods can be further increased through additional modification of standard Monte Carlo techniques for use with quasi-random sequences. For example combination of quasi-Monte Carlo with variance reduction methods, such as control variates, importance sampling and stratification, which could be combined with quasi-random sequence could lead to many improvements.

# 6 References

[1] R.E. Caflisch and W. Morokoff. Quasi-Monte Carlo computation of a finance problem. In K.T. Fang and F.J. Hickernell, editors, *Workshop on Quasi-Monte Carlo Methods and Their Applications*, pages 15–30 and UCLA CAM Report 96-16. 1996.

[2] P.E. Kloeden and E. Platen. *Numerical Solution of Stochastic Differential Equations*. Springer-Verlag, Berlin; New York, 1992.

[3] W. Morokoff and R.E. Caflisch. A Quasi-Monte Carlo approach to particle simulation of the heat equation. *SIAM Journal on Numerical Analysis*, 30:1558–1573, 1993.

[4] W. Morokoff and R.E. Caflisch. Quasi-random sequences and their discrepancies. *SIAM J. Sci. Stat. Computing*, 15:1251–1279, 1994.

[5] W. Morokoff and R.E. Caflisch. Quasi-Monte Carlo integration. *J. Comp. Phys.*, 122:218–230, 1995.

[6] B. Moskowitz and R.E. Caflisch. Smoothness and dimension reduction in quasi-Monte Carlo methods. *J. Math. Comp. Modeling*, 23:37–54, 1996.

[7] S. Ninomiya and S. Tezuka. Toward real-time pricing of complex financial derivatives. *Appl. Math. Finance*, 3:1–20, 1996.

[8] S.H. Paskov. New methodolgies for valuing derivatives. In S. Pliska and M. Dempster, editors, *Mathematics of Derivative Securities*. Isaac Newton Inst., Cambridge U. Press, 1996.

[9] S.H. Paskov and J.F. Traub. Faster valuation of financial derivatives. *J. Portfolio Manag.*, pages 113–120, 1995.

William J. Morokoff
Mathematics Dept., UCLA
and C.ATS Software, Inc.
morokoff@math.ucla.edu

Russel E. Caflisch
Mathematics Dept., UCLA
caflisch@math.ucla.edu

# Error Estimation for Quasi-Monte Carlo Methods

Giray Ökten

ABSTRACT A hybrid-Monte Carlo method designed to obtain a statistical error analysis using deterministic sequences is introduced. The method, which is called "random sampling from low-discrepancy sequences", produces estimates that satisfy deterministic error bounds yet confidence interval analysis can be applied to measure the accuracy of them.

A particular implementation of the hybrid method is applied to three problems; one from mathematical finance and others from particle transport theory. The method is compared with the pseudorandom method and two quasirandom methods. Encouraging numerical results in favor of the hybrid method are obtained.

## 1 Introduction

Recently, the term "hybrid Monte Carlo methods" has been used by several authors to describe a variety of methods in which Monte Carlo and quasi-Monte Carlo ideas are combined to obtain advantages over each of them. Mainly these ideas are studied for the reasons we now explain:

### 1.1 Improved Discrepancy and Error Bounds

One of the classical ideas used to improve the discrepancy of the Halton sequence (see [1] for the definitions of discrepancy, van der Corput sequence and Halton sequence) is due to Braaten and Weller [2]. Let us recall that each component of the $s-$dimensional Halton sequence is a van der Corput sequence in an appropriate base. In order to improve the discrepancy of the Halton sequence, the authors used permutations to permute the digits that appear in the digit expansion of each van der Corput sequence. The resulting van der Corput sequence is called a generalized van der Corput sequence and the resulting Halton sequence is called a scrambled (or, generalized) Halton sequence. The authors obtained numerical results that suggested "better" uniformity properties of the scrambled Halton sequence as well as reduced error produced upon application of it to numerical integration. Generalized van der Corput sequences are further studied by Faure [3], who showed that the best known asymptotic behavior of the discrepancy of a one-dimensional sequence is attained by a generalized van der Corput sequence.

Other types of sequences have been introduced by authors in an attempt to obtain better error reduction by making use of the specifics of the problem that is being studied. In [4], Spanier considers some particle transport problems and uses two "hybrid" sequences; "mixed sequence" and "scrambled sequence" to generate the random walk histories. The mixed sequence is a finite dimensional low-discrepancy vector sequence concatenated by a pseudorandom sequence. The "quasirandom" part of the sequence is used to simulate the initial steps of the random walk and the "pseudorandom" part is used for the subsequent steps. In this way, a low-discrepancy sequence is used for the low dimensional part of the problem, hence taking advantage of the superior error reduction properties of low-discrepancy sequences (when sample size is moderate), and a pseudorandom sequence is used for the remaining high dimensional part of the problem. A similar idea is used by Moskowitz [5] who solves a diffusion problem by simulating the stochastic process with a high dimensional pseudorandom sequence. Then the "paths" are continued for just a few steps using a low-discrepancy sequence. In this way a low-discrepancy sequence is used for a "small" part of a high dimensional integral. Both authors report numerical work that shows the superiority of their strategies over "conventional" methods. An analysis of the discrepancy of mixed sequences as well as their application to numerical integration can be found in [6].

## 1.2   Error Analysis

Suppose we want to estimate $I = \int_{[0,1]^s} f(x)dx$. In Monte Carlo methods, we draw $N$ independent random $s-$dimensional vectors $X_1, ..., X_N$ from the uniform distribution on $[0,1]^s$ to construct the estimate $\hat{I} = \frac{1}{N}\sum_{n=1}^{N} f(X_n)$. By the strong law of large numbers, $\hat{I} \to I$ as $N \to \infty$ with probability 1. Also, from the central limit theorem, $\sqrt{N}(I - \hat{I}) \Rightarrow \mathbf{N}(0, \sigma^2)$ as $N \to \infty$, where $\mathbf{N}(\mu, \sigma^2)$ denotes the normal variable with mean $\mu$ and variance $\sigma^2$. Here, $\sigma^2 = \int_{[0,1]^s} (f(X) - I)^2 dX$, and it is assumed to be finite. Hence the accuracy of the estimate of a Monte Carlo method can be measured in a practical way by the sample variance $\hat{\sigma}^2 = \frac{1}{N-1}\sum_{n=1}^{N}(f(X_n) - \hat{I})^2$.

When using quasi-Monte Carlo methods, however, we consider a "nonrandom" uniformly distributed sequence $x_1, ..., x_N$ in $[0,1)^s$ to construct the estimate $\hat{I} = \frac{1}{N}\sum_{n=1}^{N} f(x_n)$. Uniform distribution gives $\hat{I} \to I$ and an upper bound for the error is given by the Koksma-Hlawka inequality (see [1])

$$\left| I - \hat{I} \right| \leq V(f)D_N^*(x_n), \tag{1}$$

where $V(f) < \infty$ is the variation of $f$ in the sense of Hardy and Krause and $D_N^*(x_n)$ is the star discrepancy of $x_1, ..., x_N$. An advantage of the quasi-Monte Carlo method is that we get a *deterministic* error bound

$O(N^{-1}(\log N)^s)$ (as $N \to \infty$), provided $\{x_n\}$ is a low-discrepancy sequence. Using Monte Carlo methods, however, the error bounds are only *probabilistic* with order $O(N^{-1/2})$. Although the order of the error produced by the quasi-Monte Carlo method is better than that of the Monte Carlo method for sufficiently large $N$, the advantage may not be practical to obtain even if the dimension $s$ is moderate. For example, $\frac{(\log N)^9}{N} > \frac{1}{N^{1/2}}$ for $N = 10^{34}$. However numerical work shows that for functions encountered in practice, quasi-Monte Carlo methods produce more accurate estimates than the Monte Carlo method for much smaller sample sizes $N$. This may be due to the fact that the quasi-Monte Carlo error bound $O(N^{-1}(\log N)^s)$ is the order of the upper bound given by the inequality (1) *which can be a very "loose" inequality for a particular function*, although it is a "strict" inequality over the space of functions of bounded variation on $[0,1]^s$ in the sense of Hardy and Krause.

Even if we want to take the upper bound given by the Koksma-Hlawka inequality as a measure of accuracy, we still need to estimate (numerically) the factors $V(f)$ and $D_N^*(x_n)$ of it, in general. In Monte Carlo methods, however, one can practically measure the accuracy of the method by the sample standard deviation of the estimates that are produced. Therefore one of our goals is to design a technique which will produce accurate estimates with guaranteed error bounds yet whose accuracy can be measured in a practical way as in Monte Carlo methods.

To achieve a statistical error analysis using low discrepancy sequences, Owen [7] introduced randomized $(t, m, s)-$ nets and $(t, s)-$ sequences. Also see Cranley and Patterson [8] who randomized the number theoretic methods of Korobov, and Joe [9] who randomized the lattice rules.

## 2 Random Sampling from Low-Discrepancy Sequences

Our objective is to devise a method whose estimates will satisfy known deterministic bounds and yet different estimates will be used to conduct confidence interval analysis. We also want this method to be applicable to a wide range of problems, especially "infinite-dimensional" problems.

Let $P$ be a problem we want to simulate and let $I$ be a quantity to be estimated as the outcome of the simulation. Let $\Omega$ be a set of low-discrepancy sequences $\beta^{(i)} = (\beta_1^{(i)}, \beta_2^{(i)}, ...)$ that are "suitable" to simulate $P$. For example if $P$ is to estimate $I = \int_{[0,1]^s} f(x)dx$ then $\Omega$ consists of $s-$dimensional low-discrepancy vector sequences. If $P$ is an infinite-dimensional random walk problem, then $\Omega$ may contain Halton sequences in prime bases $\left(\phi_{p_{j_1}}(n), ..., \phi_{p_{j_s}}(n), ...\right)$, square root sequences that are defined by $(\{n\sqrt{p_{j_1}}\}, ..., \{n\sqrt{p_{j_s}}\}, ...)$ ($p_{j_k}$ is the $j_k$th prime number) or other "special" sequences such as mixed sequences mentioned in Section 1. The method, which we call "random sampling from low-discrepancy sequences" (or, RS-method, in short), is to use a randomly selected $\beta^{(i)} \in \Omega$ for the $i$th simulation of $P$ as $1 \leq i \leq m$, and to analyze the corresponding estimates $I_1, ..., I_m$ by statistical means.

We investigate this idea when $P$ is to estimate $I = \int_{[0,1]^s} f(x)dx$ where $f$ is of bounded variation in the sense of Hardy and Krause. Let $\Omega = \{ \beta^{(1)}, \beta^{(2)}, ..., \beta^{(r)} \}$ where each $\beta^{(i)}$ is an $s$–dimensional low-discrepancy sequence and $r$ is a sufficiently large positive integer. Put a probability measure on $\Omega$ that assigns equal "weight" to each $\beta^{(i)} \in \Omega$. For a given positive integer $N$, define the random variable $G_N$ on $\Omega$ by

$$
G_N(\beta^{(i)}) = \frac{1}{N} \sum_{n=1}^{N} f(\beta_n^{(i)})
$$

for any $\beta^{(i)} \in \Omega$. We then make the following observations: $G_N(\beta^{(i)}) \to I$ as $N \to \infty$, for every $\beta^{(i)} \in \Omega$. Hence, $|G_N(\beta^{(i)})| \leq K_i$ for every $N > 0$, for some positive number $K_i$. Put $K = \max\{K_1, ..., K_r\}$. Then, $|G_N| \leq K$ for every $N > 0$. Hence from the dominated convergence theorem, we obtain $\mathbf{E}(G_N) \to I$ as $N \to \infty$.

Now let's consider the quantity

$$
\frac{G_N(\beta^{(i_1)}) + ... + G_N(\beta^{(i_M)})}{M},
$$

which we will call an RS-estimate. For a fixed $N, M > 0$, this estimate is obtained by selecting integers $i_1, ..., i_M$ at random from the uniform distribution on $\{1, ..., r\}$ and averaging the values of the random variable $G_N$ at $\beta^{(i_1)}, ..., \beta^{(i_M)}$ of $\Omega$. For a fixed $M$, a deterministic upper bound for the RS-estimate can be given by

$$
\left| \frac{G_N(\beta^{(i_1)}) + ... + G_N(\beta^{(i_M)})}{M} - I \right|
$$

$$
\leq \frac{1}{M} \sum_{k=1}^{M} \left| G_N(\beta^{(i_k)}) - I \right|
$$

$$
\leq \frac{V(f)}{M} \sum_{k=1}^{M} D_N^*(\beta^{(i_k)}) \to 0 \text{ as } N \to \infty.
$$

If the star discrepancy of each sequence in $\Omega$ is $O(N^{-1}(\log N)^s)$, then the RS-estimate also satisfies the same order of magnitude with the implied constant being the maximum of the individual implied constants. Therefore, we expect the RS-estimate to be no worse than the estimate given by the "worst" sequence $\beta$ in $\Omega$.

Seeking a probabilistic analysis of the RS-estimates, we realize that for a fixed $N$

$$\frac{G_N(\beta^{(i_1)}) + ... + G_N(\beta^{(i_M)})}{M} \to \mathbf{E}(G_N) \ \text{as} \ M \to \infty \ \text{with probability 1,}$$

which follows from the strong law of large numbers. Then, by the central limit theorem, we can construct confidence intervals for $\mathbf{E}(G_N)$. Since $\mathbf{E}(G_N) \to I$ as $N \to \infty$, we can choose $N$ large enough to have $\mathbf{E}(G_N)$ sufficiently close to $I$ so that the confidence intervals constructed for $\mathbf{E}(G_N)$ work essentially as well as though they were confidence intervals for $I$.

Our idea raises two important questions that we are unable to answer theoretically, but investigate numerically in the rest of the paper using three model problems with known solutions. The first one is how to populate $\Omega$ in a practical implementation of the RS-method, so that the confidence interval analysis based on the central limit theorem is valid practically. One can use radical inverse type sequences, sequences obtained from multiples of irrational numbers, or both. In the preliminary numerical work, we used Halton sequences in different bases to populate $\Omega$. When we applied the RS-method to the finance problem introduced in Section 3, we obtained one sided estimates which prevented us from conducting a sound confidence interval analysis. Then we used square root sequences to populate $\Omega$ and obtained even superior results to the pseudorandom implementation. The latter are reported here.

The second question is how large $N$ should be so that the confidence intervals constructed for $\mathbf{E}(G_N)$ can be used to analyze $I$. The answer depends on the particular construction of $\Omega$ as well as the problem the method is applied to. We obtained satisfactory numerical results using numbers as small as 1000 for $N$.

# 3 Applications to Finance: Evaluation of European Call Option

We briefly explain the background of the problem: A European call option is a contract which gives the owner of the option the right to buy a prescribed asset, known as the underlying asset, at a prescribed time in the future, known as the expiry date, for a prescribed amount, known as the exercise price. The problem is to determine a fair "value" for this right. Using risk-neutral valuation (see [10]), it can be shown that the value $C(S, t)$ of such a European call option is the following expected value

$$C(S, t) = \mathbf{E}\left[e^{-r(T-t)} \max(Se^{(r-\sigma^2/2)(T-t)+\sigma\sqrt{T-t}\mathbf{Z}} - E, 0)\right]. \quad (2)$$

Here, $S$ is the price of the underlying asset at time $t$, $T$ is the expiration time, $\sigma$ is a constant called the volatility of the asset, $r$ is the risk free interest rate, $E$ is the exercise price and $\mathbf{Z}$ is a standard normal variable. An exact solution of the problem can be obtained by evaluating the right-hand side of the equation (2). We will use simulation to estimate it, when $S = 10$, $T - t = 0.5$, $\sigma = 0.4$, $r = 0.1$ and $E = 10$. The exact solution is 1.35804. To sample from $\mathbf{Z}$, we will use the Box-Muller method (see [11]), which requires generation of two numbers from the uniform distribution on $(0, 1)$.

To apply the RS-method to this problem, we need to construct a space $\Omega$ of two dimensional low-discrepancy vector sequences. Our candidate is the square root sequence, defined by $q_n = (\{n\sqrt{p_i}\}, \{n\sqrt{p_j}\})$ where $p_i$ and $p_j$ are different primes and braces about a number denote its fractional part. See [12] for numerical results showing the superiority of these sequences among a number of sequences that were applied to one infinite and several finite dimensional problems.

For a given two positive distinct integers $i_1, i_2$, we define the sequence $\{\mathbf{q}_n^{i_1,i_2}\}_{n=1}^{\infty}$ by

$$\mathbf{q}_n^{i_1,i_2} = \left(\{n\sqrt{p_{i_1}}\}, \{n\sqrt{p_{i_2}}\}\right). \tag{3}$$

Here $p_{i_j}$ denotes the $i_j$th prime number. The space $\Omega$ consists of sequences $\{\mathbf{q}_n^{i_1,i_2}\}_{n=1}^{\infty}$, as the integers $i_1, i_2$ vary between 1 and 30. There are $\binom{30}{2}$ such sequences in $\Omega$. For each simulation of the problem, we choose two distinct integers $i_1, i_2$ between 1 and 30 at random and use the sequence $\{\mathbf{q}_n^{i_1,i_2}\}_{n=1}^{\infty}$ to obtain an estimate.

## 3.1  Numerical Results

We first conduct a confidence interval analysis to compare the standard Monte Carlo method with the RS-method. Twenty 90% confidence intervals are constructed using methods based on the use of Student's $t$, bootstrap $BC_a$ and bootstrap percentile. For a discussion of bootstrap methods, see [13]. To construct each confidence interval, we simulated the problem 20 times using 4000 histories for each simulation. The following table shows the number of confidence intervals containing the true solution when the problem is simulated using pseudorandom numbers[1] and the RS-method.

---

[1]The numbers are generated using the built-in pseudorandom number generator of the software system *Mathematica*, in which language the codes used in this paper are written. The generator uses a "subtract-with-borrow" method with base $2^{31}$. For details, see G. Marsaglia and A. Zaman, "*A new class of random number generators*", The Annals of Applied Probability, 1, 462-480, 1991.

| $N = 4000$ | Pseudorandom | RS-method |
|---|---|---|
| Student's $t$ | 14 | 19 |
| Bootstrap $BC_a$ | 14 | 16 |
| Bootstrap percentile | 14 | 18 |

We expect 18 intervals to contain the true solution and the RS-method gives close results. We then compare the average width of these confidence intervals and see that the RS-method produces intervals that are narrower by approximately a factor of 3.6.

| $N = 4000$ | Pseudorandom | RS-method |
|---|---|---|
| Student's $t$ | $2.54 \times 10^{-2}$ | $0.69 \times 10^{-2}$ |
| Bootstrap $BC_a$ | $2.34 \times 10^{-2}$ | $0.67 \times 10^{-2}$ |
| Bootstrap percentile | $2.34 \times 10^{-2}$ | $0.63 \times 10^{-2}$ |

We now compare these methods based on the sample standard deviation they produce. For a given $N$ ($N = 1000, 2000, ..., 14000$), the problem is simulated 10 times using the pseudorandom and the RS-method. In Figure

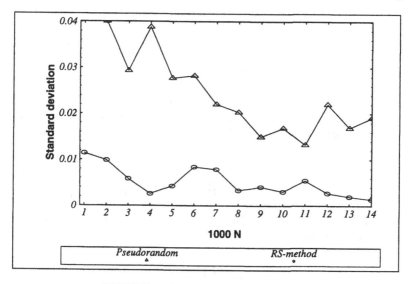

FIGURE 1. Standard deviation results

1, the sample standard deviation of the ten estimates is plotted against the number of samples $N$. We see that the standard deviation produced by the RS-method is smaller than that of the pseudorandom method by approximately a factor of 10.

We now compare the actual error produced by simulating the problem using the pseudorandom numbers, the RS-method, the Halton sequence ($\phi_2(n), \phi_3(n)$) and the square root sequence ($\{n\sqrt{3}\}, \{n\sqrt{5}\}$). For the pseudorandom and the RS-methods, we simulate the problem 10 times for a given sample size $N$ and take the average of the estimates as the result (the same strategy is used also in Figures 4 and 6). The results are displayed in Figure 2.

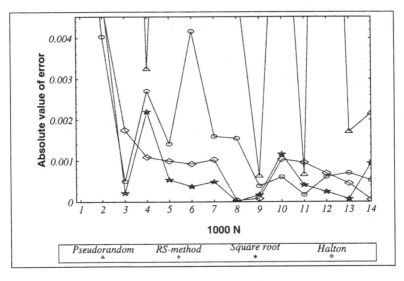

FIGURE 2. Comparison of error

As it is seen from the graph, the pseudorandom method is outperformed by the other methods and after 9000 histories it becomes difficult to distinguish between errors of the RS, square root and Halton methods.

## 4 Applications to Transport Theory

In this section, we apply the RS-method to problems from transport theory. Our objective is to estimate

$$I = \int_\Gamma g(P)\Psi(P)dP, \tag{4}$$

where $\Psi$ is the solution of the following integral equation

$$\Psi(P) = \int_\Gamma K(P, P')\Psi(P')dP' + S(P). \tag{5}$$

The equation (5) is often called the particle transport equation because of the physics it is derived from. In physical terminology, $\Gamma$ represents the phase space and $K(P, P')$ represents the probability density function of a particle making a transition from the collision point $P'$ to the point $P$. The function $\Psi(P)$ is the collision density of particles at point $P$ and $S(P)$ is the source function. The quantity $I$ that we want to estimate is often described as the reaction of the physical "detector" (described by the function $g(P)$ in (4)) to the collision density $\Psi(P)$. A detailed analysis of this problem can be found in [14].

## 4.1 Solving Matrix Equations

If we assume that the material in which the particles exist is an infinite homogeneous material and the particles can only take finitely many energy states, then (4) simplifies to (see [14] for derivations)

$$I = <g, x>, \qquad (6)$$

where, analogous to equation (5), $x$ is the solution of the following matrix equation

$$x = Hx + a. \qquad (7)$$

Here, $H = (h_{ij})$ is a known $N \times N$ matrix, $g = (g_i)$ and $a = (a_i)$ are $N$-dimensional vectors.

A detailed exposition of the solution of linear equations by simulation can be found in [11], [14] and applications of Monte Carlo, quasi-Monte Carlo and hybrid-Monte Carlo ideas can be found in [4]. If $g$ is the vector defined by $g_i = \delta_{ij}$ ($\delta_{ij} = 1$ if $i = j$, and 0 if $i \neq j$), then the quantity $I$ (see (6)) that we want to estimate becomes $x_i$; the $i$th component of the solution vector $x$ of equation (7).

In the following, we will estimate $x$ of equation (7) using the pseudo-random, RS, square root and Halton methods. We let $\mathbf{a} = (0.2, 0.3, 0.5)^T$ and $\mathbf{H} = \begin{pmatrix} .25 & .25 & .25 \\ .25 & .25 & .25 \\ .25 & .25 & .25 \end{pmatrix}$ in (7). Then the solution vector is $\mathbf{x} = (1.2, 1.3, 1.5)^T$. To generate the $n$th random walk, we will use the sequence $\{(\{n\sqrt{2}\}, \{n\sqrt{3}\}, \{n\sqrt{5}\}, ...)\}$ for the square root method, and the sequence $\{(\phi_2(n), \phi_3(n), \phi_5(n), ...)\}$ for the Halton method.

For the RS-method, the space $\Omega$ will consist of the mixed sequences that were mentioned in section 1.1. The square root sequences will be used as the quasirandom part of the mixed sequences. For a given $K$ positive distinct integers $i_1, ..., i_K$, we define the sequence $\{\mathbf{m}_n^{i_1, ..., i_K}\}_{n=1}^{\infty}$ by

$$\mathbf{m}_n^{i_1, ..., i_K} = \left(\{n\sqrt{p_{i_1}}\}, ..., \{n\sqrt{p_{i_K}}\}, \rho_1, \rho_2, ...\right). \qquad (8)$$

Here $\rho_j$ are pseudorandom numbers. A term of the sequence $\{\mathbf{m}_n^{i_1, ..., i_K}\}_{n=1}^{\infty}$ is a mixed sequence. The pseudorandom numbers $\rho_1, \rho_2, ...$ also depend on $n$ and $i_1, ..., i_K$, however, we suppress the dependence for notational convenience. $\Omega$ consists of sequences $\{\mathbf{m}_n^{i_1, ..., i_K}\}_{n=1}^{\infty}$, as the integers $i_1, ..., i_K$ vary between 1 and 30. There are $\binom{30}{K}$ sequences in $\Omega$.

For each simulation of the problem, $K$ distinct integers $i_1, ..., i_K$ between 1 and 30 are chosen at random and the sequence $\mathbf{m}_1^{i_1,...,i_K}$ is used to generate the first random walk, $\mathbf{m}_2^{i_1,...,i_K}$ to generate the second random walk and so on.

In this problem, to advance a random walk to the next state, two decisions need to be made. One of these selects either an absorption or a scattering event, while the second selects the new energy state of the scattered particle, provided that scattering does, indeed, occur. In the following, we take $K = 10$ in (8), i.e., we use low-discrepancy vectors to simulate the first 5 steps of random walks[2].

## Numerical Results

In the following, twenty 90% confidence intervals for each component of the vector $\mathbf{x}$ are constructed. To construct each confidence interval, the problem is simulated 20 times and for each simulation 1000 random walks are generated. Student's $t$, bootstrap $BC_a$ and bootstrap percentile techniques are used to construct these confidence intervals. The following table shows the number of confidence intervals that contain the true solution for each component of the vector $\mathbf{x}$, when the problem is simulated using the pseudorandom numbers.

| Pseudorandom | Component 1 | Component 2 | Component 3 |
|---|---|---|---|
| Student's $t$ | 18 | 17 | 20 |
| Bootstrap $BC_a$ | 17 | 17 | 19 |
| Bootstrap percentile | 18 | 16 | 20 |

And when the problem is simulated using the RS-method, we obtained:

| RS-method | Component 1 | Component 2 | Component 3 |
|---|---|---|---|
| Student's $t$ | 19 | 19 | 20 |
| Bootstrap $BC_a$ | 19 | 19 | 18 |
| Bootstrap percentile | 19 | 19 | 18 |

Both methods give satisfactory results. We also observe that the constructions based on the use of Student's $t$ and bootstrap techniques produce very similar results. This suggests that the sample means are approximately normally distributed in this case for both the pseudorandom and RS-methods.

We now compare the widths of the confidence intervals generated above

---

[2]Here and in Section 4.2, the parameter $K$ is not optimized but chosen to be 10 and 12. In general, the choice of $K$ is sensitive and its optimal value depends on the problem. See [4] and [6] for further discussions on this matter.

as a measure of error. The following tables give the average width of them:

| Pseudorandom | Component 1 | Component 2 | Component 3 |
|---|---|---|---|
| Student's $t$ | .0425 | .0410 | .0457 |
| Bootstrap $BC_a$ | .0398 | .0383 | .0424 |
| Bootstrap percentile | .0393 | .0380 | .0422 |

| RS-method | Component 1 | Component 2 | Component 3 |
|---|---|---|---|
| Student's $t$ | .0361 | .0377 | .0356 |
| Bootstrap-$BC_a$ | .0339 | .0347 | .0329 |
| Bootstrap percentile | .0335 | .0351 | .0327 |

The RS-method produces narrower confidence intervals than the pseudorandom implementation. We recall that these results reflect the error when the number of random walks is only 1000 for each simulation. Larger sample sizes for each simulation should produce greater relative advantages for the RS-method.

Taking the confidence interval results as the basis, we can compare the RS-method with the pseudorandom implementation by the sample standard deviations they produce. The problem is simulated 10 times, generating $N$ ($N = 1000, 2000, ..., 20000$) random walks for each simulation using the pseudorandom and the RS-method. The Euclidean norm of the resulting standard deviation is plotted in Figure 3. The RS-method gives approximately a factor of 2 improvement over the pseudorandom method.

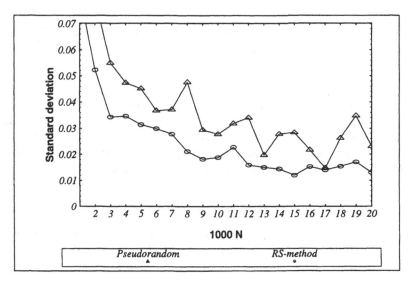

FIGURE 3. Standard deviation results

Now we compare the actual error produced by the simulations. For a given sample size $N$, each simulation produces an estimate of the vector

x. In the case of the pseudorandom and the RS-method, the problem is simulated 10 times. Therefore, for a fixed sample size $N$, we obtain 10 estimates $\mathbf{x}_1, ..., \mathbf{x}_{10}$ of $\mathbf{x}$. We define the average vector $\bar{\mathbf{x}}$ of the vectors $\mathbf{x}_1, ..., \mathbf{x}_{10}$, by $\bar{\mathbf{x}}^{(j)} = \frac{1}{10} \sum_{k=1}^{10} \mathbf{x}_k^{(j)}$, i.e., the $j$th component of the vector $\bar{\mathbf{x}}$ is the average of the $j$th components of the estimates $\mathbf{x}_1, ..., \mathbf{x}_{10}$. For the pseudorandom and RS-method, we define "$L2$ error" as the $L2$ distance between the true solution $\mathbf{x}$ and the average vector $\bar{\mathbf{x}}$. For the Halton and square root methods, there is only one estimate of $\mathbf{x}$, and $L2$ error denotes the $L2$ distance between $\mathbf{x}$ and its estimate. Figure 4 displays the $L2$ error produced by the simulation methods against a given sample size $N$.

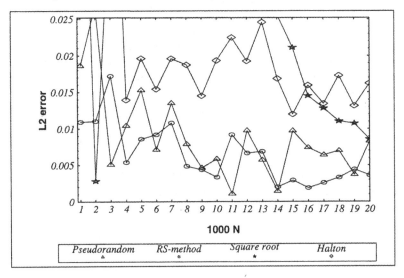

FIGURE 4. Comparison of L2 error

The RS and pseudorandom methods give better estimates than the other methods. The "poor" behavior of the Halton and square root methods may be explained by the fact that the problem has an inherently large "effective" dimension[3]. However the RS-method performs better than the quasirandom methods and slightly better than the pseudorandom method.

## 4.2   Penetration of Neutrons Through a Block

Consider a homogeneous block of thickness $d$. Suppose that neutrons that are emitted by a nuclear source, enter the block from its left face. In the block, a neutron can either be absorbed, back scattered or forward scattered

---

[3]The effective dimension of an infinite dimensional problem is an integer that is related to the rate of convergence of the Neumann series, hence the difficulty of the problem.

upon a collision. The distance between consecutive collisions of a neutron
is a random variable with the probability density function[4] $f(x) = e^{-x}$.
The problem is to find the probability of transmission of a neutron from
the right face of the block. Analytic solutions of this problem can be found
in [15]. We will estimate the probability of transmission by simulation.

In the following results, we let the probability of scattering per collision,
$p_s$, be 0.9 and the thickness, $d$, be 5 units. The events back and forward
scattering are assumed to be equally likely. The pseudorandom, RS, square
root and Halton methods are used to simulate the problem. The implementations of the methods are similar to the previous problem except that in
the RS-method we take $K = 12$ for the mixed sequences.

Numerical Results

We fix the number of random walks to be 2000 and construct twenty 90%
confidence intervals. The number of intervals that contain the true solution
is given by the following table.

| $N = 2000$ | Pseudorandom | RS-method |
|---|:---:|:---:|
| Student's $t$ | 18 | 16 |
| Bootstrap $BC_a$ | 18 | 16 |
| Bootstrap percentile | 17 | 16 |

The RS-method underestimates the expected number of confidence intervals, 18, by two. Two RS-confidence intervals missed the true solution approximately by $3 \times 10^{-5}$ and $10^{-4}$. The fact that these numbers are only a
small fraction of the average width of the RS-confidence intervals (see table
below) may keep us (at least the author!) still confident in the RS-method.

The following table shows that the RS-confidence intervals are narrower
than the pseudorandom confidence intervals by approximately a factor of
1.7 .

| $N = 2000$ | Pseudorandom | RS-method |
|---|:---:|:---:|
| Student's $t$ | .004611 | .002772 |
| Bootstrap $BC_a$ | .004285 | .002608 |
| Bootstrap percentile | .004229 | .002597 |

The sample standard deviations are compared in Figure 5. Again, the
RS-method produces estimates with smaller standard deviation than the
pseudorandom method ($N = 1000, 2000, ..., 15000$).

Figure 6 displays the absolute value of the error produced by the methods. The RS-method gives better estimates than the pseudorandom and
Halton methods for almost all history sizes $N$. The square root method

---

[4]In conventional transport applications, the intercollision probability density
function is $f_c(x) = ce^{-cx}$, where $c$ is a function of the material properties of the
block. Here we have normalized the problem by choosing $c \equiv 1$.

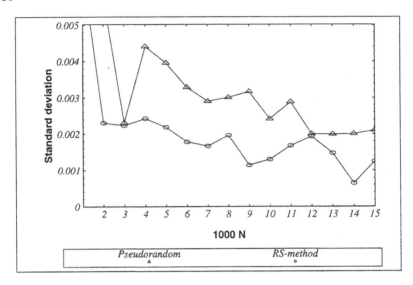

FIGURE 5. Standard deviation results

shows an irregular behavior. To obtain a better separation more histories should be generated.

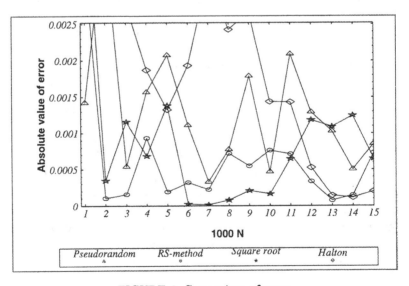

FIGURE 6. Comparison of error

# 5 Conclusions

Additional numerical work with larger sample sizes is desirable to better quantify the improvements we have observed. Although numerical results suggest the use of square root sequences and their mixed counterparts to populate $\Omega$, different constructions should also be tried in a given application.

We want to point out the common theoretical ground shared by conventional Monte Carlo error estimation and the RS-method. In one type of Monte Carlo error estimation, one makes use of the central limit theorem to construct confidence intervals using estimates that are produced by choosing different seeds in a pseudorandom number generator. Hence the space $\Omega$ consists of point sets produced from a pseudorandom number generator by selecting different seeds. Therefore, the fundamental problem we face is to develop a deterministic counterpart to the central limit theorem which will justify doing statistics with deterministic sequences, as the theory of uniform distribution corresponds to the strong law of large numbers. Such a theory will probably provide us with the "deterministic" conditions that should be satisfied by the sequences in $\Omega$ so that the estimates obtained by the sequences are subject to a "confidence interval" type analysis.

*Acknowledgments:* I thank Prof. Jerome Spanier, Prof. John Angus and the referee for their valuable remarks that improved the quality of the paper.

# 6 References

[1] H. Niederreiter, *Random Number Generation and Quasi-Monte Carlo Methods,* CBMS-SIAM, 1992.

[2] E. Braaten and G. Weller, *"An Improved Low-Discrepancy Sequence for Multidimensional Quasi-Monte Carlo Integration"*, J. Comp. Physics, 33, 249-258, 1979.

[3] H. Faure, *"Using permutations to reduce discrepancy"*, J. Comp. Appl. Math., 31, 97-103, 1990.

[4] J. Spanier, *"Quasi-Monte Carlo Methods for Particle Transport Problems"*, Proc. Conf. on Monte Carlo Methods in Scientific Computing, Univ. Las Vegas, 1994.

[5] B. S. Moskowitz, *"Application of quasi-random sequences to Monte Carlo methods"*, Ph. D. Dissertation, UCLA, 1993.

[6] G. Ökten, *"A Probabilistic Result on the Discrepancy of a Hybrid-Monte Carlo Sequence and Applications"*, Monte Carlo Methods and Applications, Vol. 2, No. 4, 255-270, 1996.

[7]   A. B. Owen, *"Randomly Permuted $(t, m, s)-Nets$ and $(t, s)-Sequences$"*, Proc. Conf. on Monte Carlo Methods in Scientific Computing, Univ. Las Vegas, 1994.

[8]   R. Cranley and T. N. L. Patterson, *"Randomization of Number Theoretic Methods for Multiple Integration"*, SIAM J. Numer. Anal., Vol. 13, No. 6, 904-914, 1976.

[9]   S. Joe, *"Randomization of lattice rules for numerical multiple integration"*, J. Comp. Appl. Math., 31, 299-304, 1990.

[10]  J. C. Hull, *Options, Futures, And Other Derivative Securities*, Prentice-Hall, Inc., 1989.

[11]  Y. R. Rubinstein, *Simulation And The Monte Carlo Method*, John Wiley & Sons, 1981.

[12]  G. Pagès and Y. J. Xiao, *"Sequences with low discrepancy and pseudo-random numbers: theoretical remarks and numerical tests"*, Technical Report, Laboratoire de Mathematiques et Modelisation, Paris, 1991.

[13]  B. Efron and R. Tibshirani, *An Introduction to the Bootstrap*, Chapman & Hall, London, 1993.

[14]  J. Spanier and E. M. Gelbard, *Monte Carlo Principles And Neutron Transport Problems*, Addison-Wesley, 1969.

[15]  J. Spanier, *"An Analytic Approach to Variance Reduction"*, SIAM J. Appl. Math., 18, 172-190, 1970.

Department of Mathematics
The Claremont Graduate School
Claremont, CA 91711
USA

# Shift–Nets: a New Class of Binary Digital $(t, m, s)$-Nets

## Wolfgang Ch. Schmid*, Salzburg

ABSTRACT  Digital nets and sequences play an important role in the theory of low-discrepancy point sets and sequences in the $s$-dimensional unit cube. This paper is devoted to the binary case. We first will improve the lower bound for the quality parameter of binary digital sequences. The main aim is to introduce a new class of binary digital nets, the so-called shift–nets, which in many cases are nets of improved quality.

## 1   Introduction

Currently, the most effective constructions of low-discrepancy point sets and sequences, which are of great importance for quasi-Monte Carlo methods in multidimensional numerical integration, are based on the concept of $(t, m, s)$-nets and $(t, s)$-sequences. A detailed theory was developed in Niederreiter [Nie87] (see also [Nie92, Chapter 4] for surveys of this theory). '

The concepts of $(t, m, s)$-nets and of $(t, s)$-sequences in a base $b$ provide point sets of $b^m$ points, respectively infinite sequences, in the half-open $s$-dimensional unit cube $I^s := [0, 1)^s$, $s \geqslant 1$, which are extremely well distributed if the quality parameters $t \in \mathbb{N}_0$ are "small". We follow [Nie92] in our basic notation and terminology.

**Definition 1** *Let $b \geqslant 2$, $s \geqslant 1$, and $0 \leqslant t \leqslant m$ be integers. Then a point set consisting of $b^m$ points of $I^s$ forms a $(t, m, s)$-net in base $b$ if every subinterval $J = \prod_{i=1}^{s} \left[ a_i b^{-d_i}, (a_i + 1) b^{-d_i} \right)$ of $I^s$ with integers $d_i \geqslant 0$ and $0 \leqslant a_i < b^{d_i}$ for $1 \leqslant i \leqslant s$ and of volume $b^{t-m}$ contains exactly $b^t$ points of the point set.*

**Definition 2** *Let $b \geqslant 2$, $s \geqslant 1$, and $t \geqslant 0$ be integers. Then a sequence $\mathbf{y}_0, \mathbf{y}_1, \ldots$ of points in $I^s$ is a $(t, s)$-sequence in base $b$ if for all $k \geqslant 0$ and $m \geqslant t$ the point set consisting of the $\mathbf{y}_n$ with $kb^m \leqslant n < (k+1)b^m$ forms a $(t, m, s)$-net in base $b$.*

So-called digital nets and sequences are of special interest due to the following two reasons. First, until now all construction methods for $(t, m, s)$-

---

*Research supported by the Austrian Research Foundation (FWF), Project P 11009 MAT nets and $(t, s)$-sequences which are relevant for applications in quasi-Monte Carlo methods are digital methods over finite commutative rings with identity. Second, digital $(t, m, s)$-nets behave extremely well for the numerical integration of functions which are representable by an in some sense rapidly converging multivariate Walsh series. In a series of papers, Larcher and several co-authors established lattice rules for the numerical integration of multivariate Walsh series by digital nets. We refer to [LSW95] for a concise introduction to the field of Larcher's lattice rules.

In the following we only consider the case where the underlying ring is a finite field $\mathbb{F}_q$ of prime-power order $q$. For information on the general case we refer to [LNS96].

**Definition 3** *Let $q$ be a prime-power and let $s \geqslant 1$ and $m \geqslant 1$ be integers. We consider the following construction principle for point sets $P$ consisting of $q^m$ points in $I^s$. We choose:*

*(i) bijections $\psi_r : Z_q = \{0, 1, \ldots, q - 1\} \to \mathbb{F}_q$ for $0 \leqslant r \leqslant m - 1$;*

*(ii) bijections $\eta_j^{(i)} : \mathbb{F}_q \to Z_q$ for $1 \leqslant i \leqslant s$ and $1 \leqslant j \leqslant m$;*

*(iii) elements $c_{jr}^{(i)} \in \mathbb{F}_q$ for $1 \leqslant i \leqslant s$, $1 \leqslant j \leqslant m$, and $0 \leqslant r \leqslant m - 1$.*

*For $n = 0, 1, \ldots, q^m - 1$ let*

$$n = \sum_{r=0}^{m-1} a_r(n) q^r \qquad \text{with all } a_r(n) \in Z_q$$

*be the digit expansion of $n$ in base $q$. We put*

$$x_n^{(i)} = \sum_{j=1}^{m} y_{nj}^{(i)} q^{-j} \qquad \text{for } 0 \leqslant n < q^m \text{ and } 1 \leqslant i \leqslant s,$$

$$\text{with} \quad y_{nj}^{(i)} = \eta_j^{(i)} \left( \sum_{r=0}^{m-1} c_{jr}^{(i)} \psi_r(a_r(n)) \right) \in Z_q$$

*for $0 \leqslant n < q^m$, $1 \leqslant i \leqslant s$, and $1 \leqslant j \leqslant m$.*
*If for some integer $t$ with $0 \leqslant t \leqslant m$ the point set*

$$\mathbf{x}_n = \left( x_n^{(1)}, \ldots, x_n^{(s)} \right) \in I^s \qquad \text{for } n = 0, 1, \ldots, q^m - 1$$

*is a $(t, m, s)$-net in base $q$, then it is called a digital $(t, m, s)$-net constructed over $\mathbb{F}_q$.*

**Remark 1** For $1 \leqslant i \leqslant s$ let $C^{(i)}$ be the $m \times m$ matrix over $\mathbb{F}_q$ with rows

$$\mathbf{c}_j^{(i)} = \left( c_{j,0}^{(i)}, \ldots, c_{j,m-1}^{(i)} \right) \qquad \text{for } j = 1, 2, \ldots, m .$$

Then we may think of the elements in (iii) either as a two-parameter system $C = \{c_j^{(i)} \in \mathbb{F}_q^m : 1 \leqslant i \leqslant s, 1 \leqslant j \leqslant m\}$ of vectors of $\mathbb{F}_q^m$ or as an $s$-tuple $(C^{(1)}, \ldots, C^{(s)})$ of $m \times m$ matrices over $\mathbb{F}_q$.

**Definition 4** *For a system* $C = \left\{ c_j^{(i)} \in \mathbb{F}_q^m : 1 \leqslant i \leqslant s, 1 \leqslant j \leqslant m \right\}$ *we define* $S(C, t, m)$ *to be the set of all subsystems*

$$\left\{ c_j^{(i)} \in \mathbb{F}_q^m : 1 \leqslant j \leqslant d_i, 1 \leqslant i \leqslant s \right\} \quad of\ C$$

*for any integers* $d_1, \ldots, d_s \geqslant 0$ *with* $\sum_{i=1}^{s} d_i = m - t$.

(For a subsystem $C \in S(C, t, m)$ with $C = \{c_1, \ldots, c_{m-t}\}$ we may also think of $C$ as an $(m - t) \times m$ matrix over $\mathbb{F}_q$, where $c_j$ is the $j$th row of $C$.)

**Lemma 1** *A system* $C$ *provides a digital* $(t, m, s)$-*net over* $\mathbb{F}_q$ *if and only if every subsystem* $C \in S(C, t, m)$ *($m-t$ vectors of* $\mathbb{F}_q^m$ *) is linearly independent over* $\mathbb{F}_q$.

**Proof** Combine [Nie92, Theorem 4.28] and [SW96, Lemma 2]. $\square$

In a quite similar way as in Definition 3 we define digital $(t, s)$-sequences constructed over $\mathbb{F}_q$.

**Definition 5** *Let* $q$ *be a prime-power and let* $s \geqslant 1$ *be an integer. We choose* $\psi_r$ *for* $r \geqslant 0$ *with* $\psi_r(0) = 0$ *for all sufficiently large* $r$, $\eta_j^{(i)}$ *for* $1 \leqslant i \leqslant s$ *and* $j \geqslant 1$, *and* $c_{jr}^{(i)}$ *for* $1 \leqslant i \leqslant s$, $j \geqslant 1$, *and* $r \geqslant 0$ *as in Definition 3. For* $n = 0, 1, \ldots$ *let*

$$n = \sum_{r=0}^{\infty} a_r(n) q^r$$

*be the digit expansion of* $n$ *in base* $q$, *where* $a_r(n) \in Z_q$ *for* $r \geqslant 0$ *and* $a_r(n) = 0$ *for all sufficiently large* $r$. *We put*

$$x_n^{(i)} = \sum_{j=1}^{\infty} y_{nj}^{(i)} q^{-j} \qquad for\ n \geqslant 0\ and\ 1 \leqslant i \leqslant s,$$

*with*

$$y_{nj}^{(i)} = \eta_j^{(i)} \left( \sum_{r=0}^{\infty} c_{jr}^{(i)} \psi_r(a_r(n)) \right) \in Z_q \qquad for\ n \geqslant 0,\ 1 \leqslant i \leqslant s,\ and\ j \geqslant 1,$$

*and we assume that for each* $n \geqslant 0$ *and* $1 \leqslant i \leqslant s$ *we have* $y_{nj}^{(i)} < q - 1$ *for infinitely many* $j$. *If for some integer* $t \geqslant 0$ *the sequence*

$$x_n = \left( x_n^{(1)}, \ldots, x_n^{(s)} \right) \in I^s \qquad for\ n = 0, 1, \ldots$$

*is a* $(t, s)$-*sequence in base* $q$, *then it is called a digital* $(t, s)$-*sequence constructed over* $\mathbb{F}_q$.

**Remark 2**

- The condition on the $y_{nj}^{(i)}$ in Definition 5 is satisfied, for instance, if $\eta_j^{(i)}(0) = 0$ and, for $r \geq 0$, $c_{jr}^{(i)} = 0$ for $1 \leq i \leq s$ and all sufficiently large $j$ (compare with [Nie92, p.72]).

- In [NX96b, Definition 4.] Niederreiter and Xing define digital sequences in a slightly refined way. They call the original definition which we use here a "digital $(t, s)$-sequence in the narrow sense constructed over $\mathbb{F}_q$".

# 2 An improved lower bound for the quality parameter of binary digital sequences

In [SW96, Proposition 1] an improved lower bound was provided for the quality parameter of digital nets over $\mathbb{F}_q$ which is the best possible result that can be obtained by the "method of linear combinations". We use this bound both to obtain an improved lower bound for the quality parameter of binary digital sequences and, in the next section, to underline the quality of the new shift method.

**Proposition 1 ([SW96, Proposition 1])** *Suppose that for some integers $s \geq 1$, $t \geq 0$, and $k \geq 2$ there exists a digital $(t, t + k, s)$-net over $\mathbb{F}_q$. Then we must have*

$$\sum_{u=1}^{\lfloor \frac{k}{2} \rfloor} \sum_{l=1}^{u} \binom{s}{l} \binom{u-1}{l-1} (q-1)^l q^{u-l} < q^{t+k} .$$

In the following we will determine lower bounds for the quality parameter of digital sequences.

**Definition 6** *For any integers $m \geq 2$, $s \geq 1$ and any prime power $q$ let $d_q(m, s)$ be the least value of $t$ such that there exists a digital $(t, m, s)$-net over $\mathbb{F}_q$, and $d_q(s)$ be the least value of $t$ such that there exists a digital $(t, s)$-sequence constructed over $\mathbb{F}_q$.*
*Furtheron we define*

$$L_q := \liminf_{s \to \infty} \frac{d_q(s)}{s} .$$

Niederreiter and Xing have shown that for any integer $s \geq 1$ and any prime power $q$ there exists a digital $(t, s)$-sequence over $\mathbb{F}_q$ with $d_q(s) = O(s)$ [NX96a, Corollary 2] and that $L_q \leq \frac{q+1}{q-1}$ [XN95, Corollary 1].

It was first shown in [LS95] that the bound $d_q(s) = O(s)$ is best possible as far as the order of magnitude in $s$ is concerned. The until now best lower bound is given by Niederreiter and Xing, due to a result of Lawrence [Law95], in the following form:

**[NX96b, Theorem 8]** *If for some integers $s \geqslant 1$, $t \geqslant 0$, and $b \geqslant 2$ there exists a general $(t, s)$-sequence in base $b$, then we must have*

$$t \geqslant \frac{s}{b} - \log_b \frac{(b-1)s + b + 1}{2} .$$

From this stronger bound for general $(t, s)$-sequences we clearly have

$$d_q(s) \geqslant \frac{s}{q} - \log_q \frac{(q-1)s + q + 1}{2}$$

and therefore $L_q \geqslant \frac{1}{q}$. For the special and for quasi-Monte Carlo applications most important case of $\mathbb{F}_2$ we will show the following improvement:

**Corollary 1** *For any dimension $s \geqslant 1$ we have*

$$d_2(s) > s \log_2 \frac{3}{2} - 4 \log_2(s-2) - 23$$

*and therefore*

$$3 \geqslant L_2 \geqslant \log_2 \frac{3}{2} .$$

*This is the best possible result for $L_2$ that can be obtained by the "method of linear combinations", i.e. from Proposition 1.*

In the proof, in several cases we make use of the following facts which will not explicitly be mentioned:

- $M(\log M - 1) + 1 \leqslant \sum_{i=1}^{M} \log i \leqslant (M+1)(\log(M+1) - 1) + \delta$ with $\delta := 2(1 - \log 2) > 0 \quad$ for $M \in \mathbb{N}$

- $\log(x - 1) \geqslant \log x - \frac{2}{x} \quad$ for $x \geqslant \frac{4}{3}$

**Proof of Corollary 1** Suppose that there exists a digital $(t, s)$-sequence over $\mathbb{F}_2$. By [NX96a, Lemma 1] this implies, for all $k \geqslant 1$, the existence of a digital $(t, t + 2k, s + 1)$-net over $\mathbb{F}_2$. Proposition 1 leads to

$$d_2(s) > \max_{k \in \mathbb{N}} \left( -2k + \log_2 \left( \sum_{u=1}^{k} \sum_{l=1}^{u} \binom{s+1}{l} \binom{u-1}{l-1} 2^{u-l} \right) \right)$$

Let now $\kappa = \kappa(s) \geqslant \frac{1}{s}$ and $\lambda = \lambda(s) \in [\frac{1}{s}, \kappa]$ be two real numbers such that $\kappa s$ is the maximal $k$ in the above inequality and $\lambda s$ is the summation index of the inner sum for which the product of the inner sum is maximal. Then we have

$$d_2(s) > -2\kappa s + \log_2\left(\binom{s+1}{\lambda s}\binom{\kappa s - 1}{\lambda s - 1}2^{\kappa s - \lambda s}\right)$$

$$\geq s(-\kappa - \lambda) + \frac{1}{\log 2}\left(\log\binom{s+1}{\lambda s} + \log\binom{\kappa s - 1}{\lambda s - 1}\right)$$

$$\geq s \cdot \left(-\kappa - \lambda + \frac{1}{\log 2}(\kappa\log\kappa - 2\lambda\log\lambda -\right.$$

$$(1-\lambda)\log(1-\lambda) - (\kappa - \lambda)\log(\kappa - \lambda)) \bigg) +$$

$$\frac{1}{\log 2}\bigg(-4\log s - \log\kappa - \log\lambda - 2\log(1-\lambda) - \log(\kappa - \lambda) -$$

$$4 - 4\delta + \frac{2}{\kappa s} - \frac{2}{\lambda s} - \frac{2}{s - \lambda s + 1} - \frac{4}{s - \lambda s} - \frac{2}{\kappa s - \lambda s}\bigg).$$

To find the optimal $\kappa$ and $\lambda$ we have to optimize the terms of $s$, and therefore to solve the following two equations:

$$0 = -1 + \frac{1}{\log 2}(\log\kappa - \log(\kappa - \lambda))$$

$$0 = -1 + \frac{1}{\log 2}(-2\log\lambda + \log(1-\lambda) + \log(\kappa - \lambda)).$$

It is easy to see that $\kappa = \frac{2}{3}$ and $\lambda = \frac{1}{3}$ are the only solutions.

Suppose now that $3|s$. From the subsequent Lemma 2b) we have that

$$d_2(s+2) \geq d_2(s+1) \geq d_2(s) > s \cdot \left(-1 + \frac{\log 3}{\log 2}\right) - 4\frac{\log s}{\log 2} - 21$$

and the result for $d_2(s)$ for every dimension $s \geq 1$ follows.

It remains to show that, by the "method of linear combinations", no better results for $L_2$ can be obtained. Let $\kappa$ and $\lambda$ be be defined like above. Then

$$\max_{k \in \mathbb{N}}\left(-2k + \log_2\left(\sum_{u=1}^{k}\sum_{l=1}^{u}\binom{s+1}{l}\binom{u-1}{l-1}2^{u-l}\right)\right)$$

$$\leq -2\kappa s + \log_2\left((2\kappa s)^2\binom{s+1}{\lambda s}\binom{\kappa s - 1}{\lambda s - 1}2^{\kappa s - \lambda s}\right)$$

$$\leq s(-\kappa - \lambda) + \frac{1}{\log 2}\left(2\log(\kappa s) + \log\binom{s+1}{\lambda s} + \log\binom{\kappa s - 1}{\lambda s - 1}\right)$$

$$\leq s \cdot \left(-\kappa - \lambda + \frac{1}{\log 2}(\kappa\log\kappa - 2\lambda\log\lambda -\right.$$

$$(1-\lambda)\log(1-\lambda) - (\kappa - \lambda)\log(\kappa - \lambda)) \bigg) +$$

$$\frac{1}{\log 2}\bigg(4\log s + 2\log\kappa + \log\lambda - \log(1-\lambda) + 2\delta +$$

$$\frac{2}{s+1} + \frac{4}{s} - \frac{2}{\lambda s}\Big).$$

We have seen above that $\kappa = \frac{2}{3}$, $\lambda = \frac{1}{3}$ is the optimal choice for $\kappa$ and $\lambda$. So

$$\max_{k \in \mathbb{N}} \left(-2k + \log_2 \left(\sum_{u=1}^{k} \sum_{l=1}^{u} \binom{s+1}{l}\binom{u-1}{l-1}2^{u-l}\right)\right)$$

$$\leq s \cdot \left(-1 + \frac{\log 3}{\log 2}\right) + 4\frac{\log s}{\log 2} + 2$$

and the result follows. $\square$

**Remark 3** For small values of $s$ the best results are obtained for a direct computation of

$$d_2(s) > -2\left\lfloor\frac{2s}{3}\right\rfloor + \log_2\left(\sum_{u=1}^{\lfloor 2s/3\rfloor} \sum_{l=1}^{u} \binom{s+1}{l}\binom{u-1}{l-1}2^{u-l}\right)$$

and therefore we have:

TABLE 1. Lower bounds for $d_2(s)$ for some small dimensions $s$

| $s$ | 1 | 2 | 3 | 4 | 5 | 6 | 7 | 8 | 9 | 10 | 11 | 12 | 13 | 14 |
|---|---|---|---|---|---|---|---|---|---|---|---|---|---|---|
| $d_2(s) \geq$ | 0 | 0 | 1 | 1 | 2 | 2 | 3 | 3 | 4 | 4 | 5 | 5 | 6 | 6 |

**Remark 4** We can generalize Corollary 1 to:
For any prime power $q$ we have

$$L_q \geq \log_q\left(1 + \frac{1}{q}\right)$$

which again is the best possible result that can be obtained by the "method of linear combinations" (the proof is the same as in the binary case). However, for $q \geq 3$, this result is worse than $L_q \geq \frac{1}{q}$ from [NX96b, Theorem 8].

# 3  Shift–nets: a new class of binary digital $(t, m, s)$-nets

The essential properties of a digital $(t, m, s)$-net depend on the properties of the first $m - t$ row vectors of each of the matrices $(C^{(1)}, \ldots, C^{(s)})$. The first row vector is the most important one, the priority, i.e. the importance for the properties of a digital net, decreases with the row number (see Lemma 1).

Therefore it is no surprise that until now, all construction methods are in some sense based on the rows of the matrices (see for example [LLNS96], [LMMS96], and [NX96b]).

In a search for new "good" row vectors for binary nets in small dimensions (for example $(t, 5, 5)$- or $(t, 7, 7)$-nets) we used the concept of $t$-configurations to construct the rows of higher priority, and we had success by using a method which we call shift method. We have extended this method to a whole system $C$ of matrices where the dimension $s$ is equal to $m$, the order of the matrices.

The essential point for the new method is, that we consider the first $m \times m$ matrix not as a system of row vectors, but as a system of $m = s$ column vectors. The other matrices are built by a shift to the left; the column vector shifted out on the left side is added on the right side.

$$
\begin{array}{ccc}
C^{(1)} & C^{(2)} & C^{(s)} \\
\begin{pmatrix} \vdots & \vdots & & \vdots \\ \mathbf{c}_1 & \mathbf{c}_2 & \cdots & \mathbf{c}_s \\ \vdots & \vdots & & \vdots \end{pmatrix} &
\begin{pmatrix} \vdots & & \vdots & \vdots \\ \mathbf{c}_2 & \cdots & \mathbf{c}_s & \mathbf{c}_1 \\ \vdots & & \vdots & \vdots \end{pmatrix} \cdots &
\begin{pmatrix} \vdots & \vdots & & \vdots \\ \mathbf{c}_s & \mathbf{c}_1 & \cdots & \mathbf{c}_{s-1} \\ \vdots & \vdots & & \vdots \end{pmatrix}
\end{array}
$$

So we have to find a proper matrix $C^{(1)}$. For a better explanation of the search for $C^{(1)}$ it is convenient to give the following definition:

**Definition 7** *Let $t \geqslant 0$, $k \geqslant 2$, $s \geqslant 1$, and $t + k \geqslant d \geqslant 1$ be integers. A system $\left\{ \mathbf{c}_j^{(i)} \in \mathbb{F}_q^{t+k} : 1 \leqslant j \leqslant t + k, 1 \leqslant i \leqslant s \right\}$ is called a $(k, t+k, d, s)$-system over $\mathbb{F}_q$ if any subsystem $\left\{ \mathbf{c}_j^{(i)} \in \mathbb{F}_q^{t+k} : 1 \leqslant j \leqslant d_i, 1 \leqslant i \leqslant s \right\}$ with $\sum_{i=1}^{s} d_i = k$ and $0 \leqslant d_1, \ldots, d_s \leqslant d$ is linearly independent over $\mathbb{F}_q$.*

**Remark 5** $C$ provides a digital $(t, t + k, s)$-net over $\mathbb{F}_q$ if and only if $C$ is a $(k, t + k, k, s)$-system over $\mathbb{F}_q$.

Now, for given integers $t \geqslant 0$ and $m \geqslant t + 2$, we want to examine, in a constructive way, if there exists a binary $(t, m, m)$-shift-net. We take an arbitrary first row vector $\mathbf{c}_1^{(1)} \in \mathbb{F}_2^m$ and construct $\mathbf{c}_1^{(2)}, \ldots, \mathbf{c}_1^{(m)}$ by the shift method. Then we examine if this is a $(m - t, m, 1, m)$-system over $\mathbb{F}_2$. If so, we will, in a same way, add second row vectors $\mathbf{c}_2^{(1)}, \ldots, \mathbf{c}_2^{(m)}$ to the system and examine if this is a $(m - t, m, 2, m)$-system over $\mathbb{F}_2$. Otherwise we have to take another first row vector and start again. We will continue this procedure until we have found a $(m - t, m, m - t, m)$-system and therefore a digital $(t, m, m)$-net over $\mathbb{F}_2$.

For the examination of a $(k, t+k, d, s)$-system over $\mathbb{F}_2$ we have to consider

$$\sum_{u=1}^{d}\sum_{l=1}^{u}\binom{s}{l}\binom{u-1}{l-1}2^{u-l}$$

linear combinations — all of them have to be linearly independent over $\mathbb{F}_2$. The rough ideas can be found in the proof of [SW96, Proposition 1], details will be given in [Sch96].

For the existence of a digital $(t, m, m)$-shift–net over $\mathbb{F}_2$, we have observed the following restrictions on the row vectors of $C^{(1)}$:

- W.l.o.g. we may choose $c_1^{(1)} = (0, 0, \ldots, 0, 1)$

- Then all of the following rows may have the form $(\ldots, 0)$

- For $2 \leqslant j \leqslant m - t$ we have: $\#1_j \geqslant m - t - (j - 1)$, where $\#1_j$ denotes the number of ones in the $j$-th row

We also have observed that, for the second row, we have $\#1_2 \leqslant t + 2$, but for growing dimension this information was not very helpful. So in most cases we restricted our search to row vectors $c_j^{(1)}$ with exactly $m - t - (j - 1)$ ones for $2 \leqslant j \leqslant m - t$.

Up to dimension $m = 16$ we have ignored this restriction and carried out an exhaustive search, i.e. for each $2 \leqslant j \leqslant m - t$ we searched over all vectors $c_j^{(1)} \in \mathbb{F}_2^m$. This means that up to dimension $m = 16$ there does not exist a binary $(t, m, m)$-shift–net with a better quality parameter than that given in Table 2.

In Table 2 we compare the quality parameters of binary $(t, m, m)$-shift–nets ("shift $t$") with the until now best known quality parameters of all (not only digital) $(t, m, m)$-nets in base 2 ("old $t$") (see the tables of net parameters given in [CLM+96]) and also with the best lower bounds for $d_2 = d_2(m, m)$ of the quality parameter $t$. Furtheron, by ⇑ • ⇓ we indicate for each dimension $s = m$ the number of parameters which were improved by the shift method and by the propagation rules of the subsequent Lemma 2 applied to the shift parameters.

**Remark 6**

- All of the best lower bounds for $d_2(m, m)$ are deduced from [SW96, Proposition 1], only for $m = 8$ (denoted by an asterisk) we have $t \geqslant 3$ by [SW96, Lemma 4] instead of $t \geqslant 2$.

- By the shift method and by the propagation rules applied to the shift parameters we have improved more than 250 of the until now best known quality parameters for $(t, m, s)$-nets (not only digital nets) in base 2. (Compare with the tables of net parameters given in [CLM+96]: tag 21 denotes the improvements by the shift method, tag 2, 3, and 16 denote three of the propagation rules.)

TABLE 2. Improvements for binary digital $(t, m, m)$-nets by the shift method

| $m$ | 3 | 4 | 5 | 6 | 7 | 8 | 9 | 10 | 11 | 12 | 13 | 14 | 15 | 16 |
|---|---|---|---|---|---|---|---|---|---|---|---|---|---|---|
| old $t$ | 0 | 1 | 1 | 2 | 3 | 4 | 5 | 5 | 6 | 7 | 8 | 9 | 10 | 11 |
| shift $t$ | 0 | 1 | 1 | 2 | 2 | 3 | 3 | 4 | 4 | 5 | 6 | 6 | 7 | 8 |
| $d_2 \geqslant$ | 0 | 1 | 1 | 2 | 2 | 3* | 3 | 3 | 4 | 4 | 5 | 5 | 6 | 6 |
| ⇑ • ⇓ | 0 | 0 | 0 | 0 | 1 | 2 | 4 | 4 | 7 | 8 | 7 | 9 | 9 | 11 |

| $m$ | 17 | 18 | 19 | 20 | 21 | 22 | 23 | 24 | 25 | 26 | 27 | 28 |
|---|---|---|---|---|---|---|---|---|---|---|---|---|
| old $t$ | 11 | 11 | 12 | 13 | 14 | 15 | 16 | 17 | 18 | 19 | 20 | 20 |
| shift $t$ | 8 | 9 | 10 | 10 | 11 | 12 | 13 | 14 | 15 | 16 | 17 | 18 |
| $d_2 \geqslant$ | 6 | 7 | 7 | 8 | 8 | 9 | 9 | 9 | 10 | 11 | 11 | 11 |
| ⇑ • ⇓ | 11 | 13 | 13 | 13 | 13 | 13 | 14 | 13 | 12 | 12 | 11 | 11 |

| $m$ | 29 | 30 | 31 | 32 | 33 | 34 | 35 | 36 | 37 | 38 | 39 |
|---|---|---|---|---|---|---|---|---|---|---|---|
| old $t$ | 20 | 21 | 22 | 23 | 24 | 25 | 26 | 27 | 28 | 29 | 30 |
| shift $t$ | 19 | 20 | 21 | 22 | 23 | 24 | 25 | 26 | 27 | 28 | 29 |
| ⇑ • ⇓ | 11 | 9 | 8 | 7 | 6 | 5 | 4 | 3 | 2 | 1 | 1 |

- Since the propagation rules for digital nets in Lemma 2 are constructive, all of these improved nets explicitly exist.

It is an important fact that the so-called propagation rules for $(t, m, s)$-nets in an arbitrary base $b \geqslant 2$ also hold for digital nets.

**Lemma 2** *Let $q$ be a prime power and $t \geqslant 0$, $m \geqslant t$, and $s \geqslant 1$ be integers.*

a) *Every digital $(t, m, s)$-net over $\mathbb{F}_q$ is a digital $(u, m, s)$-net over $\mathbb{F}_q$ for $t \leqslant u \leqslant m$.*

b) *For $1 \leqslant r \leqslant s$, every digital $(t, m, s)$-net over $\mathbb{F}_q$ can be transformed into a digital $(t, m, r)$-net over $\mathbb{F}_q$.*

c) *Every digital $(t, m, s)$-net over $\mathbb{F}_q$ can be transformed into a digital $(t, u, s)$-net over $\mathbb{F}_q$, where $t \leqslant u \leqslant m$.*

d) *Every digital $(t, m, s)$-net over $\mathbb{F}_q$ can be transformed into a digital $(t + u, m + u, s)$-net over $\mathbb{F}_q$ for any $u \in \mathbb{N}$.*

These propagation rules also hold for the general case of digital nets constructed over finite commutative rings with identity. a), b), and d) are trivial, c) was shown in [SW96, Lemma 3c)].

It is an open problem, also for general nets in base $b$, if the existence of a $(t, m, s)$-net implies the existence of a $(t+u, m+u, s+u)$-net for any $u \in \mathbb{N}$. We think that this propagation rule also will be true. For the case of binary shift–nets, numerous numerical experiments underline this presumption, so we will formulate it as a

**Conjecture** *Let $t \geqslant 0$ and $m \geqslant t$ be integers.*

*If there exists a binary $(t, m, m)$-shift–net, then for each $u \in \mathbb{N}$ there exists a binary $(t + u, m + u, m + u)$-shift–net.*

# 4 Appendix: Matrices providing the binary shift–nets

One of the main aims in improving parameters of digital nets is to provide low-discrepancy point sets of highest quality for quasi-Monte Carlo applications in scientific computing. In the appendix we will give all necessary information to implement the binary shift–nets (see Definition 3) in the following compact form: for a binary $(m - k, m, m)$-shift–net we write the first $k$ row vectors $\mathbf{c}_j^{(1)} = (c_{j,0}^{(1)}, \ldots, c_{j,m-1}^{(1)})$, $1 \leqslant j \leqslant k$, of $C^{(1)}$ as

$$\left( \sum_{r=0}^{m-1} c_{1,m-1-r}^{(1)} 2^r, \ldots, \sum_{r=0}^{m-1} c_{k,m-1-r}^{(1)} 2^r \right).$$

Note that almost all (but not exactly all) of the row vectors are of the form described in the section before: for $2 \leqslant j \leqslant k$ there are $k - (j - 1)$ ones in $\mathbf{c}_j^{(1)}$.

$$
\begin{aligned}
(0,3,3) &: (1,6,2) \\
(1,4,4) &: (1,6,2) \\
(1,5,5) &: (1,14,18,2) \\
(2,6,6) &: (1,14,18,2) \\
(2,7,7) &: (1,46,26,6,2) \\
(3,8,8) &: (1,46,26,6,2) \\
(3,9,9) &: (1,94,178,38,10,2) \\
(4,10,10) &: (1,94,170,38,10,2) \\
(4,11,11) &: (1,492,1638,86,14,18,2) \\
(5,12,12) &: (1,378,590,158,42,6,2) \\
(6,13,13) &: (1,366,572,154,14,18,2) \\
(6,14,14) &: (1,734,10652,372,46,22,10,2) \\
(7,15,15) &: (1,734,3308,372,46,22,10,2) \\
(8,16,16) &: (1,758,3132,414,46,22,10,2) \\
(8,17,17) &: (1,3562,29276,870,62,198,74,6,2) \\
(9,18,18) &: (1,3450,9814,2020,458,54,26,6,2) \\
(10,19,19) &: (1,3562,8566,1884,236,30,38,10,2) \\
(10,20,20) &: (1,10094,301178,17650,2300,314,78,22,10,2) \\
(11,21,21) &: (1,10106,54364,3974,4214,158,294,42,62,2) \\
(12,22,22) &: (1,10198,54476,18296,2674,124,150,26,6,2) \\
(m-10,m,m) &\quad \text{for } 23 \leqslant m \leqslant 39 : \\
&\quad (1,10094,51322,22822,190,626,270,22,10,2)
\end{aligned}
$$

*Acknowledgments:* We would like to thank Peter Zinterhof, the head of our Mathematics Department. In his position as head of the Research Institute for Software Technology (RIST++) he enabled us to use several DEC 3000 AXP/400 workstations for our extensive calculations. Without this computer power we never would have had that success in constructing shift–nets of such a quality.

## 5  REFERENCES

[CLM+96]  A. T. Clayman, K. M. Lawrence, G. L. Mullen, H. Niederreiter, and N. J. A. Sloane. Updated tables of parameters of $(t, m, s)$-nets. Preprint, 1996.

[Law95]  K. M. Lawrence. *Combinatorial Bounds and Constructions in the Theory of Uniform Point Distributions in Unit Cubes, Connections with Orthogonal Arrays and a Poset Generalization of a Related Problem in Coding Theory.* PhD thesis, University of Wisconsin, 1995.

[LLNS96]  G. Larcher, A. Lauß, H. Niederreiter, and W. Ch. Schmid. Optimal polynomials for $(t, m, s)$-nets and numerical integration of multivariate Walsh series. *SIAM J. Numer. Analysis*, **33**-6, 1996.

[LMMS96]  K. M. Lawrence, A. Mahalanabis, G. L. Mullen, and W. Ch. Schmid. Construction of digital $(t, m, s)$-nets from linear codes. In S. Cohen and H. Niederreiter, editors, *Finite Fields and Applications (Glasgow, 1995)*, volume 233 of *Lect. Note Series of the London Math. Soc.*, pages 189–208. Camb. Univ. Press, Cambridge, 1996.

[LNS96]  G. Larcher, H. Niederreiter, and W. Ch. Schmid. Digital nets and sequences constructed over finite rings and their application to quasi-Monte Carlo integration. *Monatsh. Math.*, **121**:231–253, 1996.

[LS95]  G. Larcher and W. Ch. Schmid. Multivariate Walsh series, digital nets and quasi-Monte Carlo integration. In H. Niederreiter and P. J.-S. Shiue, editors, *Monte Carlo and Quasi-Monte Carlo Methods in Scientific Computing (Las Vegas, 1994)*, volume 106 of *Lecture Notes in Statistics*, pages 252–262. Springer, New York, 1995.

[LSW95]  G. Larcher, W. Ch. Schmid, and R. Wolf. Digital $(t, m, s)$-nets, digital $(T, s)$-sequences, and numerical integration of multivariate Walsh series. In P. Hellekalek, G. Larcher, and P. Zinterhof, editors, *Proceedings of the 1st Salzburg Minisymposium* on *Pseudorandom Number Generation and Quasi-Monte Carlo Methods, Salzburg, Nov. 18, 1994*, volume 95–4 of *Technical Report Series*, pages 75–107. ACPC – Austrian Center for Parallel Computation, 1995.

[Nie87]  H. Niederreiter. Point sets and sequences with small discrepancy. *Monatsh. Math.*, **104**:273–337, 1987.

[Nie92]  H. Niederreiter. *Random Number Generation and Quasi-Monte Carlo Methods*. Number **63** in CBMS–NSF Series in Applied Mathematics. SIAM, Philadelphia, 1992.

[NX96a]  H. Niederreiter and C. P. Xing. Low-discrepancy sequences and global function fields with many rational places. *Finite Fields Appl.*, **2**:241–273, 1996.

[NX96b]  H. Niederreiter and C. P. Xing. Quasirandom points and global function fields. In S. Cohen and H. Niederreiter, editors, *Finite Fields and Applications (Glasgow, 1995)*, volume 233 of *Lect. Note Series of the London Math. Soc.*, pages 269–296. Camb. Univ. Press, Cambridge, 1996.

[Sch96]  W. Ch. Schmid. An algorithm to determine the quality parameter of digital nets in base 2. Preprint, University of Salzburg, 1996.

[SW97]  W. Ch. Schmid and R. Wolf. Bounds for digital nets and sequences. *Acta Arith.*, **78**:377–399, 1997.

[XN95]  C. P. Xing and H. Niederreiter. A construction of low-discrepancy sequences using global function fields. *Acta Arith.*, **73**:87–102, 1995.

Institut für Mathematik
Universität Salzburg
Hellbrunnerstraße 34
A–5020 Salzburg
Austria
e-mail: WOLFGANG.SCHMID@SBG.AC.AT

# General Sequential Sampling Techniques for Monte Carlo Simulations: Part I - Matrix Problems

Jerome Spanier
Liming Li

ABSTRACT We study sequential sampling methods based principally on
ideas of Halton. Such methods are designed to build information drawn
from early batches of random walk histories into the random walk process
used to generate later histories in order to accelerate convergence. In previously published work, such methods have been applied within the pseudorandom, rather than the quasirandom, context and have been applied
only to matrix problems. In this paper, more general sequential techniques
are formulated in an abstract space, such as Banach space. The more general formulation enables applications to linear algebraic equations and to
integral equations to be obtained as special cases through specification of
the Banach space and the operator defined on it. In this initial paper we
outline the ideas needed for consideration of the more general problem and
exhibit greatly accelerated convergence for a simple matrix problem. In a
companion paper in which similar ideas are applied to the more important class of integral equations, the need for quasirandom implementation
is stressed.

## 1 Introduction

A promising direction for improvement of Monte Carlo convergence properties is the use of sequential sampling techniques (see, for example, [1],
[2], [3] and [4]). Such techniques are designed to build information drawn
from early batches of random walk histories into the random walk process
used to generate later histories. When implemented properly, it should be
possible to achieve much faster convergence of the statistical sampling error to zero than the asymptotic $O(N^{-\frac{1}{2}})$ rate promised by the central limit
theorem.

It is to be expected, however, that implementation of any such sequential strategy will incur additional costs. Clearly, then, the error reduction
achieved must more than offset these additional costs if real improvement
is to be the result. If one defines, as is usual, the *efficiency*

$$E = \frac{1}{VT}$$

where $V$ = appropriate measure of error, such as the relative variance or the relative variation of the estimator used, and $T$ = total computation time, then the extra computational burden imposed by sequential sampling will be reflected in increased values of $T$. The reductions in $V$ per iteration that result from the use of sequential sampling may or may not offset such additional costs by enough to yield substantial gains. That is, it is not *a priori* clear that anticipated exponential increases in convergence (reductions in $V$) will result in exponential increases in overall efficiency. This is one of the central issues that arises when sequential sampling methods are attempted.

In [1], Halton discussed sequential sampling methods in the context of the matrix problem

$$\mathbf{x} = A\mathbf{x} + \mathbf{a} \tag{1}$$

where $A$ is an $s \times s$ matrix, $\mathbf{x}$ and $\mathbf{a}$ are $s$-vectors. Briefly, [1] suggests the possibility of utilizing both biased and unbiased estimators for solving Equation (1) sequentially but no specifics are offered in [1] for their use. However, in [1] and in a recent paper [2], Halton reports that the use, instead, of a sequential form of correlated sampling produces convergence superior to that achieved by using either biased or unbiased estimators, and he bases his conclusions in these papers on this latter strategy exclusively.

We summarize Halton's sequential mechanism by observing that if the solution of Equation (1) is represented as

$$\mathbf{x} = \hat{\mathbf{x}} + \mathbf{y}$$

where $\hat{\mathbf{x}}$ is an approximation to $\mathbf{x}$, then the needed correction $\mathbf{y}$ must satisfy

$$\mathbf{y} = A\mathbf{y} + \mathbf{d}$$

where

$$\mathbf{d} = \mathbf{a} + A\hat{\mathbf{x}} - \hat{\mathbf{x}}$$

is the equation error, or residual (see [5, Chapter 5, Section 5.1]), associated with using $\hat{\mathbf{x}}$ in place of $\mathbf{x}$ to solve Equation (1). Then, as $\hat{\mathbf{x}}$ approaches $\mathbf{x}$, the vectors $\mathbf{d}$ and $\mathbf{y}$ approach $\mathbf{0}$ and the effect of an error in estimating $\mathbf{y}$ is correspondingly reduced in terms of its relative impact on the solution $\mathbf{x}$.

To apply this idea sequentially, Halton begins with an initial approximation $\hat{\mathbf{y}}^{(0)}$ to the solution of Equation (1) and defines $\mathbf{x}^{(1)} = \hat{\mathbf{y}}^{(0)}$. Then with

$$\mathbf{d}^{(1)} \quad = \quad \mathbf{a} + A\mathbf{x}^{(1)} - \mathbf{x}^{(1)},$$

denote by $\hat{\mathbf{y}}^{(1)}$ an approximate solution of

$$\mathbf{y} = A\mathbf{y} + \mathbf{d}^{(1)}$$

and define

$$\begin{aligned} \mathbf{x}^{(2)} &= \mathbf{x}^{(1)} + \hat{\mathbf{y}}^{(1)} = \hat{\mathbf{y}}^{(0)} + \hat{\mathbf{y}}^{(1)} \\ \mathbf{d}^{(2)} &= \mathbf{a} + A\mathbf{x}^{(2)} - \mathbf{x}^{(2)}. \end{aligned}$$

Continuing in this fashion produces the recursions

$$\left\{ \begin{aligned} \mathbf{x}^{(k)} &= \mathbf{x}^{(k-1)} + \hat{\mathbf{y}}^{(k-1)} \\ \mathbf{d}^{(k)} &= \mathbf{a} + A\mathbf{x}^{(k)} - \mathbf{x}^{(k)} \end{aligned} \right\} \quad \mathbf{x}^{(0)} = 0, \qquad k = 1, 2, \dots \qquad (2)$$

where $\hat{\mathbf{y}}^{(k-1)}$ is an approximate solution to

$$\mathbf{y} = A\mathbf{y} + \mathbf{d}^{(k-1)}.$$

It follows easily that

$$\mathbf{d}^{(k)} = \mathbf{d}^{(k-1)} + A\hat{\mathbf{y}}^{(k-1)} - \hat{\mathbf{y}}^{(k-1)}$$

and

$$\mathbf{x}^{(k)} = \sum_{i=0}^{k-1} \hat{\mathbf{y}}^{(i)}.$$

The main result of Halton [1], [2] is that, under appropriate restrictions, geometric convergence of $\mathbf{x}^{(k)}$ to $\mathbf{x}$ is possible for this sequential method. Since Halton's implementation uses pseudorandom numbers, this geometric convergence should presumably be regarded as holding in a statistical, not a deterministic sense.

We observe that implementation of Halton's sequential algorithm requires the approximate solution of the original problem (with a reduced source, $\mathbf{d}^{(k)}$), and the computation of this reduced source itself requires application of the operator $A$ to such an approximate solution, for each iteration. These operations impose a burden of additional computation but, at least in the matrix case (which appears to have been Halton's main concern), no intrinsic difficulty. In the integral equation applications which primarily interest us, however, more fundamental difficulties arise, as we explore in [6].

These problems notwithstanding, Halton reports in both [1] and [2] the effectiveness of his sequential correlated sampling algorithm for solving very large linear systems, especially ones whose matrices are full. He provides some analytic support for the claim that sequential correlated sampling is more effective even than the best currently available deterministic methods for such matrices. We will compare his method with the one developed in this paper for very simple matrices in section 3.

## 2    Generalization to Operator Equations

The sequential matrix method described above can be extended to include very important integral transport equation problems by considering the operator transport equation

$$\phi = \mathcal{K}\phi + \sigma \tag{3}$$

where $\mathcal{K}$ is a linear operator from $\mathcal{B}$ to $\mathcal{B}$ (a Banach space). We assume that the norm of the operator $\mathcal{K}$ is less than one ($\| \mathcal{K} \| < 1$) to guarantee that the Equation (3) has exactly one solution (although the weaker assumption $\| \mathcal{K}^{n_0} \| < 1$ for some integer $n_0 \geq 1$ will suffice; this follows from rather elementary fixed point theorems in $\mathcal{B}$).

A quite general operator formulation for solving Equation (3) iteratively results by introducing a sequential dependence in both the source $\sigma$ and kernel $\mathcal{K}$ via the operator equation

$$\phi = \mathcal{K}^{(k)}\phi + \sigma^{(k)} \qquad \sigma^{(0)} = \sigma, \quad \mathcal{K}^{(0)} = \mathcal{K}. \tag{4}$$

In this way, Halton's sequential correlated sampling results by selecting $\mathcal{K}^{(k)} = \mathcal{K}$ for all $k$ and

$$\sigma^{(k+1)} = d^{(k+1)} = d^{(k)} + \mathcal{K}\hat{y}^{(k)} - \hat{y}^{(k)}, \tag{5}$$

where $\hat{y}^{(k)}$ is an approximate solution to Equation (4), while various more general biased and unbiased importance sampling sequential methods result by modifying the kernel $\mathcal{K}$ suitably from one iteration to the next. Our interest in this paper is predominantly on the latter family of sequential methods, with a view to applying these ultimately to integral equations, thus avoiding the computation of the reduced source required by Halton's correlated sampling method.

A comprehensive theory of convergence for this more general operator formulation will likely not be simple since one can expect quite different convergence characteristics depending on the specifics of the problem and the details of the implementation method. For example, if we mimic Halton's treatment for matrix problems, we would compute an initial approximation $\hat{y}^{(0)}$ to $\phi$ and define

$$\phi^{(1)} = \hat{y}^{(0)}.$$

Then with

$$d^{(1)} \quad = \quad \sigma + \mathcal{K}\phi^{(1)} - \phi^{(1)},$$

we let $\hat{y}^{(1)}$ denote an approximate solution to

$$y = \mathcal{K}y + d^{(1)}$$

and define

$$\phi^{(2)} = \phi^{(1)} + \hat{y}^{(1)} = \hat{y}^{(0)} + \hat{y}^{(1)}.$$

Continuing produces the recursions

$$\left. \begin{array}{rcl} \phi^{(k)} & = & \phi^{(k-1)} + \hat{y}^{(k-1)} \\ d^{(k)} & = & \sigma + \mathcal{K}\phi^{(k)} - \phi^{(k)} \end{array} \right\} \quad \phi^{(0)} = 0, \qquad k = 1, 2, \dots \tag{6}$$

where $\hat{y}^{(k-1)}$ is an approximate solution of

$$y = \mathcal{K}y + d^{(k-1)}.$$

It follows that

$$d^{(k)} \quad = \quad d^{(k-1)} + \mathcal{K}\hat{y}^{(k-1)} - \hat{y}^{(k-1)}$$

and

$$\phi^{(k)} = \sum_{i=0}^{k-1} \hat{y}^{(i)}.$$

The convergence of this generalized sequential method will depend on how the reduced equations

$$y = \mathcal{K}y + d^{(k)}$$

are approximately solved to produce $\hat{y}^{(k)}$. If, for example, conventional (Picard) iteration

$$\phi^{(n+1)} = \mathcal{K}\phi^{(n)} + d^{(k)}, \ \phi^{(0)} = d^{(k)},$$

is employed and restricted to $n = 0$ (i.e., a single iteration is performed to obtain $\hat{y}^{(k)}$ from $d^{(k)}$), we get

$$\hat{y}^{(k)} = \phi^{(1)} = \mathcal{K}d^{(k)} + d^{(k)}$$

with the result that

$$\phi^{(k)} = \sum_{i=0}^{k-1} \hat{y}^{(i)} = (I + \mathcal{K})\sigma + \mathcal{K}^2\phi^{(k-1)}$$

and we have arrived back at the partial sums of the Neumann series for the solution $\phi$. This series converges provided $\|\mathcal{K}^2\| < 1$. Subject to this same assumption, it is easy to show that the error $e^{(k)} = \phi^{(k)} - \phi$ can be estimated practically using the formula

$$\left\| e^{(k)} \right\| \leq \left\| \phi^{(k+1)} - \phi^{(k)} \right\| \frac{1}{1 - \|\mathcal{K}^2\|} \tag{7}$$

which depends on the error upon successive iterations and the norm of $\mathcal{K}$. While this leads to a satisfactory theory of convergence, effective sequential methods are designed to do more than advance each random walk a single step beyond the source (as in Picard iterations) and thus, properly implemented, should do substantially better than this rather naive iterative procedure. Also, permitting iterative redefinition of the operator $\mathcal{K}$ vastly complicates the analysis of convergence in this very general setting. A different convergence theory will be derived in [6] for the integral equation applications by utilizing numerical quadrature effectively to replace the integral equation by a matrix equation and applying the sequential iterations to the matrix problem.

The more general operator formulation described above has two important applications, which otherwise have to be treated separately.

I. If $\mathcal{B} = \mathcal{R}^s$ or $\mathcal{C}^s$, regarded as $s-$ dimensional real or complex vector
spaces, $\mathcal{K} = A$ is an $s \times s$ real or complex matrix, $\phi = \mathbf{x}$, $\sigma = \mathbf{a}$ are $s-$
vectors, then Equation (3) reduces to Equation (1) and may be interpreted
as describing transport in an infinite homogeneous medium with $s$ discrete
energy states available for each particle.

II. If $\mathcal{B} = L^p(\mu)$ (or $\mathcal{B} = C(X)$, the space of continuous functions on
a normed space $X$), $\mathcal{K}$ an integral operator, Equation (3) becomes the
continuous transport equation

$$\psi(P) = \int_\Gamma K(P, P')\psi(P')dP' + S(P) \tag{8}$$

where

$$\mathcal{K}\psi(P) = \int_\Gamma K(P, P')\psi(P')dP'.$$

This integral equation may be shown to be the appropriate equation to
describe the directional particle collision density, $\psi(P) = \Sigma_t(P)\phi(P)$, where
$\Sigma_t(P)$ is the total macroscopic cross section[1] at $P$ and $\phi(P)$ is the particle
flux. In Equation (8), $P$, $P'$ denote generic points in a euclidean phase
space $\Gamma$, $S(P)$ is the density of first collisions, and $K(P, P')$ is the transport
kernel describing transitions from $P'$ to $P$.

In the context of Equation (8), it is usual to apply Monte Carlo methods
to the estimation of integrals of the form

$$I = \int_\Gamma g(P)\psi(P)dP \tag{9}$$

where $g$ is some known bounded nonnegative function (usually a ratio of
nuclear cross sections) and $\psi$ satisfies Equation (8). In the more abstract
setting of Equation (3), the integral (9) is realized as either as a pairing
of the Banach space $\mathcal{B}$ and its dual space $\mathcal{B}^*$, or, if $\mathcal{B} = \mathcal{H}$ is a Hilbert
space, as an inner product on $\mathcal{H}$. That is, the function $g$ can be regarded
as defining a bounded linear functional on the space $\mathcal{B}$ via the integral (9).

Reflecting on the recursions (6) in the integral equations setting, we
now see a possibly serious drawback to their practical use. Not only is a
continuous approximation $\tilde{\psi}^{(k)}$ to the solution of Equation (8) required at
each stage, but computation of $\mathcal{K}\tilde{\psi}^{(k)}$ as a component of the reduced source
is also needed. Such a computation is tantamount to advancing by exactly
one collision each term in a Neumann series representation of the problem
solution.

---

[1]The inverse of this function is the average distance travelled between successive collisions in an infinite homogeneous medium of material with cross section
$\Sigma_t(P)$ (the cross section $\Sigma_t(P)$ has dimensions inverse length).

In an effort to avoid these difficulties, and to keep computational costs of sequential methods at a minimum, we have chosen to explore other sequential algorithms for solving Equation (3) iteratively. Indeed, since progressively improved knowledge of at least an approximation to the solution of Equation (3) seems a prerequisite for the design of an effective sequential algorithm, our thought was to see if such information alone could be used to build appropriate knowledge systematically into a rapidly converging (unbiased or biased) importance sampling scheme based on the well established theory for such that can be found, for example, in [7, Chapter 3, Section 3.7]. This theory establishes a duality between the abstract formulation based on the pair of Equations (3) and the equation

$$I = < g, \phi > \tag{10}$$

where $\phi \in \mathcal{B}$, $g \in \mathcal{B}^*$ and the equation pair

$$\phi^* = \mathcal{K}^* \phi^* + g \tag{11}$$

and

$$I^* = < \phi^*, \sigma > . \tag{12}$$

where $\phi^* \in \mathcal{B}^*$. Here, $\mathcal{K}^*$ is the operator adjoint to $\mathcal{K}$, as defined by the identity $< \mathcal{K}^* \theta^*, \theta > = < \theta^*, \mathcal{K}\theta >$ for all $\theta \in \mathcal{B}$ and all $\theta^* \in \mathcal{B}^*$. This duality permits the estimation of the inner product $I = < g, \phi >$ as $I^* = < \phi^*, \sigma >$, where $\phi, \phi^*$ satisfy Equations (3), (11), respectively. Further, through this duality, problems may be formulated either in a direct or an equivalent adjoint mode. Through duality, random variables developed with respect to a given (discrete or continuous) equation have counterpart definitions with respect to the dual equation.

The theory of importance sampling based on this duality and applied to continuous problems can be found in [7, Chapter 3, Section 3.7], and a completely parallel theory for discrete problems can easily be developed. This theory reveals that the solution $\phi^*$ of Equation (11) acts as a perfect importance function for the estimation of the quantity (10) and conversely, that the solution $\phi$ of Equation (3) acts as a perfect importance function for the estimation of the quantity (12). While exact knowledge of $\phi$ or $\phi^*$ may not be assumed in designing Monte Carlo or quasi-Monte Carlo random walk simulations, approximations such as those generated in a sequential method should play a useful role in achieving reduced error sampling

In the present paper we shall restrict our attention to applications of type I above. That is, we will apply the conventional theory of Monte Carlo simulation of random walks for the purpose of solving matrix equations as established for example, in [7, Chapter 2, Section 2.3]. Applications to integral equations will be treated in [6]. For the solution of matrix problems the random walks take place on a discrete phase space $\{1, ..., s\}$, where $s$ is the order of the linear algebraic system under consideration.

Random variables that are useful in the discrete case originated, as far as we are aware, with von Neumann and Ulam (published initially in [8]; see also [9]). For our study we chose discrete versions of the family introduced in [10], a family that includes both biased and unbiased importance sampling estimators. Our first experiments, reported here, were performed with unbiased versions of such estimators which coincide with more conventional random variables produced by importance sampling transformations.

Specifically, for the solution of the matrix problem defined by Equation (1), we will construct a new transition matrix, $(q_{ij})_{s \times s}^{(m)}$ by using the approximate solution $\tilde{x}^{(m-1)}$ as an approximate (dual) importance function from iteration $m - 1$. The new transition matrix is then used to generate the next batch of random walk histories. For the $m$ th iterative stage, discrete random walks $\omega = (i_1, i_2, ..., i_k)$ are then generated using the matrix $(q_{ij})_{s \times s}^{(m)}$ instead of the original matrix $(A_{i,j})_{s \times s}$. To construct this matrix, dual importance sampling theory suggests that we should set

$$q_{ij}^{(m)} = \frac{A_{i,j}\, \tilde{x}_j^{(m-1)}}{\sum\limits_{l=1}^{s} A_{i,l}\, \tilde{x}_l^{(m-1)} + a_i} \tag{13}$$

where $\tilde{x}_j^{(m-1)}$ is the $j$ th component of the approximate solution $\tilde{x}^{(m-1)}$ at the $(m - 1)$ st iterative stage. Then random walks $\omega = (i_0 = j, i_1, i_2, ..., i_k)$ originating in the fixed state $j$ are generated using $q_{ij}^{(m)}$ to determine successor states $j$ from the current state $i$ and weights $W_n$ are defined by:

$$W_0 = 1, \quad W_n = W_{n-1} \frac{A_{i_{n-1}, i_n}}{q_{i_{n-1}, i_n}^{(m)}}.$$

It follows [7, Chapter 2, Section 2.3] that the random variable

$$\xi_j(\omega) = W_k a_{i_k} / p_{i_k}^{(m)}, \tag{14}$$

where $p_{i_k}^{(m)} = 1 - \sum_{l=1}^{s} q_{kl}^{(m)}$, is an unbiased estimator of the $j$ th component of x. When the approximate solution $\tilde{x}$ is close to the exact solution x, we expect the variance of the estimator (14) to be small. By originating random walks in each of the $s$ discrete states at each stage, we are able to estimate all $s$ components of the unknown vector x in each stage $m$.

We recognize that the estimator $\xi$ of Equation (14) is a terminal-type estimator[2] ([7, Chapter 2, Section 2.3]) implemented in the adjoint mode simulation process. This may be regarded as a random walk process in

---

[2] Collision-type estimators [7, Chapter 2, Section 2.3] were also used in the successful construction of sequential importance sampling algorithms. These results will be reported more fully elsewhere.

which each random walk starts at index $j$ and moves backwards from state to state until arriving at its terminus. In such a backwards simulation, the original source vector $a$ is used as a scoring function at the final collision point, as in Equation (14).

In the next section we will report some numerical experience demonstrating that the suggested sequential method produces very rapid convergence when the number of (pseudorandomly generated) random walk histories per iteration is chosen properly.

## 3    A Ten State Transport Problem

To test the efficiency of our sequential sampling method, we considered an inherently finite dimensional problem first: namely, a transport problem in an infinite, homogeneous material in which particles can occupy only a finite number $s$ of energy states. Particle collisions then can result either in absorption or in transition from the current energy state to any other, but spatial variation can be ignored.

Suppose the transition matrix in this material is $A = (A_{ij})_{s \times s}$ and the source vector $\mathbf{a} = (a_1, a_2, ..., a_s)$ satisfies $a_i \geq 0$ and $\sum_{i=1}^{s} a_i = 1$, where $a_i$ is interpreted as the density of particles that undergo a first collision in the $i$ th energy state. For our test we set

$$A_{ij} = \frac{s - |i - j|}{s^2} \; ; \tag{15}$$

Then the probability that a random walk moves to the $j$ th state from the $i$ th state depends only on the distance between the indices $i$ and $j$. The goal here is to find $x_i$, the density of particles that undergo collision in the $i$ th energy state. It is not difficult to see [7, Chapter 2, Section 2.2] that the above transport problem can be exactly described by a linear matrix system. To keep the problem manageable, we look at a transport problem involving only ten energy states: $s = 10$.

The transition matrix, $A$, is symmetric and its $\mathcal{L}^\infty$−norm,

$$\|A\|_\infty = \max_i \sum_{j=1}^{10} A_{ij} = 0.75.$$ This represents a problem of moderate difficulty in terms of the rate of convergence of the matrix Neumann series. The source vector is also chosen symmetrically:

$$\mathbf{a} = (0, 0, 0, .2, .3, .3, .2, 0, 0, 0)^T$$

so that the transport problem is symmetric in its ten states.

Using the sequential algorithm described above, we solved this problem approximately by generating (using pseudorandom numbers) 100 random walk histories for each iteration and iterating 10 times. For comparison, we also solved the problem using Halton's sequential correlated sampling

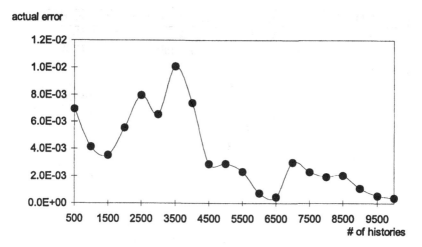

FIGURE 1. Ten state problem without sequential sampling

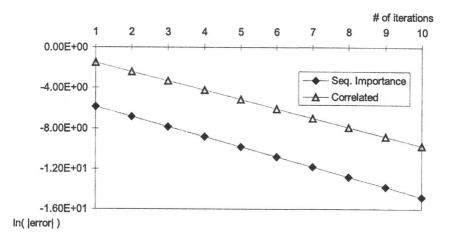

FIGURE 2. Ten state problem without sequential importance sampling and correlated sampling

method, also employing 10 sequential stages of 100 random walks each, and using plain Monte Carlo (i.e., no sequential sampling), for which we generated 10,000 random walks. The numerical results are graphed in Figures 1 and 2 in which plots of the errors in solving this $10 \times 10$ problem without and with the use of sequential sampling, respectively, are shown. A portion of the output data on which these graphs are based is displayed in Table 1. In order to emphasize the tremendous improvement obtained by using

| # of Histories | Non-Seq. Err. | Seq. Err. | Corr. Err. |
|----------------|---------------|-----------|------------|
| 100 | 2.53E-03 | 2.53E-03 | 2.97E-01 |
| 200 | 2.90E-03 | 1.08E-03 | 7.03E-02 |
| 300 | 2.17E-03 | 3.44E-04 | 4.39E-02 |
| 400 | 1.23E-02 | 1.18E-04 | 1.25E-02 |
| 500 | 6.95E-03 | 7.95E-05 | 6.82E-03 |
| 600 | 1.68E-03 | 1.78E-05 | 1.99E-03 |
| 700 | 1.40E-03 | 9.57E-06 | 6.07E-04 |
| 800 | 4.02E-03 | 3.74E-06 | 2.67E-04 |
| 900 | 5.08E-03 | 9.21E-07 | 1.59E-04 |
| 1000 | 4.14E-03 | 2.46E-07 | 8.97E-05 |

TABLE 1. Comparison of sequential & nonsequential errors

sequential sampling, Table 1 exhibits the errors obtained with and without sequential sampling for the first 1,000 random walks only, even though a total of 10,000 random walks were processed for the nonsequential run.

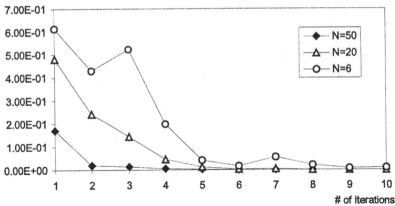

FIGURE 3. Influence on sequential convergence of the number of random walks per iteration: norm(A)=0.75

The actual error in Figure 1, and its logarithm in Figure 2, is a global measure of the error defined by

$$\text{actual error} = \frac{1}{10} \sum_{j=1}^{10} |\hat{x}_j - x_j|^2 \qquad (16)$$

where $\hat{x}_j$ is the estimated solution of the transport problem using sequential sampling, and $x_j$ is the exact solution of the transport problem.

The procedure above solves a linear system of equations whose exact solution can be found by other techniques, which makes it sensible to base our assessment of the approximation error on the actual error described above. However, for non-model problems, when $x_j$ is unknown, a different criterion is clearly needed. Instead of the actual error, we can use the average error (= norm of residual) defined by

$$\text{average error} = \frac{1}{10} \sum_{i=1}^{10} |\hat{x}_i - \sum_{j=1}^{10} A_{ij}\hat{x}_j - a_i|^2 \tag{17}$$

to examine the efficiency. The following theorem shows that the actual error (16) and the average error (17) provide equivalent measures of the accuracy of the solution.

**Theorem:** *The actual error $\to 0$ if and only if the average error $\to 0$.*
*Proof:* Because

$$|\hat{x}_i - x_i| = |\hat{x}_i - \sum_{j=1}^{10} A_{ij}\hat{x}_j - a_i + \sum_{j=1}^{10} A_{ij}\hat{x}_j + a_i - x_i|$$
$$\leq |\hat{x}_i - \sum_{j=1}^{10} A_{ij}\hat{x}_j - a_i| + \sum_{j=1}^{10} A_{ij}|\hat{x}_j - x_j|$$

and the norm of the transition matrix $||A||_\infty = \max_i \sum_{j=1}^{10} A_{ij} < 1$,

$$\max_i |\hat{x}_i - x_i| \leq \max_i |\hat{x}_i - \sum_{j=1}^{10} A_{ij}\hat{x}_j - a_i|/(1 - \max_i \sum_{j=1}^{10} A_{ij})$$

On the other hand,

$$|\hat{x}_i - \sum_{j=1}^{10} A_{ij}\hat{x}_j - a_i| = |\hat{x}_i - \sum_{j=1}^{10} A_{ij}\hat{x}_j - (x_i - \sum_{j=1}^{10} A_{ij}x_j)|$$
$$= |(\hat{x}_i - x_i) - \sum_{j=1}^{10} A_{ij}(\hat{x}_j - x_j)|$$
$$\leq \sum_{j=1}^{10} |\delta_{ij} - A_{ij}| \max_i |\hat{x}_i - x_i| \qquad \square$$

The results in Figures 1 and 2 show that convergence is greatly accelerated when sequential sampling is used. For example, to obtain absolute accuracy $10^{-4}$, our sequential sampling technique required fewer than $100 \times 5 = 500$ random walk histories (per component of solution), but without sequential sampling, more than $10,000$ random walk histories (per component) are required to achieve this same accuracy. Using Halton's correlated sampling method, this accuracy was attained after approximately 900 random walks were processed. Similar results were obtained on other sample problems using matrices of various orders; a fairly consistent advantage was seen for our algorithm over Halton's method, and both significantly outperformed plain Monte Carlo on each problem tried.

|           | N = 50    | N = 20    | N = 6     |
|-----------|-----------|-----------|-----------|
| Slope     | -1.08E+00 | -9.73E-01 | -5.23E-01 |
| Intercept | -1.01E+00 | 5.19E-01  | 1.61E-01  |

TABLE 2. Comparison of slopes and intercepts for different N values: norm(A)=0.75

|           | N = 50    | N = 20    | N = 6     |
|-----------|-----------|-----------|-----------|
| Slope     | -1.55E+00 | -1.32E+00 | -8.09E-01 |
| Intercept | 4.01E+00  | 3.27E+00  | 2.97E+00  |

TABLE 3. Comparison of slopes and intercepts for different N values: norm(A)=0.99

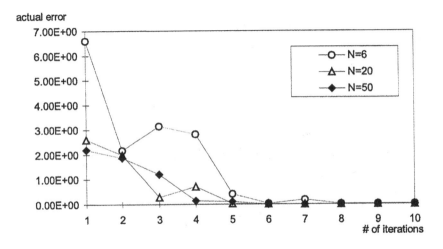

FIGURE 4. Influence on sequential convergence of the number of random walks per iteration: norm(A)=0.99

A closer analysis of our sequential sampling method for solving Equation (3) suggests that the amount of error reduction per iteration achieved depends, in general, on the operator $\mathcal{K}$, and in particular, its norm. However, to ensure that each iteration actually *reduces,* rather than *increases* the error, the number of random walk histories per iteration must be sufficiently large. In the above model problem, $\|\mathcal{K}\| = \|A\|_\infty = 0.75$ and the iterative process converges reasonably quickly. If we alter the transition matrix $A$ so that $\|A\|_\infty$ is closer to 1, then the average number of collisions per random walk is increased and the convergence of the matrix Neumann series is correspondingly slowed. For example, if the matrix $A = (A_{ij})_{10\times10}$ above is replaced by

$$\tilde{A}_{ij} = \frac{s - |i - j| + 2.4}{s^2},$$

the norm of the transition matrix becomes $\|\tilde{A}\|_\infty = \max_i \sum_{j=1}^{10} \tilde{A}_{ij} = 0.99.$

Figures 3 and 4 show the influence on convergence rate of the number $N$ of random walk histories per iteration as a function of $\|A\|_\infty$. In Figure 3 ($\|A\|_\infty = 0.75$), when $N$ is 6 (which means that only 6 random walks are generated per iteration), the factor $\lambda$ that multiplies the error may actually exceed 1 and the error will then *increase* from one iteration to the next. Choosing $N$ to be larger assures that $\lambda < 1$ and produces a steady reduction in the error for this easier of the two problems studied here. In Figure 4 ($\|A\|_\infty = 0.99$) we see that the factor $\lambda$ can exceed 1 both for $N = 6$ and $N = 20$. However, setting $N = 50$ resulted in the steady geometric reduction in error per iteration that characterizes a successful implementation of the sequential method, for this harder of the two problems studied. Determining the minimum number of random walks per sequential stage needed to guarantee monotone error reduction as a function of the problem input $A$, **a** appears to be a formidable problem.

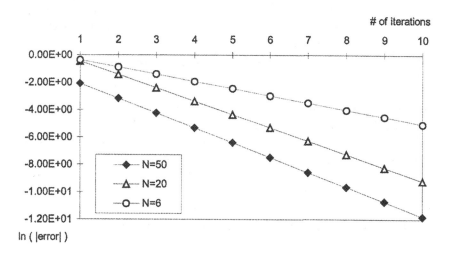

FIGURE 5. Best linear fit to data: ln error vs. number of iterations: norm(A)=0.75

In Figures 5 and 6 we have graphed three lines that were obtained by fitting the best (in the sense of least squares) straight lines to data obtained by plotting logarithms of errors versus iteration numbers for the three runs shown in Figures 3 and 4, respectively. The slopes and intercepts of these three lines are given in Table 2 and 3; the slopes provide estimates of the

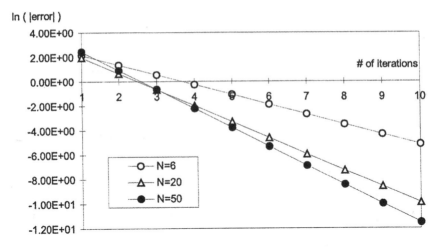

FIGURE 6. Best linear fit to data: ln error vs. number of iterations: norm(A)=0.99

error reduction per iteration[3] achieved by the three different applications of the sequential strategy. Again, the importance of generating sufficiently many random walks in each iteration to assure strict error reduction is emphasized by these data.

Experimentation with the ideas of this paper is continuing.

*Acknowledgments:* The authors gratefully acknowledge the help of Ms. Jinghong Jiang who provided invaluable assistance with the development and improvement of several computer algorithms used in this work. We also acknowledge the useful suggestions of a referee.

# 4  References

[1]  J.H. Halton, "Sequential Monte Carlo", *Proc. Camb. Phil. Soc.* **58**, (1962).

[2]  J. Halton, "Sequential Monte Carlo Techniques for the Solution of Linear Systems", *J. Sci. Comput.*, **9**, (1994).

[3]  T.E. Booth, "Exponential Convergence for Monte Carlo Particle Transport", *Trans. Amer. Nucl. Soc.*, **50**, 1986.

---

[3]In fact, the slope is an estimate of ln $\lambda$, where $\lambda$ is the average error reduction (or multiplication) factor per iteration.

[4] J. Spanier, "A New Multistage Procedure for Systematic Variance Reduction in Monte Carlo", *SIAM J. Numer. Anal.*, **8**, (1971).

[5] O. Axelsson, *Iterative Solution Methods*, Cambridge Univ. Press, 1994.

[6] L. Li and J. Spanier, "Approximation of Transport Equations by Matrix Equations and Sequential Sampling", in preparation.

[7] J. Spanier and E.M. Gelbard, *Monte Carlo Principles and Neutron Transport Problems*, Addison-Wesley, 1969.

[8] G.E. Forsythe and R.A. Leibler, "Matrix Inversion by a Monte Carlo Method", *Math. Tables Aids to Comp.*, **4**, (1950).

[9] W. Wasow, "A Note on the Inversion of Matrices by Random Walks", *Math. Tables Aids to Comp.*, **6**, (1952).

[10] J. Spanier, "A New Family of Estimators for Random Walk Problems", *J. Inst. Maths. Applics.*, **23**, (1979).

Both authors are at:
Department of Mathematics
The Claremont Graduate School
Claremont, CA. 91711
USA

# Quasi-Monte Carlo Methods for Integral Equations

Jerome Spanier
Liming Li

ABSTRACT In this paper, we establish a deterministic error bound for estimating a functional of the solution of the integral transport equation via random walks that improves an earlier result of Chelson generalizing the Koksma-Hlawka inequality for finite dimensional quadrature. We solve such problems by simulation, using sequences that combine pseudorandom and quasirandom elements in the construction of the random walks in order to take advantage of the superior uniformity properties of quasirandom numbers and the statistical (independence) properties of pseudorandom numbers. We discuss implementation issues that arise when these hybrid sequences are used in practice. The quasi-Monte Carlo techniques described in this paper have the potential to improve upon the convergence rates of both (conventional) Monte Carlo and quasi-Monte Carlo simulations in many problems. Recent model problem computations confirm these improved convergence properties.

## 1  Introduction

In the past twenty years there has been a rapid development of the subject of quadrature theory. Although there are still many unresolved issues in obtaining approximations of integrals of functions of many variables, much progress has been made, especially with deterministic methods that rely on the use of low-discrepancy nodal points. While Monte Carlo methods based on the use of pseudorandom nodal points remain the only known method capable of producing estimates whose errors are independent of the number of variables, replacement of pseudorandom points by so-called quasirandom points - which are more uniformly distributed in a certain very precise sense - can produce smaller errors, even in reasonably high dimensions.

The error analysis for these deterministic quasi-Monte Carlo quadrature formulas relies either on the Koksma-Hlawka inequality for functions of bounded variation (see [1]) or, for smooth periodic integrands, on Fourier analysis (see [2]). In the former case, the rate of convergence is limited by the rate at which $D_N^* \to 0$, where $D_N^*$ is the star discrepancy of the nodal set used. Thus, in $s$ variables, convergence rates of order $O((\log N)^s/N)$ can be achieved using sequences (e.g., $(t, s)-$ sequences, see [1]) of smallest known discrepancy. For smooth periodic functions in $s$ variables, lattice rules [2] are capable of producing even smaller errors for nodal sets of given finite cardinality $N$. However, errors uniformly (in $N$ as $N \to \infty$) as small as these are not possible, even for analytic integrands, according to a result of Korobov [3] and, in fact, errors of order $O(N^{-1})$ are then best possible. Because of limitations like these in rates of convergence, a good deal of effort is spent in trying to reduce as much as possible the constant that multiplies the $N-$ dependent factor of the approximation error or error bound. Thus, variance reduction techniques have been widely studied for conventional Monte Carlo applications, and variation reduction methods for quasi-Monte Carlo applications are increasingly of interest.

Monte Carlo and quasi-Monte Carlo methods for solving integral equations have been explored relatively more recently than the simpler problem of finite dimensional quadrature. Quasi-Monte Carlo methods are discussed, for example, in the books of Korobov [3] and Hua and Wang [4]. In general, however, improvements in the Monte Carlo solution of integral equations have been much less studied than for the estimation of finite dimensional integrals. The importance of making progress in this area is certainly clear, however, inasmuch as many vital applications - to radiation transport, semiconductor modeling, and remote geological sensing, among others - are describable in terms of integral equations. See also [5] for a recent application to computer graphics.

So far as we know, it was as recently as in the Ph.D. dissertation of Paul Chelson [6] that quasirandom methods were first applied to study importance sampling for a class of integral equations that includes the most important from the point of view of the applications. The next section of this paper reviews the salient results from [6]. In section 3 we show how Chelson's main result can be somewhat sharpened and, in section 4, we discuss recent improvements in the use of certain hybrid sequences, first introduced in [7]. These sequences combine features of both pseudorandom and quasirandom (low discrepancy) sequences with the twin objectives of achieving rapid convergence and enabling an effective error analysis based on the pseudorandom component. Section 5 discusses the very important issue of implementation of these new hybrid methods and suggests algorithms that make the new methods effective in competition with more conventional ones based exclusively on pseudorandom sequences. Finally, in section 6, we present recent numerical evidence that hybrid sequences show promise of substantially improved convergence for many practical problems.

# 2 The Mathematical Model

We review here some basic structure theory that provides the key to understanding how low discrepancy sequences can be applied to integral equation problems. We begin with the integral equation

$$\psi(\mathbf{P}) = \int_\Gamma \psi(\mathbf{P}')K(\mathbf{P},\mathbf{P}')d\mathbf{P}' + S(\mathbf{P}) \tag{1}$$

that accurately describes many problems in which particles are introduced by means of a source function $S(\mathbf{P})$ and are subsequently transported from collision events at points $\mathbf{P}'$ of the phase space $\Gamma \subset \mathcal{R}^d$ to collisions at $\mathbf{P}$ by means of the kernel $K(\mathbf{P},\mathbf{P}')$. The integral term on the right-hand side of Equation (1) is the (average) density arriving at $\mathbf{P}$ from collisions at any previous site and the solution $\psi(\mathbf{P})$ is the total (average) collision density at $\mathbf{P}$.

A typical phase space $\Gamma$ for steady-state problems is a six-dimensional space, three independent variables for the position vector $(x,y,z) = \mathbf{r}$ and three to indicate (scalar) energy (velocity) $E$ and unit direction $(\theta,\varphi) = \omega$, where $\mathbf{E} = (\theta,\varphi,E)$ , $\|\mathbf{E}\| = E$. We write $\mathbf{P} = (\mathbf{r},\mathbf{E}) \in \Gamma$. This dimensionality is sufficient to limit drastically the effectiveness of attempts to solve Equation (1) by other approximate solution procedures.

In practice, we do not solve the integral Equation (1) for the unknown function $\psi(\mathbf{P})$ for *all* $\mathbf{P}$ but rather we estimate one or more integrals of the form

$$I = \int_\Gamma g(\mathbf{P})\psi(\mathbf{P})d\mathbf{P} \tag{2}$$

where $g(\mathbf{P})$ is a known detector function describing the reaction of a detector to the collision density $\psi(\mathbf{P})$.

The conventional Monte Carlo approach to this problem is to use the Neumann series for $\psi$ and then integrate term by term to get

$$I = \int_\Gamma g(\mathbf{P})\psi(\mathbf{P})d\mathbf{P} = \int_\Gamma g(\mathbf{P})S(\mathbf{P})d\mathbf{P} + \int_{\Gamma^2} g(\mathbf{P})K(\mathbf{P},\mathbf{P}_1)S(\mathbf{P}_1)d\mathbf{P}_1 d\mathbf{P}$$
$$+ \cdots + \int_{\Gamma^{n+1}} g(\mathbf{P})K(\mathbf{P},\mathbf{P}_1)K(\mathbf{P}_1,\mathbf{P}_2)\cdots K(\mathbf{P}_{n-1},\mathbf{P}_n)S(\mathbf{P}_n)d\mathbf{P}_n\cdots d\mathbf{P}_1 d\mathbf{P}$$
$$+ \cdots \tag{3}$$

A relatively straightforward application of fixed point theory in Banach spaces shows that the series (3) converges if there exists an integer $n_0 \geq 1$ such that $\| \mathcal{K}^{n_0} \| < 1$, where $\mathcal{K}$ is the integral operator defined on the function space $L^1(\Gamma)$ by:

$$\mathcal{K}f(\mathbf{P}) = \int_\Gamma K(\mathbf{P},\mathbf{P}')f(\mathbf{P}')d\mathbf{P}' \in L^1(\Gamma),$$

where $f(\mathbf{P}) \in L^1(\Gamma)$ and $\| \cdot \|$ designates the operator norm induced by the $L^1$- norm. We note that the Neumann series representation of $I$ corresponding to Equation (3) is

$$I = \int_\Gamma g(\mathbf{P})\left(\sum_{n=0}^{\infty} \mathcal{K}^n S(\mathbf{P})\right) d\mathbf{P}$$

where $\mathcal{K}^0$ is the identity operator; i.e., $\mathcal{K}^0 f(\mathbf{P}) = f(\mathbf{P})$. Often in practice it is the case that $\| \mathcal{K} \| = 1$, but $\| \mathcal{K}^{n_0} \| < 1$ for sufficiently large $n_0$ indicating that the upper bound of the survival probability after $n_0$ collisions is strictly less than 1. This condition is sufficient to guarantee that the number of collisions made by particles in this problem is finite with probability one and that Equation (1) has a unique solution. From this point on we assume that there exists an integer $n_0 \geq 1$ such that $\| \mathcal{K}^{n_0} \| < 1$.

To solve this estimation problem by using the Monte Carlo method, one might construct a probability measure space $(\Omega, \mathcal{F}, \mu)$, making use of the source $S$ to select initial phase space coordinates for each random walk $\omega \in \Omega$, and using the transport kernel $K$, suitably normalized, to decide between termination and continuation of each random walk beyond its current phase space location and to determine the location of next collision points. There are numerous ways to accomplish this, just as there are many ways to construct samples from any probability density function of a single variable. Details concerning such constructions may be found in [6], [8] , or [9], for example, and [10] provides a useful general description of such constructions. Of course, one might prefer to base the construction of random walks on an artificial source $\hat{S}$ and kernel $\hat{K}$ whose use is designed specifically to emphasize events of greater importance. This is the general idea behind all forms of importance sampling.

# 3   Chelson's Theorems

In this section, our goal is to validate the use of quasirandom sequences to simulate the random walk process and to analyze the resulting errors. In Chelson's doctoral dissertation [6], a quasirandom perspective is adopted in studying a family of estimators that includes random variables that implement conventional importance sampling. To the authors' knowledge, these represent the first attempts to transfer variance reduction techniques developed for statistical estimators to error reduction methods in the context of quasirandom sampling. Chelson uses the notions of discrepancy and multidimensional bounded variation to establish deterministic error bounds both for estimating finite-dimensional integrals and for estimating functionals of the solution of certain integral equations.

In applying quasi-Monte Carlo methods to integral transport equations, one is concerned with the construction of sequences $\{\omega_i\}$ in the abstract space $\Omega$ of all random walk histories that guarantee the convergence

$$\lim_{N \to 0} \delta_N(\xi) = \lim_{N \to 0} \frac{1}{N} \sum_{i=1}^{N} \xi(\omega_i) - \int_{\Omega} \xi(\omega) d\mu(\omega) = 0 \qquad (4)$$

provided that the estimating function $\xi$ satisfies mild smoothness restrictions. As in [10] and [6], sequences $\{\omega_i\}$ that achieve this will be called $\mu-$

*uniformly distributed* in $\Omega$. Even for the case of finite-dimensional importance sampling applied to quasi-Monte Carlo integration, Chelson's result
providing a deterministic error bound for $\delta_N(\xi)$ is rather complicated to
describe. Here we shall state only a slightly weaker corollary of that theorem.

**Theorem 1:** *Let $f(t)$ be real-valued and integrable over $I^s$, let $g(t)$ be
a continuous probability density function on $I^s$ and assume that $f/g$ is
of bounded variation in the sense of Hardy and Krause. Let $\{x^{(n)}\}$ be any
sequence in $I^s$ and let $\{y^{(n)}\}$, $y^{(n)} \in \mathcal{R}^s$, be the sequence obtained from the
probability density function $g(y)$ by setting (see [10] for further explanation
of such constructions)*

$$y_1^{(n)} = G_1^{-1}(x_1^{(n)}), \quad y_2^{(n)} = G_2^{-1}(x_2^{(n)}), ..., \quad y_s^{(n)} = G_s^{-1}(x_s^{(n)}),$$

*where $G_k(y_k)$ are cumulative distribution functions for the conditional den-
sities*

$$g_k(y_k) = g(y_k \mid y_1, ..., y_{k-1}).$$

*Then*

$$\left| \frac{1}{N} \sum_{n=1}^{N} \frac{f(y^{(n)})}{g(y^{(n)})} - \int_{I^s} f(t)dt \right| \leq V(f/g) D_N^* \{x^{(n)}\}$$

*where $V$ denotes the Hardy-Krause variation and $D_N^*$ the star-discrepancy
of the sequence.*

Of course, the usefulness of this result derives from the possibility of
selecting the probability density function $g$ so that $V(f/g) << V(f)$. We
remark that the choice $g(y) \equiv 1$ (i.e., uniform distribution on $I^s$) produces
the classical Koksma-Hlawka inequality for the error.

In addition to this result, Chelson obtained similar deterministic error
bounds for importance sampling estimators for (infinite dimensional) random walk simulations. His particular construction of random walk histories,
which assigned special states to terminated random walks, made it awk-
ward for him to make use of the notion of discrepancy in euclidean phase
space and led him, instead, to rely on the concept of *isotropic discrepancy*
[1] over the space $\Omega$ of random walks. Once again we content ourselves here
to state a slightly weaker corollary of Chelson's infinite-dimensional result:

**Theorem 2:** *For each $\epsilon > 0$, let $T$ be such that $1 - \mu(\Omega_T) < \epsilon$, where
$\Omega_T$ is the set of all random walks $\omega \in \Omega$ making $T$ or fewer collisions before
terminating (such a $T$ exists since, with the assumption $\| K^{n_0} \| < 1$, the
random walks terminate in a finite number of steps with probability one).
Suppose $Q = \{\gamma_i\}$ is any sequence of vectors in $[0, 1)^s$ where $s = d \cdot T$
and $d = \dim(\Gamma)$, and $Q' = \{\omega_i\}$ is the sequence of random walk histories
in the space $\Omega$ generated from $Q$ by the rather natural process in which the
components of the vectors $\gamma_i$ are used to generate the individual components
of each random walk $\omega$ (see [6] for the details of Chelson's construction).*

*Let $\xi$ be a bounded random variable on $\Omega$ mapping $\Omega$ into the finite interval* $[a, b]$. *Then, for sufficiently large* $N$,

$$\delta_N\left(\xi(Q')\right) = \left|\frac{1}{N}\sum_{n=1}^{N}\xi(\omega_n) - \int_\Omega \xi(\omega)d\mu\right|$$
$$\leq (b - a)\cdot H\cdot T\cdot(4d\cdot T\sqrt{d\cdot T} + 1)(D_N^*(Q))^{\frac{1}{d\cdot T}} + (b-a)\epsilon$$

$$(5)$$

*where $H$ is a known constant.*

It is important to note in Theorem 2 that, even though the probability measure used to generate the sequence $Q' = \{\omega_i\}$ is not necessarily the measure induced by the original sequence $Q$, the discrepancy appearing in the error bound is still that of this original sequence.

Theorem 2 can be applied to importance sampling estimators as follows: Suppose $\nu$ is a probability measure on the space $\Omega$ of all random walks and that the analog probability measure $\mu$ on $\Omega$ (which is generated from the source $S$ and kernel $K$ of the Equation (1) as detailed in [6]) is absolutely continuous with respect to $\nu$. Then if $\xi$ is a random variable defined on $\Omega$ and $\zeta = \xi\frac{d\mu}{d\nu}$, one has

$$\int_\Omega \zeta d\nu = \int_\Omega \xi\frac{d\mu}{d\nu}d\nu = \int_\Omega \xi d\mu.$$

Theorem 2 is therefore directly applicable to the estimator $\zeta$ and the probability measure $\nu$ may be chosen to make the variation in $\zeta$ significantly smaller than that of $\xi$. In terms of the conclusion of Theorem 2, this would, with proper choice of $\nu$, have the effect of reducing any or all of the quantities $H, T$, and $b - a$.

Assuming that the discrepancy $D_N^*(Q) = O([(\log N)^{d\cdot T}/N])$, we see that the upper bound for the error derived from Theorem 2 is $O\left((\log N)^{d\cdot T}/N)^{\frac{1}{d\cdot T}}\right)$. For many practical problems, the maximum number $T$ of collisions required by the $N$ histories will be large if $\epsilon$ is small. The convergence rate determined by Equation (5) is, therefore, not very rapid. For example, for radiation impinging upon a thick slab in which there is very little absorption, model problem analysis (see [11]) shows that it is not unusual for $T$ to be as large as 50 or more when the number of histories $N = 10,000$. This means that the error bound is practically useless because the convergence rate in Theorem 2 is $O\left(D_N^*(Q)^{\frac{1}{d\cdot T}}\right)$ and $d\cdot T \approx 300$.

Chelson's Theorem 2 is, nevertheless, sufficient to imply the convergence $\delta_N(\xi) \to 0$ of the error as $N \to \infty$ when quasirandom vectors are used to generate the random walks provided they are $\mu-$uniformly distributed over the space $\Omega$. This is so, of course, because under those circumstances, the discrepancy $D_N^*(Q) \to 0$ as $N \to \infty$. However, as we have observed, Chelson's Theorem cannot be used to provide *effective* bounds for the error in sufficiently difficult problems. On the other hand, numerical experience [6], [12], [13] seems to produce much better results. Here we will argue that the error bound obtained by computer implementations is, in fact, not $O\left(D_N^*(Q)^{\frac{1}{d \cdot T}}\right)$ but $O\left(D_N^*(Q)\right)$. The key to this modest improvement in error bound is to refine the method for constructing random walks $\omega$ from the components of the quasirandom sequence of vectors $\{\gamma_i\}$ so that the use of isotropic discrepancy can be avoided in the proof. The details may be found in [9]; here we indicate only the main ideas.

The scheme described in [9] of constructing random walks based on the components of a quasirandom sequence of vectors $\{\gamma_i\}$ creates a mapping $\mathcal{G}$ [1] from the space $\Omega$ of all random walk histories to a sufficiently high dimensional euclidean hypercube. This mapping can be rigorously described, as in [10], in terms of conditional probability density functions constructed by using the source density $S$ and the transition kernel $K$ of an integral equation such as Equation (1). The inverse mapping $\mathcal{G}^{-1}$ maps the $d \cdot T$-dimensional unit cube to the space $\Omega$ of random walk histories. Let $\Lambda_k \subset \Omega$ be the set of all random walk histories which terminate on the $k$ th collision. Then

$$\Omega = \bigcup_{k=1}^{\infty} \Lambda_k \bigcup \Lambda_\infty$$

and we have already made assumptions on the norm of the integral operator $\mathcal{K}$ that guarantee that $\mu(\Lambda_\infty) = 0$. Since the number of random walk histories is finite in computer implementations, the maximum number $T$ of collisions made by this finite number of histories exists so that we may restrict our attention to the space $\Omega_T = \bigcup_{k=1}^{T} \Lambda_k$ instead of $\Omega$. Of course, the integer $T$ will depend on the number $N$ of random walks and will, in general, tend to $\infty$ with $N$. Even so, Chelson's *finite dimensional* theorem is applicable to this situation, as follows:

**Theorem 3:** *Suppose $Q = \{\gamma_i\}$ is a sequence of vectors in $[0, 1)^{d \cdot T}$ and $Q' = \{\omega_i\}$ is the sequence of random walk histories in $\Omega$ generated from $Q$ by the one-to-one map $\mathcal{G}$ described in detail in [8]. Then, for each $\epsilon > 0$, let $T$ be chosen as in Theorem 2; i.e., so that $1 - \mu(\Omega_T) < \epsilon$ and let $\xi$ be a bounded random variable on $\Omega$ mapping $\Omega$ into the finite interval $[a, b]$. Then, for sufficiently large $N$,*

$$\delta_N\left(\xi(Q')\right) \leq V(\xi \circ \mathcal{G}^{-1}) \cdot (D_N^*(Q)) + (b - a)\epsilon \tag{6}$$

---

[1]We choose a different notation here than was used in developing the Chelson theorems since the mapping we use is, in fact, different. The map $\mathcal{G}$ constructed by Li in [9] uses *exactly* one number in [0,1] to construct *each* coordinate of every collision point of $\omega$. It achieves this by treating random walk termination as one of a finite number of possible outcomes of each collision. This is in contrast with the treatment of Chelson in which termination or continuation of a random walk at each collision point is first decided by a binomial random variable, followed by a second (independent) decision to choose among the possible continuation states. Chelson's construction introduces non-euclidean (two point) elements into the factors of $\Omega$, and led him to the use of isotropic discrepancy.

where $\xi \circ \mathcal{G}^{-1}$ is assumed of bounded variation in the sense of Hardy and Krause considered as a function on $R^{d \cdot T}$, and $d$ is the dimension of the phase space $\Gamma$.

*Proof:* Let $\omega_i = (\omega_{i1}, \omega_{i2}, ... \omega_{ik_i}), (i = 1, 2, ..., N)$ where $\omega_{ij} \in \Gamma \subset \mathcal{R}^d$ are the phase space coordinates of the $j$ th collision for the $i$ th random walk history $1 \leq k_i \leq T$. The space $\Omega_T$ consisting of all random walks that terminate in $T$ or fewer steps is mapped in 1-1 fashion onto the space $R^{d \cdot T}$. The mapping $\mathcal{G} : \Omega_T \to [0,1)^{d \cdot T}$ is defined by

$$\mathcal{G}(\omega) = (\mathcal{G}_1(\omega_{i1}), \mathcal{G}_2(\omega_{i2}), ..., \mathcal{G}_T(\omega_{iT}))$$

and the definition of the maps $\mathcal{G}_j : R^d \to I^d$ based on the appropriate marginal and conditional probability density functions, may be found in [9]; see also [10]. Here each vector $\gamma \in I^d$ determines exactly one point $\omega_{ij}$ of the phase space $\Gamma$ for each random walk. As in the proof of Chelson's theorems, the sequence $(\omega_1, \omega_2, ...)$ of random walks will be $\mu$−uniformly distributed in $\Omega$ provided the quasirandom sequence $(\gamma_1, \gamma_2, ...)$ being used to generate $(\omega_1, \omega_2, ...)$ is uniformly distributed in the unit cube $I^{d \cdot T}$. Then

$$
\begin{aligned}
\delta_N(\xi) &= \frac{1}{N} \sum_{i=1}^{N} \xi(\omega_i) - \int_\Omega \xi(\omega) d\mu(\omega) \\
&\leq \left| \frac{1}{N} \sum_{i=1}^{N} \xi(\omega_i) - \int_{I^{d \cdot T}} \zeta(t) dt \right| + \left| \int_{I^{d \cdot T}} \zeta(t) dt - \int_\Omega \xi(\omega) d\mu(\omega) \right| \\
&\leq V(\xi \circ \mathcal{G}^{-1}) D_N^*(Q) + (b-a)\epsilon
\end{aligned}
$$

where $\zeta(t)dt = \xi(\mathcal{G}^{-1}(t)) \frac{d\mu}{dt} dt$ arises from the change of variables in the integrand. The first term in the final inequality comes from applying Theorem 2 to the finite dimensional space $I^{d \cdot T}$, while the second part is demonstrated as in [6]. □

In practice, one can make use of information gathered during a preliminary simulation (if the difficulty of the problem seems to warrant this) in order to develop a scheme for improved results based on Theorem 3. For example, by keeping track of the total number of quasirandom or pseudorandom numbers used to process the prescribed number of random walks $N$, one can derive information about the relationship between $\epsilon$ and $T$ in Theorem 3. We will have more to say about this in section 6. The interval $[a, b]$ can also be estimated from storage of the least and largest values of the estimator $\xi$, if desired.

Theorem 3 avoids using the notion of isotropic discrepancy, which was used in [6], to obtain the sharpened inequality (6). However, Theorem 3 still provides impractically large error bounds when using quasirandom sequences in the simulation of transport processes because the integer $T$ is large for many difficult problems. Since the extreme discrepancy $D_N^*$ itself goes to zero like $(\log N)^{d \cdot T}/N$ when the best presently available quasirandom sequences are used to generate the random walks, it is likely, in difficult problems, that the error bounds from Equation (6) will be inferior to the expected statistical error resulting from the use of pseudorandom sequences to generate the random walk decisions. Hybrid strategies, such as those introduced in [7], that combine pseudorandom and quasirandom sequences in an attempt to improve on the use of either alone, were designed for use on this kind of problem.

# 4 Hybrid Sequences for Simulation

The approach taken in [7] and here is to combine the best features of pseudorandom and quasirandom Monte Carlo methods, creating hybrid methods that are designed to outperform either pseudo-Monte Carlo or conventional quasi-Monte Carlo used alone. Two such methods were described in [7]:

A. A mixed strategy in which initial decisions (collisions) for each random walk are made using quasirandom vector sequences, while subsequent ones are made using either a pseudorandom or "scrambled" quasirandom sequence.

B. A scrambled quasirandom strategy in which the aim is pseudorandomly to reorder a low dimensional quasirandom vector sequence and use it, in place of a pseudorandom sequence, to make all the decisions needed to construct the mapping $\mathcal{G}$ of section 3.

The reader is asked to consult [7] for a fuller discussion of these hybrid sequences. Our concern here is primarily with their effective *implementation*. For reference later, we introduce some notation needed in implementation of the scrambled hybrid method. Let $q_1, q_2, \ldots$ denote successive elements in a fixed $d$-dimensional quasirandom sequence and let $M$ be an upper bound for the largest number of collisions required by any of the $N$ particle histories generated. The integer $M$ depends on $N$, but we suppress this dependence in our choice of notation for simplicity. Normally, $M$ would be estimated from running a small set of initial random walks and then increasing somewhat the value of $M$ observed in order to err on the side of conservatism. We denote by $P_1, P_2, \ldots, P_M$, $M$ independent random permutations of the integers $1, 2, \ldots, N$. The scrambled strategy applies the permutation $P_j$ to the elements of a matrix array of the vectors $q_i$, as indicated in the following schematic:

| *permuted quasirandom implementation* | *history 1* | *history 2* | ∘ ∘ ∘ | *history N* |
|---|---|---|---|---|
| *source* | $q_1$ | $q_2$ | | $q_N$ |
| *collision* 1 | $q_{N+P_1(1)}$ | $q_{N+P_1(2)}$ | | $q_{N+P_1(N)}$ |
| *collision* 2 | $q_{2N+P_2(1)}$ | $q_{2N+P_2(2)}$ | | $q_{2N+P_2(N)}$ |
| | ∘ | ∘ | | ∘ |
| | ∘ | ∘ | | ∘ |
| | ∘ | ∘ | | ∘ |
| *collision* M | $q_{MN+P_M(1)}$ | $q_{MN+P_M(2)}$ | | $q_{MN+P_M(N)}$ |

The manner in which the different $d-$ tuples are used to make decisions in constructing histories in the permuted quasirandom implementation is as schematically illustrated above: each individual random walk history is generated by utilizing the vectors in the indicated *column* of the matrix array.

# 5 Implementation Issues

Application of the mixed method is straightforward and relatively efficient. If, for example, initial segments of fixed dimension (determined by the number of collisions treated quasirandomly) Halton type sequences are used, quite rapid algorithms for generating the needed radical inverse functions are available and the costs per sample are only slightly higher than when pseudorandom numbers are used to make all decisions. The same is true when $\{n\theta\}$ vector sequences are used (here $\theta$ is a $d-$ tuple of irrational numbers that are linearly independent, along with the number 1, over the rationals), or a combination of the two types. However, when the scrambled strategy is employed, as was pointed out earlier, care is needed in order to avoid massive storage and running time problems.

To implement the scrambled method, one should choose a sufficiently large $N$ and scramble the sequence $q_1, q_2, \ldots$ in blocks of $N$ $d-$tuples at a time, making certain that the permutations $P_i$ used in different blocks are independent of each other. In order to avoid significant storage problems, it is important to apply the permutation $P_i$ to the $i$th batch of $d-$tuples *as needed*, <u>without</u> precomputing and storing all $(M+1)N$ $d-$ tuples in advance. This can be accomplished by making appropriate use of linear congruential algorithms to generate the needed pseudorandom numbers.

A theoretically perfect implementation of the scrambling method requires that a relatively small number, $M$, of independent permutations of the integers $\{1, \ldots, N\}$, $N =$ total sample size, be selected randomly from among the $N!$ that are theoretically available. Such a perfect implementation could, for example, be accomplished on a computer by generating blocks of $N$ pseudorandom numbers at a time and then sorting them into linear order. The mapping so induced on the integers $\{1, \ldots, N\}$ defines a permutation of those integers, and different ones would be independent of each other. This could be done for each row of the required matrix, but a great deal of computation time would be used in this process and storage of a very large $(M+1) \times N$ rectangular matrix would also be required. These requirements mean that a perfect implementation should be attempted only for small problems (i.e., problems with small $N$ ) or for an occasional larger test problem, but not for production use.

In order to make practical use of scrambling, then, an effective approximation to the perfect implementation must be worked out. In this regard, permutations may be based on the use of a linear congruential pseudorandom number generating algorithm, but the quality of the resulting permutations (in terms of their representing independent random samples from the set of all permutations) can be very adversely affected by poor parameter choices in the defining linear congruences.

Both the computer time and the storage requirements can be lessened substantially by using permutations generated by linear congruential generators, which is the stratagem we have adopted. To elaborate on this point, let $N = 2^k$ for some integer $k$ (general integer $N$ can be treated in similar fashion). Then the linear congruential generator

$$P(n + 1) \equiv aP(n) + c \qquad (\text{mod } 2^k), \qquad n = 1, 2, ..., N \qquad (7)$$

produces a full cycle permutation of $\{1,...,N\}$ if $a \equiv 1$ (mod 4) and $c$ is odd, for any chosen seed $P(1)$. Then the principal question becomes how to select the permutations $P_1, P_2, ..., P_M$ randomly, where each $P_i$ is generated by a formula such as Equation (7).

Clearly, one could randomize the choice of the multiplier, the additive constant, and the seed of the pseudorandom algorithm. We note that even if we randomized the choice of all three parameters $a, c, P(1)$, we would construct only $N^3/8$ permutations using the linear congruential method, which is still very small, when $N$ is as large as $10^6$, compared to the theoretical number $N!$ of such permutations. Model transport problems were solved to compare the effects of varying these three parameters separately. Based upon available theory concerning the qualities of randomness produced by linear congruential algorithms (see, for example, [14]), one expects that the choice of the multiplier has a more significant impact on quality than either the choice of seed or additive constant. One also anticipates that selection of multipliers that are either too small (for example, much smaller than $\sqrt{N}$) or too large (for example, larger than $N - \sqrt{N}$) results in a lower quality of randomness than does selection of intermediate choices of the multiplier. These theoretically based predictions were borne out in our experimentation.

Having settled on a range of moderate values for a randomized multiplier $a$ in Equation (7), further experimentation verified that scrambling of the multiplier gives better results than scrambling either the additive constant $c$ or the seed $P(1)$. For the present, then, our recommendation is to apply what we term "multiply scrambling", with parameter $a$ randomly chosen from an appropriate range of moderate values, for use with the scrambling method.

# 6 Numerical Results: One-Dimensional Test Problem

We now apply the previous ideas to a simple model problem similar to ones discussed in [11] and [15] to judge our simulation ideas.

A vertical slab $L$ cm thick and of unit total macroscopic cross section is assumed to contain material that may either scatter or absorb particles impinging on it from the left. Energy dependence is not included; that is, a single fixed energy or velocity is ascribed to each particle. A steady source injects particles moving perpendicular to the face at the left face of the slab. We will not consider general angular scattering distributions and will instead assume a one-dimensional model in which only directly forward scattering is permitted. The intercollision distance $d$ is assumed to have the exponential density: $\exp(-d)$ $(0 < d < \infty)$.

This is a simple radiation problem which is not hard to see can be modeled via the following integral transport equation

$$\phi(x) = \int_0^\infty K(x,y)\phi(y)dy + S(x), \quad x > 0 \tag{8}$$

where $\phi(x)$ is the particle collision density at point $x$, and the transport kernel

$$K(x,y) = \begin{cases} qe^{-(y-x)} & \text{if } x \le y \le L \\ 0 & \text{otherwise} \end{cases} \quad \text{where } q \text{ is a constant } (0 < q < 1).$$

The source function in our case is $S(x) = \delta(x)$. Our task here is to estimate the integral

$$I = \int_L^\infty \phi(x)dx,$$

which can also be expressed as an inner product

$$\phi^*(0) = (\phi^*(x), \delta(x)) = (\phi(x), g(x))$$

where

$$\phi^*(x) = \int_0^\infty K^*(x,y)\phi^*(y)dy + g(x), \quad x > 0 \tag{9}$$

$K^*(x,y) = K(y,x)$ and $g(x) = 1$ if $x > L$, $g(x) = 0$ if $x \le L$.

It is not hard to verify that

$$\phi^*(x) = \begin{cases} qe^{-(1-q)(L-x)} & \text{if } x \le L \\ 1 & \text{if } x > L \end{cases}$$

is the exact solution of (9). If the scattering probability $q = 0.9$ and the thickness of slab $L = 10$, then the fraction of particles passing through the slab, $\phi^*(0) = 0.331091$.

| # of histories | pseudo-method error | mixed-method error | scrambled-method error |
|---|---|---|---|
| 1000 | 0.002090 | 0.016909 | 0.0129092 |
| 2000 | 0.011409 | 0.013409 | 0.0139083 |
| 3000 | 0.012909 | 0.009909 | 0.0089291 |
| 4000 | 0.008409 | 0.006409 | 0.0071841 |
| 5000 | 0.004509 | 0.005109 | 0.0036908 |
| 6000 | 0.005075 | 0.005242 | 0.0047583 |
| 7000 | 0.004480 | 0.003766 | 0.0029733 |
| 8000 | 0.003909 | 0.002409 | 0.0015434 |
| 9000 | 0.003575 | 0.000575 | 0.0020773 |
| 10000 | 0.003309 | 0.000709 | 0.0013436 |

TABLE 1. Actual error of pseudo-method, mixed-method and scrambled-method for test problem

| # of histories | pseudo-method decisions | mixed-method pseudo, total | | scrambled-method decisions |
|---|---|---|---|---|
| 1000 | 6755 | 2095 | 6771 | 6833 |
| 2000 | 13523 | 4181 | 13515 | 13654 |
| 3000 | 20359 | 6198 | 20195 | 20350 |
| 4000 | 27126 | 8290 | 26940 | 26928 |
| 5000 | 33598 | 10278 | 33582 | 33621 |
| 6000 | 40390 | 12260 | 40239 | 40124 |
| 7000 | 47149 | 14280 | 46901 | 46565 |
| 8000 | 53859 | 16234 | 53511 | 53436 |
| 9000 | 60694 | 18320 | 60261 | 60123 |
| 10000 | 67317 | 20330 | 66926 | 66716 |

TABLE 2. Number of decisions using pseudo-method, mixed-method and scrambled-method

| method type | slope | $Y$-intercept |
|---|---|---|
| pseudo-method | -0.116459826 | -4.295102409 |
| mixed-method | -1.347274202 | 3.509395128 |
| scrambled-method | -1.062034185 | 5.850920323 |

TABLE 3. Slopes and $Y$-intercepts of best linear fits to log-log plots of errors

This simple problem was solved by generating 10,000 random walk histories using a FORTRAN program on the VAX/VMS$^{\text{TM}}$ (Version 5.5), and observing the fraction that cross the $x = L$ boundary. The output is shown in Table 1.

The pseudo-method referred to in Table 1 uses pseudorandom numbers to make all decisions in determining particle collision coordinates. The mixed-method in Table 1 uses a 6-dimensional quasirandom vector sequence $((t, s)$

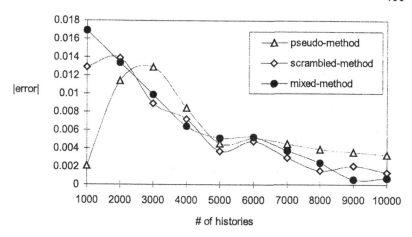

FIGURE 1. Actual errors of pseudo-method, mixed-method and scrambled-method

nets-based sequence [1]) to make the initial six decisions, while all remaining decisions are made using pseudorandom numbers. The scrambled-method in Table 1 uses the scrambling algorithm described in the previous section applied to the van der Corput-type scalar sequence $\{\phi_3(n)\}$, $n = 0, 1, \ldots$ where $\phi_b(n)$ converts the base $b$ representation of the integer $n$ to the $b-$ary fraction obtained by reflecting this representation through the $b-$ary point. The results of Table 1, which are graphed in Figure 1, show that both hybrid strategies yield better errors than the pseudo-method with the same number of samples; more than a factor of two reduction in error is obtained in both cases.

Table 2 displays the number of individual decisions required to gene- rate 10,000 random walks using each of the three different methods. The middle column of Table 2 presents first the number of pseudorandom deci- sions made for all steps after the sixth, while the second entry of the middle column gives the total number ( = pseudorandom + quasirandom) of de- cisions required to process all steps. From this information we can roughly estimate the value of $\epsilon$ corresponding to the choice $T = 6$ in Theorem 3. Because pseudorandom numbers are used for all collisions after the sixth, the ratio of the number of pseudorandom numbers used to the total num- ber used for all decisions is an estimate of $prob_T = \sum_{k=7}^{\infty} p_k$ where $p_k =$ probability that a random walk terminates in exactly $k$ collisions. From this we see that $prob_6 \approx 20330/66926 = 0.3037$.

Figure 2 employs the data from Table 1 to construct straight lines based on a least squares analysis of a semi-log plot of the errors from each of the three methods. That is, an equation $y = mx + b$ is obtained from the best

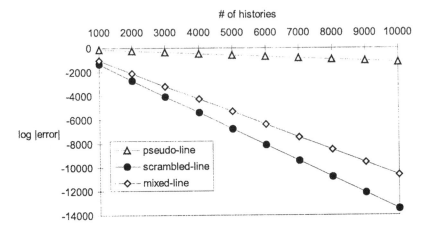

FIGURE 2. Best linear fits to semi-log plots of errors: pseudo-method, mixed-method and scrambled-method

(in the sense of least squares) linear fit to semi-log plots of the data; the slope is thus an approximation to the exponent that establishes the rate of convergence of the method used. One expects a slope of approximately $-1/2$ for the pseudorandom implementation and slopes closer to $-1$ for the other two methods. The slopes ($m$) and $Y$-intercepts ($b$) of the above three sampling methods are presented in Table 3.

From these data and from Figure 2, we can see that the slopes associated with the mixed-method and the scrambled-method are much greater than that of the pseudo-method, which means that the convergence rates of the mixed-method and the scrambled-method are much faster than that of the conventional Monte Carlo method. Of course, we notice that the pseudo-random slope is worse than expected due to the poor start during the initial several thousand histories when errors actually increased, while the slopes of the other two methods are, in fact, slightly better than one can anticipate based on the theory. Nevertheless, one can draw the preliminary conclusion than the two hybrid methods have improved upon the pseudorandom method, both in terms of error reduction achieved after 10,000 histories and in terms of estimated rate of convergence for this simple model problem.

Space limitations prevent us from reporting here on additional numerical evidence of the efficacy of using hybrid sequences in place of either pseudorandom or pure low discrepancy sequences, especially for difficult transport applications.

*Acknowledgments:* The authors gratefully acknowledge the help of Ms. Jinghong Jiang who provided invaluable assistance with the development and improvement of several computer algorithms used in this work. We also acknowledge useful suggestions of a referee.

# 7 References

[1] H. Niederreiter, *Random Number Generation and Quasi-Monte Carlo Methods*, CBMS-SIAM, 1992.

[2] I.H. Sloan and S. Joe, *Lattice Methods for Multiple Integration*, Clarendon Press, Oxford, 1994.

[3] N.M. Korobov, *Number-Theoretic Methods in Approximate Analysis*, Fizmatig, Moscow, 1963. (in Russian).

[4] L.K. Hua and Y. Wang, *Applications of Number Theory to Numerical Analysis*, Springer, Berlin, 1981.

[5] A. Keller, "A Quasi-Monte Carlo Algorithm for the Global Illumination Problem in the Radiosity Setting", *Monte Carlo and Quasi-Monte Carlo Methods in Scientific Computing* (H. Niederreiter and P. Shiue, eds.), vol. 106, Springer, 1995.

[6] P. Chelson, "Quasi-Random Techniques for Monte Carlo Methods," *Ph.D. dissertation*, The Claremont Graduate School, 1976.

[7] J. Spanier, "Quasi-Monte Carlo Methods for Particle Transport Problems", *Monte Carlo and Quasi-Monte Carlo Methods in Scientific Computing* (H. Niederreiter and P. Shiue, eds.), vol. 106, Springer, 1995.

[8] J. Spanier and E.M. Gelbard, *Monte Carlo Principles and Neutron Transport Problems*, Addison-Wesley Pub. Co., 1969.

[9] L. Li, "Quasi-Monte Carlo Methods for Transport Equations", Ph.D. dissertation, The Claremont Graduate School, 1995.

[10] E. Hlawka and R. Mück, "Uber eine Transformation von Gleichverteilen Folgen II, *Computing*, (1972).

[11] J. Spanier, "An Analytic Approach to Variance Reduction," *SIAM J. Appl. Math.*, **18**, (1970).

[12] E. Maize, "Contributions to the Theory of Error Reduction in Quasi-Monte Carlo Methods", *Ph.D. dissertation*, The Claremont Graduate School, 1981.

[13] J. Spanier and E.H. Maize, "Quasi-Random Methods for Estimating Integrals Using Relatively Small Samples", *Siam Rev.*, **36**, 18-44, (1994).

[14] D.E. Knuth, *The Art of Computer Programming, Vol. 2: Seminumerical Algorithms*, 2nd ed., Addison-Wesley, 1981.

[15] G.A. Mikhailov, *Optimization of Weighted Monte Carlo Methods*, Springer-Verlag, 1992.

Both authors are at:
Department of Mathematics
The Claremont Graduate School
Claremont, CA. 91711
USA

# Quadratic Congruential Generators With Odd Composite Modulus

## Sibylle Strandt

ABSTRACT This paper deals with quadratic congruential pseudorandom number generators with odd composite moduli. The relation between these generators and compound quadratic congruential generators is pointed out. Upper and lower bounds for the discrepancy of the generated point sets consisting of all pairs of successive pseudorandom numbers over the full period are established.

## 1 Introduction

The well known coarse lattice structure of linear congruential pseudorandom number generators leads to undesirable effects on the results in some stochastic simulations (see [EL86]). In order to avoid such misleading results one should 'run each Monte Carlo program at least twice using quite different sources of random numbers, before taking the answers of the program seriously' ([Knu81], p. 173). This state of affairs provided the motivation for studying nonlinear congruential generators which do not show such a coarse lattice structure.

The earliest nonlinear congruential method is the quadratic congruential method proposed by Knuth in 1969 ([Knu81], p. 25), which is considered in the present paper in the case of an odd composite modulus $m = p_1^{\omega_1} \cdots p_r^{\omega_r}$ with some primes $p_i \geq 3$ and integers $\omega_i \geq 2$ for $1 \leq i \leq r$. Let $\mathbb{Z}_q = \{0, 1, \ldots, q-1\}$ for integers $q \geq 1$. For integers $a, b, c, y_0 \in \mathbb{Z}_m$, a quadratic congruential sequence $(y_n)_{n \geq 0}$ of elements of $\mathbb{Z}_m$ is defined by

$$y_{n+1} \equiv ay_n^2 + by_n + c \pmod{m}, \; n \geq 0.$$

A sequence $(x_n)_{n \geq 0}$ of quadratic congruential pseudorandom numbers in the interval $[0, 1)$ can be obtained by $x_n = y_n/m$ for $n \geq 0$. These sequences are purely periodic with the maximum possible period length $m$ if and only if the conditions $a \equiv 0 \pmod{p_i}$, $b \equiv 1 \pmod{p_i}$, $c \not\equiv 0 \pmod{p_i}$ for all $1 \leq i \leq r$ and $a \not\equiv 3c \pmod 9$ if $p_i = 3$ are satisfied ([Knu81], pp. 34, 526). From now on, these conditions for the maximum possible period length are always assumed.

In the compound approach, $r$ moduli $q_i = p_i^{\omega_i}$ with pairwise different odd primes $p_i$ and integers $\omega_i \geq 2$ for $1 \leq i \leq r$ are considered. The parameters $a_i, b_i, c_i \in \mathbb{Z}_{q_i}$ have to satisfy the conditions for the maximum possible period length. Hence, one gets $r$ sequences $(x_n^{(i)})_{n \geq 0}$ of quadratic congruential pseudorandom numbers with period lengths $q_1, \ldots, q_r$, respectively. A sequence $(z_n)_{n \geq 0}$ of compound quadratic congruential pseudorandom numbers in the interval $[0, 1)$ is obtained by

$$z_n \equiv x_n^{(1)} + \cdots + x_n^{(r)} \pmod{1}, \ n \geq 0.$$

It can be shown by the Chinese Remainder Theorem that this sequence is purely periodic and has the maximum possible period length $m = q_1 \cdots q_r$. The quadratic congruential method is easy to implement and the compound approach leads to large period lengths, although exact integer calculations have to be performed only with respect to the smaller moduli $q_1, \ldots, q_r$.

Many results for the case of quadratic congruential generators with prime power modulus can be found in [Eic92], [Eic94], [Eic95a], [Eic95b], [Eic], [EN91], [EN95], and [Str94]. A review of these results can be found in [EH97] (see this volume). A characterization of the lattice structure of arbitrary quadratic congruential pseudorandom numbers is given in [EL87] and average equidistribution properties over parts of the period are treated in [Emm97]. Surveys of other nonlinear methods are given in [Nie92] and [Nie95].

In Section 2 of this paper the relation between a generator with composite modulus and a compound generator with corresponding prime power moduli is presented and upper and lower discrepancy bounds are stated and discussed. In Section 3 necessary auxiliary results are established. In Section 4 the proofs for the discrepancy bounds are worked out, and in Section 5 a proof of the relation between compound generators and generators with composite modulus is given.

## 2   Main results

The compound approach has its main advantages in the field of computing, because one can perform the necessary integer arithmetic with respect to the smaller moduli $q_i$, although a large period length $m = q_1 \cdots q_r$ is obtained. For the theoretical analysis, it is more convienient to consider ordinary generators. Fortunately, there is a one to one relation between the generated sequences of ordinary and compound quadratic congruential generators with odd moduli. For the sake of completeness the case of even moduli is included in the next theorem, too. In the case of odd moduli all terms with $\tau_i$ can be omitted. The conditions for the maximum possible period length for an even modulus can also be found in [Knu81] (pp. 34, 526).

THEOREM 1 *Let* $m = q_1 \cdots q_r$ *be a modulus with* $q_i = p_i^{\omega_i}$ *and pairwise different primes* $p_i$ *and integers* $\omega_i \geq 2$ *for* $1 \leq i \leq r$. *Let* $a$, $b$, $c \in \mathbb{Z}_m$ *be the parameters of a quadratic congruential generator with initial value* $y_0$, *and let* $a_i$, $b_i$, $c_i \in \mathbb{Z}_{q_i}$ *for* $1 \leq i \leq r$ *be the parameters of a compound quadratic congruential generator with initial values* $y_0^{(i)}$ *for* $1 \leq i \leq r$. *The corresponding sequences* $(x_n)_{n \geq 0}$ *and* $(z_n)_{n \geq 0}$ *are equal if and only if*

(i) $$ a \equiv n_i a_i + \tau_i \frac{q_i}{2} \pmod{q_i}, $$

(ii) $$ b \equiv b_i + \tau_i \frac{q_i}{2} \pmod{q_i}, $$

(iii) $$ c \equiv m_i c_i \pmod{q_i}, $$

(iv) $$ y_0 \equiv m_i y_0^{(i)} \pmod{q_i} $$

*with* $m_i = m/q_i$, $n_i \equiv m_i^{-1} \pmod{q_i}$, $\tau_i = 0$ *if* $q_i$ *is odd, and* $\tau_i \in \{0, 1\}$ *if* $q_i$ *is even for* $1 \leq i \leq r$.

The statistical independence behaviour of uniform pseudorandom numbers in the interval $[0, 1)$ is very important for their usability in a stochastic simulation. It is typically described by distribution properties of $s$-tuples of successive terms in the generated sequences. A reliable measure for the distribution of the corresponding point sets in $[0, 1)^s$ is given by the notion of discrepancy. For $N$ arbitrary points $t_0, t_1, \ldots, t_{N-1} \in [0, 1)^s$, the discrepancy is defined by

$$ D_N(t_0, t_1, \ldots, t_{N-1}) = \sup_J |F_N(J) - V(J)|, $$

where the supremum is extended over all half open subintervals $J$ of $[0, 1)^s$, $F_N(J)$ is $N^{-1}$ times the number of points among $t_0, t_1, \ldots, t_{N-1}$ falling into $J$, and $V(J)$ denotes the $s$-dimensional volume of $J$. In the present paper, the pairs

$$ \mathbf{x}_n = (x_n, x_{n+1}) \in [0, 1)^2, \ n \geq 0, $$

of successive quadratic congruential pseudorandom numbers are considered and the abbreviation

$$ D_m^{(2)} = D_m(\mathbf{x}_0, \mathbf{x}_1, \ldots, \mathbf{x}_{m-1}) $$

is used. Upper and lower bounds for the discrepancy $D_m^{(2)}$ with prime power modulus $m = p^\omega$ in [EN91] indicate that the distribution of pairs strongly depends on the maximal power of $p$ that divides the parameter $a$ and that the most 'quadratic' behaviour is obtained for $a \not\equiv 0 \pmod{p^2}$. The following presentation for composite moduli is restricted to the corresponding situation, i.e., it is assumed from now on that the parameters $a$ and $a_i$ are chosen in such a way that $a \not\equiv 0 \pmod{p_i^2}$ and $a_i \not\equiv 0 \pmod{p_i^2}$ for $1 \leq i \leq r$.

THEOREM 2 *The discrepancy $D_m^{(2)}$ of pairs of quadratic congruential pseudo-random numbers with modulus $m = p_1^{\omega_1} \cdots p_r^{\omega_r}$ with pairwise different primes $p_i \geq 3$ and integers $\omega_i \geq 2$ for $1 \leq i \leq r$ satisfies*

$$D_m^{(2)} < m^{-1/2} \prod_{i=1}^{r} \left(1 + \frac{5}{4} p_i^{-3/2}\right) \left[\frac{4}{\pi^2 p^{1/2}} \left(\log m + \frac{4\sqrt{3}\pi}{9} \log p\right)\right.$$

$$\left. \left(\log m + 0.778\right) + \frac{16}{27} p^{1/2}\right] + 2m^{-1}$$

*with $p = p_1 \cdots p_r$.*

THEOREM 3 *The discrepancy $D_m^{(2)}$ of pairs of quadratic congruential pseudo-random numbers with modulus $m = p_1^{\omega_1} \cdots p_r^{\omega_r}$ with pairwise different primes $p_i \geq 3$ and integers $\omega_i \geq 2$ for $1 \leq i \leq r$ satisfies*

$$D_m^{(2)} \geq \frac{p^{1/2}}{2(\pi + 2)} m^{-1/2}$$

*with $p = p_1 \cdots p_r$.*

THEOREM 2 and THEOREM 3 show that the discrepancy has an order of magnitude between

$$p^{1/2} m^{-1/2} \quad \text{and} \quad \left(p^{1/2} + p^{-1/2} (\log m)^2\right) m^{-1/2} .$$

Hence, for $\log m \sim p^{1/2}$, the order of magnitude of the upper bound of $D_m^{(2)}$ can be made as small as $m^{-1/2} \log m$ which coincides with the order of magnitude of the lower bound of $D_m^{(2)}$. This fits quite well the law of the iterated logarithm for discrepancies which says that the discrepancy of $m$ independent and uniformly distributed random points from $[0, 1)^2$ is almost always of the order of magnitude $m^{-1/2} (\log \log m)^{1/2}$ (cf. [Kie61]). The reader is referred to the excellent monograph [Nie92] and to the survey articles [EH97] and [Nie95] for a detailed comparison of these discrepancy estimates to corresponding results for linear and other nonlinear congruential methods.

## 3 Auxiliary results

First, some further notation is necessary. For integers $s \geq 1$ and $q \geq 2$, let

$$C_s(q) = \left\{(h_1, \ldots, h_s) \in \mathbb{Z}^s \setminus \{0\} \left| -\frac{q}{2} < h_j \leq \frac{q}{2} \quad \text{for} \quad 1 \leq j \leq s\right.\right\}$$

be a set of nonzero lattice points. For $t \in \mathbb{R}$, the abbreviation $e(t) = e^{2\pi i t}$ is used and $\mathbf{u} \cdot \mathbf{v}$ stands for the standard inner product of $\mathbf{u}, \mathbf{v} \in \mathbb{R}^s$. Define

$$r(h, q) = \begin{cases} q \sin(\pi |h|/q) & \text{for} \quad h \in C_1(q), \\ 1 & \text{for} \quad h = 0 \end{cases}$$

and

$$r(\mathbf{h}, q) = \prod_{j=1}^{s} r(h_j, q)$$

for $\mathbf{h} = (h_1, \ldots, h_s) \in C_s(q)$. Subsequently, two known results are recalled. They follow from Theorem 3.10 and from Corollary 3.17 in [Nie92], respectively.

LEMMA 1 *Let $N \geq 1$ and $q \geq 2$ be integers. Let $\mathbf{t}_n = \mathbf{y}_n/q \in [0, 1)^2$, with $\mathbf{y}_n \in \mathbb{Z}_q^2$ for $0 \leq n < N$. Then the discrepancy of the points $\mathbf{t}_0, \mathbf{t}_1, \ldots, \mathbf{t}_{N-1}$ satisfies*

$$D_N(\mathbf{t}_0, \mathbf{t}_1, \ldots, \mathbf{t}_{N-1}) \leq \frac{2}{q} + \frac{1}{N} \sum_{\mathbf{h} \in C_2(q)} \frac{1}{r(\mathbf{h}, q)} \left| \sum_{n=0}^{N-1} e(\mathbf{h} \cdot \mathbf{t}_n) \right|.$$

LEMMA 2 *The discrepancy of $N$ arbitrary points $\mathbf{t}_0, \mathbf{t}_1, \ldots, \mathbf{t}_{N-1} \in [0, 1)^2$ satisfies*

$$D_N(\mathbf{t}_0, \mathbf{t}_1, \ldots, \mathbf{t}_{N-1}) \geq \frac{1}{2(\pi + 2)|h_1 h_2| N} \left| \sum_{n=0}^{N-1} e(\mathbf{h} \cdot \mathbf{t}_n) \right|$$

*for any lattice point $\mathbf{h} = (h_1, h_2) \in \mathbb{Z}^2$ with $h_1 h_2 \neq 0$.*

In the following, the abbreviation

$$S(u, v; q) = \sum_{z \in \mathbb{Z}_q} e\left((uz^2 + vz)/q\right)$$

is used for quadratic exponential sums. The next lemma is a generalization of Lemma 6 in [EN91].

LEMMA 3 *Let $q \geq 3$ be an odd integer, let $u, v$ be integers and define $d = \gcd(u, q)$. Then*

$$|S(u, v; q)| = \begin{cases} \sqrt{qd} & \text{for} \quad v \equiv 0 \pmod{d}, \\ 0 & \text{for} \quad v \not\equiv 0 \pmod{d}. \end{cases}$$

P r o o f. First, a short calculation shows that

$$|S(u, v; q)|^2 = \sum_{w, z \in \mathbb{Z}_q} e\left((u(w^2 - z^2) + v(w - z))/q\right)$$

$$= \sum_{w, z \in \mathbb{Z}_q} e\left((u(w + z) + v)(w - z)/q\right).$$

With the substitution $(w + z, w - z) \equiv (x, y) \pmod{q}$ one obtains

$$
\begin{aligned}
|S(u, v; q)|^2 &= \sum_{x, y \in \mathbb{Z}_q} e\left((ux + v)\, y/q\right) \\
&= q \cdot \#\{x \in \mathbb{Z}_q \,|\, ux + v \equiv 0 \pmod{q}\} \\
&= \begin{cases} qd & \text{for } v \equiv 0 \pmod{d}, \\ 0 & \text{for } v \not\equiv 0 \pmod{d}, \end{cases}
\end{aligned}
$$

which yields the desired result. $\qquad\qquad\qquad\qquad\qquad\qquad\square$

The next lemma is crucial for the proof of THEOREM 2. It is a generalization of Lemma 3 in [EN95] with slightly improved estimates in its proof.

LEMMA 4 *Let $q \geq 3$ be an odd integer. Then*

$$
\sum_{\substack{\mathbf{h}=(h_1,h_2) \in C_1^2(q) \\ h_2 \not\equiv 0 \pmod{f} \\ h_1 + h_2 \equiv 0 \pmod{f}}} \frac{1}{r(\mathbf{h}, q)} < \frac{4}{\pi^2 f}\left(\log q + \frac{4\sqrt{3}\pi}{9}\log f\right)\left(\log q + 0.778\right) + \frac{16}{27}
$$

*for all divisors $f \geq 3$ of $q$.*

P r o o f. (i) First two preliminary estimates are established. Since $f/(x(f - x))$ is symmetric to $f/2$ and convex on $(0, f)$, straightforward calculations show that

$$
\sum_{d=1}^{f-1} \frac{f}{d(f - d)} < 2 \int_{1/2}^{f/2} \frac{f}{x(f - x)}\, dx < 2(\log f + \log 2).
$$

Since $[f/(x(f - x))]^2$ is symmetric to $f/2$ and convex on $(0, f)$, too, it follows that

$$
\begin{aligned}
\sum_{d=1}^{f-1} \frac{f^2}{d^2(f - d)^2} &< 2 \int_{1/2}^{f/2} \frac{f^2}{x^2(f - x)^2}\, dx \\
&< 2\left(\frac{2}{f}\log f + 2 + \frac{1}{f}(2\log 2 - 1)\right).
\end{aligned}
$$

(ii) For $f < q$ and any integer $d \in \{1, \ldots, f - 1\}$, one obtains

$$
\sum_{\substack{h=1 \\ h \equiv d \pmod{f}}}^{(q-1)/2} \frac{1}{r(h, q)} < \frac{1}{r(d, q)} + \int_0^{(q-2d)/(2f)} \frac{1}{q \sin(\pi(fx + d)/q)}\, dx
$$

$$
= \frac{1}{r(d, q)} - \frac{1}{\pi f}\log\left(\tan(\pi d/(2q))\right) < \frac{1}{r(d, q)} - \frac{1}{\pi f}\log(\pi d/(2q))
$$

$$< \frac{1}{q\sin(\pi d/q)} + \frac{1}{\pi f}\log q - \frac{0.143}{f} < \frac{2\sqrt{3}}{9d} + \frac{1}{\pi f}\log q - \frac{0.143}{f},$$

where the last inequality follows from $\sin x \geq (3\sqrt{3})/(2\pi)\cdot x$ for $x \in [0, \pi/3]$.
Therefore,

$$\sum_{\substack{h\in C_1(q)\\ h\equiv d \;(\mathrm{mod}\, f)}} \frac{1}{r(h,q)} = \sum_{\substack{h=1\\ h\equiv d \;(\mathrm{mod}\, f)}}^{(q-1)/2} \frac{1}{r(h,q)} + \sum_{\substack{h=1\\ h\equiv f-d \;(\mathrm{mod}\, f)}}^{(q-1)/2} \frac{1}{r(h,q)}$$

$$< \frac{2\sqrt{3}f}{9d(f-d)} + \frac{2}{\pi f}\log q - \frac{0.286}{f}$$

for any $d \in \{1, \ldots, f-1\}$ and $f < q$.
(iii) Finally, for $f < q$, it follows from the estimates in (ii) and (i) that

$$\sum_{\substack{\mathbf{h}=(h_1,h_2)\in C_1^2(q)\\ h_2\not\equiv 0 \;(\mathrm{mod}\, f)\\ h_1+h_2\equiv 0 \;(\mathrm{mod}\, f)}} \frac{1}{r(\mathbf{h},q)} = \sum_{d=1}^{f-1} \sum_{\substack{h_1\in C_1(q)\\ h_1\equiv d \;(\mathrm{mod}\, f)}} \sum_{\substack{h_2\in C_1(q)\\ h_2\equiv f-d \;(\mathrm{mod}\, f)}} \frac{1}{r(h_1,q)r(h_2,q)}$$

$$< \sum_{d=1}^{f-1} \left( \frac{2\sqrt{3}f}{9d(f-d)} + \frac{2}{\pi f}\log q - \frac{0.286}{f} \right)^2$$

$$< \frac{8}{27}\left( \frac{2}{f}\log f + 2 + \frac{1}{f}(2\log 2 - 1) \right)$$

$$+ \frac{8\sqrt{3}}{9}\left( \frac{2}{\pi f}\log q - \frac{0.286}{f} \right)\left( \log f + \log 2 \right)$$

$$+ f\left( \frac{2}{\pi f}\log q - \frac{0.286}{f} \right)^2$$

$$< \frac{4}{\pi^2 f}\left( \log q + \frac{4\sqrt{3}\pi}{9}\log f \right)\left( \log q + 0.778 \right) + \frac{16}{27},$$

which is the desired result for $f < q$.
(iv) For $f = q$, it follows from $\sin x > x(\pi - x)/\pi$ for $x \in (0, \pi)$ and the
second part of (i) that

$$\sum_{\substack{\mathbf{h}=(h_1,h_2)\in C_1^2(f)\\ h_1+h_2\equiv 0 \;(\mathrm{mod}\, f)}} \frac{1}{r(\mathbf{h},f)} = 2\sum_{d=1}^{(f-1)/2} \frac{1}{(f\sin(\pi d/f))^2}$$

$$< \frac{2}{\pi^2}\sum_{d=1}^{(f-1)/2} \frac{f^2}{d^2(f-d)^2} = \frac{1}{\pi^2}\sum_{d=1}^{f-1} \frac{f^2}{d^2(f-d)^2}$$

$$< \frac{4}{\pi^2 f}\left( \log f + 0.194 \right) + \frac{4}{\pi^2},$$

which completes the proof. □

Finally, one needs an estimate for the sum of a certain function over the divisors of a given integer.

LEMMA 5 *Let* $q = p_1^{\omega_1} \cdots p_r^{\omega_r}$ *be an integer with pairwise different primes* $p_i \geq 3$. *Then*

$$\sum_{d|q} \frac{1}{d^{3/2}} < \prod_{i=1}^{r} \left(1 + \frac{5}{4} p_i^{-3/2}\right) .$$

P r o o f. Straightforward calculations yield

$$\sum_{d|q} \frac{1}{d^{3/2}} = \sum_{\substack{(\beta_1,\ldots,\beta_r) \in \\ \times_{i=1}^{r} \{0,\ldots,\omega_i\}}} \frac{1}{(p_1^{\beta_1} \cdots p_r^{\beta_r})^{3/2}}$$

$$= \prod_{i=1}^{r} \sum_{\beta=0}^{\omega_i} \left(p_i^{-3/2}\right)^{\beta}$$

$$< \prod_{i=1}^{r} \frac{1}{1 - p_i^{-3/2}} < \prod_{i=1}^{r} \left(1 + \frac{5}{4} p_i^{-3/2}\right) ,$$

which is the desired result. □

With the convergence of $\sum_{i=1}^{\infty} p_i^{-3/2}$, the product $\prod_{i=1}^{\infty} \left(1 + \frac{5}{4} p_i^{-3/2}\right)$ converges, too. Therefore, the sum in LEMMA 5 is uniformly bounded for arbitrary values of $r$ and $p_i$.

## 4  Proof of the discrepancy bounds

P r o o f   o f   T H E O R E M  2. First, LEMMA 1 is applied with $N = q = m$ and $t_n = x_n$ for $0 \leq n < m$. This yields

$$D_m^{(2)} \leq \frac{2}{m} + \frac{1}{m} \sum_{h \in C_2(m)} \frac{1}{r(h,m)} \left| \sum_{n=0}^{m-1} e(h \cdot x_n) \right|$$

$$= \frac{2}{m} + \frac{1}{m} \sum_{h=(h_1,h_2) \in C_1^2(m)} \frac{1}{r(h,m)} \left| \sum_{z \in \mathbb{Z}_m} e\left((h_2 a z^2 + (h_1 + h_2 b)z)/m\right) \right|$$

$$= \frac{2}{m} + \frac{1}{m} \sum_{h=(h_1,h_2) \in C_1^2(m)} \frac{1}{r(h,m)} |S(h_2 a, h_1 + h_2 b; m)| ,$$

where in the second step $\sum_{n=0}^{m-1} e(h \cdot x_n) = 0$ for $h_1 h_2 = 0$ is used. The parameter $a$ of the quadratic congruential generator satisfies $\gcd(a,m) = p$

and therefore $\gcd(h_2 a, m) = dp$ for some divisor $d$ of $m/p$. This observation and LEMMA 3 imply that

$$D_m^{(2)} \leq \frac{2}{m} + \frac{1}{m} \sum_{d \mid (m/p)} \sum_{\substack{h=(h_1,h_2) \in C_1^2(m) \\ \gcd(h_2 a, m) = dp}} \frac{1}{r(\mathbf{h}, m)} |S(h_2 a, h_1 + h_2 b; m)|$$

$$= \frac{2}{m} + \frac{1}{m} \sum_{d \mid (m/p)} \sqrt{mdp} \sum_{\substack{h=(h_1,h_2) \in C_1^2(m) \\ \gcd(h_2 a, m) = dp \\ h_1 + h_2 \equiv 0 \;(\mathrm{mod}\, dp)}} \frac{1}{r(\mathbf{h}, m)},$$

where $b \equiv 1 \,(\mathrm{mod}\, p)$ is used in the last step. Now, the modulus $m$ can be reduced by the divisor $d$ in order to obtain

$$D_m^{(2)} \leq \frac{2}{m} + \frac{1}{m} \sum_{d \mid (m/p)} \sqrt{mdp} \sum_{\substack{k=(k_1,k_2) \in C_1^2(m/d) \\ \gcd(k_2, m/(dp)) = 1 \\ k_1 + k_2 \equiv 0 \;(\mathrm{mod}\, p)}} \frac{1}{d^2 r(\mathbf{k}, m/d)}$$

$$\leq \frac{2}{m} + \frac{1}{m} \sum_{d \mid (m/p)} \sqrt{mdp} \sum_{\substack{k=(k_1,k_2) \in C_1^2(m/d) \\ k_2 \not\equiv 0 \;(\mathrm{mod}\, p) \\ k_1 + k_2 \equiv 0 \;(\mathrm{mod}\, p)}} \frac{1}{d^2 r(\mathbf{k}, m/d)},$$

where the last step becomes obvious if the two cases $d < m/p$ and $d = m/p$ are distinguished. If $d < m/p$, then the condition $\gcd(k_2, m/(dp)) = 1$ implies that $k_2 \not\equiv 0 \,(\mathrm{mod}\, p)$. If $d = m/p$, then both conditions are satisfied for all $k_2 \in C_1(p)$. Finally, LEMMA 4 can be used to estimate

$$D_m^{(2)} < \frac{2}{m} + \frac{1}{m} \sum_{d \mid (m/p)} \sqrt{mdp} \frac{1}{d^2} \left[ \frac{4}{\pi^2 p} \left( \log(m/d) + \frac{4\sqrt{3}\pi}{9} \log p \right) \right.$$

$$\left. \left( \log(m/d) + 0.778 \right) + \frac{16}{27} \right]$$

$$< \frac{2}{m} + \frac{p^{1/2}}{m^{1/2}} \left[ \frac{4}{\pi^2 p} \left( \log m + \frac{4\sqrt{3}\pi}{9} \log p \right) \right.$$

$$\left. \left( \log m + 0.778 \right) + \frac{16}{27} \right] \sum_{d \mid (m/p)} \frac{1}{d^{3/2}} .$$

Now, LEMMA 5 yields the desired result. $\qquad \square$

Proof of THEOREM 3. First, with LEMMA 2 one gets

$$D_m^{(2)} \geq \frac{1}{2(\pi+2)|h_1 h_2| m} \left| \sum_{n=0}^{m-1} e(\mathbf{h} \cdot \mathbf{x}_n) \right|$$

$$= \frac{1}{2(\pi+2)|h_1 h_2|m} |S(h_2 a, h_1 + h_2 b; m)|$$

for any $\mathbf{h} = (h_1, h_2) \in \mathbb{Z}^2$ with $h_1 h_2 \neq 0$. The choice $\mathbf{h} = (-1, 1) \in \mathbb{Z}^2$ and an application of LEMMA 3 with $d = p$ yield

$$D_m^{(2)} \geq \frac{1}{2(\pi+2)m} \sqrt{mp},$$

which is the desired result. $\qquad\qquad\qquad\qquad\qquad\qquad\qquad$ $\square$

## 5  Proof of THEOREM 1

The equation $x_n = z_n$ is equivalent to $y_n = m z_n$ for all $n \geq 0$. According to the Chinese Remainder Theorem the latter is equivalent to $y_n \equiv m_i y_n^{(i)} \pmod{q_i}$ for all $1 \leq i \leq r$ and $n \geq 0$. In fact, both sequences have the maximum possible period length and satisfy

$$\begin{aligned}
y_{n+1} &\equiv a y_n^2 + b y_n + c \pmod{q_i}, \\
m_i y_{n+1}^{(i)} &\equiv n_i a_i (m_i y_n^{(i)})^2 + b_i (m_i y_n^{(i)}) + m_i c_i \pmod{q_i}, \quad n \geq 0,
\end{aligned}$$

so that $y_n \equiv m_i y_n^{(i)} \pmod{q_i}$ is equivalent to

$$y_0 \equiv m_i y_0^{(i)} \pmod{q_i}$$

and the property that the mappings $f_i, g_i : \mathbb{Z}_{q_i} \to \mathbb{Z}_{q_i}$ with

$$\begin{aligned}
f_i(z) &\equiv a z^2 + b z + c \pmod{q_i}, \\
g_i(z) &\equiv n_i a_i z^2 + b_i z + m_i c \pmod{q_i}
\end{aligned}$$

are identical. This last property is equivalent to

$$\begin{aligned}
2(a - n_i a_i) &\equiv 0 \pmod{q_i}, \\
a + b &\equiv n_i a_i + b_i \pmod{q_i}, \\
c &\equiv m_i c_i \pmod{q_i},
\end{aligned}$$

which yields the desired result. $\qquad\qquad\qquad\qquad\qquad\qquad\qquad$ $\square$

*Acknowledgments:* The author would like to thank the referee for valuable comments and helpful remarks.

# 6 REFERENCES

[EL86]   J. Eichenauer and J. Lehn. A non–linear congruential pseudo random number generator. *Statist. Papers*, **27**:315–326, 1986.

[EL87]   J. Eichenauer and J. Lehn. On the structure of quadratic congruential sequences. *Manuscripta Math.*, **58**:129–140, 1987.

[Eic92]  J. Eichenauer–Herrmann. A remark on the discrepancy of quadratic congruential pseudorandom numbers. *J. Comput. Appl. Math.*, **43**:383–387, 1992.

[Eic94]  J. Eichenauer–Herrmann. On the discrepancy of quadratic congruential pseudorandom numbers with power of two modulus. *J. Comput. Appl. Math.*, **53**:371–376, 1994.

[Eic95a] J. Eichenauer–Herrmann. Discrepancy bounds for nonoverlapping pairs of quadratic congruential pseudorandom numbers. *Arch. Math.*, **65**:362–368, 1995.

[Eic95b] J. Eichenauer–Herrmann. Quadratic congruential pseudorandom numbers: distribution of triples. *J. Comput. Appl. Math.*, **62**:239–253, 1995.

[Eic]    J. Eichenauer–Herrmann. Quadratic congruential pseudorandom numbers: distribution of triples, II. *J. Comput. Appl. Math.* (to appear).

[EH97]   J. Eichenauer–Herrmann and E. Herrmann. A survey of quadratic and inversive congruential pseudorandom numbers. In *this volume*, 1997.

[EN91]   J. Eichenauer–Herrmann and H. Niederreiter. On the discrepancy of quadratic congruential pseudorandom numbers. *J. Comput. Appl. Math.*, **34**:243–249, 1991.

[EN95]   J. Eichenauer–Herrmann and H. Niederreiter. An improved upper bound for the discrepancy of quadratic congruential pseudorandom numbers. *Acta Arith.*, **69**:193–198, 1995.

[Emm97]  F. Emmerich. Equidistribution properties of quadratic congruential pseudorandom numbers. *J. Comput. Appl. Math.*, **79**:207–214, 1997.

[Kie61]  J. Kiefer. On large deviations of the empiric d.f. of vector chance variables and a law of the iterated logarithm. *Pacific J. Math.*, **11**:649–660, 1961.

[Knu81]  D.E. Knuth. *The Art of Computer Programming*, Vol. 2. Addison–Wesley, Reading, MA, 2nd edition, 1981.

426

[Nie92]   H. Niederreiter. *Random Number Generation and Quasi–Monte Carlo Methods*. SIAM, Philadelphia, PA, 1992.

[Nie95]   H. Niederreiter. New developments in uniform pseudorandom number and vector generation. In *Monte Carlo and Quasi–Monte Carlo Methods in Scientific Computing, Lecture Notes in Statistics*, Vol. 106, pages 87–120, New York, 1995. Springer.

[Str94]   S. Strandt. Diskrepanzabschätzung bei quadratischen Kongruenzgeneratoren zur Erzeugung von Pseudozufallszahlen. Diplomarbeit, Technische Hochschule Darmstadt, 1994.

Sibylle Strandt
Technische Hochschule Darmstadt
Fachbereich Mathematik
Schloßgartenstraße 7
D-64289 Darmstadt
Germany
E-mail: strandt@mathematik.th-darmstadt.de

# A new permutation choice in Halton sequences

## Bruno TUFFIN[1]

ABSTRACT  This paper has several folds. We make first new permutation choices in Halton sequences to improve their distributions. These choices are multi-dimensional and they are made for two different discrepancies. We show that multi-dimensional choices are better for standard quasi-Monte Carlo methods. We also use these sequences as a variance reduction technique in Monte Carlo methods, which greatly improves the convergence accuracy of the estimators. For this kind of use, we observe that one-dimensional choices are more efficient.

## 1  Introduction

Quasi-Monte Carlo methods are deterministic analogs of Monte Carlo ones. For the latters, convergence is in $O(1/\sqrt{N})$ for an approximation with $N$ random points. It is possible to construct a sequence where the points are deterministic and "well distributed" all over the integration space, for which the convergence speed is faster (in $O(N^{-1}(\log N)^s)$ for dimension $s$). Halton sequences verify this property. In this paper we give new permutation choices for Halton sequences to improve their distribution. Next, as the major problem encountered with quasi-Monte Carlo methods is the error bound evaluation, we use the sequences in Monte Carlo methods to obtain a variance reduction. In this case, we observe that a one-dimensional choice of permutations is more efficient.

This paper is organized as follows: Section 2 describes quasi-Monte Carlo methods and Halton sequences and section 3 proposes a new choice of permutations in Halton sequences. Section 4 describes the use of such sequences as a variance reduction in Monte Carlo methods and explain why a one-dimensional choice is better for this kind of use. Finally we conclude in Section 5.

---

[1]IRISA, Campus de Beaulieu, 35042 Rennes cédex, France

# 2  Quasi-Monte Carlo methods

## 2.1  General method

Let us consider the integration of functions on $[0,1)^s$. The objective of the method is to approximate $\int_{[0,1)^s} f(u)du$ by $\frac{1}{N}\sum_{n=1}^{N} f(\xi^{(n)})$, where $(\xi^{(n)})_{n\in\mathbb{N}}$ is a deterministic sequence. Define a measure of uniform distribution over $[0,1)^s$. Let $\mathcal{P} = (\xi^{(n)})_{n\in\mathbb{N}}$ and define $A_N(z,\mathcal{P}) = \sum_{n=1}^{N} 1_{[0,z)}(\xi^{(n)})$ where $z = (z_1,\cdots,z_s) \in [0,1]^s$ and $[0,z) = \prod_{i=1}^{s}[0,z_i)$. The discrepancy in space $L^p$ of the $N$ first terms of $\mathcal{P}$ is defined by

$$T_N^{(p)*}(\mathcal{P}) = \left( \int_{[0,1)^s} \left| \frac{A_N([0,z),\mathcal{P})}{N} - \lambda_s([0,z)) \right|^p dz \right)^{1/p}.$$

This discrepancy is an expression of the mean difference between the frequency of points of $\mathcal{P}$ and the measure of each interval of form $[0,x)$. The sequence is uniformly distributed if and only if $\lim_{n\to+\infty} T_N^{(p)*}(\mathcal{P}) = 0$. An expression of $T_N^{(2)*}(\mathcal{P})$ (see [3]) is given by

$$(T_N^{(2)*}(\mathcal{P}))^2 = \frac{1}{N^2} \sum_{k,m=1}^{N} \prod_{i=1}^{s}(1-M_i^{k,m}) - \frac{2^{1-s}}{N}\sum_{k=1}^{N}\prod_{i=1}^{s}(1-\xi_i^{(k)2})+3^{-s} \quad (1.1)$$

where $\xi_i^{(n)}$ is the $i^{th}$ coordinate of the vector $\xi^{(n)}$ and $M_i^{k,m} = \max(\xi_i^{(k)},\xi_i^{(m)})$.

There exist error bounds involving this discrepancy (see [12], [3]).

Sequences with $T_N^{(p)*}(\mathcal{P}) = O(N^{-1}(\log N)^s)$ are called *low discrepancy sequences*.

A new notion of discrepancy, $T_N^{(2)}(\mathcal{P})$, is described by Morokoff and Caflish in [3]. It is shown that the computation of $T_N^{(2)*}(\mathcal{P})$ takes more into account the points $x \in [0,1)^s$ near the origin $0 = (0,\cdots,0)$ than the others. The definition is the following: let $x < y$ denote the inequality for each coordinate of vectors, i.e. $x_i < y_i$ $(i = 1,\cdots s)$. Then

$$T_N^{(2)}(\mathcal{P}) = \left[ \int_{x,y\in[0,1)^s;x<y} \left| \frac{A_N([x,y),\mathcal{P})}{N} - \lambda_s([x,y)) \right|^2 dxdy \right]^{1/2}.$$

With this new definition, all the points of the space $[0,1)^s$ have the same importance. It is proven in [3] that, if $m_i^{k,m} = \min(\xi_i^{(k)},\xi_i^{(m)})$,

$$(T_N^{(2)}(\mathcal{P}))^2 = \frac{1}{N^2} \sum_{k,m=1}^{N} \prod_{i=1}^{s}(1-M_i^{k,m})m_i^{k,m} - \frac{2^{1-s}}{N}\sum_{k=1}^{N}\prod_{i=1}^{s}(1-\xi_i^{(k)})\xi_i^{(k)}+12^{-s}.$$

$$(1.2)$$

Unfortunately, even if this definition is more representative of equi-distribution, there exists yet no error bound using this discrepancy.

There exist many examples of low discrepancy sequences [4]. In this paper, we focus on an important family, Halton sequences, and their properties.

## 2.2 Halton sequences and improvements

Let $p \in \mathbb{N}$. Let us denote the digit expansion of $n \in \mathbb{N}$ in base $p$ by

$$n = a_j p^j + \cdots + a_1 p + a_0 \text{ with } a_i \in \{0, \cdots, p-1\}.$$

The radical-inverse function of $n$ is then defined by

$$\Phi_p(n) = a_0/p + a_1/p^2 + \cdots + a_j/p^{j+1}.$$

If $p_1, \cdots, p_s$ are $s$ mutually prime integers, the sequence $\mathcal{P} = (\xi^{(n)})_{n \in \mathbb{N}}$ defined by

$$\xi^{(n)} = (\Phi_{p_1}(n), \cdots, \Phi_{p_s}(n)),$$

is a Halton sequence and it verifies

$$T_N^{(2)*}(\mathcal{P}) = O(N^{-1}(\log(N))^s). \tag{1.3}$$

In practice, we will always let $p_i$ equal to the $i^{th}$ prime number. In spite of the asymptotic low discrepancy of such sequences, it is observed in [1], [3], [8] that, in a large dimension, a good distribution needs many iterations to occur (i.e. we obtain bad distributions for a small number of iterations). As a matter of fact, the monotone cycles of length $p_i$ for the $i^{th}$ projection introduce a great regularity between coordinates. An important research effort has been made to improve this. In [3] the sequence is scrambled independently for each coordinate: for the $s$ sequences (corresponding to the $s$ coordinates), the $N$ points are ranged from the smallest to the largest and then they are randomly permuted. It is possible to show that this operation do not change the bound on the discrepancy given by (1.3).

Braaten and Weller [1] have improved the distribution of the sequence with the introduction, for each prime number $p_i$ $(1 \leq i \leq s)$, of a permutation $\pi_{p_i}$ on $\{0, \cdots, p_i-1\}$ satisfying $\pi_{p_i}(0) = 0$ and leading to a more chaotic sequence. In this case again, for each choice of permutation the convergence speed (1.3) remains unchanged. This scrambling gives sequences that are the most efficient for certain dimensions and for certain applications [6].

If, from the digit expansion of $n$ in base $p_i$, we set

$$S_{p_i}(n) = \pi_{p_i}(a_0)/p_i + \pi_{p_i}(a_1)/p_i^2 + \cdots + \pi_{p_i}(a_j)/p_i^{j+1},$$

the new Halton sequence is $\mathcal{P} = (\xi^{(n)})_{n \in \mathbb{N}}$ with $\xi^{(n)} = (S_{p_1}(n), \cdots, S_{p_s}(n))$. The problem is to choose good permutations. Braaten and Weller [1] have built the permutations $\pi_{p_i}$ as follows. For each $p_i$, $\pi_{p_i}$ is chosen in an unidimensional way: if we know $\pi_{p_i}(1), \cdots, \pi_{p_i}(j)$, we choose $\pi_{p_i}(j+1)$ as the element minimizing the mean square discrepancy $T_{j+1}^{(2)*}$ of the $j+1$ points $\{\pi_{p_i}(1)/p_i, \cdots, \pi_{p_i}(j)/p_i, \pi_{p_i}(j+1)/p_i\}$.

Although this choice gives good results, we construct here four multi-dimensional algorithms which should give better results than Braaten and Weller's one, which is built uni-dimensionally.

# 3 New choice of permutations

Two algorithms, MCT* and MCL*, based on $T_N^{(2)*}(\mathcal{P})$, are already described in [8] (where they are called respectively MC1 and MC2). We recall them here and give two new choices, called MCT and MCL, based on $T_N^{(2)}(\mathcal{P})$. We will use expression (1.2), instead of (1.1) in [8], to make the choice of permutations in Halton sequences. The algorithms we generate are the same as those in [8], but for $T_N^{(2)}(\mathcal{P})$ instead of $T_N^{(2)*}(\mathcal{P})$.

## 3.1 Algorithm MCL

As for Braaten and Weller's one, this method for the choice of permutations gives a table which is available for any dimension. But in our case, the table is generated line per line instead of element per element. Thus the whole $(j+1)^{th}$ line is chosen knowing the $j$ previous lines. As the only restriction for the $j^{th}$ permutation is $\pi_{p_j}(0) = 0$, there are $(p_j - 1)!$ possibilities. For example, for $j = 10$, we have $28! = 3.049e29$ choices. Since it is impossible to compute at each time the discrepancy, we make our choice in a random manner.

Given a $K$-sample $(\pi^{(1)}, \cdots, \pi^{(K)})$ of permutations $\pi$ of $\{0, \cdots, p_{j+1} - 1\}$ such that $\pi(0) = 0$, an estimator of $\pi_{p_{j+1}}$ is the permutation which minimizes $\{T_{p_{j+1}-1}^{(2)}(\mathcal{P}_{\pi^{(k)}}) | 1 \leq k \leq K\}$ for $p_{j+1} - 1$ points. That is, $T_{p_{j+1}-1}^{(2)}(\mathcal{P}_\pi) = \min_{1 \leq k \leq K} T_{p_{j+1}-1}^{(2)}(\mathcal{P}_{\pi^{(k)}})$ where $\mathcal{P}_\pi$ is the sequence in dimension $j+1$ issued from $\pi$ (with permutations $\pi_{p_1}, \cdots, \pi_{p_j}$ fixed for the $j$ first coordinates). We call this algorithm MCL. The algorithm with the same technique, but for $T_N^{(2)*}(\mathcal{P})$ is called MCL* [8].

## 3.2 Algorithm MCT

Let us show now an algorithm creating a whole table for each dimension: in a fixed dimension $s$, we search for a table $(\pi_{p_1}, \cdots, \pi_{p_s})$, with $\pi_{p_i}$ permutation of $\{0, \cdots, p_i - 1\}$ such that $\pi_{p_i}(0) = 0$ and $\pi_{p_i}$ minimizes $T_{p_s-1}^{(2)}(\mathcal{P}_{p_1, \cdots, p_s})$ where $\mathcal{P}_{p_1, \cdots, p_s}$ is the sequence in dimension $s$ associated with permutations $(\pi_{p_1}, \cdots, \pi_{p_s})$. As the number of possible permutations is even larger than before, we use again a random approach.

Let

$$\Omega_s = \{(\pi_1, \cdots, \pi_s) \mid \quad \forall 1 \leq i \leq s \; \pi_i \text{ is a permutation of}$$
$$\{0, \cdots, p_i - 1\} \text{ satisfying } \pi_i(0) = 0\}.$$

Let $\Pi$ be a random variable uniformly distributed on $\Omega_s$ and

$$(\pi_1^{(k)}, \cdots, \pi_s^{(k)})_{1 \leq k \leq K}$$

a $K$-sample from $\Pi$, with $\pi_i^{(k)}$ $i^{th}$ coordinate (i.e. permutation) of $k^{th}$ variable $\pi^{(k)}$. An estimator of $(\pi_{p_1}, \cdots, \pi_{p_s})$ is the table of permutations $(\pi_1^{(j)}, \cdots, \pi_s^{(j)})$ such that

$$T_{p_s-1}^{(2)}(\mathcal{P}_{p_1,\cdots,p_s}^{j}) = \min_{1 \leq k \leq K} T_{p_s-1}^{(2)}(\mathcal{P}_{p_1,\cdots,p_s}^{k}),$$

with $\mathcal{P}_{p_1,\cdots,p_s}^{k}$ the sequence corresponding to permutations $(\pi_1^{(k)}, \cdots, \pi_s^{(k)})$. We call this algorithm MCT. The algorithm with the same techniques, but for $T_N^{(2)*}(\mathcal{P})$ is called MCT* in [8].

## 3.3 Results

| Dim | Halton | B W | MCT* (#iter.) | MCL* (#iter.) |
|-----|--------|-----|---------------|----------------|
| 2 | 4.34e-2 | 4.86e-2 | 4.34e-2 | 4.34e-2 |
| 3 | 1.67e-2 | 9.72e-3 | 8.33e-3 | 1.35e-2 |
| 4 | 6.97e-3 | 6.15e-3 | 3.02e-3 | 4.51e-3 |
| 5 | 2.49e-3 | 1.42e-3 | 9.54e-4 (5 $10^6$) | 1.16e-3 1.79e-3 |
| 6 | 1.91e-3 | 6.79e-4 | 4.58e-4 ($10^6$) | 5.85e-4 ($10^6$) |
| 7 | 1.18e-3 | 2.78e-4 | 1.79e-4 ($10^6$) | 1.82e-4 ($10^6$) |
| 8 | 9.86e-4 | 1.81e-4 | 7.95e-5 ($10^6$) | 8.06e-5 ($10^6$) |
| 9 | 6.76e-4 | 1.03e-4 | 3.13e-5 ($10^6$) | 3.10e-5 ($10^6$) |
| 10 | 4.18e-4 | 1.76e-5 | 1.16e-5 ($10^6$) | 1.09e-5 ($10^6$) |
| 11 | 3.63e-4 | 1.13e-5 | 4.59e-6 (5 $10^5$) | 4.25e-6 (5 $10^5$) |
| 12 | 2.46e-4 | 5.34e-6 | 1.75e-6 (2 $10^5$) | 1.64e-6 (5 $10^5$) |
| 13 | 1.94e-4 | 2.34e-6 | 6.43e-7 ($10^5$) | 6.16e-7 (5 $10^5$) |
| 14 | 1.71e-4 | 1.60e-6 | 2.46e-7 (6 $10^5$) | 2.27e-7 (2 $10^5$) |
| 15 | 1.38e-4 | 4.07e-7 | 8.87e-8 (2 $10^5$) | 7.87e-8 (2 $10^5$) |
| 16 | 1.04e-4 | 2.11e-7 | 3.31e-8 (3.5 $10^5$) | 2.73e-8 (2 $10^5$) |

TABLE I. $(T_{p_s-1}^{(2)*})^2$ for the different algorithms in dimension $s$.

The square of discrepancies $(T_{p_s-1}^{(2)*}(\mathcal{P}))^2$ obtained for each method and each dimension are given in Table I and for $(T_{p_s-1}^{(2)}(\mathcal{P}))^2$ in Table II. The number of iterations used ($K$ in the previous subsections) is indicated after the value of discrepancy when we are not sure to obtain the real minimum. For algorithms MCL* and MCL, it indicates the number of iterations for the choice of the line. We can make the following remarks on both Tables: the improvements of our algorithms increases with the dimension with respect to standard Halton and Braaten and Weller permutations (due to the multi-dimensional heuristic). Moreover, the best results are given by algorithm

| Dim | Halton | MCT (#iter.) | MCL (#iter.) |
|-----|--------|--------------|--------------|
| 2 | 1.39e-2 | 1.39e-2 | 1.39e-2 |
| 3 | 1.15e-3 | 9.93e-4 | 1.13e-3 |
| 4 | 1.39e-4 | 1.16e-4 | 1.22e-4 |
| 5 | 1.13e-5 | 8.85e-6 ($5\ 10^6$) | 9.16e-6 |
| 6 | 1.62e-6 | 1.15e-6 ($10^6$) | 1.11e-6 ($10^6$) |
| 7 | 1.83e-7 | 1.15e-7 ($10^6$) | 1.11e-7 ($10^6$) |
| 8 | 2.81e-8 | 1.55e-8 ($10^6$) | 1.36e-8 ($10^6$) |
| 9 | 3.59e-9 | 1.89e-9 ($10^6$) | 1.78e-9 ($10^6$) |
| 10 | 4.03e-10 | 2.08e-10 ($10^6$) | 1.75e-10 ($10^6$) |
| 11 | 6.26e-11 | 3.09e-11 ($5\ 10^5$) | 2.60e-11 ($10^6$) |
| 12 | 7.57e-12 | 3.48e-12 ($2\ 10^5$) | 2.83e-12 ($5\ 10^5$) |
| 13 | 1.07e-12 | 4.28e-13 ($2\ 10^5$) | 3.26e-13 ($5\ 10^5$) |
| 14 | 1.75e-13 | 6.96e-14 ($6\ 10^5$) | 4.37e-14 ($2\ 10^5$) |
| 15 | 2.66e-14 | 9.34e-15 ($3\ 10^5$) | 5.25e-15 ($2\ 10^5$) |
| 16 | 3.73e-15 | 1.31e-15 ($3.5\ 10^5$) | 6.46e-16 ($2\ 10^5$) |

TABLE II. $(T_{p_s-1}^{(2)})^2$ for the different algorithms.

MCL* for $(T_{p_s-1}^{(2)*}(\mathcal{P}))^2$ and MCL for $(T_{p_s-1}^{(2)}(\mathcal{P}))^2$, although it should be given by respectively MCT* and MCT (as they are algorithms which, for a sufficiently large number of iterations, give the smallest discrepancy). Then it should be better to use the permutations given by MCL* or MCL, and it should be more practical because we get only one table for every dimension. Furthermore, the larger the dimension, the longer the cycles dependence between successive coordinates in standard Halton sequences [8]. If we compare the discrepancies for large dimensions (in this way, if the number of points $p_s - 1$ is large and the dimension $s$ is small, standard Halton gives good results), then our improvements are significant.

Let us compare in Table III, on the example of the function in dimension 16 defined by $\prod_{i=1}^{16} 12(x_i - 1/2)^2$, the approximation of the integral

$$\int_{[0,1)^{16}} \prod_{i=1}^{16} 12(x_i - \frac{1}{2})^2 dx_i, \text{ which value is 1, given by } \frac{1}{N} \sum_{n=1}^{N} \prod_{i=1}^{16} 12(\xi_i^{(n)} - \frac{1}{2})^2.$$

We see that Braaten and Weller's permutation choice improves the quality of the approximation in comparison with Halton's one, and that ours improve the one of Braaten and Weller. In this case, for a standard quasi-Monte Carlo method, the best choices are given by algorithms MCL and MCT.

| It. | Halton | B W | MCT* | MCL* | MCT | MCL |
|-----|--------|-----|------|------|-----|-----|
| $10^3$ | 0.48939 | 1.69686 | 0.76792 | 0.18115 | 0.29996 | 0.65553 |
| $10^4$ | 0.42182 | 2.95772 | 0.53146 | 1.11605 | 0.64943 | 1.29340 |
| $10^6$ | 0.90964 | 1.06985 | 0.95899 | 1.05608 | 1.00286 | 1.02135 |

TABLE III. Test on function $\prod_{i=1}^{16} 12(x_i - 0.5)^2$ for the different algorithms.

## 4 On the use of low discrepancy sequences in Monte Carlo methods

Unfortunately, the known error bounds are generally impossible to evaluate in practice. Then, to obtain a useful error bound, we use low discrepancy sequences to reduce variance in Monte Carlo methods.

Let $X$ be a random variable uniformly distributed on $[0, 1)^s$ and $(\xi^{(k)})_{k \in \mathbb{N}}$ a low discrepancy sequence as described in the previous section. Instead of simulating the random variable $f(X)$, we study

$$Z = \frac{1}{n} \sum_{k=1}^{n} f(\{X + \xi^{(k)}\}), \qquad (1.4)$$

where $\{x\}$ is the fractional part for each coordinate of $x \in \mathbb{R}^s$. To our knowledge, this type of technique has been used for the first time in [2] where $(\xi^{(k)})_{k \leq n}$ is a lattice developed by Korobov and then for Bayesian integration in [7] with low discrepancy sequences. Owen [5] uses a slightly different technique, where the randomness is introduced on permutations for Niederreiter sequences. In our applications, we will use the permuted Halton sequences described in the previous section.

**Theorem 1** *[9] If $(\xi^{(k)})_{k \in \mathbb{N}}$ is a low discrepancy sequence and $f$ a (bounded) Riemann integrable function, we have*

$$\sigma^2 \left( \frac{1}{n} \sum_{k=1}^{n} f(\{X + \xi^{(k)}\}) \right) = O(n^{-2}(\log n)^{2s}).$$

Then, for a sufficiently large $n$, we are sure to obtain a variance reduction with respect to $n$ independent random variables $f(X)$, which variance is in $n^{-1}$. Efficient applications of this method to the analysis of product-form multi-class queuing networks and of a cellular system with dynamic resource sharing can be found respectively in [11] and in [10].

To compare permutation choices for this kind of method, we use the function $f(x_1, \cdots, x_{12}) = \prod_{i=1}^{12} \frac{\pi}{2} \sin(\pi x_i)$. We take $n = 10^4$ elements of the low discrepancy sequences and make $I = 100$ independent iterations of the random variable to estimate the variance. We compare in Table IV the variance of this estimator for each permutation choice with the one of the standard Monte Carlo estimator for $10^6$ iterations, to have the same number of calls to the function.

| Method | Variance |
|--------|----------|
| Monte Carlo | 1.130e-2 |
| Halton | 7.228e-4 |
| B& W | 1.645e-5 |
| MCL* | 7.188e-5 |
| MCT* | 3.361e-5 |
| MCL | 1.019e-4 |
| MCT | 6.753e-5 |

TABLE IV. Variance of different estimators of $\int_{[0,1)^s} \prod_{i=1}^{12} \frac{\pi}{2} \sin(\pi x_i) dx_i$: the Monte Carlo one and those using permuted Halton sequences as a variance reduction.

An important remark is that the best choice for this application is the one-dimensional sequence of Braaten and Weller (except for MCT, but our experiments tell us that this is marginal). As a matter of fact, the introduction of an additional term in (1.4) breaks the multi-dimensional building of the sequence. Another remark is that the Braaten and Weller sequence generally outperforms the $(0, s)$ sequence of Niederreiter, usually used in quasi-Monte Carlo methods, for dimensions smaller than 10 (see [9]).

# 5 Conclusion

We give here new permutation choices in Halton sequences. The new choices are based on $T_N^{(2)}$ or on $T_N^{(2)*}$ and are multi-dimensional, which is an advantage with respect to previous proposals. These choices are better in a standard quasi-Monte Carlo integration. We also use low-discrepancy sequences as variance reduction techniques in Monte Carlo methods. Whereas classical Monte Carlo algorithms do not change the convergence speed, ours does it. For this utilization of Quasi-Monte Carlo in a Monte Carlo scheme (and because of the additional term) a one dimensional choice for the permutations is recommended. It is commonly known that quasi-Monte Carlo techniques give better accuracy than Monte Carlo ones. Nevertheless in practice, to obtain an error bound, a combination of Monte Carlo and quasi-Monte Carlo methods can be efficiently used.

# 6 References

[1] E. Braaten and G. Weller. An Improved Low-Discrepancy Sequence for Multidimensional Quasi-Monte Carlo Integration. *J. Comput. Phys.*, 33:249–258, 1979.

[2] R. Cranley and T.N.L. Patterson. Randomization of number theoretic methods for multiple integration. *SIAM J. Numer. Anal.*, 13(6):904–

914, December 1976.

[3] W. J. Morokoff and R. E. Caflisch. Quasi-Random Sequences and their Discrepancies. *SIAM Journal on Scientific Computing*, pages 1571–1599, December 1994.

[4] H. Niederreiter. *Random Number Generation and Quasi-Monte Carlo Methods*. CBMS-SIAM 63, Philadelphia, 1992.

[5] A. B. Owen. Randomly permuted $(t, m, s)$-nets and $(t, s)$-sequences. In Harald Niederreiter and Peter Jau-Shyong Shiue, editors, *Monte Carlo and Quasi-Monte Carlo Methods in Scientific Computing*, volume 106 of *Lecture Notes in Statistics*, pages 299–315. Springer, 1995.

[6] G. Pagès and Y.J. Xiao. Sequences with low discrepancy and pseudo-random numbers: theoretical results and numerical tests. *to appear in J. Statist. Comput. Simulat.*, 1996.

[7] J. E. H. Shaw. A quasirandom approach to integration in Bayesian statistics. *Ann. Statist.*, 16:895–914, 1988.

[8] B. Tuffin. Improvement of Halton sequences distribution. Technical Report 998, IRISA, March 1996.

[9] B. Tuffin. On the Use of Low Discrepancy Sequences in Monte Carlo Methods. *Monte Carlo Methods and Applications*, 2(4):295–320, 1996.

[10] B. Tuffin. Variance reduction technique for a cellular system with dynamic resource sharing. In *Proceedings of the 10th European Simulation Multiconference*, pages 467–471, Budapest, June 1996.

[11] B. Tuffin. Variance Reductions applied to Product-Form Multi-Class Queuing Network: antithetic variates and low discrepancy sequences. Technical Report 1005, IRISA, April 1996.

[12] S. K. Zaremba. Some applications of multidimensionnal integration by parts. *Ann. Pol. Math*, XXI:85–96, 1968.

# Optimal U–Type Designs[*]

Peter Winker
Dept. of Economics and Statistics
University of Konstanz
P.O. Box 5560
D-78434 Konstanz
Peter.Winker@uni-konstanz.de

Kai–Tai Fang
Dept. of Mathematics
Hong Kong Baptist University
Kowloon Tong
Hong Kong
ktfang@hkbu.edu.hk

ABSTRACT Designs with low discrepancy are of interest in many areas of statistical work. U–type designs are among the most widely studied design classes. In this paper a heuristic global optimization algorithm, Threshold Accepting, is used to find optimal U–type designs (uniform designs) or at least good approximations to uniform designs. As the evaluation of the discrepancy of a given point set is performed by an exact algorithm, the application presented here is restricted to small numbers of experiments in low dimensional spaces. The comparison with known optimal results for the two–factor uniform design and good designs for three to five factors shows a good performance of the algorithm.

## 1 Introduction

The interest in finding experimental designs with low discrepancy can be motivated from their application for the evaluation of multidimensional integrals by numerical methods.[1] Quasi–Monte Carlo methods based on some designs with low discrepancy were developed to face the "curse of dimensionality" caused by the large error bounds of the straightforward generalization of classical one dimensional integration rules.[2]

One widely studied class of quasi–Monte Carlo methods is based on U–type designs. A U–type design in dimension $d$ with $n$ points ($d$ factors each with $n$ levels) denoted by $U_{n \times d}$ is given by a $n \times d$ matrix of full

---

[*]Research was supported by the Deutsche Forschungsgemeinschaft, Sonderforschungsbereich 178, "Internationalisierung der Wirtschaft" at the University of Konstanz and a Hong Kong UGC grant. We are indebted to an anonymous referee for helpful comments. All remaining shortcomings are our owns.

[1]Cf. Bates, Buck, Riccomagno and Wynn (1996), Fang and Wang (1994), p. 237.
[2]Cf. Hua and Wang (1981), Niederreiter (1992a), and Fang and Wang (1994).

rank with any column being a permutation of $\{1, 2, \ldots, n\}$. Without loss of generality it can be assumed that the first column of any U–type design is given by $(1, 2, \ldots, n)'$. However, for the second column there are $n! - 1$ possible choices, $n! - 2$ for the third column, and so on. Thus, if $d$ and $n$ are large, the number of U–type designs is very large. Let $\mathcal{U}_{n \times d}$ be the set of all U–type designs in dimension $d$ with $n$ points.

For the application in quasi–Monte Carlo methods it is necessary to find optimal or at least very good U–type designs for given $d$ and $n$ in the sense that they should minimize the discrepancy function or any other reasonable measure of uniformity on the set of all U–type designs $\mathcal{U}_{n \times d}$.[3] An optimal U–type design in this sense is called uniform design. For given $d$ and $n$, in general, the uniform design is not unique. There exists a relationship between uniform designs and other concepts used to obtain high quality designs. In particular, Fang (1995) shows that any orthogonal design can be obtained from a U–type design by the so–called pseudo–level technique. Furthermore, the U–type designs used to obtain orthogonal designs are $(t, m, s)$–nets as described by Niederreiter (1992a).

As the discrepancy of a set of points in dimension $d$ is defined on the unit cube $[0, 1]^d$ the transformation

$$
\begin{array}{ccc}
\rho: & \mathcal{U}_{n \times d} & \longrightarrow & [0, 1]^d \\[2mm]
& U_{n \times d} = (u_{ij})_{\substack{1 \leq i \leq n \\ 1 \leq j \leq d}} & \longmapsto & \left(\frac{u_{ij} - 0.5}{n}\right)_{\substack{1 \leq i \leq n \\ 1 \leq j \leq d}}
\end{array}
$$

is used. Then, the discrepancy $D(U_{n \times d})$ of any element $U_{n \times d}$ of $\mathcal{U}_{n \times d}$ is given by

$$
D(U_{n \times d}) \equiv D(\rho(U_{n \times d})) = \sup_{\mathbf{x} \in [0,1]^d} \left| \frac{M(\mathbf{x})}{n} - F(\mathbf{x}) \right|,
$$

where $M(\mathbf{x})$ is the number of points $\mathbf{u}$ in $\rho(u_{.1}), \ldots, \rho(u_{.d})$ satisfying $\mathbf{u} \leq \mathbf{x}$, i.e. $u_j \leq x_j$ for all $j = 1, \ldots d$, and $F(\mathbf{x})$ is the distribution function of the uniform distribution on $[0, 1]^d$.

The mathematical programming problem to be solved in order to find uniform designs can be expressed in the form

$$
\text{minimize } D(U_{n \times d}) \text{ over } U_{n \times d} \in \mathcal{U}_{n \times d}. \tag{1}
$$

Unfortunately, the calculation of the discrepancy for a given U–type design already proves to be very complex. In fact, the algorithms proposed by Niederreiter (1973) for the one dimensional case, by De Clerk (1986) for dimension two and by Bundschuh and Zhu (1993) for the general case are exact, but they use computing time growing exponentially in the problem size as given by $d$ and $n$.

---

[3]Cf. Koksma–Hlawka inequality in Niederreiter (1992b), p. 26, equation (2).

Furthermore, as indicated above the cardinality of $\mathcal{U}_{n \times d}$ itself grows at a tremendous rate in $d$ and $n$. Thus, although there is a trivial algorithm for solving (1) by enumeration of all elements of $\mathcal{U}_{n \times d}$, calculating the corresponding discrepancy and choosing the minimum, it is not feasible even for quite modest instance sizes. In fact, one might be tempted to assume that (1) is NP–complete or even NP–hard.

There exist several possibilities to tackle the problem despite its high computational complexity. Instead of using a global optimization heuristic as presented in this paper, one could restrict the search to some subset of $\mathcal{U}_{n \times d}$. For example, number–theoretic methods use good lattice point sets as candidates for good U–type designs. Using such a priori selections out of the large set of U–type designs the double optimization problem (1) is essentially reduced to a one step optimization problem, namely the calculation of the discrepancy.

For smaller problem instances the exact algorithm proposed by Bundschuh and Zhu (1993) offers an opportunity to find the exact value of the discrepancy for a given set of points. When $n$ and $d$ are large, their algorithm – which essentially is an enumeration algorithm – uses a tremendous amount of computing time growing at the order $O(n^d)$.[4] For larger problem instances ($d > 6$ and $n > 1000$), the use of the optimization heuristic Threshold Accepting supplies high quality lower bounds for the discrepancy. In fact, Winker and Fang (1997) show that for those problem instances for which the exact calculation is still feasible the optimization heuristic almost always could give the exact value for the discrepancy.

This paper deals with the "outer loop" optimization problem in (1), the choice of a general good or optimal U–type design for given $d$ and $n$. Again, a heuristic optimization approach using Threshold Accepting will be used. This algorithm generates a sequence of candidate solutions to the problem. The (exact) discrepancy for these solutions is calculated by the algorithm of Bundschuh and Zhu (1993). Therefore, the application is restricted to smaller instances. Alternatively to the discrepancy, the $L_2$–discrepancy and the notion of D–optimality are used as measures of uniformity.[5] The U–type designs obtained by minimizing different measures of uniformity are compared by the corresponding values for the different objective functions.

The final step, to combine both heuristic optimization approaches to generate high quality solutions even for large $d$ and $n$ will be subject of further work.

The paper is organized in the following way. The next section will give a description of the Threshold Accepting heuristic and its implementation for the problem of uniform designs. In section 3 some implementation details

---

[4]Some results presented in Winker and Fang (1997) might give an impression of the order of magnitude one has to expect.

[5]Cf. Fang and Hickernell (1995) for an overview on such measures and their interrelationships.

will be discussed somewhat deeper before section 4 presents results for optimized U–type designs for $d \leq 5$ and $n \leq 30$. For $d = 2$ these results are compared with the exact solutions provided by Li and Fang (1995) and good approximations provided by GLP–sets, incomplete Latin squares and U–type designs based on orthogonal designs.[6] The paper concludes with the main findings and an outlook to the two loop optimization problem.

## 2   The Threshold Accepting Algorithm

The problem of finding uniform designs for given dimension $d$ and number of points $n$ falls under the heading of large scale integer programming problems. Even if the discrepancy function could be evaluated easily for any given design, the selection of an optimal design remains a complex task. As the set of U–type designs $\mathcal{U}_{n \times d}$ is finite a simple enumeration algorithm would give the exact result. However, as pointed out above this algorithm is infeasible for higher dimensional problems with more than just a few points.

Li and Fang (1995) propose an algorithm for finding uniform designs in the two–dimensional case. It seems possible to find generalizations of their approach for $d > 2$. However, the time complexity of this algorithm might grow exponentially in the dimension – the course of dimensionality reappears. Nevertheless, the results provided by Li and Fang (1995) can be used as a benchmark of the heuristic optimization algorithm proposed in the sequel.

As integer optimization heuristic the Threshold Accepting algorithm (TA) introduced by Dueck and Scheuer (1990) is used. The TA algorithm may be characterized as refined local search or evolutionary algorithm such as the more widely used Simulated Annealing algorithm.[7]

The TA implementation used for the search of uniform designs starts with a randomly generated U–type design $U^0_{n \times d}$. Afterwards, in each iteration the algorithm tries to replace the current solution $U^c_{n \times d}$ with a new one. The new design $U^t_{n \times d}$ is randomly generated in a given neighbourhood of the current solution. The discrepancy is calculated for the new design and the result is compared with the value of the current design.

In order to avoid getting stuck in a local minimum and to ensure the convergence to a global minimum with the number of iterations tending to infinity the right acceptance criterium has to be choosen. Whereas a trivial local search algorithm accepts only improvements of the objective function, TA allows for a temporary worsening up to a given threshold value. As this threshold value is non–zero, i.e. positive during most of the iteration steps,

---

[6]Cf. Fang and Hickernell (1995).
[7]Cf. Kirkpatrick et al. (1983) and Fox (1995).

440

bad local minima can be left again. The threshold sequence decreases to zero as the algorithm proceeds. Consequently, the TA algorithm will end up in a local minimum though of high quality with good chances of being a global minimum at least for small problem sizes.[8]

Figure 1 presents a flow chart of the TA implementation for minimizing the discrepancy of U–type designs for given dimensions $d$ and $n$.

FIGURE 1. Threshold Accepting Algorithm for Uniform Designs

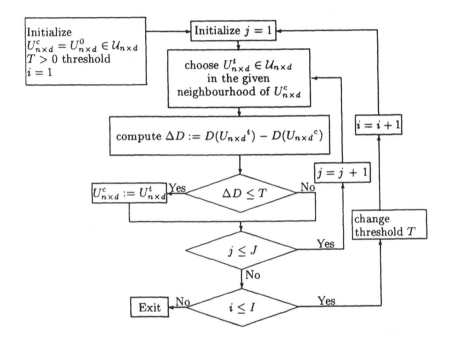

In analogy to Simulated Annealing the sequence of threshold values $T$ might be interpreted as a cooling schedule. Consequently, for a given value of $T$ a number of $J$ iterations has to be performed in order to bring the system in a stable state with regard to this threshold parameter. Afterwards, the threshold value is decreased until it reaches zero after $I$ steps of the outer loop.

Although the structure of the algorithm is quite simple its implementation to the approximation of uniform designs has to take into consideration several aspects. In the next section three central aspects will be discussed,

---

[8]Althöfer and Koschnick (1991) proved that TA will eventually converge to a global optimum for a good choice of the threshold sequence as the number of iterations tends to infinity.

namely *the definition of local neighbourhoods, the generation of the thresh-old sequence*, and updating procedures to *reduce the computing time* needed for the exact evaluation of the discrepancy for the new design $U_{n \times d}^t$.

# 3   Implementation for Uniform Design

Let us start with the introduction of some local structure on the domain $\mathcal{U}_{n \times d}$. Using the transformation $\rho$ defined in the introduction, $\mathcal{U}_{n \times d}$ is a (finite) subset of the $d$–dimensional unit cube $C^d$. Therefore, the projections of some $\varepsilon$–spheres with regard to the Euclidian metric in $\mathbf{R}^d$ would be a natural choice to define local neighbourhoods. Then, two elements of $\mathcal{U}_{n \times d}$ would belong to the same $\varepsilon$–neighbourhood if and only if their Euclidian distance is smaller than $\varepsilon$. However, this choice would generate at least two problems related to those described in Winker and Fang (1997). The first problem applies to the cardinality of the neighbourhoods. For large $d$, one would have to choose $\varepsilon$ very large ($> 0.5$) to make sure that a $\varepsilon$–neighbourhood contains more than one element of $\mathcal{U}_{n \times d}$. The second problem consists of generating and storing all the information about these neighbourhoods.

As the problem is related to the approximation of the discrepancy, it is not too surprising that the solution used in Winker and Fang (1997) can be carried over to the uniform design problem. Two U–type designs will be regarded as close to each other if they differ in at most $k \leq d$ columns. The difference for a given column is restricted to the exchange of two elements. Given the natural ordering of all elements in this column for any design in $\mathcal{U}_{n \times d}$, the second element should be one of the $l \leq n$ next neighbours of the first one.[9]

Using this definition of local neighbourhoods the local structure of the problem can be simulated in the following sense. A large number $\tilde{I}$ of elements in $\mathcal{U}_{n \times d}$ is randomly generated. For each element another element out of its predefined neighbourhood is also randomly choosen. For all pairs of U–type designs the difference of their discrepancy is calculated. Then, a frequency plot of those local deviations contributes besides the insights to the local structure of the discrete optimization problem a possibility to generate the threshold sequence endogenously.

To this end, the calculated differences are sorted in decreasing order. The performance of the algorithm can be improved using only a lower fraction $\alpha$ of this sequence, i.e. $I := \alpha \tilde{I}$. As the steps between two consecutive thresholds $T$ in this sequence become smaller as the sequence approaches zero, a fixed number of iterations $J$ for each step can be used. Nevertheless,

---

[9]Cf. Winker and Fang (1997) for a formal definition of this maximum order norm.

the total number of iterations for a given interval of thresholds will increase due to the shape of the distribution of deviations.

Li and Fang (1995) observe that there is a lower border to the minimal difference between the discrepancy for two U–type designs in the two-dimensional case. The endogenous generation of threshold values described above automatically profits from such features of the problem as it will never use a threshold value smaller than the smallest step to be expected with the sole exception of the value zero at the very end of the iteration process used to make sure that a local minimum is achieved.

A last detail of the implementation is concerned with reducing the number of necessary evaluations of the discrepancy for a U–type design. To this end, the observation is used, that the discrepancy function achieves its maximum on a finite subset of $[0, 1]^d$. Now, when moving from a current U–type design $U_{n \times d}^c$ to a new one $U_{n \times d}^t$ the maximum of the discrepancy function for the new one might be achieved close to the old one. Hence, before using the algorithm of Bundschuh and Zhu (1993) for the calculation of the discrepancy it is tested whether the discrepancy function for the new U–type design grows by more than the current threshold near the old maximum. Since in that case the new solution is not accepted anyway, a superfluous evaluation can be avoided.

# 4   Some Results

The evaluation of the discrepancy for a given U–type design using the algorithm of Bundschuh and Zhu (1993) is quite computer intensive.[10] For example, to calculate the discrepancy of a U–type design with $d = 5$ and $n = 122$, it takes 1,226.64 seconds of CPU–time on an IBM RS 6000/3AT workstation (SPECfp: 187.2). As the heurisitc optimization algorithm requires a huge number of evaluations – at least 10,000 –, the first tests of the TA implementation were restricted to $d = 2, \ldots, 5$ and $n \leq 30$. For this range of values, the discrepancy can be calculated quite fast. Furthermore, the testing procedure described above reduces the number of necessary evaluations considerably. Probably, a more careful analysis of the local structure of the problem would allow for an even more efficient implementation.

For each pair $(d, n)$ 25 different values of the parameter $\alpha$ for the endogenous generation of the threshold sequence were used, and for each value of $\alpha$ 10 different seeds for the random number generator were selected. Consequently, a total number of 250 runs was performed for each combination of $d$ and $n$. All trials were repeated for 10,000 iterations and 100,000 iterations,

---

[10] We are indebted to Mr. Vincent Chin and Mr. J.X. Pan for leaving us the C–code for our calculations.

respectively. The following tables 1 – 2 summarize the results.

TABLE 1. Discrepancy of Best U–type Designs by TA $d = 2, 3$

| | $d = 2$ | | | | $d = 3$ | | | |
|---|---|---|---|---|---|---|---|---|
| | 10,000 iter. | | 100,000 iter. | | 10,000 iter. | | 100,000 iter. | |
| $n$ | best | mean | best | mean | best | mean | best | mean |
| 05 | $.25000^{a)}$ | .25184 | $.25000^{a)}$ | .25000 | $.32900^{b)}$ | .33562 | $.32900^{b)}$ | .34412 |
| 06 | $.20139^{a)}$ | .22300 | $.20139^{a)}$ | .21778 | $.29688^{b)}$ | .30149 | $.29688^{b)}$ | .29921 |
| 07 | $.17857^{a)}$ | .21629 | $.17857^{a)}$ | .19106 | $.24891^{b)}$ | .27037 | $.24891^{b)}$ | .26005 |
| 08 | $.16016^{a)}$ | .18711 | $.16016^{a)}$ | .20544 | $.22925^{b)}$ | .25060 | $.22925^{b)}$ | .24775 |
| 09 | $.14506^{a)}$ | .15401 | $.14506^{a)}$ | .14859 | $.21142^{b)}$ | .21264 | $.21142^{b)}$ | .21429 |
| 10 | $.13750^{a)}$ | .13950 | $.13750^{a)}$ | .14088 | $.18663^{b)}$ | .19578 | $.19438^{b)}$ | .19772 |
| 11 | $.13017^{a)}$ | .13182 | $.13017^{a)}$ | .13208 | $.17740^{b)}$ | .18295 | $.17665^{b)}$ | .18454 |
| 12 | $.11979^{a)}$ | .12326 | $.11979^{a)}$ | .12644 | $.16385^{b)}$ | .16967 | $.16385^{b)}$ | .17249 |
| 13 | $.11391^{a)}$ | .11494 | $.11391^{a)}$ | .12021 | $.15379^{b)}$ | .16125 | $.15379^{b)}$ | .16198 |
| 14 | $.10587^{a)}$ | .11033 | $.10587^{a)}$ | .10898 | $.14573^{b)}$ | .15064 | $.14573^{b)}$ | .15255 |
| 15 | $.09889^{a)}$ | .10144 | $.09889^{a)}$ | .10288 | $.13604^{b)}$ | .14376 | $.13426^{b)}$ | .14573 |
| 16 | $.09473^{a)}$ | .09833 | $.09473^{a)}$ | .09813 | $.12869^{b)}$ | .13570 | $.12869^{b)}$ | .13452 |
| 17 | $.08737^{a)}$ | .09170 | $.08737^{a)}$ | .08992 | $.12454^{b)}$ | .12966 | $.12454^{b)}$ | .13096 |
| 18 | $.08565^{a)}$ | .08904 | $.08565^{a)}$ | .09047 | $.11885^{b)}$ | .12416 | $.11885^{b)}$ | .12489 |
| 19 | $.08102^{a)}$ | .08483 | $.08102^{a)}$ | .08359 | $.11115^{b)}$ | .11870 | $.11312^{b)}$ | .11979 |
| 20 | $.07813^{b)}$ | .08038 | $.07563^{a)}$ | .07856 | $.10976^{b)}$ | .11464 | $.10836^{b)}$ | .11515 |
| 21 | $.07540^{a)}$ | .07676 | $.07540^{a)}$ | .07744 | $.10459^{b)}$ | .11009 | $.10601^{b)}$ | .11112 |
| 22 | $.07076^{a)}$ | .07608 | $.07076^{a)}$ | .07326 | $.10193^{b)}$ | .10625 | $.10020^{b)}$ | .10483 |
| 23 | $.06947^{b)}$ | .07212 | $.06758^{a)}$ | .07141 | $.09597^{b)}$ | .10263 | $.09613^{b)}$ | .10105 |
| 24 | $.06641^{b)}$ | .07044 | $.06641^{b)}$ | .06849 | $.09399^{b)}$ | .09957 | $.09282^{b)}$ | .09767 |
| 25 | $.06280^{b)}$ | .06772 | $.06280^{b)}$ | .06691 | $.09000^{b)}$ | .09485 | $.09196^{b)}$ | .09769 |
| 26 | $.06176^{b)}$ | .06701 | $.06176^{b)}$ | .06323 | $.08942^{b)}$ | .09347 | $.08931^{b)}$ | .09414 |
| 27 | $.05933^{b)}$ | .06471 | $.05864^{b)}$ | .06159 | $.08639^{b)}$ | .09124 | $.08723^{b)}$ | .09151 |
| 28 | $.05644^{b)}$ | .06180 | $.05580^{b)}$ | .06051 | $.08428^{b)}$ | .08899 | $.08352^{b)}$ | .08912 |
| 29 | $.05797^{b)}$ | .06100 | $.05499^{b)}$ | .05776 | $.08179^{b)}$ | .08662 | $.07958^{b)}$ | .08454 |
| 30 | $.05417^{b)}$ | .05808 | $.05306^{b)}$ | .05684 | $.07933^{b)}$ | .08466 | $.07979^{b)}$ | .08497 |

a) Equal to known optimum (Li and Fang (1995)).

b) Improvement to best U–type designs from Fang and Hickernell (1995).

Table 1 gives the results for the two– and three–dimensional case. For $d = 2$ and $n \leq 23$ the results can be compared with the exact results provided by Li and Fang (1995) and the results provided by Fang and

Hickernell (1995) for GLP–sets, incomplete Latin squares and U–type designs based on orthogonal designs. As can be seen, in the two–dimensional case, for almost all numbers of levels the exact optimum is reached at least for one trial already for 10,000 iterations. The column with the mean discrepancy for all trials indicates that one cannot be sure always to find the exact minimum, but in general one will find at least a very good approximation, often superior to any U–type design generated by other methods. Furthermore, for the two–dimensional case suboptimal results can be used as input to the Li and Fang (1995) algorithm and reduce the necessary computing time of this algorithm considerably. In fact, the algorithm by Li and Fang (1995) can be used to check the results generated by TA for optimality.

As so far no algorithm is known to solve the uniform design problem for dimensions larger than 2, it is not possible to compare the results for $d = 3, 4, 5$ presented in the other tables with any known results. However, it is still possible to compare the achieved values with the best gained from GLP–sets, incomplete Latin squares or U–type designs based on orthogonal designs.[11] The improvement achieved by the TA implementation reaches an order of magnitude of 10 to 20% for the considered instances, already when using only 10,000 iterations per trial.

The computation time for the trials with 10,000 iterations ranges from 0.53 CPU–seconds for $d = 2$ and $n = 5$ to more than 4,000 CPU–seconds for $d = 5$ and $n = 30$. Hence, it might be assumed that the time consumption of our TA implementation is some orders of magnitude smaller than for the (deterministic) algorithm of Li and Fang (1995). Of course, the difference will increase for larger problem instances.

The results show that the TA implementation gives a reasonable mean approximation to uniform designs for small problem instances without an excessive tuning of the optimization parameters. Some more tuning probably could increase the quality of the results as first trials for the two–dimensional case indicate, when the true optimum could be reproduced already with 10,000 iterations for all instances up to $n = 23$. Of course, one cannot expect that this TA implementation will give uniform designs for all problem instances. Nevertheless, it can provide very good U–type designs with regard to minimizing the discrepancy.

As the calculation of the discrepancy $(D)$ for a given set of points is very complex, it might be justified to use alternative measures of uniformity. The TA implementation described above allows also the optimization of other objective functions such as the $L_2$–discrepancy $(D_2)$ and the notion of D–optimality (D–opt. to be maximized). The advantage of these alternative measures is the fact that they are easy to compute for a given set of points. However, they lead to different error bounds for simlulation and

---

[11] Cf. Fang and Hickernell (1995).

TABLE 2. Discrepancy of Best U-type Designs by TA $d = 4, 5$

| | $d = 4$ | | | | $d = 5$ | | | |
|---|---|---|---|---|---|---|---|---|
| | 10,000 iter. | | 100,000 iter. | | 10,000 iter. | | 100,000 iter. | |
| $n$ | best | mean | best | mean | best | mean | best | mean |
| 05 | .40310 | .40310 | .40310 | .40462 | .44279 | .44348 | .44279 | .44680 |
| 06 | .36068[a] | .36070 | .36068[a] | .36074 | .40007[a] | .40007 | .40007[a] | .40008 |
| 07 | .31489[a] | .32218 | .31489[a] | .32073 | .36286[a] | .36349 | .36286[a] | .36325 |
| 08 | .27660[a] | .29085 | .27660[a] | .28922 | .33105[a] | .33277 | .33105[a] | .33212 |
| 09 | .25757[a] | .25873 | .25757[a] | .26133 | .30387[a] | .30411 | .30387[a] | .30387 |
| 10 | .24303[a] | .24419 | .24303[a] | .24314 | .27378[a] | .28074 | .27378[a] | .27858 |
| 11 | .21753[a] | .22580 | .21753[a] | .22186 | .26038[a] | .26285 | .26038[a] | .26118 |
| 12 | .20315[a] | .20976 | .20315[a] | .20692 | .24281[a] | .25048 | .24281[a] | .24722 |
| 13 | .19366[a] | .19962 | .18980[a] | .19760 | .22740[a] | .23704 | .22740[a] | .23245 |
| 14 | .18197[a] | .18861 | .18197[a] | .18484 | .21729[a] | .22545 | .21378[a] | .22052 |
| 15 | .17315[a] | .17890 | .17083[a] | .17916 | .20943[a] | .21454 | .20943[b] | .21454[b] |
| 16 | .16602[a] | .17135 | .16445[a] | .16801 | .19765[a] | .20658 | .20974[a] | .21995 |
| 17 | .15633[a] | .16466 | .15593[a] | .16105 | .19185[a] | .19902 | .18648[a] | .19412 |
| 18 | .15033[a] | .15825 | .14665[a] | .15430 | .18240[a] | .19069 | .17999[a] | .18608 |
| 19 | .14534[a] | .15237 | .14249[a] | .14862 | .17453[a] | .18431 | .17446[a] | .17951 |
| 20 | .14015[a] | .14704 | .13845[a] | .14309 | .16966[a] | .17829 | .16640[a] | .17307 |
| 21 | .13576[a] | .14248 | .13240[a] | .13781 | .16398[a] | .17206 | .16149[a] | .16689 |
| 22 | .13170[a] | .13716 | .13170[a] | .13708 | .15968[a] | .16690 | .15807[a] | .16184 |
| 23 | .12875[a] | .13348 | .12529[a] | .12948 | .15680[a] | .16283 | .15280 | .15735 |
| 24 | .12173[a] | .12953 | .12171[a] | .12521 | .15274[a] | .15839 | .14848[a] | .15312 |
| 25 | .12046[a] | .12597 | .11823[a] | .12201 | .15023[a] | .15620 | .15023[b] | .15620[b] |
| 26 | .11846[a] | .12321 | .11420[a] | .11823 | .14431[a] | .14992 | .14103[a] | .14492 |
| 27 | .11415[a] | .11953 | .11154[a] | .11505 | .14073[a] | .14659 | .13343[a] | .13760 |
| 28 | .11305[a] | .11713 | .10882[a] | .11235 | .13783[a] | .14293 | .13783[b] | .14293[b] |
| 29 | .10983[a] | .11423 | .10665[a] | .10981 | .13485[a] | .14046 | .13485[b] | .14046[b] |
| 30 | .10710[a] | .11155 | .10324[a] | .10699 | .13179[a] | .13734 | .13179[b] | .13734[b] |

a) Improvement to best U-type designs from Fang and Hickernell (1995).

b) Results for 10,000 iterations.

numerical integration. Therefore, it is interesting to compare the U-type designs obtained by optimizing different objective functions.

Table 3 gives the best out of 250 trials for $d = 2, \ldots, 5$, $n = 30$ and 10,000 iterations for each of the three measures of uniformity $D$, $D_2$ and D-opt. It should be taken into account that due to the high complexity of calculating the discrepancy for a given set of points, a run with 10,000 iterations for $D$ takes more CPU–time than 50 runs with 10,000 iterations each for $D_2$. With the implementation of a double loop TA optimization for the discrepancy this difference will become much smaller.

TABLE 3. Uniformity of Best U–type Designs by TA

| | Discrepancy | $L_2$–Discrepancy | D–optimality |
|---|---|---|---|
| Objective | | $d = 2$ | |
| $D$ | .054167 | .000279 | .997679 |
| $D_2$ | .054167 | .000220 | .997778 |
| D–opt. | .096944 | .000594 | .997779 |
| | | $d = 3$ | |
| $D$ | .079329 | .000391 | .995760 |
| $D_2$ | .085764 | .000283 | .983994 |
| D–opt. | .161366 | .000695 | .996670 |
| | | $d = 4$ | |
| $D$ | .101710 | .000455 | .939160 |
| $D_2$ | .126803 | .000282 | .929029 |
| D–opt. | .177859 | .000646 | .995520 |
| | | $d = 5$ | |
| $D$ | .131788 | .000338 | .864948 |
| $D_2$ | .165550 | .000218 | .863392 |
| D–opt. | .210430 | .000516 | .994286 |

The results in table 3 are consistent in the sense that the optimal values in each row are obtained for the elements on the diagonal, i.e. in order to obtain a design with low $D$ it is best to minimze $D$. Furthermore, one might conclude that a low $D$ design can be better proxied by minimzing $D_2$ than be maximizing the D–opt. criterium. In fact, in dimension two the global optimal for $D$ can be found by minimizing $D_2$. Finally, the differences in $D_2$ for the designs optimized with regard to $D$ and $D_2$, respectively, are smaller than the differences in $D$. Hence, designs with low discrepancy $D$ exhibit a better performance for other measures of uniformity than viz.

# 5   Conclusion and Outlook

The results generated by the simple TA implementation presented in this paper proved to be superior to most of the standard methods used for the generation of "good" U–type designs. In fact, for the two–dimensional case, where exact optima are known, they could be replicated always by the TA implementation. Thus, it might be concluded that this method might be useful for the generation of "good" U–type designs.

In this paper, the search for optimal or at least "good" U–type designs has been restricted to low dimensional problems with few points and to the discrepancy as objective function. Thus, some natural extensions for future research are given. The most straightforward extension is given by the combination of the TA implementation for the approximation of the discrepancy presented in Winker and Fang (1997) with the TA implementation given in this paper. Then, the search for "good" U–type designs would become possible for much larger problem instances, for example $d = 11$ and $n \leq 1000$.

A further topic for future research is the comparison of U–type designs generated by minimizing different objective functions besides the discrepancy. Of course, the TA implementation given here can easily be adopted to other objective functions. Then, it will be possible to see whether easier to calculate functions might also result in good U–type designs.

# 6  References

[1] I. Althöfer, and K.-U. Koschnick (1991). *On the Convergence of 'Threshold Accepting'.* Applied Mathematics and Optimization, 24, 183–195.

[2] R. Bates, R. Buck, E. Riccomagno, and H. P. Wynn (1996). *Experimental Design and Observation for Large Systems.* Journal of the Royal Statistical Society B, 57, 77–94.

[3] P. Bundschuh, and Y. C. Zhu (1993). *A method for exact calculation of the discrepancy of low-dimensional finite point set (I).* Abhandlungen aus dem Mathematischen Seminar der Universität Hamburg 63, 115–133.

[4] L. De Clerk (1986). *A method for exact calculation of the stardiscrepancy of plane sets applied to the sequences of Hammersley.* Monatshefte für Mathematik, 101, 261–278.

[5] G. Dueck, and T. Scheuer (1990). *Threshold Accepting: A General Purpose Algorithm Appearing Superior to Simulated Annealing.* Journal of Computational Physics, 90, 161–175.

[6] K.-T. Fang, and F. J. Hickernell (1995). *The uniform design and its applications*, Bulletin of the International Statistical Institute, 50th session, book 1, 11, pp. 49–65.

[7] K.-T. Fang, and Y. Wang (1994). *Applications of Number Theoretic Methods in Statistics.* Chapman and Hall, London.

[8] Y. Fang (1995). *Relationships between Uniform and Orthogonal Designs.* The 3rd ICSA Statistical Conference, Beijing.

[9] B. L. Fox. *Simulated Annealing: Folklore, Facts, and Directions.* In: H. Niederreiter and P. J.-S. Shiue (eds.). *Monte Carlo and Quasi-Monte Carlo Methods in Scientific Computing.* Lecture Notes in Statistics 106, Springer, New York 1995.

[10] L. K. Hua, and Y. Wang (1981). *Applications of Number Theory to Numerical Analysis.* Springer, Berlin.

[11] S. Kirkpatrick, C. Gelatt, and M. Vecchi (1983). *Optimization by Simulated Annealing.* Science, 220, 671–680.

[12] W. Li, and K.-T. Fang (1995). *A Global Optimum Algorithm on Two Factor Uniform Design.* In: K.-T. Fang and F. J. Hickernell (eds.). *Proceedings Workshop on Quasi-Monte Carlo Methods and Their Applications.* Hong Kong Baptist University, 189–201.

[13] H. Niederreiter, (1973). *Application of diophantine approximations to numerical integration.* In: C. F. Osgood (ed.). *Diophantine Approximations and Its Applications.* Academic Press, New York, 129–199.

[14] H. Niederreiter, (1992a). *Random Number Generation and Quasi-Monte Carlo Methods.* SIAM CBMS–NSF Regional Conference Series in Applied Mathematics, Philadelphia.

[15] H. Niederreiter, (1992b). *Lattice Rules for Multiple Integration.* In: Marti, K. (ed.). *Stochastic Optimization.* Lecture Notes in Economics and Mathematical Systems 379, Springer–Verlag, Berlin.

[16] P. Winker, and K.-T. Fang (1997). *Application of Threshold Accepting to the Evaluation of the Discrepancy of a Set of Points.* SIAM Journal on Numerical Analysis, forthcoming.

# Lecture Notes in Statistics

For information about Volumes 1 to 51
please contact Springer-Verlag